Hepatic Encephalopathy
Pathophysiology and Treatment

Experimental Biology and Medicine

Hepatic Encephalopathy: *Pathophysiology and Treatment,* edited by *Roger F. Butterworth and Gilles Pomier Layrargues,* 1989
Molecular Actions of Alcohol: *Neurobiology and Metabolism,* edited by *Grace Y. Sun, P. Kevin Rudeen, W. Gibson Wood, Yau Huei Wei, and Albert Y. Sun,* 1989
Macromolecular Biorecognition: *Principles and Methods,* edited by *Irwin Chaiken, Emilia Chiancone, Angelo Fontana, and Paolo Neri,* 1987
Molecular Basis of Lymphokine Action, edited by *David R. Webb, Carl W. Pierce, and Stanley Cohen,* 1987
Biology of Copper Complexes, edited by *John R. J. Sorenson,* 1987
Epstein–Barr Virus and Human Disease, edited by *P. H. Levine, D. V. Ablashi, M. Nonoyama, G. R. Pearson, and R. Glaser,* 1987
Methods of Protein Sequence Analysis • 1986, edited by *Kenneth A. Walsh,* 1987
Regulation of Immune Gene Expression, edited by *Marc Feldmann and Andrew McMichael,* 1986
Biological Methylation and Drug Design, edited by *Ronald T. Borchardt, Cyrus R. Creveling, and Per Magne Ueland,* 1986
Retroviruses and Human Pathology, edited by *Robert C. Gallo, Dominique Stehelin, and Oliviero E. Varnier,* 1985
Directed Drug Delivery, edited by *Ronald T. Borchardt, Arnold Repta, and Valentino Stella,* 1985
Immune Regulation, edited by *Marc Feldmann and N. A. Mitchison,* 1985
Human T Cell Clones, edited by *Marc Feldmann, Jonathan R. Lamb, and James N. Woody,* 1985
Inositol and Phosphoinositides, edited by *John E. Bleasdale, Joseph Eichberg, and George Hauser,* 1985
Growth, Cancer, and the Cell Cycle, edited by *Philip Skehan and Susan J. Friedman,* 1984
Ir Genes, edited by *Carl W. Pierce, Susan E. Cullen, Judith A. Kapp, Benjamin D. Schwartz, and Donald C. Shreffler,* 1983
Methods in Protein Sequence Analysis, edited by *Marshall Elzinga,* 1982
Inflammatory Diseases and Copper, edited by *John R. J. Sorenson,* 1982

Hepatic Encephalopathy

Pathophysiology and Treatment

Edited by

Roger F. Butterworth, PhD

and ***Gilles Pomier Layrargues,*** MD

Hôpital St. Luc, Montréal, Quebec, Canada

Humana Press • Clifton, New Jersey

Library of Congress Cataloging-in-Publication Data

Hepatic encephalopathy: pathophysiology and treatment/edited by
Roger F. Butterworth, Gilles Pomier Layrargues.
p. cm. – (Experimental biology and medicine)
The results of a symposium, held in Val David, Quebec from Oct.
30–Nov. 1, 1988.
ISBN 0-89603-164-0
1. Hepatic encephalopathy–Congresses. I. Butterworth, Roger F.
II. Layrargues, Gilles Pomier. III. Series : Experimental biology
and medicine (Clifton, N.J.)
[DNLM: 1. Hepatic Encephalopathy—physiopathology—congresses.
2. Hepatic Encephalopathy—therapy—congresses. WI 700 H5257 1988]
RC848.H4H463 1989
616.3'62—dc20
DNLM/DLC
for Library of Congress 89-20039
CIP

Crescent Manor
PO Box 2148
Clifton, New Jersey 07015

The results of a symposium, held in Val David, Quebec from Oct. 30–Nov. 1, 1988.

Preface

Hepatic Encephalopathy (HE) is a neuropsychiatric disorder resulting from liver failure. HE may be associated with fulminant (acute) hepatic failure or chronic liver disease with portal–systemic shunting. The latter condition is characterized neuropathologically by astrocytic rather than neuronal changes (Alzheimer Type II astrocytosis). The former is frequently accompanied by cerebral edema.

Several hypotheses have been proposed to explain the pathogenesis of HE. These include:

1. A toxic action of a substance (or substances) such as ammonia on brain function
2. A deficit of cerebral energy metabolism
3. Neurotransmitter changes, and, more recently
4. The role of "endogenous benzodiazepines."

This volume summarizes the results of a symposium held in Val David, Quebec from October 30–November 1, 1988, that was devoted to an evaluation of the evidence for and against the various hypotheses of HE. Data from studies in patients, in experimental (animal) models of HE, and in cultured cell preparations were discussed. In addition, a review of available approaches to the treatment and management of HE was included. The therapeutic use of lactulose, antibiotics, dietary treatment, and branched-chain amino acid treatments were included, as well as the results of preliminary studies of the therapeutic use of the benzodiazepine antagonist, flumazenil.

Roger F. Butterworth, PhD
Gilles Pomier Layrargues, MD

Acknowledgments

The symposium was made possible by the generous financial assistance of:

Hoffman-La Roche Ltd.
Fonds de la recherche en santé du Québec
The Medical Research Council of Canada
Département de médecine, Université de Montréal
The Canadian Liver Foundation
Janssen Pharmaceutica
Interfalk Canada
Squibb Canada

We would like to express our gratitude to Dominique Roy for her help with the organization of the meeting and with the preparation of manuscripts as well as to other members of the André-Viallet Clinical Research Centre for their invaluable assistance.

Above all, we would like to thank all participants at the meeting for their invaluable input and lively criticism. Thanks to their efforts, the meeting was able to establish some degree of order in a complex area of research.

Contents

C. Treatment of Hepatic Encephalopathy

Update of Existing Therapeutic Approaches

Potential Therapeutic Use of Benzodiazepine Antagonists

Contributors

GEORG BANSKY • *Division of Clinical Pharmacology, Department of Medicine, University Hospital, Zurich, Switzerland*

M. BARALDI • *Clinica Medica III e Cattedra di Farmacognosia, Universita' di Modena, Modena, Italy*

ANTHONY S. BASILE • *Laboratory of Neuroscience, NIDDK, National Institutes of Health, Bethesda, MD*

MARK L. BASSETT • *Liver Diseases Section, NIDDK, National Institutes of Health, Bethesda, MD*

KEVIN L. BEHAR • *Department of Molecular Biophysics and Biochemistry, Yale University, New Haven, CT*

FINN BENGTSSON • *Department of Clinical Pharmacology, Laboratory for Experimental Brain Research, Lund University Hospital, Lund, Sweden*

MARCELLE BERGERON• *Laboratory of Neurochemistry, André-Viallet Clinical Research Centre, Hôpital St-Luc (University of Montreal), Montreal, Quebec, Canada*

ANNE-MARIE BESNARD • *Laboratory of Neurochemistry, André-Viallet Clinical Research Centre, Hôpital St-Luc (University of Montreal), Montreal, Quebec, Canada*

ANDRES T. BLEI • *Department of Medicine, Lakeside VA Medical Center and Northwestern University, Chicago, IL*

MOGENS BUGGE • *Department of Surgery, Lund University Hospital, Lund, Sweden*

ROGER F. BUTTERWORTH • *Laboratory of Neurochemistry, André-Viallet Clinical Research Centre, Hôpital St-Luc (University of Montreal), Montreal, Quebec, Canada*

HAROLD O. CONN • *VA Medical Center West Haven, Yale University School of Medicine, New Haven, CT*

ARTHUR J. L. COOPER • *Departments of Biochemistry and Neurology, Cornell University Medical College, New York, NY*

CARL W. COTMAN • *Department of Psychobiology, University of California at Irvine, CA*

J. F. S. CROCKER • *Department of Pediatrics, Izaak Walton Killam Hospital for Children, Halifax, Nova Scotia, Canada*

S. C. DIGOUT • *Department of Pediatrics, Izaak Walton Killam Hospital for Children, Halifax, Nova Scotia, Canada*

PETER R. DODD • *Department of Pharmacology, University of Sydney, N. S. W., Australia*

LJUSK SIW ERIKSSON • *Huddinge University Hospital, Karolinska Institute, Huddinge, Sweden*

PETER FERENCI • *1st Department of Gastroenterology, University of Vienna, Austria*

MARIA R. FERREIRA • *Liver Diseases Section, NIDDK, National Institutes of Health, Bethesda, MD*

SUSAN M. FITZPATRICK • *Department of Molecular Biophysics and Biochemistry, Yale University, New Haven, CT*

H. CHRIS FROMM • *Liver Diseases Section, NIDDK, National Institutes of Health, Bethesda, MD*

SERGIO H. GAMMAL • *Liver Diseases Section, NIDDK, National Institutes of Health, Bethesda, MD*

ALAN S. GELBARD • *Biophysics Laboratory, Memorial Sloan Kettering Cancer Center, New York, NY*

JEAN-FRANÇOIS GIGUÉRE • *Laboratory of Neurochemistry, André-Viallet Clinical Research Centre, Hôpital St-Luc (University of Montreal), Montreal, Quebec, Canada*

GUYLAINE GIRARD • *Laboratory of Neurochemistry, André-Viallet Clinical Research Centre, Hôpital St-Luc (University of Montreal), Montreal, Quebec, Canada*

CATHERINE GIROD • *Unité de Recherches de Physiopathologie Hépatique (INSERM U-24), Hôpital Beaujon, France*

MURIEL GODARD • *Pediatric Research Center, Sainte-Justine Hospital (University of Montreal), Montreal, Quebec, Canada*

A. GOSWANI • *Medical College of Wisconsin, Milwaukee, WI*

GEORG GRIMM • *1st Department of Medicine, University of Vienna, Austria*

K. GYR • *Department of Internal Medicine, Division of Gastroenterology, Kantonsspital Liestal, University of Basel, Switzerland*

RICHARD A. HAWKINS • *Department of Physiology and Biophysics, University of Health Sciences, The Chicago Medical School, North Chicago, IL*

H. HORTNAGL • *University of Vienna, Austria*

H. O. Jauregui • *Department of Pathology, Rhode Island Hospital, Providence, RI*

J. Jessy • *Department of Physiology and Biophysics, University of Health Sciences, Chicago Medical School, North Chicago, IL*

Kaj Johansen • *Department of Surgery, Harborview Medical Center, University of Washington, School of Medicine, Seattle, WA*

Graham A. R. Johnston • *Department of Pharmacology, University of Sydney, N. S. W., Australia*

E. Anthony Jones • *Liver Diseases Section, NIDDK, National Institutes of Health, Bethesda, MD*

K. Kaminsky • *Cleveland Metropolitan General Hospital, Case Western Reserve University, Cleveland, OH*

James C. K. Lai • *Departments of Biochemistry and Neurology, Cornell University Medical College, New York, NY*

Marie Lambert • *Pediatric Research Center, Sainte-Justine Hospital (University of Montreal), Montreal, Quebec, Canada*

Joël Lavoie • *Laboratory of Neurochemistry, André-Viallet Clinical Research Centre, Hôpital St-Luc (University of Montreal), Montreal, Quebec, Canada*

Didier Lebrec • *Unité de Recherches de Physiopathologie Hépatique (INSERM U-24), Hôpital Beaujon, France*

Samuel S. Lee • *Unité de Recherches de Physiopathologie Hépatique (INSERM U-24), Hôpital Beaujon, France*

Dorothy K. Leong • *Department of Pharmacology, University of Sydney, N. S. W., Australia*

Jacques Letarte • *Pediatric Research Center, Sainte-Justine Hospital (University of Montreal), Montreal, Quebec, Canada*

Emile Levy • *Pediatric Research Center, Sainte-Justine Hospital (University of Montreal), Montreal, Quebec, Canada*

Alan H. Lockwood • *Department of Neurology and Positron Diagnostic and Research Center, University of Texas Health Science Center, Houston, TX*

Jill E. Maddison • *Department of Veterinary Clinical Sciences, University of Sydney, N. S. W., Australia*

Anke M. Mans • *Department of Physiology and Biophysics, University of Health Sciences, The Chicago Medical School, North Chicago, IL*

Paul Martin • *Liver Diseases Section, NIDDK, National Institutes of Health, Bethesda, MD*

A. J. McCullough • *Cleveland Metropolitan General Hospital, Case Western Reserve University, Cleveland, OH*

Peter J. Meier • *Division of Clinical Pharmacology, Department of Medicine, University Hospital, Zurich, Switzerland*

R. Meier • *Department of Internal Medicine, Division of Gastroenterology, Kantonsspital Liestal, University of Basel, Liestal, Switzerland*

Kevin D. Mullen • *Cleveland Metropolitan General Hospital, Case Western Reserve University, Cleveland, OH*

M. G. Murphy • *Department of Physiology and Biophysics, Dalhousie University, Halifax, Nova Scotia, Canada*

Ch. R. K. Murthy • *School of Life Sciences, University of Hyderabad, Hyderabad, India*

S. Naik • *Department of Pathology, Rhode Island Hospital, Providence, RI*

Anders Nobin • *Department of Surgery, Lund University Hosptial, Lund, Sweden*

Michael D. Norenberg • *Department of Pathology, VA Medical Center and University of Miami, School of Medicine/Jackson Memorial Hospital, Miami, FL*

Christine Peterson • *Department of Psychobiology, University of California at Irvine, CA*

Ch. Pifl • *University of Vienna, Austria*

Gilles Pomier Layrargues • *Hepatology Unit, André-Viallet Clinical Research Centre, Hôpital St-Luc (University of Montreal), Montreal, Quebec, Canada*

Ijaz Qureshi • *Pediatric Research Center, Sainte-Justine Hosptial (University of Montreal), Montreal, Quebec, Canada*

Winfried A. Raabe • *Department of Neurology, VA Medical Center and University of Minnesota, Minneapolis, MN*

Tomas A. Reader • *Centre de recherche en sciences neurologiques (CRSN), University of Montreal, Montreal, Quebec, Canada*

A. M. Russo • *Clinica Medica III e Cattedra di Farmacoligia, Universita' di Modena, Modena, Italy*

Karin Rimpel-Lamhaouar • *Departments of Biochemistry and Neurology, Cornell University Medical College, New York, NY*

H. Santangini • *Department of Pathology, Rhode Island Hospital, Providence, RI*

J. W. Schenck • *Medical College of Wisconsin, Milwaukee, WI*

STEVEN SCHENKER • *Department of Gastroenterology and Nutrition, University of Texas Health Science Center, and Audie L. Murphy Memorial Veterans Hospital, San Antonio, TX*

ROBERT G. SHULMAN • *Department of Molecular Biophysics and Biochemistry, Yale University, New Haven, CT*

E. SINGER • *University of Vienna, Austria*

PHIL SKOLNICK • *Laboratory of Neuroscience, NIDDK, National Institutes of Health, Bethesda, MD*

K. M. SZAUTER • *Cleveland Metropolitan General Hospital, Case Western Reserve University, Cleveland, OH*

JOHN C. SZERB • *Department of Physiology and Biophysics, Dalhousie University, Halifax, Nova Scotia, Canada*

P. D. TOLENTINO • *Cleveland Metropolitan General Hospital, Case Western Reserve University, Cleveland, OH*

PETER G. TRABER • *Ann Arbor VA Medical Center and University of Michigan, Ann Arbor, MI*

MISAEL URIBE • *Liver Unit, National Institute of Nutrition, Mexico*

R. R. VARMA • *Hepatology Unit, Division of Gastroenterology, Medical College of Wisconsin, Milwaukee, WI*

E. VENTURA • *Clinica Medica III e Cattedra di Farmacoligia, Universita' di Modena, Modena, Italy*

C. VEZZELLI • *Clinica Medica III e Cattedra di Farmacoligia, Universita' di Modena, Modena, Italy*

FREDRICK L. WEBER, JR. • *Case Western Reserve University School of Medicine, VA Medical Center, Cleveland, OH*

BERNARD WILLEMS • *Hepatology Unit, André-Viallet Clinical Research Centre, Hôpital St-Luc (University of Montreal), Montreal, Quebec, Canada*

C. YURDAYDIN • *University of Vienna, Austria*

P. ZANOLI • *Clinica Medica III e Cattedra di Farmacognosia, Universita' di Modena, Modena, Italy*

M. L. ZENEROLI • *Clinica Medica III e Cattedra di Farmacoligia, Universita' di Modena, Modena, Italy*

LESLIE ZIEVE • *VA and Hennepin County Medical Centers, University of Minnesota, Minneapolis, MN*

CH. ZIMMERMAN • *University of Vienna, Austria*

A. Introduction

HEPATIC ENCEPHALOPATHY: THE PRESENT AND THE FUTURE

Steven Schenker, M.D.
University of Texas Health Science Center
and Audie Murphy Veterans Hospital,
San Antonio, Texas 78287

INTRODUCTION

Hepatic encephalopathy may be defined as an alteration in mental state accompanying and due to liver disease. Hepatic dysfunction may be of the chronic type (i.e. cirrhosis) with spontaneous or surgically-induced portal-systemic shunting of blood (portal-systemic encephalopathy, PSE), or caused by acute liver failure (viral or toxic). Encephalopathy, in the former disorder (PSE), is usually insidious in onset, precipitated by some specific insult (i.e. gastrointestinal hemorrhage, azotemia, hypokalemic alkalosis) and is often rapidly reversible with appropriate therapy. In this regard, PSE is unique since treatment is relatively effective despite lack of complete understanding of the pathogenesis of this syndrome. Some 75 percent of such patients respond to rather simple and innocuous therapy within 72 hours and, in fact, lack of improvement is cause to reevaluate the diagnosis.

It is the other type of hepatic coma (a term used hereafter interchangeable with encephalopathy), that which occurs with acute liver failure, for which our treatment is poor. Until recently, the survival has varied from 20-35 percent, worsening with older age. In this connection, two key questions come up. They are: 1) what is the pathogenesis of this disorder and is it the same as for PSE and 2) what progress has been made in the treatment of hepatic coma accompanying acute liver failure? The answer to the first question is unknown. Are the events of PSE telescoped over a shorter time during acute coma or are they quite different? Until this information is available, one must be very careful about extrapolating the pathogenesis of one disorder to another, especially with studies based on experimental animal models. As regards the second issue, a substitute liver (i.e. transplantation) has been the most promising therapeutic approach with more than 60 percent salvage in several small series involving about 60 patients (1-3). Unfortunately, such data have not been strictly controlled and utilize either historic comparisons or patients who had not been transplanted for various reasons as a basis for comparisons. Nevertheless, the published results are very impressive. There has also been some modest progress made, both at the level of pathogenesis (4-6) and possibly the treatment (7) of cerebral edema. Finally, there has been an explosion of data regarding liver regeneration with identification of a myriad of factors (8-10) and further unravelling of the complex process of control of the growth process (11-13). By contrast, there has been little progress in the important area of liver support systems, an issue which may be critical because of the logistic difficulties in getting the patient with fulminant liver failure to a liver transplant center and obtaining the new organ in a timely fashion. Hopefully, advances in prolonging the preservation of donor liver with newer solutions (14) may help in this regard. Another approach which seems to hold promise for these patients is allogenic hepatocyte transplantation. It has prolonged life in animals with partial hepatectomy (15).

The thrust of this discussion will be to assess the current status of the pathogenesis of hepatic encephalopathy. Despite research spanning many decades and many exciting hypotheses which have been extensively tested, the area is very controversial and the final answer still uncertain.

Pathogenesis of hepatic encephalopathy

a. Research problems:

Table 1 lists the main reasons for the difficulty in assessing the pathogenesis of hepatic encephalopathy. These include 1) poor understanding of the neurochemical mechanisms of normal mental state (i.e. awareness), 2) the heterogeneity of the brain at functional, cellular and biochemical levels, 3) applicability of various animal models of encephalopathy to the disease state in humans and finally 4) the perennial problem of assessing causal vs. casual relationships of cerebral biochemical findings to mental state changes. Perhaps more longitudinal studies in precomatose, comatose and postcomatose (therapeutic reversal of coma) states can help to resolve some of the cause-effect problems.

Table 1

PROBLEMS IN STUDYING THE PATHOGENESIS OF HEPATIC ENCEPHALOPATHY

1. Poor understanding of mechanism(s) of normal mental state (consciousness)

2. Heterogeneity of brain:
   - functional
   - cellular (astroglial cells)
   - biochemical (compartmentation)

3. Applicability of animal models to diverse types of human hepatic encephalopathy

4. Cause-effect relationship proof

The concept of brain heterogeneity deserves special emphasis as it may have an important bearing on the study of and, indeed, the mechanism(s) of hepatic encephalopathy.

Initial studies in experimental animals and clinical observations in patients with various brain lesions led to the concept that the reticular activating system in the brainstem may be critical for the off and on aspects of consciousness and the cortex for the content of consciousness (16). This view is clearly simplistic, as other areas of the brain have important modifying effects on awareness, but it served to focus attention initially on functional differences in various brain regions and the need to study these individually.

An example of regional differences in brain metabolism is illustrated well by ammonia processing. In rats studied over 4 weeks after a portacaval shunt there was a higher ammonia concentration in brain cortex than in the brainstem (17). Similar variation in data have been obtained when exogenous ammonia is given to animals (18) or when human brain from controls and patients with hepatic encephalopathy is analyzed postmortem for glutamate (19). It would appear, therefore, that various regions of the brain metabolize ammonia, a putative neurotoxin in hepatic encephalopathy, in a quantitately different manner. The explanation for this variable processing of ammonia is not known.

Moreover, in rats with portacaval shunts, regional metabolic rate for glucose is five-fold greater in the reticular formation of the brain as compared to the cortex (20). While the concept of increased glucose consumption in brain with shunting has been disputed on methodologic grounds (21), the variability in diverse brain regions is indisputable. This heterogeneity extends further to brain energy metabolism and to small brain areas. As shown in Figure 1, when rats are given ammonia, ATP and phosphocreatine fell transiently in the reticular activating formation, but not in the posterior colliculi of the brainstem (22).

Identical results were obtained by McCandless et al. (23) with octanoate, using both rapid brain freezing with liquid nitrogen and microwave irradiation.

A further indication of heterogeneity is at the cellular level, as shown best in astroglial cells. The elegant studies of Norenberg (24), Albrecht (25) and others, have shown that these cells which are uniquely altered in patients with PSE and hyperammonemia metabolize ammonia selectively and may exhibit a different biochemical profile than other brain cells, and their foot processes adjoin and form a part of the blood brain barrier. Recent reports by Norenberg of the protective effects of cyclic AMP on ammonia-induced changes in astrocytes in tissue culture (26) are particularly intriguing and may have both mechanistic and therapeutic implications. These cells deserve more study. Thus, regionally, microregionally, and at the cellular level brain metabolism may vary. This even extends to intracellular pools where metabolic compartmentation of glutamate and gamma-aminobutyric acid (GABA) is well documented (27). In conclusion, brain heterogeneity is an important aspect of cerebral function and requires consideration in studies of hepatic encephalopathy.

b. Theories of Pathogenesis:

Inasmuch as most forms of PSE (excluding hepatocerebral degeneration and spastic paraparesis) and hepatic coma ensuing from acute liver disease are potentially fully reversible, it has been generally agreed that most forms of this disorder are on a metabolic rather than structural basis. This implies that the astrocytic changes and cerebral edema seen in the chronic and more acute forms of hepatic encephalopathy are not primary etiologic events. This view is consistent with the current belief that the edema is of cytotoxic origin and is not on a vasogenic basis. The response of many patients with cerebral edema to mannitol is consonant with such a pathogenesis. A metabolic (neurochemical/neurophysiologic) basis for PSE and its full reversibility with therapy has been

1) <u>False neurotransmitters</u>.

The false neurotransmitter concept stipulates an excess of weak neurotransmitters (i.e. octopamine) in lieu of true ones, such as dopamine and norepinephrine (30). The basis for this innovative idea was high octopamine levels in brain of animals in coma due to liver damage, as well as in serum and urine of patients with PSE. Moreover, a low branched chain/aromatic amino acid ratio was

Table 2

THEORIES OF HEPATIC ENCEPHALOPATHY

1. False (weak) neurotransmitter excess

2. Gamma-aminobutyric acid (GABA) effect:
   - excess GABA in brain
   - increased GABA receptor binding and spillover sensitivity to other sedative ligands
   - endogenous sedative (? nitrogenous)

3. Ammonia excess (? synergism with other toxins)
   - various mechanisms (neuronal membrane effects, neurotransmitter (glutamate, aspartate) depletion, NADH/NAD shuttle impairment, energy loss (? secondary)

noted in patients with PSE and this would form a deficit of norepinephrine and dopamine, and an excess of false neurotransmitters in brain (30). Unfortunately, as detailed elsewhere, there are many problems with accepting this concept (31). Perhaps the most damaging are lack of effect on consciousness of administering large doses of false neurotransmitters intraventricularly and elevating brain levels of these false neurotransmitters by pharmacologic means (32). Moreover, very low concentrations of dopamine and norepinephrine in the brain of experimental animals did not induce coma and no changes in these neurotransmitters were found in the brain of patients dying with PSE. Controlled trials of L-Dopa in patients have been of no benefit (33). It is evident that this concept is unproven. It has, however, stimulated an exciting and novel research area and, thus, has benefited the field.

2) GABA: As regards the GABA hypothesis, each of the GABA subconcepts (see above) deserve individual consideration. The original view that GABA generated in the gut, poorly metabolized by the damaged liver, penetrates a deranged blood/brain barrier and inhibits neurotransmission in the brain (34) seems to be untenable on the basis of currently available data. First, the initial assays for blood GABA appear to have measured other amino acids; hence the increment of true GABA in plasma of animals and patients with hepatic coma is increased to a considerably lesser degree, if at all (35,36). This issue, however, is still controversial (37). Changes in blood/brain barrier with hepatic coma have also been questioned (38-39). In any event, the usual concentration of GABA in brain is much higher than in blood, hence it is difficult to see how small increments in blood can affect brain GABA levels. Brain GABA concentrations, and GABA metabolizing enzymes in experimental animal with liver damage, and in patients with PSE, also are normal (40,41). Conceivably a subcompartment of GABA, not measurable by available techniques, may be increased, but this view is conjectural. One may even ask how much GABA excess is needed to induce coma since 3-4-fold increases of GABA in cortex cerebellum of rats given the GABA transaminase inhibitor, ethanolamine sulfate, over 4 weeks, failed to inhibit consciousness (42). Again, perhaps, pools of GABA in key synaptic sites are crucial. Still, the data give one pause.

What about the concept of increased GABA receptor binding? It is difficult to reconcile such a view with increased GABA in brain since this would be expected to cause down-regulation of the appropriate receptors. A detailed analysis of this concept is outside of the scope of this brief review, but in essence, binding data are variable in different animal models and most lately show no change. The latest count is increased binding in 5 studies, unchanged binding in 11 and decreased binding in 2 reports (43). Also, why should lactulose be of therapeutic value if the problem is purely central?

In the final analysis, one would hope for an answer from human postmortem material or by in vivo studies using newer techniques, such as positron emission tomography (PET), to assess GABA receptor status in patients with PSE. The issue is important as it might explain the apparent sensitivity of patients with liver disease to benzodiazepines (44) and would imply that treatment of PSE could be directed at the brain (45). Unfortunately, postmortem human brain studies are not easy to carry out and may be confounded by the antemortem condition of the patient, drugs taken or other variables of stress, electrolytes, alcohol intake, convulsions, etc (46). In the most comprehensive report available, Butterworth et al. (47) found no change in GABA agonist binding to 2 brain regions in cirrhotics with PSE as compared to controls. Benzodiazepine binding, likewise, was not altered in the cirrhotics. By contrast, in a preliminary study, Ferenci obtained an apparent increase in GABA binding to postmortem human brain obtained from patients with mild coma, while with severe coma there was a decrease (48). Does the degree of coma matter? What about subclasses of receptors and variations in methodology used (48a). Another issue is the basis for upregulation of the GABA receptor. One group has suggested that this is due to "denervation" (49). The precise meaning of this is uncertain, but it would have to account for the rapid reversal of coma due to PSE with therapy. Thus, the GABA receptor upregulation concept is clearly unproven.

Perhaps the most exciting and plausible aspect of the GABA hypothesis relates to the presence of an endogenous benzodiazepine-like sedative in patients with PSE. Teleologically it would seem reasonable for such a substance to exist. Otherwise, what would be the purpose of receptors for benzodiazepines in brain? In fact, such substances have been extracted from various body fluids and it has been speculated that they may be hypoxanthine, inosine, desmethyldiazepam, or other agents. The presence of such substances in plants or milk has also been reported (50), and it has been suggested that they are ingested rather than synthesized in the body. Such a theory would also fit the scattered reports of therapeutic benefit in experimental animals and patients of benzodiazepine antagonists (51-54) and of increased benzodiazepine binding to brain of some animal models with coma due to liver damage (55). Unfortunately, the animal data, both as to binding and therapeutic success, have been inconsistent (40,43), no increase in benzodiazepine binding has been noted in some human studies (47) and the therapeutic studies in patients have been surprisingly uncontrolled. Nevertheless, the evidence for such a putative sedative continues to accumulate. Some have shown the presence of an inhibitor in plasma and CSF of patients with PSE, although the standard error was large and the relationship of the sedative to coma vs. just liver failure remains to be validated (56). Other studies have utilized cerebellar Purkinje neurons in tissue culture from animals with coma and controls, and have shown increased sensitivity of the former to benzodiazepine antagonists, implying possible occupancy of the receptor site by an endogenous inhibitor (57). Finally, using positron emission tomography and a benzodiazepine antagonist as a labeled probe, greater uptake of the label was seen in the brain of patients with PSE than in controls (44). This has been attributed as possibly due to displacement of an endogenous ligand, although other interpretations (i.e. greater transfer of the label to brain of PSE patients) are also possible. Using the same technique, decreased binding to epileptic foci in vivo has been documented. Clearly the specificity of this technique will have to be documented. Perhaps studies prior, during and after reversal of coma may better establish its significance. Overall, this endogenous sedative concept likewise has not been proven, but it is presently buffeted by fewer negatives. It could also account for the apparent sensitivity of such patients to other exogenous sedatives, especially the benzodiazepines (58). Moreover, if it turned out to be nitrogenous in nature it might explain the beneficial effects of lactulose and other similar therapeutic approaches.

3. Ammonia. The last hypothesis relates to ammonia toxicity, likely aided and abetted by other synergistic toxins. This view has long been championed by Zieve (59). The concept fits well experimental animal data and is consistent with clinical findings in children with urea cycle abnormalities and ammonia elevation. Problems with it are lack of good correlation between mental state and blood ammonia in some patients with liver disease. This, however, is probably not surprising as the methods for measuring blood ammonia are not carried out well in many laboratories, blood ammonia (even arterial and fasting) may not reflect brain ammonia well and, of course, it is likely that factors other than elevated ammonia play a role in hepatic coma. The role of brain vs. blood ammonia is illustrated well by better correlation of cerebrospinal fluid glutamine (the end product of ammonia metabolism in brain) with the clinical state (60,61). It is this reviewer's opinion that on balance ammonia must continue to be considered a prime (but not the only) villain in the scenario of hepatic encephalopathy.

Lest it be said that there is nothing new to the ammonia toxicity concept, it should be emphasized that data showing the importance of small bowel glutamine as a major source of ammonia (62) and of skeletal muscle as a detoxifying site are indications of progress in the field (63).

The mechanism(s) of ammonia toxicity to the brain are uncertain - possibly because there are so many. They include a direct neurophysiological effect on the neuronal membrane altering chloride extrusion, impaired formation of the excitatory neurotransmitters - glutamate and aspartate, and impairment of the hydrogen shuttle into mitochondria with a secondary decrease in energy formation (64,67). These various effects are not mutually exclusive. Three points are worth making. First, the neurophysiological/membrane mechanism requires further testing at various stages of coma-preceding, during and after recovery from it - to establish the functional significance of changes reported. This is not easy to do in the setting wherein such measurements are made. Second, the decrease in glutamate (and other excitatory amino acid neurotransmitters) has been gaining adherents. The evidence for this view has been summarized elsewhere (43,65), and briefly includes 1) consistent, low glutamate in brain of patients and animals with hepatic coma using various techniques, including nuclear magnetic resonance, 2) possible decrease in glutamate release from brain with ammonia, 3) increase in glutamate binding to brain of some, but not all, animals with liver damage and 4) the key role of glutamate in ammonia metabolism. The protective effect of methionine sulfoximine on ammonia toxicity (66), may bear on this, as well. This has been attributed to inhibition of glutamine synthesis (hence, elevation of glutamate) by the drug, although other effects of this agent cannot be ruled out. The fall of glutamate with ammonia excess may be not only due to formation of glutamine, but also the inhibition of glutaminase by ammonia. This may link to interference with the malate/aspartate shuttle of H^+ equivalents from cytoplasm to mitochondria, which is necessary to reduce intramitochondrial NAD to NADH (67). This brings up the third point, effects of ammonia on energy metabolism. A fall in mitochondrial NADH may secondarily reduce energy formation, as discussed earlier, and perhaps corroborated recently by the finding of high spinal fluid lactate in patients with PSE (68). It should be emphasized that changes in energy level may be localized to key areas of the brain (22,23), may be seen only at high ammonia levels and may not play a primary role in the coma. Thus, normal energy stores in cortex of such animals using NMR are not surprising (69), as they have been reported earlier. Moreover, not all data relating to glutamate binding to brain receptors is consistent. This may relate again to animal model, stage of coma, technique used and/or subclass of receptors assayed.

With these caveats, probably the best consensus as to pathogenesis, at this time, is that 1) ammonia is still _a_ primary toxin, 2) the possibility of an endogenous neuroinhibitor acting on the GABA receptor supercomplex needs to be explored further and 3) the mechanism of hepatic encephalopathy is a disturbed balance of excitatory and inhibitory neurotransmitters at key sites. This lumped concept may not only decrease the nonconstructive controversy in the field, it may even be correct. It is a composite concept that currently makes the most sense to me.

FUTURE NEEDS

What about future research needs (Table 3)? These include, at the basic level, 1) better validated animal models, 2) more correlative studies of neurochemistry and neurophysiology, 3) more emphasis on microheterogeneity (i.e. study of astroglial cells) and 4) a better pathophysiological definition of cerebral edema. At the clinical level, the needs are for 1) support systems for acute liver failure, 2) studies of hepatic regeneration, 3) controlled studies of GABA and benzodiazepine antagonists, 4) better definition of minimal PSE and its reversal and 5) use of newer (properly validated) techniques such as PET scanning and NMR for _in vivo_ assessment of cerebral function. Perhaps the most important need is for a critical _but_ open mind. About 100 years ago Aldous Huxley stated, "It is the customary fate of new truths to begin as heresies and to end as superstitions". It is important to remember this as we assess new concepts in the field of hepatic encephalopathy. Where will we be in another 100 years?

Plus ça change, plus c'est la même chose.

Table 3

FUTURE RESEARCH NEEDS

<u>Basic</u>

. Better validated animal models, especially of PSE.

. More correlative studies of neurochemistry and neurophysiology.

. More emphasis on brain microheterogeneity- i.e. astroglial cells.

. Better definition of blood/brain barrier and cerebral edema.

<u>Clinical</u>

. Support systems for acute liver failure.

. Studies of hepatic regeneration.

. <u>In vivo</u> controlled studies of agonists and antagonists with proper end points.

. Definition of minimal PSE.

. Use of newer <u>in vivo</u> techniques in patients- i.e. positron emission tomography and nuclear magnetic resonance.

REFERENCES

1. Iwatsuki, S., Esquivel, C.O., Gordon, R.D., et al. (1985). Liver transplantation for fulminant hepatic failure. Semin. Liver Dis. 5, 325-328.

2. Ringe, B., Pilchmayr, R., Lancchart, W., et al. (1986). Indications and results of liver transplantation in acute hepatic failure. Transplant. Proc. 18, 86-88.

3. Bismuth, H., Samuel, D., Gugenheim, J. (1987). Emergency liver transplantation for fulminant hepatitis. Ann. Int. Med. 107, 337-341.

4. Lo, W.D., Ennis, S.R., Goldstein, G.W., et al. (1987). The effects of galactosamine-induced hepatic failure upon blood/brain barrier permeability. Hepatology 7, 452-456.

5. Gantz, R., Swain, M., Traber, P., et al. (1988). A role for ammonia in the brain edema of fulminant hepatic failure (FHF) in the rat. Studies in cortical brain slices. Hepatology 8, 1247.

6. Seda, H.W.M., Hughes, R.D., Gove, C.D., et al. (1984). Inhibition of rat brain Na+, K+-ATPase activity by serum from patients with fulminant hepatic failure. Hepatology 4, 74-79.

7. Canalese, S., Gimson, A.E.S., Davis, C. et al (1982). Controlled trial of dexamethasone and mannitol for the cerebral edema of fulminant hepatic failure. Gut 23, 625-629.

8. LaBrecque, D.R., Steele, G., Fogerty, S. et al. (1987). Purification and physical-chemical characterization of hepatic stimulator substance. Hepatology 7, 100-106.

9. Michalopaulos, G., Houck, K.A., Dolan, M.L. et al. (1984). Control of hepatocyte replication by two serum factors. Cancer Res. 44, 4414-4419.

10. Francavilla, A., DiLeo, A., Polineno, L. et al. (1986). The effect of hepatic stimulatory substance isolated from regenerating hepatic cytosol and 50,000 and 300,000 subfractions in enhancing survival in experimental acute hepatic failure in rats treated with galactosamine. Hepatology 6, 1346-1351.

11. Baskin, G.S., Henderson, G.I., Schenker, S. (1988). Ethanol and hepatic regeneration. Hepatology 8, 408-411.

12. Alsion, M.R. (1986). Regulation of hepatic growth. Physiol. Rev. 66, 499-541.

13. Roper, S.E., Burwen, S.J., Barker, M.E., Jones, A.L. (1987). Translocation of epidermal growth factor to the hepatocyte nucleus during rat liver regeneration. Gastroenterology 92, 1243-1250.

14. Jamieson, N.V., Sandburg, R., Lindell, S. et al. (1988). Successful 24-30 hour preservation of the canine liver: a preliminary report. Transplant. Proceed. 20, 945-947.

15. Demetriou, A.A., Reisner, A., Sanchez, J. et al. (1988). Transplantation of microcarrier-attached hepatocytes into 90% partially hepatectomized rats. Hepatology 8, 1006-1009.

16. Magoun, H.W. (1963). The Waking Brain, 2nd Ed., C.S. Thomas, Springfield, IL.

17. Butterworth, R.F., Giguère, J.F., Michaud, J. et al. (1987). Ammonia: key factor in the pathogenesis of hepatic encephalopathy. Neurochem. Path. 6, 1-12.

18. Hindfelt, B. (1973). The effect of acute ammonia intoxication upon the brain energy state in rats pretreated with L-methionine D-L-sulphoximine. Scan. J. Clin. Lab. Invest. 31, 289-299.

19. Butterworth, R.F., Lavoie, J., Pomier Layrargues, G., Giguère, J-F. (1988). Neurotransmitter changes in autopsied brain tissue from cirrhotic patients with hepatic encephalopathy. In Advances in Ammonia Metabolism and Hepatic Encephalopathy, P.B. Soeters, J.H.P. Wilson, A.J. Meijer, E. Holm (Eds.). Elsevier, pp. 319, 324.

20. Lockwood, A.H., Ginsberg, M.D., Rhoades, H.M., Gutierrez, M.T. (1986). Cerebral glucose metabolism after portacaval shunting in the rat. J. Clin. Invest. 78, 86-95, 1986.

21. Hawkins, R.A., Mans, A.M., Biebuyck, J.F. (1987). Changes in brain metabolism in hepatic encephalopathy. Neurochem. Pathol. 6, 35-66.

22. McCandless, D.W., Schenker, S. (1981). Effect of acute ammonia intoxication on energy stores in the cerebral reticular activating system. Exp. Brain Res. 44, 325-330.

23. McCandless, D.W. (1985). Octanoic acid induced coma and reticular formation energy metabolism. Brain Res. 335, 131-137.

24. Norenberg, M.D. (1987). The role of astrocytes in hepatic encephalopathy. Neurochem. Pathol. 6, 13-34, 1987.

25. Albrecht, J., Wysmyk-Cybula, U., Rafalowska, V. (1985). Na+/K+-ATPase activity and GABA uptake in astroglial cell-enriched fractions and synaptosomes derived from rats in the early stages of experimental hepatogenic encephalopathy. Acta Neurol. Scand. 72, 317-320.

26. Gregorios, J.B., Mozes, L.W., Norenberg, L.O.B., Norenberg, M.D. (1986). Effect of cyclic AMP on ammonia-induced alterations in primary astrocyte cultures. J. Neuropathol. Exp. Neurol. 45, 396-407.

27. Cooper, A.J.L., Plum, F. (1987). Biochemistry and physiology of brain ammonia. Physiol. Rev. 67, 440-519.

28. Bernthal, P., Hays, A., Tarter, R.F. et al. (1987). Cerebral CT scan abnormalities in cholestatic and hepatocellular diseases and their relationship to neuropsychological test performance. Hepatology 7, 107-114.

29. Gilberstadt, S., Gilberstadt, H., Zieve, L. et al. (1980). Psychomotor performance defects in cirrhotic patients without overt encephalopathy. Arch. Int. Med. 140, 519-521.

30. Fischer, J.E., Baldessarini, R.J. (1976). Pathogenesis and therapy of hepatic coma. Prog. Liver Dis. 1, 363.

31. Hoyumpa, A.M., Schenker, S. (1982). Perspectives in hepatic encephalopathy. J. Lab. Clin. Med. 100, 477-487.

32. Zieve, L., Olsen, E.L. (1977). Can hepatic coma be caused by a reduction of brain noradrenaline or dopamine? Gut 18, 688.

33. Michel, H., Solere, M., Granier, P., et al. (1980). Treatment of cirrhotic hepatic encephalopathy with L-Dopa. A controlled trial. Gastroenterology 79, 207.

34. Schafer, D.F., Jones, E.A. (1982). Potential neural mechanisms in the pathogenesis of hepatic encephalopathy. In: Progress in Liver Disease, H. Popper, F. Schaffner (Eds.), New York: Grune & Stratton, vol. 7, pp. 615-627.

35. Maddison, J.E., Dodd, P.R., Morrison, M. et al. (1987). Plasma GABA, GABA-like activity and the brain GABA-benzodiazepine receptor complex in rats with chronic hepatic encephalopathy. Hepatology 7, 621-628.

36. Moroni, F., Riggio, O., Carla, V. et al. (1987). Hepatic encephalopathy: lack of changes of gamma-aminobutyric acid content in plasma and cerebrospinal fluid. Hepatology 7, 816-820.

37. Levy, L.J., Leek, J., Losowsky, M.S. (1987). Evidence for gamma-aminobutyric acid as the inhibitor of gamma-aminobutyric acid binding in the plasma of humans with liver disease and hepatic encephalopathy. Clin. Sci. 73, 531-534.

38. Lo, W.D., Ennis, S.R., Goldstein, G.W. et al. (1987). The effects of galactosamine-induced hepatic failure upon blood/brain barrier permeability. Hepatology 7, 452-456.

39. Traber, P.G., Canto, M.D., Granger, D.R. et al. (1987). Electron microscopic evaluation of brain edema in rabbits with galactosamine-induce fulminant hepatic failure: ultrastructure and integrity of the blood-brain barrier. Hepatology 7, 1272-1277.

40. Maddison, J.E., Dodd, P.R., Johnston, G.A.R. et al. (1987). Brain gamma-aminobutyric acid receptor binding is normal in rats with thioacetamide-induced hepatic encephalopathy despite elevated plasma gamma-aminobutyric acid-like activity. Gastroenterology 93, 1062-1068.

41. Lavoie, J., Giguere, J.F., Layrargues, G.P., Butterworth, R.F. (1987). Amino acid changes in autopsied brain tissue from cirrhotic patients with hepatic encephalopathy. J. Neurochem. 49, 692-697.

42. Sykes, C., Prestwich, S., Horton, R. (1984). Chronic administration of the GABA-transaminase inhibitor ethanolamine o-sulphate leads to up-regulation of GABA binding sites. Biochem. Pharmacol. 33, 387-393.

43. Schenker, S., Brady, C.E. (1988). Pathogenesis of hepatic encephalopathy. In: Hepatic Encephalopathy: Management with Lactulose and Related Carbohydrates. H.O. Conn, J. Bircher (Eds.), Medi-Ed. Press, East Lansing, MI, pp. 15-30.

44. Samson, Y., Bernau, J., Pappata, S. et al. (1987). Cerebral uptake of benzodiazepine measured by positron emission tomography in hepatic encephalopathy (correspondence). N. Engl. J. Med. 316, 414-415.

45. Schafer, D.F. (1987). Hepatic coma: studies on the target organ. Gastroenterology 93, 1131-1134.

46. Evoniuk, G., Skolnick, P. (1988). Anion regulation of agonist and inverse agonist binding to benzodiazepine receptors. J. Neurochem. 51, 1169-1175.

47. Butterworth, R.F., Lavoie, J., Giguere, J.-F., Pomier-Layrargues, G. (1988). Affinities and densities of high affinity [^{3}H]-muscimol (GABA-A) binding sites and of central benzodiazepine receptors are unchanged in autopsied brain tissue from cirrhotic patients with hepatic encephalopathy. Hepatology 8, 1084-1088.

48. Ferenci, P., Riederer, P., Jellinger, K. et al. (1988). Changes in cerebral receptors for gamma-aminobutyric acid in patients with hepatic encephalopathy. Liver 8, 225-230.

48a. Watkins, J.C., Evans, R.H. (1981). Excitatory amino acid transmitters. Ann. Rev. Pharmacol. Toxicol. 21, 165-204.

49. Zeneroli, M.L., Iuliano, E., Racogni, G. et al (1982). Metabolism and brain uptake of gamma-aminobutyric acid in galactosamine-induced hepatic encephalopathy in rats. J. Neurochem. 38, 1219-1222.

50. Wildman, J., Mohler, H., Vetler, W. et al. (1987). Diazepam and N-desmethyldiazepam are found in rat brain and adrenal and may be of plant origin. J. Neural. Transm. 70, 282-288.

51. Baraldi, M., Zeneroli, M.L., Ventura, E. et al. (1984). Supersensitivity of benzodiazepine receptors in hepatic encephalopathy due to fulminant hepatic failure in the rat: reversal by a benzodiazepine antagonist. Clin. Sci. 67, 167-175.

52. Bassett, M.L., Mullen, K.D., Skolnick, P. et al. (1987). Amelioration of hepatic encephalopathy by pharmacologic antagonism of the $GABA_A$-benzodiazepine receptor complex in a rabbit model of fulminant hepatic failure. Gastroenterology 93, 1069-1077.

53. Scollo-Lavizzari, C., Steinmann, E. (1985). Reversal of hepatic coma by benzodiazepine antagonist (Ro15-1788) (correspondence). Lancet 1, 1324.

54. Bansky, G., Meier, P.J., Ziegler, W.H. et al. (1985). Reversal of hepatic coma by benzodiazepine antagonist (Ro15-1788) (correspondence). Lancet 1, 1324.

55. Baraldi, M., Zeneroli, M.L. (1982). Experimental hepatic encephalopathy: changes in the binding of gamma-aminobutyric acid. Science 216, 427.

56. Mullen, K.D., Szauter, K.M., Kaminsky, K., Matuszak, S.M. (1988). Characterization of "endogenous" benzodiazepine (BZ) activity detected in human hepatic encephalopathy (HE). Hepatology 8, 1352.

57. Basile, A.S., Gammal, S.H., Mullen, K.D. et al. (1988). Differential responsiveness of cerebellar Purkinje neurons to GABA and benzodiazepine receptor ligands in an animal model of hepatic encephalopathy. J. Neurosci. 8,2414-2421.

58. Bakter, G., Fisch, H.V., Garkoganes, G. et al. (1987). Mechanism of the excessive sedative response to cirrhotics to benzodiazepines: model experiments with trazolan. Hepatology 7, 629-638.

59. Zieve, L. (1987). Pathogenesis of hepatic encephalopathy. Metab. Brain Dis. 2, 147-165.

60. Watanabe, A., Takei, N., Higashi, T. et al. (1984). Glutamic acid and glutamine levels in serum and cerebrospinal fluid in hepatic encephalopathy. Biochem. Med. 32, 225-231.

61. Plum, F. and Hindfeldt, B. (1986). The neurological complications of liver disease. In: Handbook of Clinical Neurology. P.J. Vinken,G. W.Bruyn (Eds.), Elsevier, New York,pp. 349-377.

62. Soeters, P.B., van Leeuwen, P.A.M., van Berlo, C.L.H. (1988). Nitrogen metabolism in the gut. In: Hepatic Encephalopathy: Management with Lactulose and Related Carbohydrates. H.O. Conn, J. Bircher (Eds.), Medi-Ed, East Lansing, MI, pp. 31-39.

63. Lockwood, A.M., McDonald, J.M., Reiman, R.E., Gelbard, A.S. et al. (1979). The dynamics of ammonia metabolism in man. Effects of liver disease and hyperammonemia. J. Clin. Invest. 63, 449-460.

64. Schenker, S., Hoyumpa, A.M. (1984). Pathophysiology of hepatic encephalopathy. Hospital Practice 19, 99-121.

65. Butterworth, R.F., Lavoie, J., Giguere, J.F., et al. (1987). Cerebral GABA-ergic and glutamatergic function in hepatic encephalopathy. Neurochem. Pathol. 6, 131-144.

66. Warren, K.S., Schenker, S. (1964). Effect of an inhibitor of glutamine synthesis (methionine sulfoximine) on ammonia toxicity and metabolism. J. Lab. Clin. Med. 64, 442.

67. Hindfelt, B., Plum, F., Duffy, T.E. (1977). Effect of acute ammonia intoxication or cerebral metabolism in rats with portacaval shunts. J. Clin. Invest. 59, 386-396.

68. Yao, H., Sadoshima, S., Fujii, K. et al. (1987). Cerebrospinal fluid lactate in patients with hepatic encephalopathy. Eur. Neurol. 27, 182-187.

69. Dentz, N.E.P., Chamuleau, R.A.F.M. (1988). In vivo ^{31}NMR spectroscopy of the rat cerebral cortex during acute hepatic encephalopathy. NMR Biomed. 1, 101-106.

B. Pathophysiology of Hepatic Encephalopathy

The Role of Ammonia and Other Toxins

AMMONIA IN LIVER AND EXTRAHEPATIC TISSUES: AN OVERVIEW OF METABOLISM AND TOXICITY IN MAMMALS

Arthur J.L. Cooper*, James C.K. Lai* and Alan S. Gelbard+
*Departments of Biochemistry and Neurology, Cornell University Medical College, New York, NY 10021 and +Biophysics Laboratory, Memorial Sloan Kettering Cancer Center, New York, NY 10021

INTRODUCTION

Ammonia is a major byproduct of systemic and cerebral nitrogen metabolism and is generated in at least 20 enzymatic reactions within the major organs of the body. Ammonia is thought to be generated in the gastrointestinal tract by the action of bacteria on nitrogenous substrates and by deamidation of glutamine in the large and small intestine. Substantial amounts of ammonia are generated in the liver from glutamate and in the kidney by deamidation of glutamine. The principal fate of systemic blood ammonia, in the brain and other organs, is incorporation into glutamine (amide). Portal vein ammonia (which is present at a much higher concentration (∿0.5-1.0 mM) than in the peripheral arterial or venous (20-110 μM) blood), on the other hand, is largely detoxified as urea in the liver. The glutamine derived from brain, muscle and other tissues acts as an energy source for the gut and at the same time releases ammonia for urea synthesis. Thus, ultimately, most extrahepatic ammonia is incorporated into urea by temporary storage in glutamine (amide). Despite the central importance of ammonia in nitrogen homeostasis, its concentration must be kept low. At elevated concentrations, ammonia is a potent neurotoxin. In liver disease, the normal ammonia detoxification mechanisms may be compromised. Alternatively, as may occur in chronic cirrhosis or surgical construction of a portacaval shunt, portal systemic shunting of blood past the liver will greatly increase systemic ammonia concentrations. Indeed, excess circulating ammonia has long been implicated as a contributing factor to the neurological dysfunction that is a frequent complication of liver failure and a number of metabolic diseases. Many theories for ammonia neurotoxicity have been put forward, including 1) disruption of energy metabolism, 2) changes in blood-brain barrier function, 3) changes in neurotransmitter metabolism, 4) alteration of protein/nucleic acid synthesis, and 5) interference of normal electrophysiological properties. With regard to the first hypothesis, excess ammonia has been shown to stimulate glycolysis and inhibit brain α-ketoglutarate dehydrogenase complex *in vitro*. *In vivo* ammonia increases lactate and the lactate/pyruvate ratio in brain (possibly by stimulating glycolysis, depressing tricarboxylic acid cycle activity and blocking the malate-aspartate shuttle). A rational basis for the treatment of patients with liver disease and other hyperammonemic syndromes is to stabilize nitrogen homeostasis to bring ammonia levels back into the normal range.

CEREBRAL AMMONIA METABOLISM

In a recent survey Cooper and Plum (1987) listed 16 enzymatic pathways known to generate ammonia in brain, of which the purine nucleotide cycle (net: L-aspartate + H_2O + GTP $\longrightarrow$ fumarate + NH_3 + ADP + P_i), glutaminase and glutamate dehydrogenase are quantitatively the most important. Some controversy exists as to which pathway is the most important source of cerebral ammonia (see for example, Benjamin, 1982). Benjamin and Quastel (1975) have formulated a theory of glutamine cycling wherein glutamate released from the nerve endings is "neutralized" by uptake into the astrocytes; carbon is returned to the nerve endings in the form of electrophysiologically neutral glutamine. Within the nerve terminals, glutamine is then a source of glutamate and possibly GABA (cf. Van Den Berg *et al*., 1978); indeed, nerve terminals are rich in glutaminase (Bradford and Ward, 1975; Kvamme, 1983). Thus, it is highly likely that glutaminase may locally generate appreciable quantities of ammonia. Nevertheless, Schultz and Lowenstein (1978) have argued strongly that the purine nucleotide cycle is the major source of metabolically generated whole brain ammonia.

A solution of ammonia contains ammonium (NH_4^+) plus ammonia free base (NH_3). The pK_a depends strongly on temperature and ionic strength but under physiological conditions (i.e., 37°C, moderate ionic strength) the pK_a is ∿9.2 (see the discussion by Cooper and Plum, 1987). Thus, at normal pH values only ∿1% of ammonia in the tissues is in the form of NH_3. The blood-brain barrier is largely impervious to ammonium. The on-off rate for the proton is exceedingly rapid compared to the transit time of blood through the brain so that ammonia can easily enter the brain from the blood largely by diffusion of the free base. Nevertheless, there is also some limitation to entry of NH_3; the single pass extraction of [^{13}N]ammonia (^{13}N is a positron-emitting isotope, $t_{1/2}$ of 9.96 min) is less than 100%. Values ranging from 20-50% have been obtained, depending on species and physiological state (Phelps _et al_., 1977; Cooper _et al_., 1979; Lockwood _et al_., 1979, 1980; Raichle and Larson, 1981).

Compartmentation of brain metabolism has been known for many years. In pioneering ^{15}N tracer studies, by Berl _et al_. (1962), it was shown that blood-derived ammonia is largely converted in brain to glutamine in a small compartment of rapidly turning over, glutamate that is distinct from a larger, more slowly turning over glutamate compartment. Over the years much evidence has accumulated to suggest that the astrocytes represent (or contribute to) the small compartment (see Cooper and Plum, 1987) and, indeed, much of the glutamine synthetase of the brain is found in the astrocytes (Norenberg and Martinez-Hernandez, 1979). Work from our laboratory emphasizes just how effective the small pool glutamine synthetase is at detoxifying blood-borne ammonia. Using [^{13}N]ammonia we showed that blood-borne ammonia on entering the brain is rapidly ($t_{1/2}$= 3 sec) and overwhelmingly (≧98%) converted to glutamine (amide) (Cooper _et al_., 1979, 1985). [^{13}N]Ammonia administered to the CSF of rats was also shown to be efficiently metabolized to L-[_amide_-^{13}N]glutamine in the small compartment (Cooper _et al_., 1979). This finding is a consequence of the anatomical localization of astrocytes; astrocytes underlie the epindymal layer lining the CSF side of the ventricles.

Cerebral glutamine synthetase is evidently quite effective at detoxifying blood-derived ammonia under normal physiological conditions. However, the enzyme appears to be working at near maximal capacity and cannot prevent the build-up of brain ammonia during chronic hyperammonemia. In a model of chronic liver disease (i.e. 14-week portacaval shunted rat), whole brain glutamine synthetase was not increased (Cooper _et al_., 1985) compared to controls. In a more recent study, Butterworth _et al_. (1988) verified these findings for the portacaval-shunted rats and showed that, in the brain stem and cortex, glutamine synthetase activity is actually significantly decreased compared to controls. Since the brain has no alternative to glutamine synthetase for detoxifying blood-borne ammonia (the urea cycle is incomplete in the brain), these findings emphasize, once again, the importance of maintaining regimens designed to keep blood ammonia levels low in patients with liver disease.

Effective removal of incoming ammonia by the brain suggests that there should be an arterio-venous (A-V) difference for ammonia across normal brain. This has not been demonstrated for the normal rat brain, possibly because of technical difficulties in measuring low concentrations of ammonia; a positive A-V difference, however, has been found across the brains of humans and cows (see discussion by Cooper and Plum, 1987) and of portacaval shunted rats (Duffy _et al_., 1983). Several studies have shown a net uptake of amino acids across the human brain (e.g., Felig _et al_., 1973). In order for the brain to maintain nitrogen homeostasis, some mechanism must exist for exit of nitrogen from the brain. Some studies have shown no output of glutamine whereas other studies, including one from our laboratory (Duffy _et al_., 1983), have shown a net output of glutamine. The output was particularly marked in portacaval-shunted rats. Thus, glutamine egress may serve to remove waste nitrogen from the brain. In this regard, it is interesting to note that the concentration of glutamine in CSF is more than twice that of all the other amino acids combined (Ferraro and Hare, 1984).

In summary, a simplified working hypothesis for brain nitrogen homeostasis may be posited as follows: glutamine synthetase plays a pivotal role in brain metabolism. Blood-derived and endogenous ammonia is detoxified as glutamine in astrocytes. This glutamine may serve as a souce of neurotransmitter glutamate (or GABA). At the same time, egress of glutamine to blood or the CSF may provide a means of removal, at least in part, of accumulated ammonia nitrogen (as the amide) or of some amino acid nitrogen (as the amine) and some carbon (Duffy *et al.*, 1983; Cooper and Plum, 1987). The area of brain nitrogen homeostasis is little understood, and somewhat controversial. Obviously, more work is needed to further establish the validity of the above working hypothesis.

LIVER AMMONIA METABOLISM

The theoretical capacity of liver glutamine synthetase to remove ammonia is similar to that of the urea cycle (see discussion by Cooper et al., 1987), yet it is generally assumed that the urea cycle is the more important process for the removal of portal vein ammonia. We have investigated the short-term metabolic fate of $[^{13}N]$ammonia in rat liver *in vivo* (Cooper *et al.*, 1987; 1988a,b). $[^{13}N]$Ammonia administered as a bolus to the portal vein of anesthetized rats was rapidly (order of seconds) incorporated into liver urea and glutamine (amide) in the ratio of 14:1 (Cooper *et al.*, 1987 (see below)). The periportal hepatocytes are enriched in the urea cycle enzymes whereas glutamine synthetase is largely located downstream in the sinusoid in a small ring of perivenous hepatocytes (cf. Häussinger and Gerok, 1986). Ammonia is largely removed from the portal vein before it reaches the glutamine synthetase of the perivenous hepatocytes. The glutamine synthetase in these cells may 1) act as a back-up system under conditions in which the urea cycle may be overloaded, or 2) metabolize endogenously-produced ammonia.

Label very rapidly appeared in glutamate, aspartate, alanine and various urea cycle components. As noted above, only a small amount of label appeared in glutamine. The very rapid labeling of glutamate and aspartate showed that 1) there is little diffusional limitation for entry of ammonia into the mitochondria compared to rates of enzymatic removal of ammonia, and 2) the rate of attainment of equilibrium distribution of label among components of the mitochondrial glutamate dehydrogenase and aspartate aminotransferase reactions was reached very quickly ($t_{1/2} \leq 2$ sec). We were unable to detect label corresponding to AMP in a deproteinized liver homogenate. Since aspartate was heavily labeled, the findings suggested that the purine nucleotide cycle (as others had previously proposed (Rognstad, 1977; Krebs *et al.*, 1978)) is not an important source of ammonia in the liver. No evidence for [^{13}N]AMP formation in the liver could be obtained when [^{13}N]glutamate or [^{13}N]alanine were administered (see below). Of interest was the finding that, although label was readily incorporated into citrulline, argininosuccinate, arginine, and urea, no label could be detected in carbamyl phosphate (limit of detection $\leq$ 0.2% of the label in the liver). Estimates of carbamyl phosphate in liver have varied widely, from $\sim$ 1.20 nmol/g (Tatibana and Shigesada, 1976) to 150 nmol/g (Raijman, 1974). It is often assumed that the major pool of carbamyl phosphate in the liver is in the mitochondria . Our data suggest that, regardless of whether cytosol or mitochondria represent the major pool of carbamyl phosphate, the concentration of carbamyl phosphate cannot be >20 µM within the mitochondria (Cooper *et al.*, 1987) (see also Meijer *et al.*, 1985). Ornithine transcarbamylase activity is very high in the liver mitochondria; it is possible that carbamyl phosphate, once formed, is rapidly channeled to citrulline.

The short-term metabolic fate of L-[^{13}N]alanine, L-[^{13}N]glutamate and L-[*amide*-^{13}N]glutamine was also investigated in rat liver *in vivo*. Label was found to equilibrate very rapidly among components of the cytosolic aspartate aminotransferase and alanine aminotransferase reactions. The time required to reach equilibrium distribution of label was of the order of seconds (Cooper *et al.*, 1988,a,b). The relative importance of glutamine as a source of urea nitrogen in the liver has been disputed in the literature. However, based on the results of experiments with isolated perfused livers, Haussinger and Gerok (1986) have stressed the importance of glutamine in this regard and have proposed a glutamine cycle as follows: Approximately two-thirds of portal vein-derived ammonia is converted to urea and one-third to glutamine in the liver. The glutamine released from the perivenous cells (and some glutamine released to the blood by other organs) is returned to the liver wherein it acts as a source of ammonia for urea synthesis in the periportal cells. The capacity of isolated hepatocytes to hydrolyze glutamine is high (Ochs, 1984) and Meijer (1985) has suggested that ammonia released from the glutaminase reaction is efficienty channeled to urea _in vivo_. As mentioned above, our findings definitely establish that portal vein ammonia is metabolized in the liver _in vivo_ in part to glutamine, but the proportion is smaller than that envisaged by Häussinger _et al_. (∿ 6% versus 33%). We also found that portal vein-administered L-[_amide_-^{13}N]glutamine is indeed converted to [^{13}N]ammonia and [^{13}N]urea in the intact rat liver _in vivo_ but that contribution of glutamine to ammonia for urea synthesis was rather modest - ∿ 7-13% of the ammonia used for urea synthesis. Thus, our data firmly establish that the processes involved in the glutamine cycle do indeed occur in the rat liver _in vivo_ but that, at least in the fed, anesthetized rat, the cycle is only working at a modest capacity. The cycle may be of more quantitative significance under altered physiological states, but this needs to be further investigated.

It is becoming increasingly apparent that the liver possesses a remarkable segregation of enzymatic activities (Häussinger and Gerok, 1986). We have already alluded to the periportal localization of the urea cycle enzymes and to the perivenous localization of glutamine synthetase. In addition, alanine aminotransferase is preferentially located in periportal hepatocytes (Quistorff, 1985). Häussinger and Gerok (1984) suggested that ∿ 65% of glutamate taken up by the liver from the portal vein enters the perivenous hepatocytes. Our ^{13}N experiments have provided direct evidence for the zonal uptake of glutamate and alanine in rat liver _in vivo_.

It has long been appreciated that alanine is a major gluconeogenic/ureogenic precursor in the liver (e.g. Krebs _et al_., 1978). The concentration of alanine in the portal vein is high (1.0-1.2 mM) and 25-30% of this alanine is removed on passage through the liver (Cooper _et al_., 1988a,b). Following a bolus injection of L-[^{13}N]alanine into the portal vein, label rapidly appeared in liver metabolites. Glutamate was quickly labeled because of the high activity of alanine aminotransferase. However, by 1 min, appreciable label was found in urea (∿ 24%); very little label was found in glutamine (∿ 1.6%) despite the fact that, by 30 sec, 48% of the label was in glutamate. The data suggest that ∿ 95% of the alanine was taken up in the urea-cycle-enriched periportal hepatocytes and only ∿5% in the glutamine-synthetase-containing perivenous hepatocytes (see below). Gluconeogenesis from portal vein-derived alanine is facilitated by preferential uptake into the periportal hepatocytes where the process is further enhanced by the high activity of alanine aminotransferase.

Previous studies have shown limited permeability of liver slices (possibly at the cell/plasma membrane) to penetration of L-[U-^{14}C]glutamate (Hems *et al.*, 1968). Nevertheless, we (Cooper *et al.*, 1988a,b) and others (Pardridge, 1977; Taylor and Rennie, 1987) have shown that a sizable (∿ 20%) fraction of portal vein glutamate is taken up by the liver. Our data further show that, on entry into the liver cells, this glutamate is rapidly metabolized. By 30 sec after a bolus administration of L-[^{13}N]glutamate, ∿ 50% of the label in the liver was in a metabolized form; ∿ 15% was in glutamine. This labeling pattern was very different from that found for L-[^{13}N]alanine despite comparable labeling of the glutamate pool by 30 sec. The data suggest that, of the glutamate nitrogen taken up by the liver, some (25-40%) is metabolized in the periportal hepatocytes (to urea) and the remainder (60-75%) in the perivenous hepatocytes (Cooper *et al.*, 1988a) (see below). The latter value is in remarkable agreement with the value estimated by Häussinger and Gerok (1984) of ∿65%. (It should be pointed out, however, that we cannot rule out the possibility of some portal-vein glutamate entering the parenchymal cells of the liver.)

Our estimates for partitioning of uptake of alanine and glutamate are based on the fact that label is diverted from L-[^{13}N]glutamate to glutamine about _10-12 times more slowly_ when L-[^{13}N]alanine is the precursor of liver L-[^{13}N]glutamate than when portal vein L-[^{13}N]glutamate is the precursor. On the other hand, liver L-[^{13}N]glutamate, derived from portal vein L-[^{13}N]alanine, is converted to [^{13}N]urea ∿ _3 times faster_ than is liver L-[^{13}N]glutamate, derived from portal vein L-[^{13}N]glutamate. A reasonable depiction of the uptake of these two amino acids into liver and the metabolic fate of the nitrogen in periportal and perivenous cells is as follows:

	[^{13}N]Ala uptake				[^{13}N]Glu uptake
			3/1		
periportal	95% →	Glu →	urea	←	∿35%
perivenous	5% →	Glu →	Gln	←	∿65%
			1/12		

ON THE SOURCE OF AMMONIA FOR UREA SYNTHESIS

Following the characterization of glutamate dehydrogenase by von Euler and the discovery of enzymatic transamination by Braunstein in the late 1930s, it became apparent that these two enzymes were important for the conversion of amino acid nitrogen to urea. In 1955, Braunstein pointed out that glutamate-utilizing aminotransferases could be reversibly linked to aspartate aminotransferase and glutamate dehydrogenase to provide a means of incorporating ammonia into an amino acid ("transreamination") or removing ammonia from an amino acid ("transdeamination"). Indeed, for some time now, biochemistry textbooks have shown the glutamate dehydrogenase reaction as a major source of ammonia for urea synthesis. However, this conclusion has not been universally accepted. In vitro, the reaction is thermodynamically strongly in favor of glutamate synthesis, but, because of the rapid reversibility of the reaction, it is difficult to determine from tracer studies whether glutamate dehydrogenase catalyzes net synthesis or breakdown of glutamate. However, Krebs, Braunstein and others have emphasized that, in the liver, the glutamate dehydrogenase reaction will be "pulled" toward glutamate breakdown by the continuous, irreversible removal of ammonia as urea. As noted above, our ^{13}N data (Cooper et al., 1987, 1988b) show just how rapidly label exchange is catalyzed by glutamate dehydrogenase. Our data suggest that, at equilibrium, nitrogen is partitioned 80% to liver ammonia and 20% to mitochondrial glutamate. Because this reversible exchange of nitrogen ($t_{1/2}$ to reach equilibrium $\leq$ 2 sec) is much faster than irreversible loss of ammonia nitrogen as urea ($t_{1/2}$ = 10-11 sec), urea formation from ammonia must result in a "pulling" of nitrogen from the mitochondrial pool of glutamate to ammonia and hence to urea. On the other hand, we were able to show that, at one minute after administration of [^{13}N]ammonia, some of the label in the liver was in *cytosolic* glutamate. Thus, the glutamate dehydrogenase was responsible for "pulling" away some ammonia from urea synthesis (at least temporarily) by the following sequence: $NH_3 \longrightarrow Glu_{mit} \longrightarrow Asp_{mit} \longrightarrow Asp_{cyt} \longrightarrow Glu_{cyt}$. For glutamate dehydrogenase to participate in *net* ammonia production in liver, irreversible nitrogen removal from glutamate ($NH_3 \longrightarrow$ urea) must be faster than glutamate nitrogen removal from the mitochondria. Because over the first minute, ∿ 5 times as much label was directed toward urea (plus urea cycle components) than toward glutamate (plus aspartate plus alanine) in the cytosolic compartment, our data suggest that glutamate dehydrogenase does, indeed, catalyze a net removal of ammonia from glutamate in rat liver *in vivo*. However, it must be borne in mind that glutamate dehydrogenase is not the only source of ammonia for urea synthesis. We have estimated that, in the fed, anesthetized rat, ammonia destined for urea synthesis in the liver is obtained very approximately as follows: 1) portal vein ammonia (33%), 2) portal vein glutamine(amide) (∿ 6-13%), 3) mitochondrial glutamate via the glutamate dehydrogenase reaction (∿ 20%), and 4) other sources including hepatic vein ammonia and glutamine, and the many enzymes in the liver capable of generating ammonia (∿ 33-40%) (Cooper *et al*., 1988a).

AMMONIA METABOLISM IN OTHER ORGANS

Because of the lack of sensitivity of the usually employed glutamine synthetase assay (i.e., the γ-glutamylhydroxamate assay), it has not always been possible to detect glutamine synthetase activity in organs other than liver, brain and kidney (see Cooper (1988) for a discussion). However, the presence of glutamine synthetase has definitely now been established in skeletal muscle, heart, adipocytes and lung (cf. Cooper, 1988). Skeletal muscle, particularly under conditions of vigorous exercise, can generate ammonia from AMP (see Cooper and Plum (1987) for a discussion). On the other hand, resting skeletal muscle can take up circulating ammonia and metabolize it to glutamine. Although the specific activity of skeletal muscle glutamine synthetase is fairly low compared to that of liver or brain, the large bulk of skeletal muscle ensures that muscle is a major organ for the removal of circulating ammonia. Lockwood *et* al. (1979) estimate that, in adult humans, skeletal muscle may remove as much as 50% of circulating ammonia. The authors further point out that, since muscle wasting is a common feature of liver disease, the inability of such patients to utilize the normal ammonia detoxification machinery of the liver may be further exacerbated by a compromise in extrahepatic removal of circulating ammonia. In addition to uptake of ammonia, the skeletal muscle also accumulates branched-chain amino acids and some other amino acids. Nitrogen balance is maintained, in part, by a net efflux of glutamine and alanine (see, for example, Harper et al., 1984).

In studies of ^{13}N-organ distribution of label following administration of a bolus of [^{13}N]ammonia via the femoral vein of rats, Freed and Gelbard (1982) noticed very rapid appearance of label in lung. In a later study, we noted that ∿ 30% of the dose accumulated in the lungs within the first few seconds of administration of the dose. Subsequently, we found that much of this label was incorporated into glutamine and that lung contains glutamine synthetase activity (Freed B.R. and Cooper A.J.L., unpublished results; see Cooper 1988). We have proposed that the lungs may be a (previously unappreciated) source of blood glutamine. A similar conclusion has also been reached recently by T. Welbourne (1988; and personal communication).

The kidney has long been recognized as an important organ for maintaining nitrogen homeostasis. Most waste nitrogen is eliminated in the urine as urea but smaller amounts are released as ammonium, amino acids, protein, creatinine and other nitrogenous compounds. The kidney may also contribute ammonia to the blood. Much of the kidney ammonium is derived from glutamine(amide) (Pitts, 1964) although glutamine(amine) and glutamate may also contribute. Glutamine synthetase is present in the kidneys of rats, mice, guinea pigs, rabbits and some other species but appears to be of low activity or absent in the kidneys of dogs and humans (see discussion by Cooper, 1988).

WHOLE BODY NITROGEN HOMEOSTASIS

Interorgan pathways of nitrogen flux are complex and will depend, on age, nutritional state and other factors. Nevertheless, a few generalizations may be pointed out. Tissues other than liver either do not have a functioning urea cycle or have very limited capacity to make urea from ammonia. Extrahepatic tissues therefore must rely on glutamine synthesis for the detoxification of ammonia. At the same time, waste amino acid nitrogen can also be incorporated into glutamine amine via α-keto-glutarate-linked transamination (amino acid (N) ⟶ glu (N) ⟶ glutamine (α-N)). Glutamine egress occurs from muscle, adipose tissue, and, probably, brain and lung. This glutamine is transported to the gut wherein it acts as a major energy source (Windmueller and Spaeth, 1974). Ammonia released by this cellular process, together with ammonia derived from the degradation of nitrogenous substances in the lumen of the gut, is transferred via the portal vein to the liver wherein it is converted to urea. Thus, extrahepatic ammonia is temporarily detoxified as glutamine (amide); ultimately, the nitrogen is incorporated mostly into urea.

Another amino acid present in blood in high concentrations is alanine and this amino acid also plays a special role in the maintenance of nitrogen homeostasis. Alanine is liberated from skeletal muscle in large amounts. The alanine nitrogen is obtained via linked transamination reactions; the branched chain amino acids are particularly effective nitrogen donors to pyruvate via linked transaminases in muscle (Harper et al., 1984). The α-keto acid analogues of the branched chain amino acids are oxidized to CO_2 to provide some energy or are released to the blood. The alanine liberated to the blood together with alanine in the portal vein (derived, in part, from the breakdown of glutamine in the small intestine) acts as an excellent source of nitrogen for urea synthesis. Felig (1973) has pointed out that the loss of carbon from the muscle as alanine may be replenished by glucose orginally released by the liver. Some of this glucose is, in turn, derived by gluconeogenesis from pyruvate within the liver. This "alanine cycle", in which C and N are transported between muscle and liver, is probably not stoichiometric, but the cycle does emphasize the interorgan flow of C and N in the maintenance of whole body homeostasis.

ON THE MECHANISM OF AMMONIA TOXICITY

Ammonia was first considered to be a major toxin in hepatic failure over 100 years ago. However, from time to time, the relative importance of ammonia in hepatic encephalopathy has been challenged. Blood ammonia levels do not always correlate with the degree of encephalopathy. Occasionally, a patient with encephalopathy has normal blood ammonia levels. However, ammonia is a potent neurotoxin in experimental animals, and patients with liver disease are susceptible to an ammonia challenge. Moreover, excess ammonia is directly toxic to infants with inborn errors of the urea cycle; neurological damage in such children is minimized by aggressive treatment to reduce hyperammonemia as soon as possible after birth. Ammonia is a complicating factor in patients with valproate poisoning who have normal liver function. A blood ammonia in the normal range in patients with liver disease does not preclude a previous hyperammonemic episode. Moreover, the bloods drawn are almost always venous and the ammonia levels in venous blood do not necessarily reflect the concentration of ammonia delivered to the brain in the arterial supply. CSF glutamine and α-ketoglutaramate (the α-keto acid analogue of glutamine) concentrations correlate much better with the degree of encephalopathy than do blood ammonia concentrations. Presumably, a high CSF glutamine concentration reflects a previous excess of ammonia within the CNS. Thus, the overall consensus at this time has swung back in favor of ammonia. Ammonia is now regarded as a contributor to the neuropathology of liver disease, but ammonia is, by no means, the only pathological factor in the disease (for original references see Cooper and Plum, 1987).

As mentioned in the introductory statement, several theories on the toxicity of ammonia have been put forward. (See Raabe (this volume) for a discussion on the electrophysiological effects of excess ammonia and Hawkins (this volume) for a discussion of the changes in glucose metabolism and blood-brain barrier amino acid transport in chronically hyperammonemic (i.e., portacaval-shunted) rats.) In this section we will focus on ammonia-induced changes in cerebral energy metabolism.

Ammonia stimulates glycolysis in brain extracts, possibly by stimulating/disinhibiting phosphofructokinase

(Lowry and Passonneau, 1966). Such an activation could account for the stimulation of glycolysis sometimes observed in the brains of acutely hyperammonemic rats. On the other hand, high concentrations of ammonia inhibit cerebral respiration *in vitro*; elevated ammonia levels in animals injected with [U-^{14}C]glucose result in reduction in the amount of label incorporated into brain glutamate, aspartate and GABA (see discussion by Cooper and Plum, 1987). Hindfelt *et al.* (1977) have proposed that excess ammonia interferes with the malate-aspartate shuttle. Bessman and Bessman (1955) suggested that excess ammonia interferes with brain energy metabolism by promoting the reductive amination of α-ketoglutarate to glutamate, draining the TCA cycle. However, brain α-ketoglutarate levels are normal or elevated and glutamate levels are generally depressed in the brains of hyperammonemic animals. The TCA cycle depletion theory is now discounted, but excess ammonia *can* affect the TCA cycle in other ways. McKhann and Tower (1961) noted that excess ammonia interferes with oxidation of pyruvate and α-ketoglutarate *in vitro*. In keeping with these findings we showed that ammonia, in the pathophysiological range (0.5-1 mM), strongly inhibits rat brain α-ketoglutarate dehydrogenase complex (Lai and Cooper, 1986). Thus, it seems that excess ammonia may interfere with the TCA cycle by impeding flow through the cycle rather than by draining carbon from it.

A possible scheme for interference of brain energy metabolism in acute hyperammonemia is as follows: Excess ammonia stimulates glycolysis, but at the same time inhibits both the TCA cycle and the malate-aspartate shuttle. These selective effects should result in an elevation of the lactate/pyruvate ratio in the brains of rats and, indeed, this is found to be the case. Despite these changes in overall energy metabolism, whole-brain changes in ATP and PCr in the hyperammonemic rat are often only modest (but see below), leading some authors to question whether disruption of energy metabolism is a major factor in ammonia intoxication. However, acute ammonia intoxication in the mouse is accompanied by transient falls in PCr and ATP within the paramedian structures of the brain stem, whereas the energy state of the neighboring structures remains unaffected (McCandless and Schenker, 1981). Hindfelt (1983) has summarized the effects of excess ammonia on brain ATP

PCr as follows: In both acute and sustained moderate hyperammonemia, the PCr level falls whereas the sum total of ATP + ADP + AMP remains unchanged. In more persistent low-grade hyperammonemia, the adenylate pool declines whereas energy charge (ATP + 0.5 X ADP/(ATP + ADP + AMP)) is not affected. At higher concentrations of ammonia in the brain (1-3 mmol/kg), ATP falls while AMP and ADP correspondingly rise and the energy charge is decreased; the changes are especially marked in the brainstem. In portacaval-shunted rats subjected to an acute ammonia challenge, brain ATP falls slightly but significantly by 60 min whereas PCr falls markedly (Hindfelt *et al.*, 1977).

Some controversy exists over the effects of chronic hyperammonemia on brain glucose utilization. One study showed a slight drop following portacaval shunting in rats but, by 12 wk a tendency appeared toward increased glucose utilization (Cruz and Duffy, 1983; see also Lockwood, this volume). However, Mans *et al.* (1983) noted that brain glucose utilization was decreased by 20% in 4- to 8-wk portacaval-shunted rats. In encephalopathic patients, the cerebral metabolic rate for O_2 ($CMRO_2$) and cerebral blood flow (CBF) declined roughly in parallel with the severity of the neurological changes (Posner and Plum, 1960). In alert, 8-wk portacaval-shunted rats with chronic low grade hyperammonemia, the addition of a superimposed ammonia load promptly induced stupor and EEG abnormalities accompanied by substantial reductions in $CMRO_2$ and CBF (Gjedde *et al.*, 1978). Recently, Ross (1988) studied the ^{31}P-NMR spectra of the brains of a small number of patients with liver disease (clinical status not given) and reported a significant decrease in cerebral Pi and Pi/ATP ratio in such patients. ^{31}P-NMR spectroscopy seems promising. Perhaps, as the methodology improves, ^{31}P-NMR might afford a non-invasive means of assessing energy metabolism in the brains of patients with liver disease and of correlating energy deficit with degree of cerebral impairment.

In conclusion, hyperammonemia seems to affect cerebral energy metabolism via a variety of mechanisms. CBF and $CMRO_2$ are generally decreased; changes in whole brain "high-energy" phosphates are generally modest except in severe hyperammonemia, but the changes may be more pronounced in regions of the brain stem.

CONCLUSION

Ammonia is an integral part of nitrogen homeostasis, yet its concentration must be kept low. Hyperammonemia is a contributing factor to many diseases including liver cirrhosis, inborn errors of the urea cycle and valproate toxicity. The mechanism whereby excess ammonia is neurotoxic is poorly understood. Undoubtedly, excess ammonia affects many physiological functions of the brain, including energy metabolism. The effect on energy metabolism is complicated; probably, several individual enzymatic steps of glycolysis and the TCA cycle are affected. In experimental animals, changes in cerebral "high-energy" phosphates are generally modest, but become more pronounced under conditions of severe hyperammonemia. Whether excess ammonia associated with liver disease directly affects cerebral energy metabolism in patients with liver disease remains to be seen; however, $CMRO_2$ and CBF are decreased in such patients. The cerebral "energy status" of patients with liver disease and a correlation (if any) with degree of encephalopathy may perhaps be ascertained in future studies by the use of *in vivo* ^{31}P-NMR spectroscopy.

Acknowledgments: These studies were supported, in part, by NIH grants DK-16739 and NS-24592, CA-34603 and Department of Energy contract EE-77-5-5268. We thank Karen Morgan for preparing this manuscript.

REFERENCES

Benjamin, A.M. (1982) Ammonia. In: *Handbook of Neurochemistry* (2nd Ed) (A. Lajtha, ed); Plenum Press; New York; pp. 117-137.

Benjamin, A.M., and Quastel, J.H. (1975) Metabolism of amino acids and ammonia in rat brain cortex slices *in vitro*: A possible role of ammonia in brain function. *J. Neurochem.* *25*, 197-206.

Berl, S., Takagaki, G., Clarke, D.D., and Waelsch, H. (1962) Mebabolic compartments *in vivo*: Ammonia and glutamic acid metabolism in brain and liver. *J. Biol. Chem.* *237*, 2562-2569.

Bessman, S.P., and Bessman, A.N. (1955) The cerebral and peripheral uptake of ammonia in liver disease with an hypothesis for the mechanism of hepatic coma. J. Clin. Invest. 34, 622-628.

Bradford, H.F. and Ward, H.K. (1975) Glutamine as a metabolic substrate for isolated nerve endings: inhibition by ammonium ion. Biochem Soc. Trans. 3, 1223-1226.

Braunstein, A.E. (1957) Les voies principales de l'assimilation et dissimilation de l'azote chez les animaux. Adv. Enzymol. Rel. Areas Mol. Biol. 19, 335-389.

Butterworth, R.F., Girard G., and Giguère J.-F. (1988) Regional differences in the capacity for ammonia removal by brain following portocaval anastomosis. J. Neurochem. 51, 486-490.

Cooper, A.J.L. (1988) Glutamine synthetase. In: Glutamine and Glutamate in Mammals, Vol 1. (E. Kvamme, ed) CRC Press, Inc.; Boca Raton, Florida; pp 7-31.

Cooper, A.J.L., and Plum, F. (1987) Biochemistry and Physiology of Brain Ammonia. Physiol. Rev. 67, 440-519.

Cooper, A.J.L., McDonald, J.M., Gelbard, A.S., Gledhill, R.F., and Duffy, T.E. (1979) The metabolic fate of ^{13}N-labeled ammonia in rat brain. J. Biol. Chem. 254, 4982-4992.

Cooper, A.J.L., Mora, S.N., Cruz, N.F., and Gelbard, A.S. Cerebral ammonia metabolism in hyperammonemic rats (1985) J. Neurochem. 44, 1716-1723.

Cooper, A.J.L., Nieves, E., Coleman, A.E., Filc-DeRicco, S., and Gelbard, A.S. (1987) Short-term metabolic fate of [^{13}N]ammonia in rat liver. J. Biol. Chem. 262, 1073-1080.

Cooper, A.J.L., Nieves, E., Filc-DeRicco, S., and Gelbard, A.S. (1988a) Short-term metabolic fate of [^{13}N]ammonia, L-[^{13}N]alanine, L-[^{13}N]glutamate and L-[amide-^{13}N]glutamine in normal rat liver in vivo. In: Advances in Ammonia Metabolism and Hepatic Encephalopathy (P.B. Soeters, J.H.P. Wilson, A.J. Meijer and E. Holm, eds) Elsevier; Amsterdam; pp. 11-25.

Cooper, A.J.L., Nieves, E., Rosenspire, K.C., Filc-DeRicco, S., Gelbard, A.S., and Brusilow, S.W. (1988b) Short-term metabolic fate of ^{13}N-labeled glutamate, alanine and glutamine(amide) in rat liver. J. Biol. Chem. 263, 12268-12273.

Cruz, N.F., and Duffy, T.E. (1983) Local cerebral glucose metabolism in rats with chronic portacaval shunts. J. Cereb. Blood Flow Metab. 3, 311-320.

Duffy, T.E., Plum, F., and Cooper, A.J.L. (1983) Cerebral ammonia metabolism in vivo. In: Glutamine, Glutamate and GABA in the Central Nervous System (L. Hertz, E. Kvamme, E.G. McGeer, and A. Schousboe, eds) Alan R. Liss; New York; pp. 371-388.

Felig, P. (1973) The glucose-alanine cycle. Metabolism 22, 179-207.

Felig, P., Wahren, J., and Ahlborg, G. (1973) Uptake of individual amino acids by the human brain. Proc. Soc. Exp. Biol. Med. 142, 230-231.

Ferraro, T.N., and Hare, T.A. (1984) Triple-column ion-exchange physiological amino acid analysis with fluorescent detection: Baseline characterization of human cerebrospinal fluid. Anal. Biochem. 143, 82-94.

Freed, B.R., and Gelbard, A.S. (1982) Distribution of ^{13}N following intravenous injection of [^{13}N]ammonia in the rat. Can J. Pharmacol. 60, 60-67.

Gjedde, A., Lockwood, A.H., Duffy, T.E., and Plum, F. (1978) Cerebral blood flow and metabolism in chronically hyperammonemic rats: Effect of an acute ammonia challenge. Ann. Neurol. 3, 325-330.

Harper, A.E., Miller, R.H., and Block, K.P. (1984) Branched-chain amino acid metabolism. Ann. Rev. Nutr. 4, 409-545.

Häussinger, D. and Gerok, W. (1984) Hepatocyte heterogeneity in ammonia metabolism: impairment of glutamine synthesis in CCl_4 induced liver cell necrosis with no effect on urea synthesis. Chem-Biol. Interactions 48, 191-194.

Häussinger, D., and Gerok, W. (1986) Metabolism of amino acids and ammonia. In: Regulation of Hepatic Metabolism: Intra- and Intercellular Compartmentation (R.G. Thurman, F.C. Kauffman, and K. Jungermann, eds) Plenum Press; New York; pp 253-291.

Hems, R., Stubbs, M., and Krebs, H.A. (1968) Restricted permeability of rat liver for glutamate and succinate. Biochem J. 107, 807-815.

Hindfelt, B. (1983) Ammonia intoxication and brain energy metabolism. In: New Aspects of Clinical Nutrition (G. Kleinberger, and E. Deutsch, eds) Karger; Basel; pp. 474-484.

Hindfelt, B., Plum, F., and Duffy, T.E. (1977) Effects of acute ammonia intoxication on cerebral metabolism in rats with portacaval shunts. J. Clin. Invest. 59, 386-396.

Krebs, H.A., Hems, R., Lund, P., Halliday, D., and Read, W.W.C. (1978) Sources of ammonia for urea synthesis Biochem. J. 176, 733-737.

Kvamme, E. (1983) Ammonia metabolism in the CNS. Prog. Neurobiol. 20, 109-132.

Lai, J.C.K., and Cooper, A.J.L. (1986) Brain α-ketoglutarate dehydrogenase complex: Kinetic properties, regional distribution and effects of inhibition. J. Neurochem. 47, 1376-1386.

Lockwood, A.H., Finn, R.D., Campbell, J.A., and Richman, T.B. (1980) Factors that affect the uptake of ammonia by the brain: the blood-brain pH gradient. Brain Res. 181, 259-266.

Lockwood, A.H., McDonald, J.M., Reiman, R.E., Gelbard, A.S., Laughlin, J.S., and Duffy, T.E. (1979) The dynamics of ammonia metabolism in man. Effects of liver disease and hyperammonemia. J. Clin. Invest. 63, 449-460.

Lowry, O.H., and Passonneau, J.V. (1966) Kinetic evidence for multiple binding sites on phosphofructokinase. J. Biol. Chem. 241, 2268-2279.

Mans, A.M., Biebuyck, J.F., Davis, D.W., Bryan, R.M., and Hawkins, R.A. (1983) Regional cerebral glucose utilization in rats with portacaval anastamosis. J. Neurochem. 40, 986-991.

McCandless, D.W., and Schenker, S. (1981) Effect of acute ammonia intoxication on energy stores in the cerebral reticular activating system. Exp. Brain Res. 44, 325-330.

McKhann, G.M. and Tower, D.B. (1961) Ammonia toxicity and cerebral oxidative metabolism. Am. J. Physiol. 200 420-424.

Meijer, A. (1985) Channeling of ammonia from glutaminase to carbamoyl-phosphate synthetase in liver mitochondria. FEBS Lett. 191, 249-251.

Meijer, A.J., Lof, C., Ramos, I.C. and Verhoeven, A. (1985) Control of ureogenesis. Eur. J. Biochem. 148, 189-196.

Norenberg, M.D., and Martinez-Hernandez, A. (1979) Fine structural localization of glutamine synthetase in astrocytes of rat brain. Brain Res. 161, 303-310.

Ochs, R.S. (1984) Glutamine metabolism of isolated rat hepatocytes. Evidence for catecholamine activation of α-ketoglutarate dehydrogenase. J. Biol. Chem. 259, 13004-13010.

Pardrige, W.M. (1977) Unidirectional influx of glutamine and other neutral amino acids into liver of fed and fasted rat in vivo. Am. J. Physiol. 232, E492-E496.

Pardridge, W.M. (1983) Brain Metabolism: A perspective from the blood-brain barrier. Physiol. Rev. 63, 1481-1535.

Phelps, M.E., Hoffman, E.J., and Raybaud, C. (1977) Factors which affect cerebral uptake and retention of $^{13}NH_3$. Stroke 8, 694-702.

Pitts, R.F. (1964) Renal production and excretion of ammonia. Am. J. Med. 36, 720-742.

Posner, J.B., and Plum, F. (1960) The toxic effects of

carbon dioxide and acetazolamide in hepatic encephalopathy. J. Clin. Invest. 39, 1246-1258.

Quistorff, B. (1985) Gluconeogenesis in periportal and perivenous hepatocytes in rat liver, isolated by a new high-yield/collagenase perfusion technique. Biochem. J. 229, 221-226.

Raichle, M.E., and Larson, K.B. (1981) The significance of the NH_3-NH_4^+ equilibrium on the passage of ^{13}N-ammonia from blood to brain. A new regional residue detection model. Circ. Res. 48, 913-937.

Raijman, L. (1974) Citrulline synthesis in rat tissues and liver content of carbamoyl phosphate and ornithine. Biochem. J. 138, 225-232.

Rognstad, R. (1977) Sources of ammonia for urea synthesis in isolated rat liver cells. Biochim. Biophys. Acta 496, 249-254.

Ross, B.D. (1988) The current state of clinical magnetic resonance spectroscopy with phosphorus-31: An overview from Hammersmith. Mag. Res. Medicine Biol. 1, 81-98.

Schultz, V., and Lowenstein, J.M. (1978) The purine nucleotide cycle. Studies of ammonia production and interconversions of adenine and hypoxanthine nucleotides and nucleosides by rat brain in vivo. J. Biol. Chem. 253, 1938-1943.

Tatibana, M., and Shigesada, K (1976) Regulation of urea biosynthesis by the acetylglutamate-arginine system. In: The urea cycle (S. Grisolia, R. Baguena and F. Mayor, eds) John Wiley and Sons, Inc; New York; pp. 301-313.

Taylor, P.M., and Rennie, M.J. (1987) Perivenous localization of Na-dependent glutamate transport in perfused rat liver. FEBS Lett. 221, 370-374.

Van Den Berg, C.J., Matheson, D.F., and Nijemanting, W.C. (1978) Compartmentation of amino acids in brain: The GABA-glutamine-glutamate cycle. In: Amino Acids and Chemical Transmitters (F. Fonnum, ed) Plenum Press; New York; pp. 709-723.

Welbourne, T.C. (1988) Role of the lung in glutamine homeostasis. In: New Aspects of Renal Ammonia Metabolism. Contributions to Nephrology. Vol 63 (G. Baverel, A.C.Schoolwerth, H. Endou, M. Rengel, and A. Tizcanello, eds) Karger; Basel; pp. 178-182.

Windmueller, H.G. and Spaeth, A.E. (1974) Uptake and metabolism of plasma glutamine by the small intestine. J. Biol. Chem. 249, 5070-5079.

NEUROPHYSIOLOGY OF AMMONIA INTOXICATION

Winfried A. Raabe

Departments of Neurology, VA Medical Center and

University of Minnesota, Minneapolis, MN 55417

Hepatic encephalopathy is a potentially reversible disease. Therefore, a structural abnormality of the CNS cannot be the cause of the encephalopathy. The encephalopathy is the expression of an abnormal function, or dysfunction, of the CNS. Since the CNS consists of neuronal circuits, a dysfunction of these neuronal circuits underlies the dysfunction of the CNS. Neuronal circuits consist of individual neurons which must function normally to properly interact with each other. Hence it can be presumed that the cause of encephalopathy is a dysfunction of individual neurons, which disturbs interactions between individual neurons in neuronal circuits, causes a dysfunction of neuronal circuits, and causes a dysfunction of the CNS.

Ammonia intoxication has been implicated in the pathogenesis of hepatic encephalopathy. This paper reviews the effects of NH_4^+ on the function of neurons and on neuronal interactions. The abnormal neuronal interactions due to NH_4^+ are related to the encephalopathy due to acute ammonia intoxication and to the encephalopathy due to acute ammonia intoxication after portasystemic shunt, an experimental model of hepatic encephalopathy.

POSTSYNAPTIC INHIBITION

In the mammalian CNS, postsynaptic inhibition is mainly mediated by GABA or glycine (13). In brief, an action potential in the presynaptic neuron travels down the axon and opens in the presynaptic terminal voltage-gated Ca^{++}-channels. The subsequent influx of Ca^{++} into the presynaptic terminal initiates the release of the inhibitory transmitter, GABA or glycine into the synaptic cleft. GABA or glycine cross the synaptic cleft, combine with GABA or glycine receptors on the postsynaptic neuron and initiate the opening of Cl^- channels in the postsynaptic neuron. Because of a gradient between the extra- and intracellular concentrations of Cl^-, the extracellular Cl^--concentration being higher than the intracellular concentration, the transmitter mediated opening of Cl^--channels triggers an influx of Cl^- into the postsynaptic neuron. This influx of Cl^- adds negative

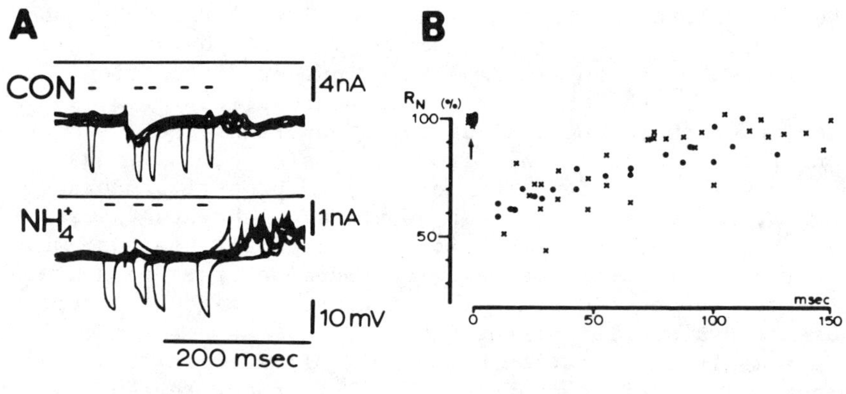

Fig. 1. Effects of NH_4^+ on recurrent collateral IPSP in pyramidal tract cell of cerebral cortex. **(A)** *CON* - Control records show hyperpolarizing IPSP (several tracings superimposed). *NH_4^+* - ammonium acetate i.v. abolishes the hyperpolarizing action of the IPSP. Upper traces indicate intracellular hyperpolarizing current injection to monitor the change in neuron input resistance, lower traces are intracellular records, positivity upwards. **(B)** Neuron input resistance (R_N) before (•) and after (×) abolition of the hyperpolarizing action of the IPSP by ammonium acetate i.v. Arrow indicates neuron input resistance before begin of IPSP. Cat, pentobarbital anesthesia. (From ref. 26, with permission.)

charges to the intracellular space and produces a hyperpolarization of the membrane potential in the postsynaptic neuron, i.e., the inhibitory postsynaptic potential (IPSP), Fig. 1. When the transmitter action on postsynaptic receptors is terminated, the Cl^--channels in the postsynaptic neuron close. Excess Cl^- in the postsynaptic neuron is removed by a mechanism extruding Cl^- from the intracellular to the extracellular space (15). This mechanism for Cl^--extrusion is responsible for maintaining the concentration gradient of Cl^- across the neuronal membrane. The mechanism for Cl^--extrusion may be a transport system supported by a Cl^--dependent ATPase (12).

Ammonia intoxication impairs postsynaptic inhibition in cerebral cortex (28), thalamus (36), brainstem (14) and spinal cord (15, 16, 29, 30) *in vivo*. NH_4^+ produces this effect by inactivating the extrusion of Cl^- from neurons (15). The inactivation of Cl^--extrusion abolishes the concentration gradient for Cl^- across the neuronal membrane. As a consequence, the opening of Cl^- channels by the inhibitory transmitters GABA and glycine does no longer cause an influx of Cl^- into neurons. This lack of the addition of negative charges to the intracellular space is shown by the absence of a hyperpolarizing IPSP during ammonia intoxication, Figs. 1, 2, 3, 5, 9 and 13.

The effects of GABA- or glycine-mediated inhibitory synaptic transmission on the postsynaptic neuron are not impaired when NH_4^+ inactivates Cl^--extrusion and abolishes the hyperpolarizing IPSP. The inhibitory transmitters GABA and glycine still open Cl^--channels when NH_4^+ abolishes the hyperpolarizing IPSP. This can be shown in two ways. (i) The opening of Cl^--channels in the postsynaptic neuron by the inhibitory synaptic transmitter temporarily decreases the neuron input resistance. This decrease in neuron input resistance is unchanged when NH_4^+ inactivates Cl^--extrusion and abolishes the hyperpolarizing IPSP, Fig. 1. (ii) Before and after NH_4^+, the opening of Cl^--channels by inhibitory synaptic transmission triggers de- or hyperpolarizing IPSPs when the actual membrane potential is different from the equilibrium potential of the IPSP (E_{IPSP}), Fig. 2.

[The E_{IPSP} is the membrane potential at which the opening of Cl^--channels by inhibitory synaptic transmission produces no net exchange of Cl^- across the neuronal membrane. Normally, E_{IPSP} is about 10 mV more negative than the resting membrane potential and inhibition produces a hyperpolarizing IPSP, cf. Fig 3 *Control*. When the neuron is

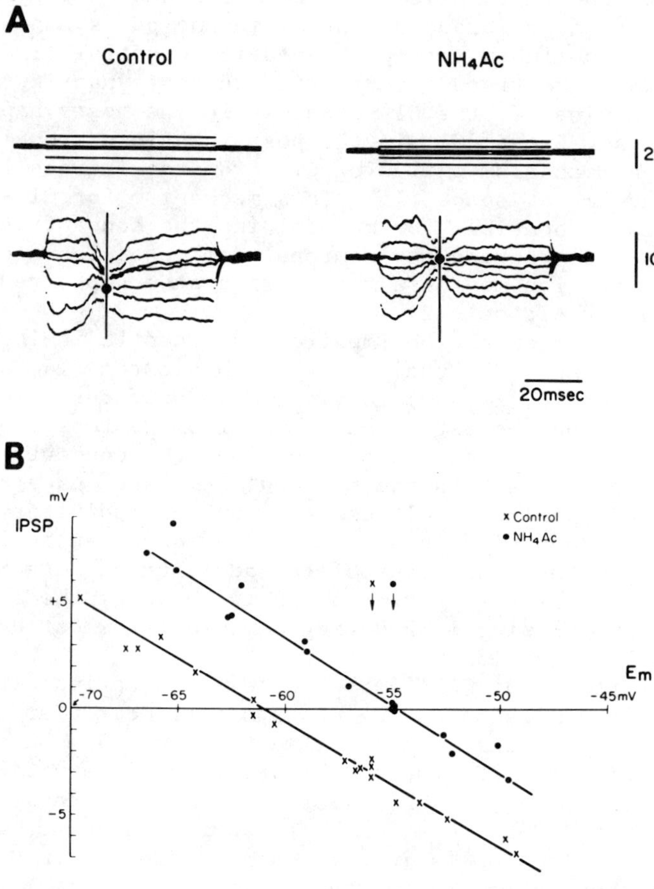

Fig. 2. Effect of NH_4^+ on E_{IPSP} in spinal motoneuron. (A) *Control* - Stimulation of femoral nerve elicits hyperpolarizing IPSP; variation of the resting membrane potential by intracellular current injection shows that E_{IPSP} (•) is 5.5 mV more negative than the resting membrane potential. *NH_4Ac* - Ammonium acetate i.v., 1.9 mmol/kg, abolishes the hyperpolarizing action of the IPSP and shifts E_{IPSP} (•) to the level of the resting membrane potential. Upper traces indicate intracellular current injection, lower traces are intracellular records, positivity upwards. (B) Plot of IPSP-amplitude vs. resting membrane potential during applied current steps before (×) and after (•) ammonium acetate i.v. →

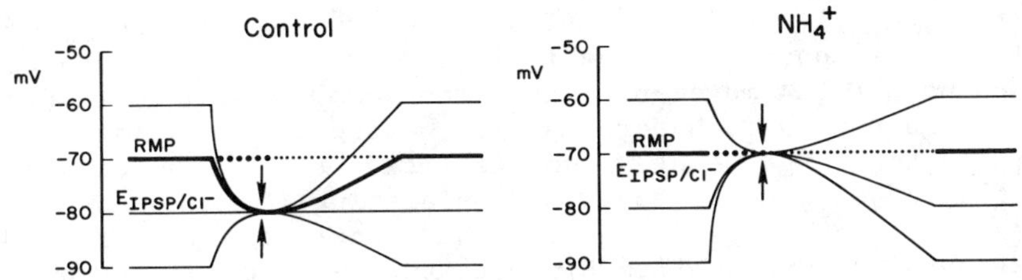

Fig. 3. Schematic representation of the effects of NH_4^+ on E_{IPSP}. *Control* - In control conditions, E_{IPSP} is about 10 mV more negative than the resting membrane potential. At resting membrane potential (RMP) and at more depolarized membrane potentials, the opening of Cl^--channels by the inhibitory transmitter, indicated by large dots (•), causes an influx of Cl^- into the neuron, see downward arrow, and hyperpolarizing IPSPs. When the membrane potential is hyperpolarized, e.g., by intracellular current injection, to the level of E_{IPSP}, the opening of Cl^--channels by the inhibitory transmitter does not produce a net Cl^-flux across the neuronal membrane. When the membrane potential is hyperpolarized beyond the level of E_{IPSP}, the opening of Cl^--channels by the inhibitory transmitter produces an efflux of Cl^- from the neuron, see upward arrow, and a depolarizing IPSP. *NH_4^+* - The inactivation of Cl^--extrusion by NH_4^+, shifts E_{IPSP} to the level of the resting membrane potential. The opening of Cl^--channels does not produce a net Cl^--flux across the neuronal membrane. When the membrane potential is depolarized, Cl^--channel opening produces again a Cl^--influx and a hyperpolarizing IPSP. When the membrane potential is hyperpolarized, i.e., more negative than the resting membrane potential and E_{IPSP}, Cl^--channel opening by the inhibitory transmitter produces an efflux of Cl^- and depolarizing IPSPs.

Fig. 2, cont. Vertical arrows indicate resting membrane potential. Lines are linear regression of plotted data. Ammonium acetate i.v. shifts E_{IPSP} from -61.5 mV to -55 mV. The parallel shift of the plotted lines indicates that ammonium acetate had no effect on the postsynaptic conductance change associated with the IPSP. Cat, pentobarbital anesthesia. (From ref. 29, with permission.)

hyperpolarized by intracellular current injection to the level of E_{IPSP}, the negative charges inside the neuron counteract the Cl^--gradient across the neuronal membrane. Cl^--channel opening by the inhibitory transmitter produces then no net Cl^- movement across the membrane and no IPSP. When the neuron is hyperpolarized beyond the level of the E_{IPSP}, the large negative charge inside the neuron causes an efflux of Cl^- and produces a depolarizing IPSP when Cl^--channels open, Fig. 3 *Control*. The inactivation of Cl^--extrusion by NH_4^+ shifts E_{IPSP} to the level of the resting membrane potential, Fig. 3 *NH_4^+*. The opening of Cl^--channels by the inhibitory transmitter does not produce a change of the membrane potential, i.e., does not produce an IPSP. After NH_4^+, depolarization of the neuron recreates a difference between membrane potential and E_{IPSP}, the latter being more negative than the actual membrane potential. Inhibition produces then again a hyperpolarizing IPSP, i.e., hyperpolarizing from a depolarized membrane potential to E_{IPSP} at the resting membrane potential. Similarly, hyperpolarization creates a difference between membrane potential and E_{IPSP}, the latter being more positive than the actual membrane potential. The opening of Cl^--channels produces then a depolarizing IPSP, Fig. 3 *NH_4^+*.]

The observations on the lack of an effect of NH_4^+ on the opening of Cl^--channels by inhibitory synaptic transmission indicate that NH_4^+ does not significantly affect action potential conduction into presynaptic terminals, Ca^{++}-influx into presynaptic terminals, the release of transmitter from presynaptic terminals, as well as the action of the inhibitory transmitters GABA and glycine on receptors on the postsynaptic neuron.

Recently, Ca^{++}- and voltage-dependent Cl^--currents have been described (21). At present, the physiological function of these Cl^--currents is unknown. Since these Ca^{++}- and voltage-dependent Cl^--currents also depend on the Cl^--gradient across the membrane, the inactivation of Cl^--extrusion by NH_4^+ will also affects these currents.

INHIBITORY NEURONAL INTERACTIONS

To be pathophysiologically significant, the inactivation of Cl^--extrusion by NH_4^+ has to disturb inhibitory neuronal interactions. Figure 4 shows that the inactivation of Cl^--extrusion and the subsequent shift of E_{IPSP} to the level

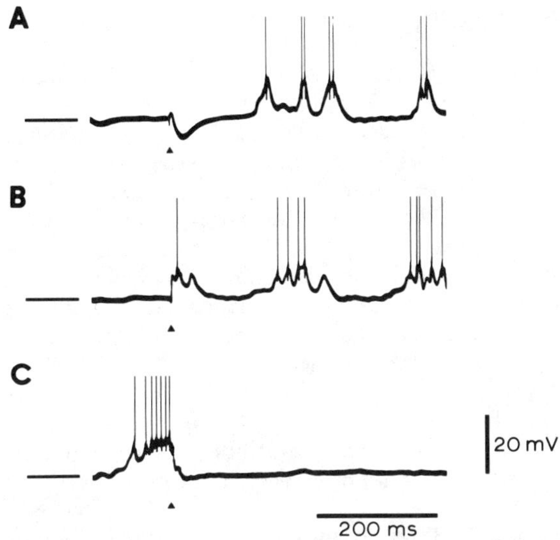

Fig. 4. Effect of NH_4^+ on hyperpolarizing Cl^--mediated postsynaptic inhibition in pyramidal tract cell of cerebral cortex. Resting membrane potential of -69 mV is indicated by line at left of tracings. Stimulation of nucleus ventralis lateralis thalami, see triangles, elicits a large hyperpolarizing IPSP and three EPSPs, which are largely hidden by the IPSP and can normally only be seen with membrane hyperpolarization (7). **(A)** Before ammonium acetate i.v., stimulation produces a small EPSP followed by a large, hyperpolarizing IPSP which suppresses all neuronal excitation. **(B)** Ammonium acetate, 2.14 mmol/kg, abolishes the hyperpolarizing IPSP. The three EPSPs, which are normally hidden by the IPSP, can now be clearly discerned. The second EPSP elicits an action potential. Stimulus occurs while the neuron is at resting membrane potential. **(C)** After ammonium acetate, 2.14 mmol/kg; stimulus occurs during transient depolarization of membrane potential by EPSPs. The inhibitory transmitter opens Cl^--channels and produces an influx of Cl^- into the neuron, which repolarizes the neuron to the level of the resting membrane potential and suppresses all neuronal excitation. Tracing (C) was obtained 20 s after tracing (B). Cat, pentobarbital anesthesia, action potentials are truncated, electropositivity upwards. (From ref. 26, with permission.)

of the resting membrane potential affects the efficacy of postsynaptic inhibition. After NH_4^+ inhibition no longer suppresses neuronal excitation by excitatory postsynaptic potentials (EPSPs), Fig. 4 B.

However, it should be noted that this disinhibition is only seen when postsynaptic inhibition occurs at the level of the resting membrane potential, i.e., at E_{IPSP}, Fig 4 B. In contrast, when postsynaptic inhibition occurs in a neuron which is temporarily depolarized by EPSPs, inhibition is still effective, Fig. 4 C. In this latter case, inhibitory synaptic transmission opens Cl^--channels at a point in time at which the actual membrane potential is depolarized and does not correspond to E_{IPSP}. Therefore, the opening of Cl^--channels by the inhibitory synaptic transmitter produces an influx of Cl^- into the neuron and an IPSP. This IPSP repolarizes the neuron to the level of E_{IPSP} (= resting membrane potential). The currents underlying this repolarizing IPSP are sufficient to suppress neuronal excitation and to inhibit or "break" any ongoing action potential generation by EPSPs.

IPSPs occurring in neurons at resting membrane potential probably serve to "gate out" action potential generation by EPSPs occurring during the IPSP. Therefore, IPSPs occurring at the level of the resting membrane potential may be called "gate-IPSPs". IPSPs occurring during a transient depolarization of the resting membrane potential by EPSPs, which break action potential generation by EPSPs, may be called "break-IPSPs". The inactivation of Cl^--extrusion by NH_4^+ does not produce a dysfunction of all Cl^--dependent postsynaptic inhibition in the CNS but rather a partial dysfunction of postsynaptic inhibition. NH_4^+ affects the inhibitory efficacy of gate-IPSPs, but NH_4^+ does not affect the inhibitory efficacy of break-IPSPs (28).

Gate-IPSPs occur frequently in spontaneously active neurons (2, 8, 22, 23). Therefore, the inactivation of Cl^--extrusion by NH_4^+ can be presumed to affect inhibitory neuronal interactions in the spontaneously active CNS.

POSTSYNAPTIC EXCITATION

In general, the mechanisms of postsynaptic excitation correspond to those of postsynaptic inhibition, see above. Exceptions are the nature of the excitatory transmitter, which is presumed to be glutamate (or aspartate), and the ion channels opened by postsynaptic receptors, Na^+- and K^+-permeable channels instead of Cl^--permeable channels.

Ammonia intoxication impairs excitatory synaptic transmission (1, 10, 25, 27, 37, 38). NH_4^+ blocks the conduction of action potentials into the presynaptic terminals of excitatory afferents, Fig. 5. There is indirect evidence that the conduction block may be caused by a depolarization (27). The latter may be due to the K^+-like properties of NH_4^+ as well as extracellular K^+ accumulation because of an inhibition of K^+-uptake into astrocytes by NH_4^+ (27).

The conduction block for action potentials into presynaptic terminals decreases the number of synaptic terminals which excrete excitatory transmitter onto the postsynaptic neuron. Subsequently, the excitatory synaptic potential

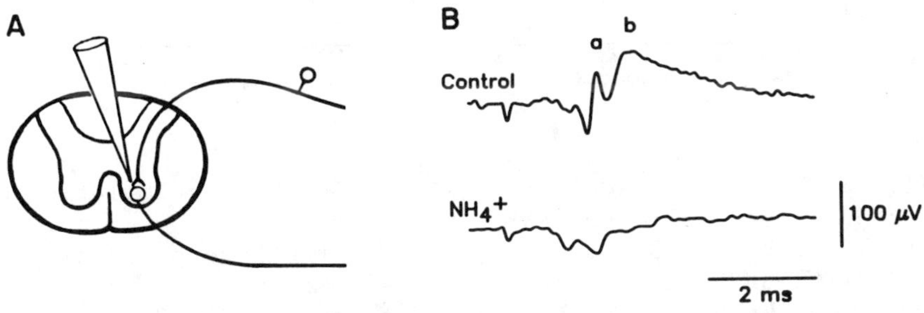

Fig. 5. Block of action potential conduction into presynaptic terminals by NH_4^+. **(A)** Experimental arrangement: extracellular recording electrode in spinal motoneurone nucleus simultaneously records afferent fiber spike and postsynaptic potential. **(B)** *Control* - Stimulation of the nerve to posterior biceps-semitendinous muscle produces an action potential in presynaptic afferent fibers ("a") followed by monosynaptic EPSP in posterior biceps-semitendinous motoneuron nucleus ("b"). Cat, pentobarbital anesthesia, electronegativity upwards. *NH_4^+* - Ammonium acetate i.v., 4.97 mmol/kg, abolishes the action potential in presynaptic afferent fibres and the EPSP. (Raabe, W., unpublished data.)

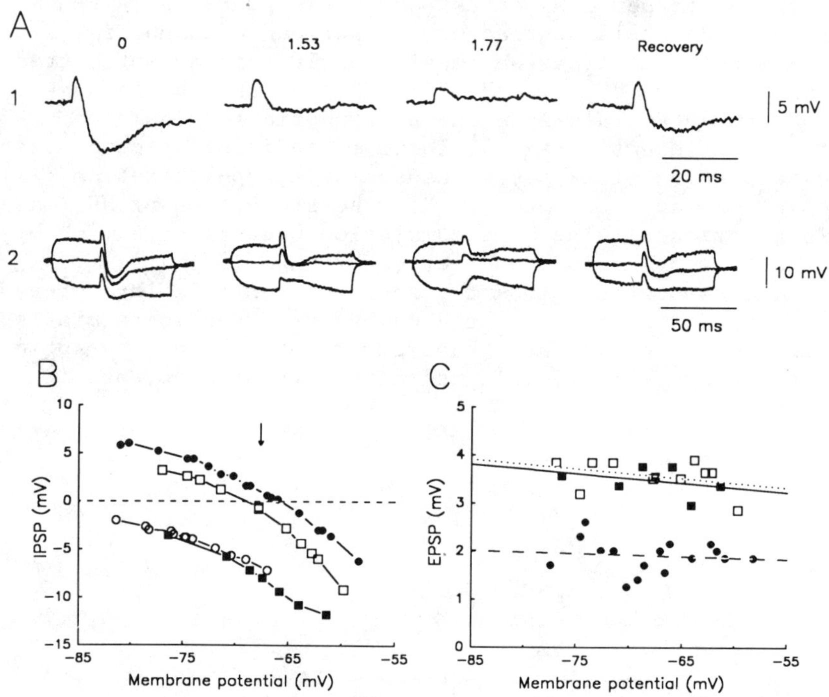

Fig. 6. NH_4^+ decreases EPSP. **(A)** *0* - Control, stimulation of peroneal nerve elicits EPSP-IPSP sequence. *1.53* - Ammonium acetate i.v., 1.53 mmol/kg, decreases the hyperpolarizing IPSP without effect on EPSP. *1.77* - Ammonium acetate i.v., 1.77 mmol/kg abolishes IPSP and decreases EPSP. *Recovery* - 35 min after end of ammonium acetate i.v. **(B)** Plot of IPSP-amplitude vs. membrane potential during applied current steps. (■) - before ammonium acetate, (□) - 1.53 mmol/kg ammonium acetate, (•) - 1.77 mmol/kg ammonium acetate, (○) - Recovery. Arrow indicates resting membrane potential. **(C)** Plot of EPSP-amplitude vs. membrane potential during applied current steps. Symbols as in (B). Lines are regression of plotted points. Cat, pentobarbital anesthesia, intracellular record from spinal motoneuron, electropositivity upwards, third infusion of ammonium acetate i.v. (A)1 and (A)2 are the same except for time calibration, and intracellular current injection to de- or hyperpolarize the membrane potential in (A)2. (Raabe, W., unpublished data.)

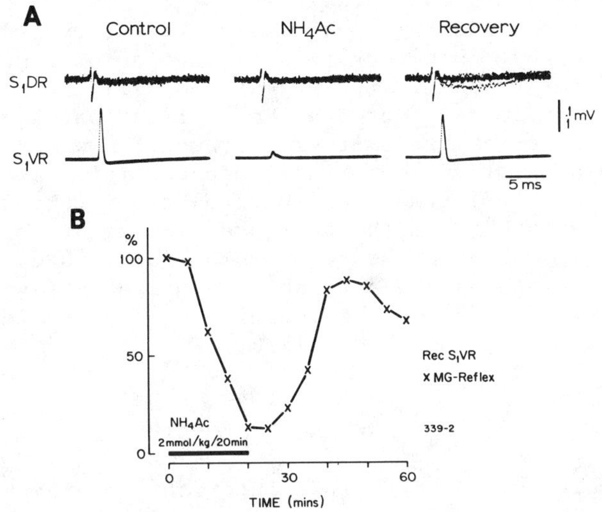

Fig. 7. Effect of NH_4^+ on the spinal monosynaptic reflex. **(A)** *Control* - before ammonium acetate i.v., stimulation of the medial gastrocnemius nerve produces a large reflex discharge in the S_1 ventral root. *NH_4OAc* - ammonium acetate i.v., 2 mmol/kg (second of two infusions), markedly decreases the reflex discharge without changing the spike of afferent fibers recorded from the S_1 dorsal root. *Recovery* - 25 min after termination of ammonium acetate i.v., the reflex discharge has almost completely recovered. S_1DR - record from S_1 dorsal root at entry into spinal cord, negativity downwards; S_1VR - record from S_1 ventral root, negativity upwards. **(B)** Plot of the area between reflex discharge and baseline in relation to ammonium acetate i.v. Cat, cerveau isolé preparation, spinal cord section at L_2. (From ref. 26, with permission.)

(EPSP) decreases, Figures 5 and 6. The decrease of the EPSP decreases action potential generation in the postsynaptic neuron, i.e., decreases excitatory synaptic transmission, Fig. 7.

The excitatory synaptic transmitter glutamate is thought to be derived from glutamine via cleavage by glutaminase. NH_4^+ inhibits glutaminase (3, 17). Therefore, it had been suggested that NH_4^+ may impair excitatory synaptic transmission by inhibiting the utilization of glutamine as a precursor for transmitter glutamate (4, 11).

NH_4^+ may indeed decrease the utilization of glutamine as a precursor for transmitter glutamate. However, because NH_4^+ blocks the invasion of action potentials into presynaptic terminals (27), the contribution of the inhibition of glutaminase by NH_4^+ to the decrease of excitatory synaptic transmission cannot be neurophysiologically investigated. At present, no methods are available to differentiate a transmitter depletion due to the inhibition of glutaminase from the effects of a conduction block.

NH_4^+ AND ENCEPHALOPATHY

An acute systemic ammonia intoxication progressively increases NH_4^+ in the CNS. This progressive increase of NH_4^+ in the CNS affects postsynaptic inhibition before affecting postsynaptic excitation. Figure 6 A shows an EPSP-IPSP sequence recorded intracellularly from a spinal motoneuron. The progressive infusion of ammonium acetate i.v. initially markedly decreases the hyperpolarizing IPSP without decreasing the EPSP. Only when the infusion of ammonium acetate i.v. is continued beyond the decrease of the IPSP, the EPSP decreases too. Accordingly, the doses of ammonium acetate i.v. necessary to affect IPSPs, 2.5 - 3.8 mmol/kg (15, 16, 29, 30, 31), are smaller than those necessary to affect EPSPs, 6 mmol/kg (27). And, the CNS tissue level of NH_4^+ necessary to affect postsynaptic inhibition, 0.74-0.92 μmol/g (24, 32, 33), is lower than that necessary to affect postsynaptic excitation, ≥ 1.4 μmol/g (calculated from ref. 25 with the assumption of a brain/blood ratio of ≥ 1, cf. ref. 9).

These observations indicate that an acute progressive systemic ammonia intoxication at first inactivates Cl^--extrusion and produces a dysfunction of inhibitory neuronal interactions mediated by gate-IPSPs. When CNS tissue levels of NH_4^+ exceed those necessary to affect postsynaptic inhibition, NH_4^+ produces a conduction block in presynaptic terminals and disturbs excitatory synaptic transmission (31). At this point, NH_4^+ may then also produce a dysfunction of

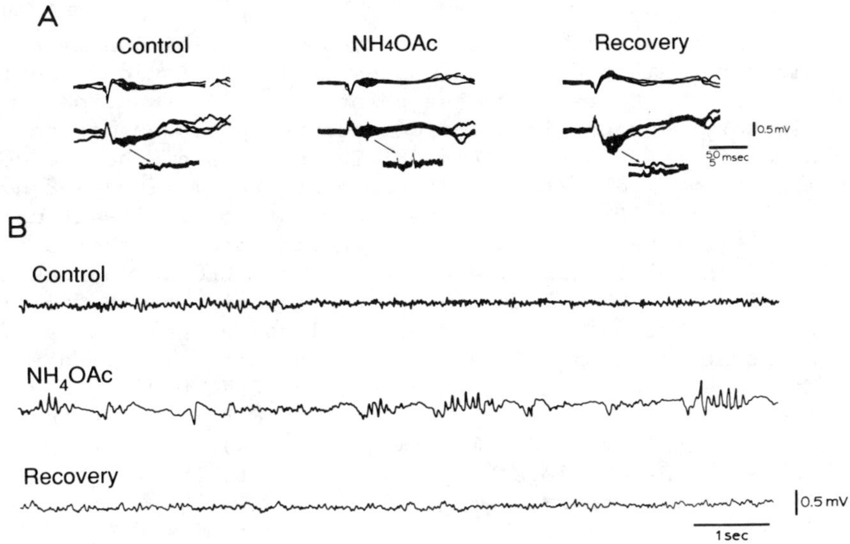

Fig. 8. Effect of NH_4^+ on inhibition and electrocorticogram. **(A)** Effect of ammonium acetate i.v. on inhibition of action potential generation in a pyramidal tract cell by recurrent postsynaptic inhibition, cf. ref. 28. *Control* - A brief train of stimuli to the pyramidal tract evokes at the cortical surface a negative wave and intracortically a positive wave, indicating recurrent collateral postsynaptic inhibition of pyramidal tract cells. The generation of an action potential in response to a single stimulus to the pyramidal tract is suppressed when the single stimulus is applied 25 ms after the conditioning stimulus train to the pyramidal tract, inset at bottom. *NH_4OAc* - Ammonium acetate i.v., 3.3 mmol/kg, decreases the amplitude of evoked potentials and abolishes the suppression of antidromic action potentials by recurrent postsynaptic inhibition, see inset. *Recovery* - Eleven minutes after termination of ammonium acetate i.v., the evoked potentials and the suppression of antidromic action potential by recurrent postsynaptic inhibition have recovered. Upper tracings are surface electrocorticogram; lower tracings are extracellular record from pyramidal tract cell and depth electrocorticogram recorded by the same →

inhibition mediated by break-IPSPs because of a conduction block in presynaptic terminals of afferents to inhibitory interneurons and/or of presynaptic terminals of inhibitory interneurons.

The effect of NH_4^+ on postsynaptic inhibition is sufficient to initiate encephalopathy. The infusion of ammonium acetate i.v., titrated to the point that a dysfunction of inhibitory neuronal interactions occurs, produces marked changes in the EEG (31), Fig. 8. The changes of the EEG develop in parallel with the decrease of the amplitude of hyperpolarizing IPSPs due to the inactivation of Cl^--extrusion by NH_4^+ (31), Fig. 9. The EEG is an indicator of the function of the whole CNS. The changes of the EEG may be taken as a sign that NH_4^+ changed the function of the CNS from normal to abnormal, i.e., from normal to encephalopathy. The concentration of CNS tissue NH_4^+ which is necessary to affect inhibition in the cerebral cortex, 0.74-0.92 μmol/g, corresponds to the concentration observed during the onset of encephalopathy, 0.8 - 1.2 μmol/g (9, 35). Furthermore, encephalopathy develops before NH_4^+ affects excitatory synaptic transmission (25), Fig. 10.

A dysfunction of the reticular formation due to NH_4^+ (18, 35) is not necessary to develop encephalopathy. NH_4^+ directly affects the cerebral cortex and does not need and abnormal input from the brainstem reticular formation to develop dysfunction. In animals in which the reticular activating system was disconnected from the cerebral cortex, ammonia intoxication produces significant changes of the EEG (31), Fig. 11.

Fig. 8, cont. electrode; time calibration for electrocorticogram is 50 ms, for insets 5 ms. **(B)** spontaneous surface electrocorticogram before, during and after recovery from effect of ammonium acetate i.v. on recurrent postsynaptic inhibition of pyramidal tract cell shown in (A); *Control* - low amplitude 12-20 Hz activity dominates the electrocorticogram. *NH_4OAc* - during the effect of ammonium acetate i.v. on recurrent postsynaptic inhibition, the electrocorticogram shows high amplitude slow waves and superimposed spindles. *Recovery* - After recovery of recurrent postsynaptic inhibition from ammonium acetate i.v., the electrocorticogram is again dominated by low amplitude 12-20 Hz activity. Cat, cerveau isolé preparation, negativity upward in all records. (From ref. 26, with permission.)

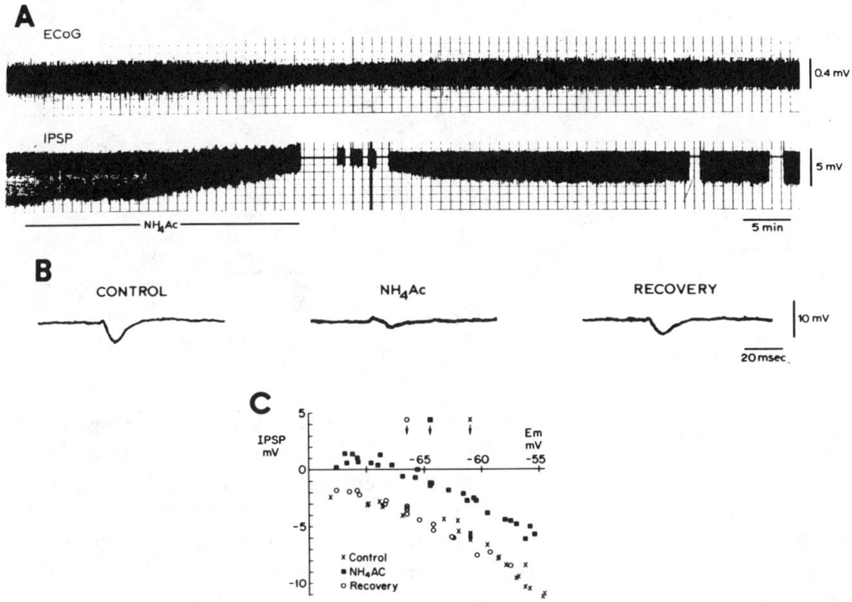

Fig. 9. NH_4^+ simultaneously affects electrocorticogram and IPSP. **(A)** Simultaneous record of electrocorticogram (ECoG) and IPSP (AC records). Ammonium acetate i.v. (NH_4Ac), 3.34 mmol/kg, reversibly decreases the amplitudes of ECoG and IPSP. **(B)** Effect of ammonium acetate i.v. on IPSP (DC-coupled record). *Control* - Stimulation of peroneal nerve elicits small EPSP which is followed by a large IPSP. *NH_4Ac* - Ammonium acetate i.v., 3.34 mmol/kg, decreases IPSP. *Recovery* - 40 min after end of ammonium acetate i.v., the IPSP has recoverd. **(C)** Plot of IPSP-amplitude vs. membrane potential during applied current steps before, during and after ammonium acetate i.v. The resting membrane potential, indicated by downward arrows, slightly increased during the experiment. Ammonium acetate i.v. decreased the amplitude of the IPSP by shifting the E_{IPSP} (IPSP = 0 mV) to a less negative level. Cat, cerveau isolé preparation, electrocorticogram: negativity upwards, IPSP: negativity downwards, gastrocnemius motoneuron. (From ref. 31, with permission.)

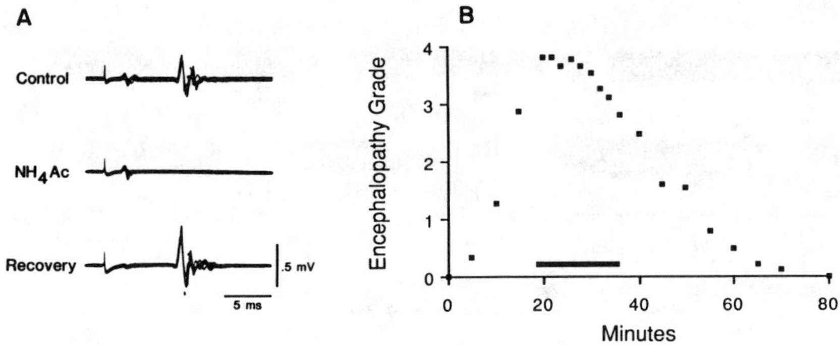

Fig. 10. (A) Effects of NH_4^+ on H-reflex and M-response. *Control* - Stimulation of the posterior tibial nerve elicits a small M-response and a large H-reflex in the plantar muscles. *NH_4Ac* - Ammonium acetate i.v., 6.45 mmol/kg, abolishes the H-reflex and does not affect the M-response. *Recovery* - After termination of ammonium acetate i.v., the H-reflex reappeared. M-response and H-reflex are slightly changed compared to Control because spontaneous movement of the animal slightly changed the position of stimulation and recording electrodes. **(B)** H-reflex during the encephalopathy due to ammonia intoxication. Rat, N=9, no anesthesia, ammonium acetate i.p., 5.2 mmol/kg. Plot of encephalopathy grade vs. time after ammonium acetate administration. Bar marks the average period of disappearance of the H-reflex. (From ref. 25, with permission.)

The dominant inhibitory synaptic transmitter in the cerebral cortex is presumed to be GABA. The predominant inhibitory synaptic transmitter in the brainstem is presumed to be glycine (13). Ammonia intoxication affects both, postsynaptic inhibition mediated by GABA and glycine. Therefore, it can be presumed that NH_4^+ initiates the **encephalopathy due to acute ammonia intoxication** by simultaneous effects of NH_4^+ on gate-IPSPs in the cerebral cortex, thalamus, brainstem and other structures with Cl^--dependent IPSPs in the CNS. Of

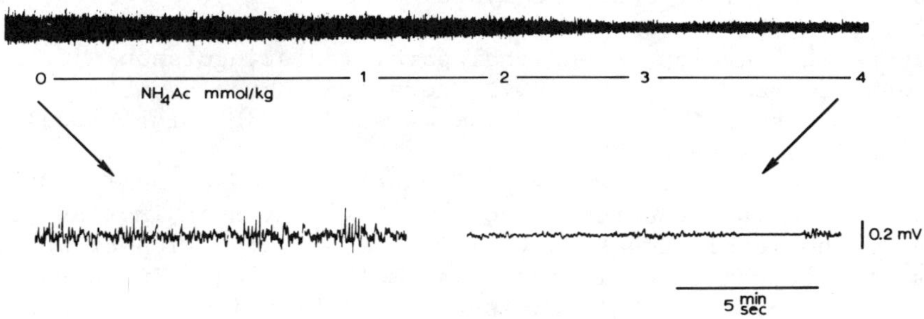

Fig. 11. NH_4^+ changes the electrocorticogram after destruction of pathways from the reticular activating system in the brainstem to cerebral cortex. Upper tracing - Ammonium acetate i.v., 4 mmol/kg, changes the electrocorticogram; this change becomes apparent with the infusion of 2-3 mmol/kg. Lower tracings - Electrocorticogram in detail before and after ammonium acetate i.v., 4 mmol/kg. Cat, cerveau isolé preparation; electronegativity upward; time calibration for upper tracing is 5 min, for lower tracing is 5 s. (From ref. 31, with permission.)

course, a faulty input from the thalamus and the brainstem reticular activating system to the cerebral cortex may contribute to the dysfunction of the cerebral cortex in addition to the action of NH_4^+ on cortical inhibition. Only when CNS tissue levels of NH_4^+ rise beyond those necessary to affect Cl^--extrusion and postsynaptic inhibition, a dysfunction of excitatory neuronal interactions and break-IPSPs will also contribute to the signs and symptoms of encephalopathy.

Clinical evidence corroborates this hypothesis, specifically, the dysfunction of excitation only in progressed stages of encephalopathy. In acutely occurring hyperammonemias, the muscle stretch reflexes vanish only in progressed stages of the encephalopathy, when patients are already comatose and have blood (and brain) levels of NH_4^+ beyond 1.4 mmol/l (cf. ref. 25).

NH_4^+, PORTASYSTEMIC SHUNT AND ENCEPHALOPATHY

Many patients with hepatic encephalopathy have cirrhosis of the liver and portasystemic shunting of blood (6). In addition, patients with a healthy liver and a portasystemic shunt may develop an encephalopathy indistinguishable from hepatic encephalopathy (20). Therefore, chronic portasystemic shunting of blood may be used as an animal model of hepatic encephalopathy.

A portacaval shunt significantly increases NH_4^+ and glutamine in the cerebral cortex. NH_4^+ may even increase beyond the level necessary to initiate a dysfunction of postsynaptic inhibition in normal animals, Table I. In spite of this increase of NH_4^+, postsynaptic inhibition in portacaval shunted animals may be normal, Fig. 12. This observation indicates that after portasystemic shunting of blood Cl^--extrusion develops a tolerance to chronically increased levels of NH_4^+. However, a portacaval shunt also significantly decreases the exogenous load of NH_4^+ and the increase

TABLE I

Concentrations of NH_4^+ and glutamine (average ± S.E.M., N=6) in the cerebral cortex of normal and PC-shunted (PCS) cats before and during disinhibition of pyramidal tract cells by NH_4Ac i.v.

	PCS (days)	NH_4Ac i.v. (mmol/kg)	NH_4^+ (μmol/g)	Glutamine (μmol/g)
Normal				
Control	-	-	0.24±0.02	2.92±0.13
Disinhibition	-	2.43±0.61	0.77±0.11[b]	5.54±0.58[b]
PCS				
Control	154±3	-	1.37±0.17[c]	14.28±1.40[d]
Disinhibition	163±12	0.70±0.31[c]	1.61±0.16[a,d]	13.63±1.19[a,d]

[a] NS vs. PCS/Control
[b] P <0.005 vs.Normal/Control
[c] P <0.05 vs. Normal/Disinhibition
[d] P <0.005 vs. Normal/Disinhibition

(From ref. 33)

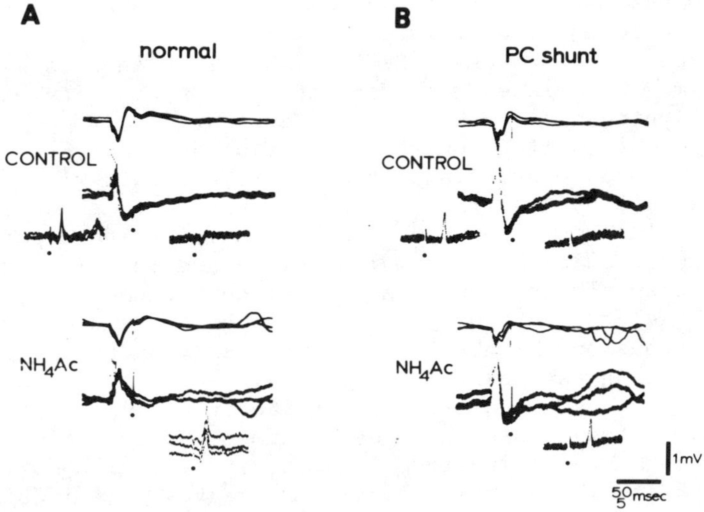

Fig. 12. Effect of ammonium acetate i.v. on recurrent postsynaptic inhibition of pyramidal tract cells in normal and portacaval shunted cats. **(A)** Normal cats. *Control* - A brief train of stimuli to the pyramidal tract evokes at the cortical surface a negative potential and intracortically a positive potential indicating recurrent postsynaptic inhibition of pyramidal tract cells. A single stimulus to the pyramidal tract triggers antidromic an action potential in a pyramidal tract cell, inset left; when the single stimulus is applied 25 ms after the brief train of stimuli producing recurrent inhibition, the generation of antidromic action potentials is suppressed, inset right. Dots (•) mark single stimuli. *NH_4Ac* - Ammonium acetate i.v., 2.5 mmol/kg, decreases the amplitude of the evoked potentials and abolishes the suppression of antidromic action potentials by recurrent postsynaptic inhibition, inset right. **(B)** Portacaval shunted cats. *Control* - As in (A), Control. *NH_4Ac* - Ammonium acetate i.v., 0.23 mmol/kg, decreases the amplitudes of the evoked potentials and abolishes the suppression of antidromic action potentials by recurrent postsynaptic inhibition, inset right. (From ref. 33, with permission.)

of NH_4^+ in the cerebral cortex necessary to affect postsynaptic inhibition (33), Fig. 12 and Table I.

These observations indicate that in spite of the tolerance to NH_4^+ after a portasystemic shunt, postsynaptic inhibition in cerebral cortex develops an increased sensitivity to an acute NH_4^+-load to the CNS (33). This increased sensitivity may be related to a decrease in the activity of glutamine synthetase in the cerebral cortex (5).

The observations on the effects of a portasystemic shunt on postsynaptic inhibition correlate well with the observations on the effects of a portasystemic shunt on the behavioral signs of encephalopathy. Animals with a portacaval shunt require a much smaller exogenous NH_4^+-load to develop encephalopathy than animals without a shunt (5, 9).

The electrophysiological, neurochemical and behavioral observations on animals with portasystemic shunts suggest the following pathogenesis of **portasystemic encephalopathy**, i.e., the encephalopathy precipitated in patients with a chronic portasystemic shunt by a sudden and systemic NH_4^+-load such as resulting from a gastro-intestinal hemorrhage. A portasystemic shunt chronically increases NH_4^+ in blood and CNS. Since postsynaptic inhibition develops a tolerance to a chronic increase of NH_4^+ in the CNS, patients are not necessarily encephalopathic in spite of increased NH_4^+ in blood and CNS. However, because a portasystemic shunt also increases the sensitivity of postsynaptic inhibition to an acute NH_4^+-load, a small and acute increase of NH_4^+ in the blood, such as from a gastro-intestinal hemorrhage, may increase NH_4^+ in the CNS sufficiently to affect postsynaptic inhibition mediated by gate-IPSPs and to initiate encephalopathy (33). Because of the decrease in glutamine synthetase activity after portasystemic shunt (5), portasystemic encephalopathy may affect the cerebral cortex earlier and more severely than other areas of the CNS, e.g., thalamus and brainstem reticular formation.

INHIBITION, NH_4^+, GABA AND BENZODIAZEPINE RECEPTORS

Recently, it has been suggested that hepatic encephalopathy is caused by an increase of GABA-mediated postsynaptic inhibition, either due to gut-derived GABA (34) or due to an endogenous ligand acting on the GABA-benzodiazepine receptor complex (19). This hypothesis seems to be supported by the observation that Ro15-1788, a benzodiazepine

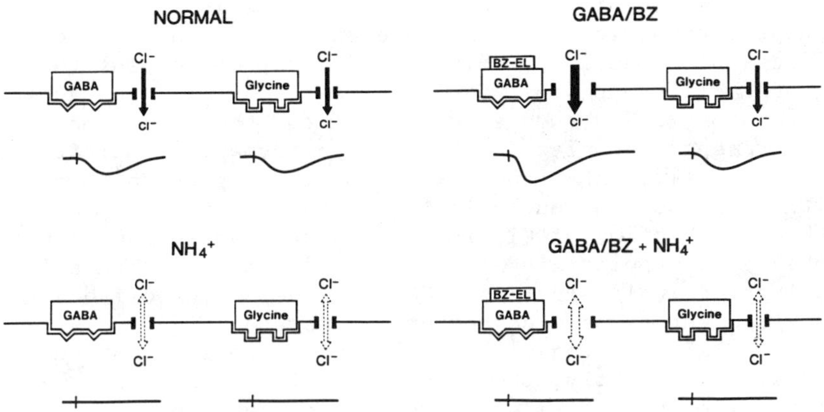

Fig. 13. *NORMAL* - Schematic representation of GABA- and glycine-mediated postsynaptic inhibition. After release by presynaptic terminals, GABA and glycine occupy receptors on the postsynaptic membrane. The receptors open Cl^--channels and permit Cl^- to enter the postsynaptic neuron. The force driving the entry of Cl^- is a concentration gradient between extra- and intracellular Cl^-, the extracellular concentration being higher (large Cl^-) than the intracellular concentration (small Cl^-). The result of the Cl^--influx into the neuron is a hyperpolarizing IPSP. It does not matter whether GABA or glycine is the inhibitory synaptic transmitter. *NH_4^+* - Effect of NH_4^+ on GABA and glycine mediated postsynaptic inhibition. NH_4^+ abolishes the concentration gradient between extra- and intracellular Cl^- (equal size of Cl^- extra- and intracellular); there is no net flux of Cl^- across the neuronal membrane (dashed arrows) and no hyperpolarizing IPSPs are generated, although GABA and glycine still open Cl^--channels in the postsynaptic membrane. *GABA/BZ* - Schematic representation of the effects of an endogenous ligand (BZ-EL) acting on the GABA-benzodiazepine receptor complex. GABA+BZ-EL opens Cl^--channels either more frequently or for longer period of time than normal (large pore in postsynaptic membrane). More Cl^- can enter the neuron (thick downward arrow) and a larger and/or longer hyperpolarizing IPSP is generated. BZ-EL has no effect on glycine mediated →

antagonist, can improve hepatic encephalopathy (cf. ref. 19).

Since GABA is considered to be the major inhibitory synaptic transmitter in the cerebral cortex, and since NH_4^+ affects GABA-mediated inhibition, it may be helpful to clarify the differences between the effects of NH_4^+ and gut-derived GABA, or an endogenous ligand for the GABA-benzodiazepine receptor complex (BZ-EL), on postsynaptic inhibition.

Normally, GABA- and glycine-mediated postsynaptic inhibition open Cl^--channels in the postsynaptic neuron. The subsequent influx of Cl^- into the postsynaptic neuron produces a hyperpolarizing IPSP, Fig. 13 *Normal*. Ammonia intoxication, which increases NH_4^+ in the CNS, abolishes the Cl^--gradient across the neuronal membrane, Fig. 13 *NH_4^+*, cf., the size of the Cl^- symbol on the two sides of the membrane. Subsequently, the opening of Cl^--channels by the inhibitory transmitters GABA and glycine does not produce an influx of Cl^- into the postsynaptic neuron and hyperpolarizing IPSPs are not generated, Fig. 13 *NH_4^+*.

The GABA-benzodiazepine hypothesis suggests that in hepatic encephalopathy gut derived GABA or BZ-EL occupy the GABA-benzodiazepine receptor complex. GABA or BZ-EL increase GABA-mediated inhibition by either opening more Cl^--channels, or opening Cl^--channels longer, allowing more Cl^- than normal to enter the postsynaptic neuron. This increases GABA-mediated IPSPs but does not affect IPSPs mediated by glycine, Fig. 13 *GABA/BZ*.

Since there is much evidence for a role of ammonia intoxication in the pathogenesis of hepatic encephalopathy, GABA or BZ-EL and NH_4^+ may simultaneously affect postsynaptic inhibition. This case is schematically shown in Fig. 13 *GABA/BZ + NH_4^+*. Although, gut-derived GABA or BZ-EL increases the opening of Cl^--channels by GABA, the lack of a Cl^--gradient across the membrane prevents the generation of hyperpolarizing IPSPs by GABA- and glycine-mediated postsynaptic inhibition.

Fig. 13, cont. inhibition. *GABA/BZ+NH_4^+* - Effect of NH_4^+ on GABA-mediated inhibition in the presence of an endogenous ligand (BZ-EL) for the GABA-benzodiazepine receptor complex. The presence of BZ-EL does not change the effects of NH_4^+ on postsynaptic inhibition mediated by GABA and glycine.

The potential effects of NH_4^+, GABA, BZ-EL on inhibitory neuronal interactions mediated by GABA are shown in Fig. 14. Normally, postsynaptic inhibition suppresses the generation of action potentials by EPSPs occurring during IPSPs, Fig. 14 *Normal - Control*. When NH_4^+ increases in the CNS, gate-IPSPs loose their inhibitory efficacy and do not suppress action potential generation, Fig. 14 *Normal - NH_4^+/"Gate"*. NH_4^+ does not affect the inhibitory efficacy of break-IPSPs, Fig. 14 *Normal - NH_4^+/"Break"*. Gut derived GABA or BZ-EL prolong the inhibitory efficacy of GABA-mediated postsynaptic inhibition, Fig. 14 *GABA/BZ-EL - Control*. Although, gut derived GABA or BZ-EL may increase GABA-mediated inhibition, NH_4^+ abolishes the inhibitory efficacy of gate-IPSPs mediated by GABA, Fig. 14 *GABA/BZ-EL - NH_4^+/"Gate"*. However, NH_4^+ does not affect the inhibitory efficacy of break-IPSPs mediated by GABA; inhibition is still increased by GABA or BZ-EL, Fig. 14 *GABA/BZ-EL - NH_4^+/ "Break"*.

The simultaneous actions of gut-derived GABA or BZ-EL and NH_4^+ on postsynaptic inhibition result in increases and decreases of postsynaptic inhibition. GABA-mediated inhibition by gate-IPSPs looses its inhibitory efficacy and is decreased, whereas GABA-mediated inhibition by break-IPSPs is increased when compared to control, Fig. 14. Glycine-mediated inhibition by gate-IPSPs is decreased, whereas glycine-mediated inhibition by break-IPSPs retains its normal inhibitory efficacy (not shown).

Ro15-1788 is presumed to be a benzodiazepine antagonist, which by itself may have no effect on GABA-mediated inhibition, however, reverses the effects of benzodiazepines or BZ-EL on GABA-mediated inhibition. The question arises whether Ro15-1788 could possibly reverse also the effects of NH_4^+ on postsynaptic inhibition. This is most unlikely. In the simultaneous presence of NH_4^+ and BZ-EL, Ro15-1788 can be suggested to reverse the effect of BZ-EL on GABA-mediated inhibition. Ro15-1788 probably reverses the increased inhibitory efficacy of break-IPSPs to a normal inhibitory efficacy, Fig. 14 *BZ-EL+Ro15-1788 - NH_4^+/"Break"*. Ro15-1788 probably has no effect on the inactivation of Cl^--extrusion by NH_4^+ and the loss of the inhibitory efficacy of gate-IPSPs, Fig. 14 *BZ-EL+Ro15-1788 - NH_4^+/"Gate"*. Of course, Ro15-1788 is unlikely to affect the loss of inhibitory efficacy by NH_4^+ of glycine mediated gate-IPSPs. Thus, it is most unlikely that Ro15-1788 counteracts the effects of NH_4^+ on inhibitory neuronal interactions in cerebral cortex, thalamus, brainstem and other areas of the CNS.

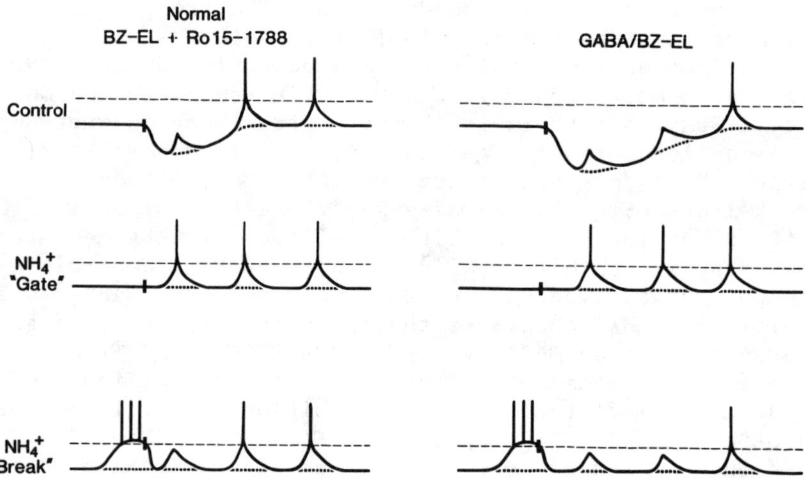

Fig. 14. Schematic representation of the effects of NH_4^+, an endogenous ligand for the GABA-benzodiazepine receptor complex (BZ-EL) and Ro15-1788 on efficacy of GABA-mediated postsynaptic inhibition. **(NORMAL)** *Control* - In control conditions hyperpolarizing postsynaptic inhibition suppresses the generation of action potentials by the first of three EPSPs. *NH_4^+/"Gate"* - NH_4^+ abolishes the hyperpolarizing action and efficacy of postsynaptic inhibition mediated by gate-IPSPs; the generation of action potentials is not prevented. *NH_4^+/"Break"* - Although NH_4^+ abolishes the hyperpolarizing action of postsynaptic inhibition, inhibition mediated by break-IPSPs still prevents the generation of action potentials. **(GABA/BZ-EL)** *Control* - An endogenous ligand for the GABA-benzodiazepine receptor complex (BZ-EL) increases the efficacy of postsynaptic inhibition and the first two EPSPs do not generate action potentials. *NH_4^+/"Gate"* - Although BZ-EL acts on the GABA-benzodiazepine receptor complex, NH_4^+ abolishes the efficacy of inhibition mediated by gate-IPSPs. *NH_4^+/"Break"* - NH_4^+ does not affect the inhibitory efficacy of break-IPSPs when BZ-EL acts on GABA-mediated postsynaptic inhibition. The inhibitory efficacy of GABA-mediated IPSPs is still increased as in GABA/BZ - Control. **(BZ-EL + Ro15-1788)** If Ro15-1788 has a pure antagonist action on the GABA-benzodiazepine receptor complex, the increased inhibition due to BZ-EL is returned to NORMAL. →

CONCLUSIONS

A systemic ammonia intoxication increases NH_4^+ in the CNS. The increase of NH_4^+ in the CNS affects at first inhibitory neuronal interactions mediated by gate-IPSPs. When NH_4^+ in the CNS increases beyond the level necessary to affect inhibition, excitatory synaptic transmission and inhibition mediated by break-IPSPs will be disturbed by NH_4^+. Thus, ammonia intoxication and the subsequent increase of NH_4^+ in the CNS disturb neuronal interactions, disturb the proper function of neuronal circuits and produce encephalopathy in experimental animals. NH_4^+ can be directly demonstrated to do that what is necessary to develop encephalopathy, i.e., to alter the function of neurons and to disturb neuronal interactions.

The effects of NH_4^+ on postsynaptic inhibition and excitation can be inferred to cause in man the encephalopathy due to acute ammonia intoxication such as seen with defects of the urea cycle. The effects of NH_4^+ on inhibition and excitation are also sufficient to cause portasystemic encephalopathy, i.e., the hepatic encephalopathy precipitated in patients with portasystemic shunts by an acute ammonia load such as from a gastro-intestinal hemorrhage. In portasystemic encephalopathy, however, NH_4^+ may not be the sole cause of the encephalopathy. Other toxins, e.g., short chain fatty acids and mercaptans, may act in synergism with NH_4^+ to initiate encephalopathy (39).

The recent observations on the improvement of hepatic encephalopathy after administration of Ro15-1788 led to the hypothesis that an endogenous ligand for the GABA-benzodiazepine receptor is involved in the pathogenesis of hepatic encephalopathy. The discussion of the effects of NH_4^+ and BZ-EL on postsynaptic inhibition outlines the differences between their actions on postsynaptic inhibition. Since NH_4^+ can be presumed to be increased in patients with portasystemic shunts, and Ro15-1788 does not reverse the effect of NH_4^+ on inhibition, much research is needed to understand the mechanism by which Ro15-1788 may improve encephalopathy in spite of increased NH_4^+ in the CNS.

Fig. 14, cont. Ro15-1788 does not affect the dysfunction of postsynaptic inhibition due to the effect of NH_4^+ on Cl^--extrusion from neurons.

ACKNOWLEDGMENTS

I thank Dr. J. C. Szerb for his helpful discussion of the effects of NH_4^+ on the CNS during this symposium. This research was supported by grants from the Veterans Administration.

REFERENCES

1. Alger, B. E. & Nicoll, R. (1983). Ammonia does not selectively block IPSPs in rat hippocampal pyramidal cells. J. Neurophysiol. 49, 1381-1391.

2. Andersen, P. & Sears, T. A. (1964). The role of inhibition in the phasing of spontaneous thalamo-cortical discharge. J. Physiol. 173, 459-480.

3. Bradford, H. F. & Ward, H. K. (1975). Glutamine as a substrate for isolated nerve-endings: inhibition by ammonium ions. Biochem. Soc. Trans. 3, 1223-1226.

4. Bradford, H. F., Ward, H. K., & Thomas, A. J. (1978). Glutamine - A major substrate for nerve endings. J. Neurochem. 30, 1453-1459.

5. Butterworth, R. F., Girard, G., & Giguere, J.-F. (1988). Regional differences in the capacity for ammonia removal by brain following portocaval anastomosis. J. Neurochem. 51, 486-490.

6. Conn, H. O. (1988) The hepatic encephalopathies. In: Conn, H. O. & Bircher, J. (Eds.), Hepatic Encephalopathy: Management with Lactulose and related Carbohydrates. Medi-Ed Press, East Lansing, Michigan, pp. 3-14.

7. Creutzfeldt, O. D., Watanabe, S., & Lux, H. D. (1966) Relations between EEG phenomena and potentials of single cortical cells. I. Evoked responses after thalamic and epicortical stimulation. Electroenceph. Clin. Neurophysiol. 20, 1-18.

8. Deschenes, M., Paradis, M., Roy, J. P., & Steriade, M. (1984). Electrophysiology of neurons of lateral thalamic nuclei in cat: Resting properties and burst discharges.

J. Neurophysiol. 51, 1196-1219.

9. Ehrlich, M., Plum, F., & Duffy, T. E. (1979). Blood and brain ammonia concentrations after portacaval anastomosis. Effects of acute ammonia loading. J. Neurochem. 34, 1538-1542.

10. Fan, P. & Szerb, J. C. (1988). Concentration-dependent effects of ammonium ions on synaptic transmission in the rat hippocampal slice. Abstract, this symposium.

11. Hamberger, A. C., Chiang, G. H., Nylén, E. S., Scheff, S. W. & Cotman, C. W. (1979). Glutamate as a CNS transmitter. I. Evaluation of glucose and glutamine as precursors for the synthesis of preferentially released glutamate. Brain Res. 168, 513-530.

12. Inagaki, C., Oda, W., Kondo, K., & Kusumi, M. (1987). Histochemical demonstration of Cl--ATPase in rat spinal motoneurons. Brain Res. 419, 375-378.

13. Krnjevic, K. (1974). Chemical nature of synaptic transmission in vertebrates. Physiol. Rev. 54, 418-540.

14. Llinas, R. & Baker, R. (1972). A chloride-dependent inhibitory postsynaptic potential in cat trochlear motoneurons. J. Neurophysiol. 35, 484-492.

15. Lux, H. D. (1971). Ammonium and chloride extrusion: Hyperpolarizing synaptic inhibition in spinal motoneurons. Science 173, 555-557.

16. Lux, H. D., Loracher, C., & Neher, E. (1970). The action of ammonium ion on postsynaptic inhibition in spinal motoneurons. Exp. Brain Res. 11, 431-447.

17. Matheson, D. F. & van den Berg, C. J. (1975). Ammonia and brain glutamine: inhibition of glutamine degradation by ammonia. Biochem. Soc. Trans. 3, 525-528.

18. McCandless, D. W. & Schenker, S. (1981). Effect of acute ammonia intoxication on energy stores in the cerebral reticular activating system. Exp. Brain Res. 44, 325-330.

19. Mullen, K. D., Martin, J. V., Mendelson, W. B., Bassett,

M. L. & Jones, E. A. (1988). Could an endogenous benzodiazepine ligand contribute to hepatic encephalopathy? The Lancet I, 457-459.

20. Ohnishi, K., Saito, M., Sato, S., Nakayama, T., Takashi, M., Iida, S., Nomura, F., Koen H. & Okuda, K. (1985). Direction of splenic venous flow assessed by pulsed doppler flowmetry in patients with a large splenorenal shunt. Relation to spontaneous hepatic encephalopathy. Gastroenterology 89, 180-185.

21. Owen, D. G., Segal, M., & Barker, J. L. (1986). Voltage-clamp analysis of a Ca2+- and voltage-dependent chloride conductance in cultured mouse spinal neurons. J. Neurophysiol. 55, 1115-1135.

22. Purpura, D. P. & Cohen, B. (1962). Intracellular recording from thalamic neurons during recruiting responses. J. Neurophysiol. 25, 621-635.

23. Raabe, W. (1974). Simultaneous and independent IPSPs in nearby neurons in cat motor cortex. Brain Res. 72, 153-157.

24. Raabe, W. (1981). Ammonia and disinhibition in cat motor cortex by ammonium acetate, monofluoroacetate and insulin induced hypoglycemia. Brain. Res. 210, 311-322.

25. Raabe, W. (1987). The H-reflex in the encephalopathy due to ammonia intoxication. Exp. Neurol. 96, 601-611.

26. Raabe, W. (1987). Synaptic transmission in ammonia intoxication. Neurochem. Pathol. 6, 145-166.

27. Raabe, W. (1988). Ammonium decreases excitatory synaptic transmission in cat spinal cord in vivo. Submitted for publication.

28. Raabe, W. & Gumnit, R. J. (1975). Disinhibition in cat motor cortex by ammonia. J. Neurophysiol. 38, 347-355.

29. Raabe, W. & Lin, S. (1983). Ammonia intoxication and hyperpolarizing inhibition. Exp. Neurol. 82, 711-715.

30. Raabe, W. & Lin, S. (1984). Ammonia, postsynaptic inhibition and CNS energy state. Brain Res. 303, 67-76.

31. Raabe, W. & Lin, S. (1985). Pathophysiology of ammonia intoxication. Exp. Neurol. 87, 519-532.

32. Raabe, W. & Onstad, G. (1982). Ammonia and methionine sulfoximine intoxication. Brain Res. 242, 291-298.

33. Raabe, W. & Onstad, G. (1985). Porta-caval shunting changes neuronal sensitivity to ammonia. J. Neurol. Sci. 71, 307-314.

34. Schafer, D. F. & Jones, E. A. (1982). Hepatic encephalopathy and the gamma-aminobutyric-acid neurotransmitter system. Lancet I, 18-20.

35. Schenker, S., McCandless, D. W., Brophy, E. & Lewis, M. S. (1967). Studies on the intracerebral toxicity of ammonia. J. Clin. Invest. 46, 838-848.

36. Singer, W. & Creutzfeldt, O. (1974). Unpublished results quoted by Creutzfeldt, O. and Houchin, J. in Neuronal Basis of EEG waves, in Handbook of Electroencephalography and Clinical Neurophysiology (Rémond, A., Ed.), Elsevier, Amsterdam, Vol. 2, Part C, pp. 2C3-2C55.

37. Théoret, Y., Davies, M. F., Esplin, B., & Capek, R. (1985). Effects of ammonium chloride on synaptic transmission in the rat hippocampal slice. Neuroscience 14, 798-806.

38. Théoret, Y. & Bossu, J.-L. (1985). Effects of ammonium salts on synaptic transmission to hippocampal CA1 and CA3 pyramidal cells in vivo. Neuroscience 14, 807-821.

39. Zieve, L. (1982) Hepatic encephalopathy. In: Schiff. L. & Schiff, E. (Eds.), Diseases of the Liver. Lippincott, Philadelphia, PA, pp. 433-460.

EFFECT OF PORTACAVAL ANASTOMOSIS ON AMMONIA METABOLISM IN BRAIN AND LIVER

Guylaine Girard, Jean-François Giguère and Roger F. Butterworth

Lab. of Neurochemistry, André-Viallet Clinical Res. Centre, Hôpital St-Luc (University of Montreal), Montreal, Qué. Canada H2X 3J4

SUMMARY

Construction of an end-to-side portacaval anastomosis (PCA) in the rat results in liver atrophy and increased blood and brain ammonia. Since brain is devoid of an effective urea cycle, removal of ammonia involves glutamine synthesis and PCA results in increased brain glutamine. Glutamine synthetase (GS) activities are decreased by 15% in cerebral cortex following PCA but are unchanged in brainstem. Administration of ammonium acetate to rats with a PCA results in severe encephalopathy progressing to coma. Glutamine content of brainstem of comatose rats is increased a further 2-fold whereas that of cerebral cortex is unchanged. Consequently, ammonia levels in cerebral cortex rise rapidly to attain levels of the order of 5 mM. These findings suggest a limitation in the capacity of cerebral cortex to remove additional blood-borne ammonia following PCA. GS activities of liver are decreased by 80% following PCA. This latter effect probably results from the selective loss of perivenous hepatocytes following PCA.

INTRODUCTION

There is substantial clinical and experimental evidence to suggest that ammonia toxicity is a major contributing factor in the pathogenesis of hepatic encephalopathy associated with subacute and chronic liver disease. Ammonia levels in patients with severe liver disease and encephalopathy are frequently found to be elevated in both blood and csf and hepatic encephalopathy results in neuropathological damage similar to that encountered in patients with congenital hyperammonemias (recent reviews, Zieve, 1982; Cooper and Plum, 1987; Butterworth et al., 1987).

Construction of an end-to-side portacaval shunt in the rat results in liver atrophy, sustained hyperammonemia (Table 1) and subtle neurological symptoms of encephalopathy which include abnormal locomotor activity (Holmin and Siesjo, 1974), altered sleep patterns (Beaubernard et al., 1977) and modifications of neuromuscular coordination (Butterworth and Giguère, 1984). Feeding of ammonium salts or resins to shunted rats leads to severe encephalopathy progressing to coma (Hindfelt et al., 1977; Norenberg, 1977) and neuropathological examination of brains from ammonia-treated shunted rats reveals Alzheimer type II astrocytosis, the morphological change characteristic of hepatic encephalopathy associated with chronic liver disease.

Table 1: EFFECTS OF PORTACAVAL ANASTOMOSIS IN THE RAT

	Sham operated	Portacaval shunt
Body weight (g)	358.0 ± 7.0	301.0 ± 11.1*
Liver weight (g)	13.1 ± 0.4	6.2 ± 0.33*
Blood ammonia (mM)	0.17 ± 0.01	0.33 ± 0.02*

*$p<0.01$ compared to sham operated

Effect of portacaval anastomosis on brain ammonia and amino acid levels

Following portacaval shunting, ammonia content of brain increases 2 to 3-fold with increases in cerebral cortex being significantly greater than those observed in brainstem (Figures 1 and 2) suggesting that ammonia removal by brain following portacaval shunting proceeds in a region-selective manner. Since brain is devoid of an effective urea cycle, removal of excess ammonia relies on glutamine formation. Portacaval shunting results in region-selective increases of brain glutamine (Giguère and Butterworth, 1984 and Figures 1 and 2). Similarly, increased brain glutamine is a consistent finding in patients with hepatic encephalopathy associated with both acute and chronic liver diseases (Record et al., 1976; Giguère and Butterworth, 1982; Lavoie et al., 1987a).

Administration of ammonium salts to rats with portacaval shunts results in severe encephalopathy progressing to coma (Hindfelt et al., 1977; Butterworth and Giguère, 1984). Brain glutamate levels are decreased (Hindfelt et al., 1977; Giguère and Butterworth, 1984 and Figures 1 and 2) and brain ammonia levels rise precipitously to attain levels of 3-5 mM in shunted rats administered ammonium salts. Glutamine content of brainstem is increased a further 2-fold in comatose shunted rats but no further increases are observed in cerebral cortex (Figures 1 and 2).

In dogs with portacaval shunts, ammonia infusions resulted in decreased incorporation of ^{15}N-ammonia into glutamine (Ukida et al., 1988). These findings suggest a limitation in the capacity of cerebral cortex to remove additional blood-borne ammonia by formation of glutamine following portacaval shunting. Such observations have parallels in clinical and electrophysiological studies. Patients with portal-systemic shunting (of either a spontaneous or surgical nature) are particularly susceptible to develop severe encephalopathy following the ingestion of ammonia-

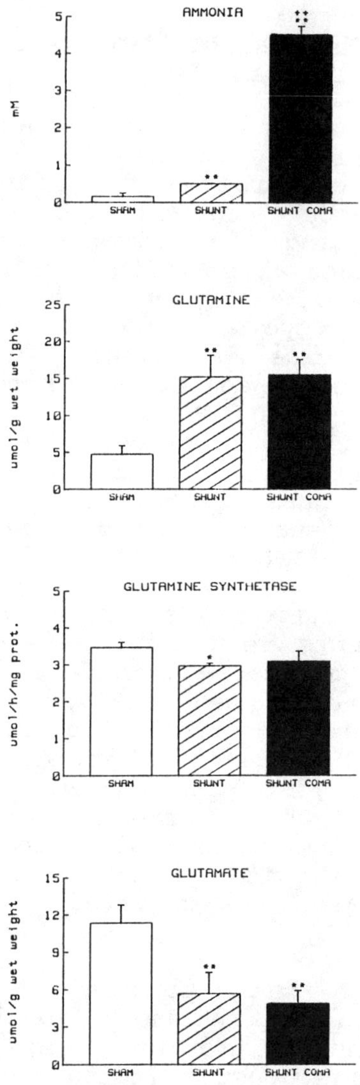

Figure 1. Effect of PCA and subsequent administration of ammonium acetate on ammonia, glutamine, glutamine synthetase and glutamate in cerebral cortex of rats.
* $p<0.05$ compared to sham
**$p<0.01$ compared to sham
++$p<0.01$ compared to shunt

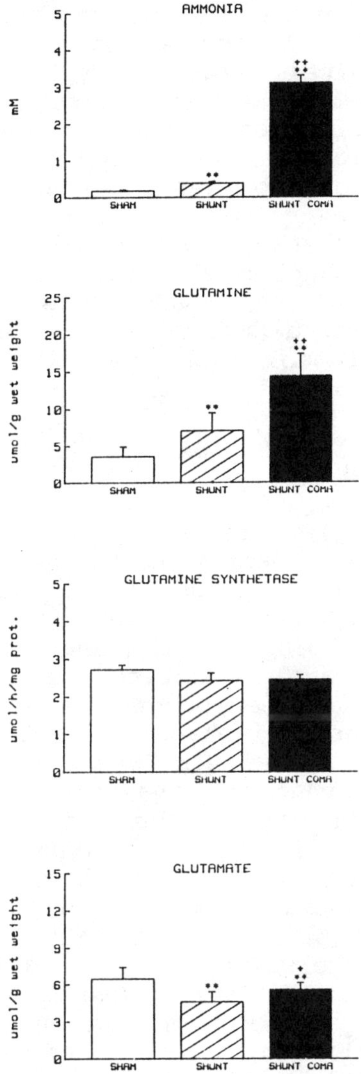

Figure 2. Effect of PCA and subsequent administration of ammonium acetate on ammonia, glutamine, glutamine synthetase and glutamate in brainstem of rats.
**p<0.01 compared to sham
+p<0.05 compared to shunt
++p<0.01 compared to shunt

containing substances (McDermott and Adams, 1954; Norenberg, 1981). Furthermore, in a series of electrophysiological studies on the effects of ammonia on synaptic transmission in the cerebral cortex of the cat, Raabe and coworkers showed that infusions of ammonium acetate led to inactivation of chloride extrusion from neurons resulting in a shift in the E_{IPSP} and altered inhibitory neurotransmission (Raabe and Gumnit, 1975; Raabe and Lin, 1985 and chapter by Raabe in this volume). Following portacaval anastomosis, animals were extremely sensitive to small doses of ammonium acetate but glutamine levels did not increase further (Raabe and Onstad, 1985).

Ammonia removal by brain; key role of astrocytic glutamine synthetase

The enzyme responsible for glutamine synthesis in brain, namely glutamine synthetase (GS) is highly enriched (and possibly uniquely localized) in astrocytes (Norenberg and Martinez-Hernandez, 1979). Neuropathological examination of the brain following portacaval shunting reveals astrocytic changes and it has been suggested that such changes reflect increased metabolic activity directed towards ammonia removal by these cells (Zamora et al., 1973). Portacaval anastomosis results in a small but significant loss of GS activity in cerebral cortex (Figure 1) but no change in brainstem (Figure 2). It has been suggested that the loss of GS activity in cerebral cortex may result from end-product (glutamine) inhibition of the enzyme (Butterworth et al., 1988). Similar reductions of GS activity have been observed in autopsied brain tissue from cirrhotic patients with hepatic encephalopathy (Lavoie et al., 1987b and Giguère et al., this volume).

GS in brain appears to operate at near-maximal capacity (Cooper et al., 1985). Cremer et al., 1975, were unable to detect any subtantial increase of glutamine synthesis in rat brain following portacaval anastomosis, again suggesting a limitation in the capacity of brain to remove additional blood-borne ammonia.

Effect of portacaval anastomosis on liver glutamine synthetase

Portacaval shunting results in a massive loss of GS activity in liver (Figure 3). This loss of GS activity is not due to the postoperative loss of weight associated with the shunting procedure as shown by studies in which animals were food-restricted (pair fed)

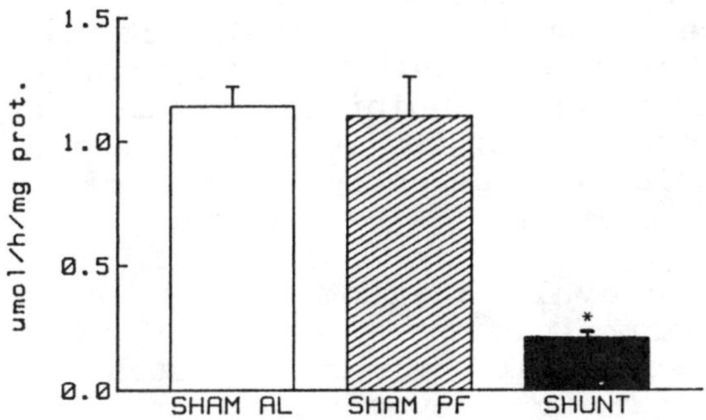

Figure 3. Effect of portacaval anastomosis or food restriction on glutamine synthetase activities in liver AL ad libitum fed; PF: pair fed controls.

Loss of hepatic GS activity most likely results from the selective loss of perivenous hepatocytes following portacaval shunting. Immunohistochemical studies have shown that GS is localized in the perivenous area of the liver lobule (Gerbhardt and Mecke, 1983). Following perivenous liver cell necrosis by carbon tetrachloride pretreatment of the rat, Haussinger and Gerok (1984) observed a significant decrease of ammonia uptake by the perfused liver. Glutamine synthesis (but not urea synthesis) from added ammonia was inhibited following perivenous cell loss. It was suggested that the hyperammonemia resulting from acute severe liver disease associated with predominantly perivenous liver cell necrosis such as that observed in CCl_4 poisoning, paracetamol overdose or viral hepatitis results from impaired GS activity rather than diminished urea synthesis (Haussinger and Gerok, 1984).

CONCLUSIONS

1. Glutamine synthetase activites in both brain (cerebral cortex) and liver are decreased following portacaval anastomosis. In the case of brain, reduced GS activities result either from loss of astrocytic integrity or from product inhibition of the enzyme. In liver, GS activity is decreased probably as a result of selective loss of perivenous hepatocytes.

2. Portacaval anastomosis results in an impairment in the already-limited capacity of certain brain structures such as cerebral cortex to remove additional blood-borne ammonia.

REFERENCES

Beaubernard, D., Salomon, F., Grange, D., Thangapregassam, M.J., and Bismuth, J. (1977). Experimental hepatic encephalopathy: changes of the level of wakefulness in the rat with portocaval shunt. Biomedicine 27, 169-171.

Butterworth, R.F. and Giguère, J.F. (1984). Region-selective glutamine changes in the CNS in relation to function in experimental subacute hepatic encephalopathy. In Advances in Hepatic Encephalopathy and Urea Cycle Disease . Kleinberger, P. Ferenci, P., P. Riederer and H. Thaler, eds). pp. 394-401.

Butterworth, R.F., Giguère, J.F., Michaud, J., Lavoie, J. and Pomier Layrargues, G. (1987). Ammonia: key factor in the pathogenesis of hepatic encephalopathy. Neurochem. Pathol. 6, 1-12.

Butterworth, R.F., Girard, G. and Giguère, J.F. (1988). Regional differences in the capacity for ammonia removal by brain following portacaval anastomosis. J. Neurochem, 51, 486-490.

Cooper, A.J.L. and Plum, F. (1987). Biochemistry and physiology of brain ammonia. Physiol. Rev., 67, 4410-519.

Cooper, A.J.L., Mora, S.N., Cruz, N.F. and Gerlbard, A.S. (1985). Cerebral ammonia metabolism in hyperammonemic rats. J. Neurochem. 44, 1716-1723.

Cremer, J.E., Heath, D.F., Teal, H.M., Woods, M.S. and Cavanagh, J.B. (1975). Some dynamic aspects of brain metabolism in rats given a portacaval anastomosis. Neuropathol. Appl. Neurobiol. 3, 293-311.

Gerbhardt, R. and Mecke, D. (1983). Heterogeneous distribution of glutamine synthetase among rat liver parenchymal cells in situ and in primary culture. Embo. J., 2, 567.

Giguère, J.F. and Butterworth, R.F. (1982). Glutamic acid abnormalities in the central nervous system in hepatic encephalopathy. Clin. Biochem., 15, 95.

Giguère, J.F. and Butterworth, R.F. (1984). Amino acid changes in regions of the CNS in relation to function in experimental portal-systemic encephalopathy. Neurochem. Res., 9, 1309-1321.

Haussinger, D. and Gerok, W. (1984). Hepatocyte heterogeneity in ammonia metabolism: impairment of glutamine synthesis in CCl_4 induced liver cell necrosis with no effect on urea synthesis. Chem.-Biol. Interactions, 48, 191-194.

Hindfelt, B., Plum, F. and Duffy, T.E. (1977). Effect of acute ammonia intoxication on cerebral metabolism in rats with portacaval shunts. J. Clin. Invest., 59, 386-396.

Holmin, T. and Siesjo, B.K. (1974). The effects of portacaval anastomosis upon the energy state and upon acid-base parameters of the rat brain. J. Neurochem., 22, 403-412.

Lavoie, J., Giguère, J.F., Pomier Layrargues, G., and Butterworth R.F. (1987a). Amino acid changes in autopsied brain tissue from cirrhotic patients with hepatic encephalopathy. J. Neurochem., 49, 692-697.

Lavoie, J., Giguère, J.F., Pomier Layrargues, G. and Butterworth, R.F. (1987b). Activities of neuronal and astrocytic marker enzymes in autopsied brain tissue from patients with hepatic encephalopathy. Metab. Brain Dis., 2, 283-290.

McDermott, W.V. and Adams, R.D. (1954). Episodic stupor associated with an Eck fistula in the human with particular reference to the metabolism of ammonia. J. Clin. Invest. 33, 1-9.

Norenberg, M.D. (1977). A light and electron microscopic study of experimental portal-systemic (ammonia) encephalopathy. Lab. Invest., 36, 618-627.

Norenberg, M.D. (1981). The astrocyte in liver disease. Adv. Cell. Neurobiol., 2, 303-352.

Norenberg, M.D. and Martinez-Hernandez, A. (1979). Fine structural localization of glutamine synthetase in astrocytes of rat brain. Brain Res., 161, 303-310.

Raabe, W. and Gumnit, R.J. (1975). Disinhibition in rat motor cortex by ammonia. J. Neurophysiol., 38, 347-355.

Raabe, W. and Lin, S. (1985). Pathophysiology of ammonia intoxication. Exp. Neurol., 87, 519-532.

Raabe, W. and Onstad, G. (1985). Porta-caval shunting changes neuronal sensitivity to ammonia. J. Neurol. Sci., 71, 307-314.

Record, C.O., Buxton, B., Chase R.A., Curzon, G., Murray-Lyon, I.M., and Williams, R. (1976). Plasma and brain amino acids in fulminant hepatic failure and their relationship to hepatic encephalopathy. Eur. J. Clin. Invest., 6, 387-394.

Ukida, M., Morishita, H., Morimoto, Y., Usui, H. and Nagashima, H. (1988). Limited glutamine synthesis in brains of dogs with a portacaval anastomosis after ^{15}N-ammonium chloride loading. In Adv. Ammonia Metab. and Hep. Enceph., (P.B. Soeters, J.H.P. Wilson, A.J. Meijer and E. Holm, eds), Excerta Medica, Amsterdam, pp. 433-438.

Zamora, A.T., Cavanagh, J.B., and Kyu, M.H. (1973). Ultrastructural responses of the astrocytes to portocaval anstomosis in the rat. J. Neurol. Sci. 18, 25-45.

Zieve, L. (1982). Hepatic encephalopathy. In Diseases of the Liver, 5th ed. (L. Schiff and E.R. Schiff, eds), pp. 433-459, Lippincott, Philadelphia.

SELECTIVE INHIBITION OF MITOCHONDRIAL DEHYDROGENASES BY AMMONIA AND FATTY ACYL COENZYME A DERIVATIVES

James C.K. Lai, Karin Rimpel-Lamhaouar, and Arthur J.L. Cooper
Departments of Biochemistry and Neurology, Cornell University Medical College, New York, NY 10021, USA

INTRODUCTION

Hyperammonemia and organic acidemia are consistently found in hepatic encephalopathy, Reye's syndrome, organic acidemias, and several other metabolic encephalopathies (Lai et al., 1977; Cooper and Plum, 1987; Cooper and Lai, 1987; Butterworth et al., 1987). However, the biochemical mechanisms underlying the neurotoxicity of these compounds have not been fully elucidated. Administration of fatty acids and ammonia can induce reversible coma in experimental animals (see Lai et al. (1987) and Cooper and Plum (1987) for discussion). It is known that once the fatty acids enter tissues and cells, they are metabolically converted to CoA derivatives. An excess of these CoA derivatives may lead to metabolic derangements (see Patel et al. (1983) and Stumpf et al. (1985) for discussion).

We have investigated the hypothesis that one mechanism underlying the toxicity of ammonia and fatty acids is the selective inhibition of key, rate-limiting and/or regulated mitochondrial dehydrogenases by ammonia and CoA derivatives of fatty acids.

PREPARATION OF BRAIN, LIVER AND HEART MITOCHONDRIA AND ASSAY OF ACTIVITIES OF VARIOUS DEHYDROGENASES

Rat brain (nonsynaptic), liver, and heart mitochondria were isolated as described previously (Lai and Cooper, 1986). The activities of pyruvate dehydrogenase complex (PDHC), NAD^+- and $NADP^+$-linked isocitrate dehydrogenases (NAD^+-ICDH & $NADP^+$-ICDH), α-ketoglutarate dehydrogenase complex (KGDHC), and NAD^+-linked malate dehydrogenase (MDH) were determined by procedures detailed elsewhere (Lai and Clark, 1976; Lai and Cooper, 1986).

SELECTIVE INHIBITION OF MITOCHONDRIAL DEHYDROGENASES BY AMMONIA

Excess ammonia stimulates glycolysis and increases lactate production, and the lactate/pyruvate ratio (Cooper and Lai, 1987). Assuming that ammonia disrupts mitochondrial oxidative metabolism, the ammonia-induced decrease in the flux into and through the tricarboxylic acid (TCA) cycle of glycolytically-derived carbon would favor increased lactate production and accumulation (Cooper and Lai, 1987). The early finding of McKhann and Tower (1961) that very high, toxic levels (> 15 mM) of ammonia inhibited α-ketoglutarate and pyruvate-supported oxygen uptake by crude brain mitochondrial preparations prompted us to postulate that one biochemical mechanism underlying the toxicity of ammonia may be the selective inhibition of key, rate-limiting and/or regulated mitochondrial dehydrogenases. We have, therefore, undertaken studies to address this possibility.

Our initial study focused on the α-ketoglutarate dehydrogenase complex (KGDHC), a rate-limiting and regulated multienzyme complex of the TCA cycle. At pathophysiological levels of 0.2 - 2 mM, ammonia inhibits KGDHC in brain (but not in heart) mitochondria, with an IC_{50} of 2 mM (Lai and Cooper, 1986).

To further investigate the possibility that ammonia may selectively inhibit key, rate-limiting and/or regulated dehydrogenases, the effect of ammonia on the activities of NAD^+- and $NADP^+$-linked isocitrate dehydrogenases (NAD^+-ICDH and $NADP^+$-ICDH) and malate dehydrogenase (MDH) was monitored. At pathological and toxicological levels of 10 mM or higher, ammonia inhibits both brain mitochondrial and liver mitochondrial (see also Katunuma et al. (1966) for additional discussion) NAD^+-ICDH in a concentration-related manner (Table 1). The liver NAD^+-ICDH is more sensitive to inhibition by ammonia than the brain enzyme; 20 mM ammonia inhibits the liver and the brain enzyme by 42% and 31%, respectively ($P < 0.05$; Table 1). However, neither brain mitochondrial $NADP^+$-ICDH nor brain mitochondrial MDH is affected by ammonia despite the high levels (10-20 mM) employed (Table 1).

Table 1. Inhibition of Mitochondrial Dehydrogenases by Ammonium Chloride and Acetate

Enzyme	Ammonium Chloride or Acetate(mM) 5	10	15	20	30
Brain mitochondrial NAD^+-ICDH	104	88*	78*	69*	74*
Liver mitochondrial NAD^+-ICDH	103	80*	ND	58*	ND
Brain mitochondrial $NADP^+$-ICDH	ND	ND	93	84	100
Brain mitochondrial MDH	ND	108	ND	94	ND

Values are the averages of enzymatic activities derived from three or more experiments and are expressed as % of the control mean value (100%). Values marked with * are significantly different ($P < 0.05$ by ANOVA) from corresponding control values. The abbreviations are: NAD^+- and $NADP^+$-ICDH, NAD^+- and $NADP^+$-linked isocitrate dehydrogenase; MDH, malate dehydrogenase; ND, not determined. The control activity values are: brain mitochondrial NAD^+-ICDH, 120 mU/mg protein; brain mitochondrial $NADP^+$-ICDH, 23 mU/mg protein; brain mitochondrial MDH, 8 U/mg protein; liver mitochondrial NAD^+-ICDH, 4 mU/mg protein.

The results of our studies, thus far, are consistent with the notion that, at pathophysiological levels (0.2 - 2 mM), ammonia selectively inhibits brain α-ketoglutarate dehydrogenase complex (KGDHC) (Lai and Cooper, 1986). At the higher and patho-toxicological levels of 10 - 20 mM, ammonia selectively inhibits the regulated NAD^+-linked isocitrate dehydrogenase in brain and liver (Table 1). However, the more active and less tightly regulated malate dehydrogenase in brain is not affected by ammonia at these high levels (Table 1).

DIFFERENTIAL INHIBITION OF MITOCHONDRIAL DEHYDROGENASES BY FATTY ACYL COENZYME A DERIVATIVES

To investigate the possibility that fatty acyl CoAs may selectively inhibit key, rate-limiting and/or regulated mitochondrial dehydrogenases, we have examined the effects of butyryl- (short-chain), octanoyl- (medium-chain), and palmitoyl (long-chain) CoAs on several brain mitochondrial dehydrogenases (Table 2). The CoA derivatives inhibit several dehydrogenases: The susceptibility of the enzymes to inhibition by these compounds is in the order: KGDHC >> NAD^+-ICDH >> MDH (Table 2). For example, at pathological levels (250 μM or lower), the CoAs inhibit brain mitochondrial KGDHC by more than 70% (Table 2). By contrast, at the higher level of 1 mM, the CoAs inhibit brain mitochondrial NAD^+-ICDH by ≧ 20% (Table 2). However, even at the higher level of 1 mM, the CoAs only inhibit brain mitochondrial MDH to a much smaller extent (by ≦ 20%; Table 2).

Of the fatty acyl CoAs investigated, the inhibitory potency increases with increasing chain length. The medium-chain and long-chain acyl CoAs are significantly more potent than the short-chain acyl CoAs (Table 2). Moreover, because the enzymatic activities are usually measured with optimal concentrations of substrates, the inhibitory effects of the acyl CoAs may be more pronounced *in vivo* where substrate concentrations may not be saturating.

Table 2. Inhibition of Brain Mitochondrial Dehydrogenases by Fatty Acyl CoAs

	Brain Mitochondrial Enzymatic Activity (% of control)		
	KGDHC	NAD^+-ICDH	MDH
Butyryl CoA:			
250 µM	21	ND	ND
1 mM	5	80	92^+
Octanoyl CoA:			
12.5 µM	63	ND	ND
250 µM	0	93^+	ND
1 mM	6	54	74
Palmitoyl CoA:			
12.5 µM	58	ND	ND
62.5 µM	38	ND	ND
250 µM	9	96^+	ND
1 mM	9	53	83

Values are the means of three or more experiments and are significantly ($P < 0.05$ by ANOVA) different from corresponding control activity (100%) values, except those marked with $^+$. The control values are: α-ketoglutarate dehydrogenase complex (KGDHC), 50 mU/mg protein; NAD^+-linked isocitrate dehydrogenase (NAD^+-ICDH), 120 mU/mg protein; malate dehydrogenase (MDH), 8 U/mg protein; ND, not determined.

Results from our ongoing studies also indicate that the fatty acyl CoAs are inhibitors of purified bovine heart KGDHC and pyruvate dehydrogenase complex (PDHC). For example, at 1 mM, butyryl CoA inhibits bovine heart PDHC and KGDHC by 90% and 50%, respectively. At 1 mM, octanoyl CoA inhibits bovine heart PDHC and KGDHC by 95% and 60%, respectively. Even more potent than butyryl- and octanoyl CoA, palmitoyl CoA, at 60 μM, inhibits bovine heart PDHC and KGDHC by 90% and 70%, respectively. Thus, our preliminary data obtained with the purified bovine heart enzyme complexes demonstrate that the bovine heart PDHC is more sensitive than the bovine heart KGDHC to inhibition by the fatty acyl CoAs. Moreover, our results suggest that brain mitochondrial KGDHC is more sensitive to inhibition by the CoAs than the bovine heart enzyme complex.

IMPLICATIONS OF OUR FINDINGS IN HEPATIC ENCEPHALOPATHY, REYE'S SYNDROME, AND OTHER METABOLIC ENCEPHALOPATHIES

Since (a) hyperammonemia and organic acidemia are persistent features in hepatic encephalopathy, Reye's syndrome, and several other metabolic encephalopathies (Stumpf et al., 1985; Cooper and Plum, 1987; Cooper and Lai, 1987), and (b) an excess of these compounds has been shown to be toxic to experimental animals, it is not unreasonable to speculate that the neurotoxicity of ammonia and short- and medium-chain fatty acids may play pathophysiological roles in these metabolic diseases. We have proposed that one mechanism underlying the toxicity of ammonia and short- and medium-chain fatty acids is the interference with brain energy metabolism through the selective inhibition of key, rate-limiting and/or tightly regulated mitochondrial dehydrogenases by ammonia and by fatty acyl CoAs. Consistent with our hypothesis is the observation that, at pathophysiological levels, ammonia and fatty acyl CoAs selectively inhibit brain mitochondrial α-ketoglutarate dehydrogenase complex (Tables 1 and 2), a rate-limiting and tightly regulated enzyme complex in the TCA cycle. The fact that pathophysiological levels of fatty acyl CoAs also significantly inhibit the purified bovine heart pyruvate dehydrogenase complex (PDHC), a rate-limiting and highly regulated multi-enzyme complex, is also consistent with our hypothesis. Since PDHC regulates the entry of glycolytically derived carbon into the tricarboxylic acid (TCA) cycle, our results strongly suggest that pathophysiological levels of fatty acyl CoAs inhibit both the entry of glycolytically-derived carbon into and the flux of the carbon through the TCA cycle.

Additionally, our data suggest that, when the levels of ammonia and fatty acyl CoAs reach pathological and/or toxicological ranges, these toxins will also inhibit other mitochondrial dehydrogenases (e.g., $NAD^+$and $NADP^+$-linked isocitrate dehydrogenases).

SUMMARY

At pathophysiological levels, ammonia and fatty acyl CoAs inhibit brain mitochondrial α-ketoglutarate dehydrogenase complex. At these levels, the fatty acyl CoAs also inhibit purified bovine heart α-ketoglutarate dehydrogenase and pyruvate dehydrogenase complexes. At pathological and/or toxicological levels, these toxins also inhibit brain mitochondrial NAD^+- and $NADP^+$-linked isocitrate dehydrogenases (J.C.K. Lai, unpublished data). These results are consistent with our hypothesis that one mechanism underlying the disruption of brain energy metabolism by ammonia and by short- and medium-chain fatty acids is the selective inhibition of key mitochondrial rate-limiting and/or tightly regulated dehydrogenases by ammonia and fatty acyl CoAs. Our data suggest that the selective inhibition of rate-limiting and/or regulated TCA cycle enzymes may assume pathophysiological importance in hepatic encephalopathy, Reye's syndrome, organic acidemias, and several other metabolic encephalopathies.

ACKNOWLEDGEMENTS

Our studies were supported, in part, by NIH grants NS 24592 and DK 16739. We thank Ms. Karen Morgan for typing this manuscript.

REFERENCES

Butterworth R.F., Giguere J.-F., Michaud J., Lavoie J., and Layrargues G.P. (1987) Ammonia: key factor in the pathogenesis of hepatic encephalopathy. Neurochem. Pathol. 6, 1-12.

Cooper A.J.L. and Lai J.C.K. (1987) Cerebral ammonia metabolism in normal and hyperammonemic rats. Neurochem. Pathol. 6, 67-95.

Cooper A.J.L. and Plum F. (1987) Biochemistry and physiology of brain ammonia. Physiol. Rev. 67, 440-519.

Katunuma N., Okada M., and Nishii Y. (1966) Regulation of the urea cycle and TCA cycle by ammonia, in Advances in Enzyme Regulation, Vol. 4, (Weber G., ed.), pp. 317-335. Pergamon Press, Oxford.

Lai J.C.K. and Clark J.B. (1976) Preparation and properties of mitochondria derived from synaptosomes. Biochem. J. 154, 423-432.

Lai J.C.K. and Cooper A.J.L. (1986) Brain α-ketoglutarate dehydrogenase complex: kinetic properties, regional distribution, and effects of inhibitors. J. Neurochem. 47, 1376-1386.

Lai J.C.K., Silk D.B.A., and Williams R. (1977) Plasma short-chain fatty acids in patients with fulminant hepatic failure. Clin. Chim. Acta 78, 305-310.

McKhann G.M. and Tower D.B. (1961) Ammonia toxicity and cerebral oxidative metabolism. Am. J. Physiol. 200, 420-424.

Patel T.B., DeBuysere M.S., and Olson M.S. (1983) The effect of propionate on the regulation of the pyruvate dehydrogenase complex in the rat liver. Arch. Biochem. Biophys. 220, 405-414.

Stumpf D.A., Parker Jr. W.D., and Angelini C. (1985) Carnitine deficiency, organic acidemias, and Reye's syndrome. Neurology 35, 1041-1045.

ACUTE ACTION OF AMMONIA ON LEUCINE METABOLISM IN ISOLATED ASTROCYTES, NEURONS AND OLIGO CELLS OF RAT BRAIN

J. JESSY and Ch. R. K. MURTHY
Dept. Physiology and Biophysics, University of Health Sciences, Chicago Medical School, N. Chicago 60064, Il. USA,(JJ); School of Life Sciences, University of Hyderabad, Hyderabad, INDIA

It is well known that in brain ammonia is detoxified by way of glutamine synthesis and this process requires glutamate as a precursor (Cooper et al., 1979, 1985; Cooper and Plum, 1987). Results obtained in the studies on metabolic compartmentation led to the belief that the glutamate required for this process is generated in the astrocytes by the reductive amination of α-ketoglutarate (KG) in the reaction mediated by glutamate dehydrogenase (Berl and Clarke, 1983). However, it was observed recently that the glutamate generated in this reaction may not be serving as the substrate for the synthesis of glutamine in brain either in normal or in hyperammonemic states (Cooper et al., 1979, 1985; Subbalakshmi and Murthy, 1983; Yudkoff et al., 1983; Yu et al., 1984). Hence, it was proposed that the pool of glutamate required for glutamine synthesis in brain may be generated from the transamination of branched-chain amino acids (leucine, isoleucine and valine; BCAA) with KG (Cooper et al., 1979, 1985; Duffy and Plum, 1982; Jessy and Murthy, 1985, 1988).

Most of the earlier studies with BCAA were on their transport across the blood-brain barrier (BBB) and their incorporation into cerebral proteins. Besides being incorporated into proteins, BCAA are also oxidized to CO_2 in brain. It has been demonstrated that the carbon and nitrogen of BCAA are incorporated into glutamate and glutamine (Cremer et al., 1975, Yudkoff et al., 1983). Though it was believed earlier that BCAA are metabolized exclusively in the astrocytes (Cremer et al., 1975), recent studies indicated that neuronal components *may be involved in their metabolism (Chaplin et al., 1976; Murthy and Hertz, 1987 a,b). Incorporation of BCAA carbons into myelin lipids suggests the involvement of oligodendroglia in their metabolism (Wiggins et al., 1979). An enhancement in the transport of BCAA across BBB and their incorporation into glutamate and glutamine was shown in hyperammonemic states (Cremer et al., 1975; Mans et al., 1982; Cangiano et al., 1984). In contrast to the above results, studies with astrocytes in primary cultures indicated that the production of CO_2 from these amino acids is suppressed in the presence of pathological concentrations of ammonia (Murthy and Hertz, 1987 a,b).*

In the present report, we describe the uptake, metabolism of BCAA and the activities of BCAA-aminotransferase (BCAA-T) and branched chain keto acid decarboxylase (BCKA-DC) in the neurons, astrocytes and oligo cells isolated from the cerebral cortex of normal and hyperammonemic rats.

METHODS

Hyperammonemia was induced in rats by an intraperitoneal injection of 2.5 mmoles of ammonium acetate/100 g body weight and after 25 min animals were killed by decapitation. Blood was collected from the neck wound for ammonia determination. Astrocytes and neurons were isolated by the method of Farooq and Norton (1978) and oligo cells by the method of Snyder et al.,(1980). Viability of the isolated cells was determined by trypan blue dye exclusion and the cell number with a hemocytometer. Cell protein content was determined by the method of Lowry et al., (1951). Methods suggested by Jessy and Murthy (1985) and Sullivan et al., (1976) were adopted for assaying BCAA-T with leucine as substrate and BCKA-DC using α-ketoisocaproic acid (KIC) as substrates respectively. Metabolism of these amino acids was determined by trapping $^{14}CO_2$ from [U-^{14}C]leucine in 30 min and uptake by determining the accumulated radioactivity in the cells incubated for 3 min in Krebs-Ringer glucose phosphate. Labelled KIC formed during the metabolism of labelled leucine was converted to 2,4-dinitrophenylhydrazone and was extracted into cyclohexane before determining the radioactivity. Purity of commercially obtained radioactive leucine and the extracted keto acid were determined by paper chromatography and autoradiography. Ammonia levels in the serum and in brain (frozen in liquid nitrogen) were determined by Berthelot's reaction using phenol-hypochlorite reagent.

RESULTS AND DISCUSSION

Intraperitoneal administration of a single dose of ammonium acetate elevated ammonia levels in serum from 0.1±0.01 (3) to 1.8±0.04 (3) μmoles/ml (mean±s.d) and in the brain from 0.4±0.09 (3) to 2.6±0.52 (3) μmoles/gm wet wt. Earlier studies in this laboratory established that the purity of the isolated cells is about 80-85% (Subbalakshmi and Murthy, 1985).

Results presented in Table I indicate the kinetic parameters, Km and Vm, for leucine uptake into astrocytes, neurons and oligo cells. When a comparison was made among the different cell types, Km for leucine uptake was similar in astrocytes and oligo cells which was higher than that in neurons. The Vm value for the uptake of leucine was higher in astrocytes than in the other two cell types.

Though the theory of metabolic compartmentation predicts that leucine is metabolised in astrocytes, it is interesting to note that this amino acid is taken up by all the three cell types. This is understandable as this amino acid is also required

TABLE I

Kinetics of leucine uptake into cellular fractions isolated from normal and hyperammonemic rat cerebral cortex

		Astrocytes	Neurons	Oligo Cells
Control	Km	2.4±0.9 (3)	1.2±0.03 (3)	2.9±0.2 (3)
	Vm	72±9 (3)	22±2 (3)	29±2 (3)
Amm. Ac	Km	2.6±0.3 (3)	6.6±1.6 (3)$^{+}$	2.7±0.5 (3)
	Vm	168±15 (3)$^{++}$	142±24 (3)**	112±12 (3)$^{++}$

Uptake studies were carried out in Krebs-Ringer glucose phosphate and 100 μg of cell protein at 37°C for 3 min. Km is expressed in mM and Vm as μmoles of amino acid taken up/mg protein/hr. Numbers in parenthesis indicate number of experiments. All the values are mean ± s.d and the comparisons are with their respective controls. $^{+}$ $p<0.01$; **$p<0.005$; $^{++}$$p<0.001$.

for protein synthesis. However, greater Vm values for the leucine uptake in the astrocytes is keeping with the postulates of the theory of metabolic compartmentation. As the rate of oxidation of leucine is higher than its incorporation into proteins (Murthy and Hertz, 1987 a), it is possible that these amino acids are preferentially oxidized in these cells. No changes were obser ved in the Km for the uptake of leucine in the astrocytes and oligo cells isolated from the cerebral cortex of hyperammonemic rats. However, there was a two to six fold increase in the Vm for the uptake of leucine in these cells in hyperammonemic states. Under these conditions, both the Km and Vm for the uptake of leucine was enhanced in the neurons.

Though no reports are available on the uptake of BCAA into brain cells in hyperammonemic states, there are evidences to indicate that the transport of these amino acids across BBB is increased under these conditions (Mans et al., 1982). In fact, Cangiano et al., (1984) demonstrated that this phenomenon is associated with an enhancement in the synthesis of glutamine in brain in hyperammonemic states. It is obvious that enhanced transport of BCAA across BBB would elevate the extracellular concentrations of these amino acids in brain in hyperammonemic conditions. Unless a corresponding change occurs in the uptake of these amino acids in different cell types in brain, they may not be efficiently utilized for the synthesis of glutamine. Results obtained in the present study are in agreement with the above concept.

As Vm is usually taken as a representative of the number of transport sites, changes in this parameter would represent a change in the number of carrier molecules in the cell membrane. This is usually achieved by the synthesis of new proteins or degradation of the existing molecules. An increase in protein content of neurons and astrocytes, observed in the present study in hyperammonemic states (Table II), suggests that the observed increase in the leucine transport might be due to enhanced protein synthesis. In addition, an increase in the transport is also possible by an unmasking of the already existing sites in the membrane and this may be achieved by changes in the physical properties of the membrane such as membrane fluidity. There are reports on the alterations in the fluidity of neuronal membranes by ammonium ions (O'Connor et al., 1982).

Thus, in hyperammonemic states there is an increase in the transport of leucine not only into the brain but also into different cell types. It would be interesting to study the metabolic fate of leucine in the cells under these conditions. Such an attempt is made in the present study by measuring the production of carbon dioxide in the cells isolated from normal and hyperammonemic rats. As was mentioned earlier, besides being incorporated into proteins, BCAA are also oxidized to CO_2 *in brain. Though Patel and Balazs (1970) reported that the incorpo-*

TABLE II

Protein content in Neurons, Astrocytes and Oligo Cells in Normal and Hyperammonemic States

	Control	Amm. Ac. treated	
Neurons	391±54 (4)	561±37 (4)	$p < 0.005$
Astrocytes	614±32 (4)	794±54 (4)	$p < 0.001$
Oligo Cells	167±35 (4)	131±31 (6)	N.S.

Protein in the trichloroacetic acid precipitate of the cells was determined by the method of Lowry et al., (1951). Values are pg protein/cell. Values are mean ± s.d. Numbers in the parenthesis indicate the number of experiments. N.S.: Not significant.

ration into proteins is higher than that of metabolism, recent studies indicate that this is not true (Chaplin et al.,1976; Murthy and Hertz, 1987 a,b). Leucine metabolism is initiated by its transamination with KG resulting in the formation of glutamate (GLU) and KIC. The keto acid is decarboxylated at C-1 position resulting in the formation of CO_2 and branched-chain keto acyl CoA. The latter is metabolized in a series of reactions and the end products are acetoacetyl CoA and acetyl CoA which are finally oxidized in the citric acid cycle (TCA cycle). Thus, CO_2 production from leucine occurs in two stages viz., when the KIC is decarboxylated at C-1 position and when the end products of leucine metabolism are oxidized in the TCA cycle. Though these two processes can be distinguished by using [1-^{14}C]- and [U-^{14}C]-leucine, presently we have used only [U-^{14}C]leucine.

Rate of production of CO_2, in the present study, is expressed in two modes viz., per mg protein and per cell. This becomes essential as the protein content varies from cell to cell and even within the same cell in normal and hyperammonemic conditions (Table II). As the enzymes involved in the metabolism of leucine and of TCA cycle constitute only a small fraction of the total protein in the cell, these results on the protein content do not provide any information on the changes in the contents of enzymes involved. If these changes in the protein content are not due to the changes in the amount of enzymes involved in the leucine metabolism, then activity/mg protein gives an underestimate of the process and this complication is avoided if the activity is represented per cell.

All the three cell types in brains of normal animals are capable of oxidizing leucine (Table III), an observation not in agree-

TABLE III
Production of $^{14}CO_2$ from [U-^{14}C]leucine in Cell Fractions in Normal and Hyperammonemic Rats

	Astrocytes	Neurons	Oligo Cells
Control	9±1.3 (6)	10±1.5 (6)	45±13 (5)
Amm. Ac.	18±1.0 (6)*	N.D. (4)*	23±6 (4)+
Control	57±8 (6)	40±5 (6)	83±7 (4)
Amm. Ac.	140±8 (5)*	N.D. (4)*	30±7 (4)*

Values are mean ± s.d. Numbers in parenthesis indicate the number of experiments. N.D.: Not detectable. Top panel:pmoles/mg protein/min; Lower panel: pmoles x 10^{-7} /cell/min.
*: $p < 0.001$; + $p < 0.02$.

ment with the earlier proposal that astrocytes are the sites of leucine metabolism. However, these results with neurons are in agreement with the results of Chaplin et al.,(1976) and of Murthy and Hertz (1987 a,b). Similarly, oxidation of BCAA in oligo cells suggests the possibility of incorporation of leucine carbons into lipids (Wiggins et al., 1979). Though all the three cell types are capable of oxidizing leucine, differences were noticed in the rates of oxidation in them. Rate of leucine oxidation was found to be higher in oligo cells than in astrocytes or neurons irrespective of mode of expression of activity. In the latter two cell types, the rate of leucine oxidation was more or less same when it was expressed per mg protein. However, when the rates were expressed per cell, astrocytes exhibited a higher rate of leucine oxidation than neurons. This suggests that the differences in the utilization of leucine might be due to the differences in the protein content of different cells. Moreover, these results also suggest that neurons and oligo cells also participate in the cerebral BCAA metabolism and are capable of producing GLU from leucine at rates comparable to astrocytes. However, net contribution of these cells to the total BCAA metabolism in brain might be less than that of astrocytes as the latter outnumber the former two cell types.

Following the administration of an acute dose of ammonium acetate, there was a suppression of leucine oxidation in neurons and oligo cells. Irrespective of mode of expression, leucine oxidation was enhanced by two fold in the astrocytes isolated from hyperammonemic rats.

As a result of enhanced oxidation of leucine in astrocytes, there would be an increase in the carbon flow through TCA cycle and incorporation of both nitrogen and carbon of leucine into GLU. This concept is in agreement with the results of Cremer et al., (1975) and the suggestion of Duffy and Plum (1982). Suppression of leucine oxidation in neurons and oligo cells in hyperammonemic states might be due to alterations in (a) transport of leucine and/or (b) activities of BCAA-T and BCKA-DH and/or (c) operational rates of TCA cycle. The first possibility may be ruled out as leucine transport was enhanced in both these cell types in hyperammonemic states (Table I). So also the last possibility, as there are enough evidences to indicate that operational rate of TCA cycle is either unaltered or slightly increased in brain preparations in the presence of pathological concentration of ammonium ion (Hawkins et al., 1973; Ratnakumari et al., 1986). This is suggestive of changes in the activities of the above two enzymes might be responsible for the alterations observed in leucine oxidation in hyperammonemic states. A similar argument is applicable for astrocytes where leucine oxidation was enhanced except that the change in leucine transport is a valid reason. Hence, we have measured the activities of BCAA-T and of BCKA decarboxylase (Table VI).

Results obtained with the enzymes indicated a suppression of BCAA-T and an elevation in the decarboxylase activity in the neurons isolated from hyperammonemic rats. Changes in the activities of these enzymes in the astrocytes and oligo cells isolated from the hyperammonemic rats varied with the mode of expression of activity. These results suggest that the decrease in the production of CO_2 *might be due to the suppression of BCAA-T activity in hyperammonemic states. However, reduction in BCAA-T activity does not explain the observed increase in the production of* CO_2 *from leucine in astrocytes. Moreover, increased activity of the decarboxylase in the neurons and oligo cells isolated from the hyperammonemic rats fails to explain the decrease in leucine oxidation in these cells. However, this anomaly could be an experimental artifact as Triton X-100 was added to the assay mixtures to obtain maximal activities. With the abolition of the permeability barriers by Triton, KIC formed in the BCAA-T reaction is freely available for the decarboxylase. With the increase in the decarboxylase activity, very little KIC will be left over which may give an underestimate of transaminase activity. To resolve this discrepancy production of [^{14}C] KIC from [U-^{14}C] leucine was studied in the cells under these conditions (Table IV).*

TABLE IV

[^{14}C]KIC Content in the Astrocytes, Neurons and Oligo Cells isolated from Normal and Hyperammonemic Rats

		Astrocytes	Neurons	Oligo Cells
A	Control	21±3 (4)	17±3 (4)	22±6 (4)
	Amm. Ac.	32±3 (4)**	22±3 (4)*	21±3 (4)
B	Control	13±2 (4)	6.5±1 (4)	3.6±0.9(4)
	Amm. Ac.	25±2 (4)+	12.0±2 (4)**	2.7±0.3(4)

*Values are mean ± s.d. Values in panel A are pmoles/mg protein/min and in panel B are pmoles x 10^{-6}/cell/min. Numbers in parenthesis indicate the number of experiments. *$p < 0.05$; ** $p < 0.005$; + $p < 0.001$.*

An increase in KIC content was noticed in the neurons and astrocytes isolated from the hyperammonemic rats but not in the oligo cells. This suggests that the suppression in transaminase might be due to the increased utilization of the KIC in the decarboxylase reaction. Moreover, suppression in the leucine oxidation in neurons might not be due to the nonavailability of the KIC. A similar conclusion might be applicable for oligo cells even though the KIC content was unaltered. However, these results fail to explain as to why the leucine oxidation was suppressed in the neurons despite an increase in the decarboxylase activity and the availability of KIC. Such a situation would arise if the KIC formed during the transamination is not made available for the decarboxylase either by its utilization in an alternate metabolic pathway or its efflux from the cell. As there are no known alternate metabolic pathways for KIC utilization, we determined its efflux from the cells (Table VI).

From these results, it is obvious that the KIC content in the medium is many fold higher than that in the cells isolated from both normal and hyperammonemic rats. This suggests that the rate of KIC formation is far in excess of its utilization and that not all the KIC formed is made available to the decarboxylase. Further, it appears that (i) even under normal conditions the amount of leucine nitrogen incorporated into glutamate might be more than the incorporation of its carbon; and (ii) there might be an efflux of KIC from the cells. An increase in medium content of KIC in the neurons isolated from hyperammonemic rats supports the concept that the

TABLE V

Tissue and Medium content of KIC in Astrocytes, Neurons and Oligo cells of Normal and Hyperammonemic rats

		Control	Amm. Ac. treated	% over control
Astro-cytes	P	0.67±0.1 (4)	1.00±0.2 (4)	+ 49 p < 0.05
	M	19.2 ±3.4 (4)	30.3 ±2.3 (4)	+ 58 p < 0.005
Neurons	P	0.33±0.07 (4)	0.50±0.07 (4)	+ 50 p < 0.02
	M	16.3±2.7 (4)	21.5 ±3.00 (4)	+ 32 p < 0.05
Oligo Cells	P	0.55±0.12(4)	0.62±0.17 (4)	+ 13 N.S.
	M	18.8±2.70(4)	20.5 ±2.30 (4)	+ 9 N.S.
Astro-cytes	P	4.1±0.6 (4)	7.9 ±1.6 (4)	+ 93 p < 0.005
	M	118±21 (4)	241±19 (4)	+104 p < 0.001
Neurons	P	1.3±0.3 (4)	2.8±0.4 (4)	+115 p < 0.001
	M	64±10 (4)	121±17 (4)	+ 89 p < 0.005
Oligo Cells	P	0.9±0.1 (4)	0.8±0.2 (4)	- 12 N.S.
	M	32±5 (4)	27±3 (4)	- 16 N.S.

Legend same as in Table V except that the values in Panel B are pmoles x 10^{-7}/cell/min. P = Pellet; M = Medium.

suppression of leucine oxidation might not be due to the decrease in BCAA-T but due to an enhanced efflux of the KIC from the cells. In these cells, the amount of GLU formed due to transamination would be higher in hyperammonemic states than in normal conditions. However, a similar increase in the KIC content in the astrocytes in hyperammonemic states would not explain the increased oxidation of leucine. This may be due to the fact that the oxidation of acetyl CoA (formed from KIC) requires oxaloacetate (OAA). If the OAA of the TCA cycle is utilized, then pyruvate oxidation and the operation of malate-aspartate shuttle would be affected (Murthy and Hertz, 1988). As astrocytes are capable of generating OAA by CO_2 fixation, they can utilize the KIC more efficiently than the neurons. Increased leucine oxidation in hyperammonemic states in astrocytes might be due to the reported enhancement in the anaplerotic replenishment of OAA (Berl and Clarke,1983). Inhibition of leucine oxidation in neurons might spare the OAA for TCA cycle and for the operation of malate-aspartate shuttle. Any acetyl CoA formed during the oxidation of

TABLE VI

ACTIVITIES OF BCAA-T AND BCKA-DC IN CELLULAR FRACTIONS IN HYPERAMMONEMIC STATES.

		Astrocytes	Neurons	Oligo Cells
I		Leucine aminotransferase		
A	Control	11.3±2.8 (5)	5.7±0.5 (5)	10.0±1.5 (7)
	Amm.Ac	8.3±1.0 (6)	2.7±0.2 (6)	10.0±1.7 (6)
		p<0.05 -27%	p< 0.001 -53%	
B	Control	7.0±1.7 (5)	2.2±0.1 (5)	1.7±0.3 (7)
	Amm.Ac	6.6±0.8(6)	1.6±0.1 (6)	1.3±0.2 (6)
			p< 0.001 -27%	p<0.02 -22%
II		Branched-chain keto acid decarboxylase		
A	Control	43±10 (5)	28±05 (5)	35±3 (5)
	Amm.Ac	82±20 (5)	58±13 (5)	45±3 (5)
		p<0.001 +91%	p<0.005 +107%	p<0.001 + 29%
B	Control	26±06 (5)	11±2 (5)	5.9±0.5 (5)
	Amm.Ac	65±16 (5)	33±7 (5)	5.9±0.4 (5)
		p< 0.001 +150%	p<0.001 +200%	

Values are mean±s.d. A:nmoles/mg protein/min. B:pmoles X 10^{-3}/cell/min. I:KIC produced; II:$Fe(CN)_6$ reduced. Number in parenthesis indicates the number of experiments.

KIC would accumulate under these conditions and exert a feed back inhibition on BCKA-DH. Moreover, differences in other parameters, such as the kinetics of BCKA transport across membranes and the affinities of the enzymes involved in the metabolism of the BCKA, in these cells might also be responsible for such responses.

In conclusion, it is suggested that the GLU required for glutamine synthesis in hyperammonemic states might be generated by the transamination of BCAA such as leucine. The keto acids formed during this process might be utilized to a limited extent and the excess keto acids are exported out of the cells. These results also suggest that all the major three cell types of brain i.e., neurons, astrocytes and oligo cells are capable of generating glutamate from leucine and thus they may participate in the process of ammonia detoxification.

ACKNOWLEDGEMENTS

Financial assistance was given by C.S.I.R., India.

REFERENCES

Berl S. and Clarke D.D. (1983) Metabolic compartmentation concept., in Glutamine,Glutamate and GABA in Central Nervous System (Hertz L., Kvamme E., McGeer E. and Schousboe A. eds.,) pp 205-217, Alan R. Liss, Inc., New York.

Cangiano P.C., Cangiano C., James J.H., Ceci F., Fischer J.E. and Strom R.(1984) Effect of ammonia on amino acid uptake by brain microvessels. J.Biol.Chem 259, 5295-5300.

Chaplin E.R., Goldberg A.L. and Diamond I. (1976) Leucine oxidation in brain slices and nerve endings. J.Neurochem. 26, 701-707.

CooperA.J.L., McDonald J.M., Gelbard A.S., Gledhill R.F. and Duffy T.E. (1979) The metabolic fate of ^{13}N labelled ammonia in rat brain. J.Biol.Chem. 254, 4982-4992.

Cooper A.J.L., Mora S.N., Cruz N.F. and Gelbard A.S. (1985) Cerebral ammonia metabolism in hyperammonemic rats. J. Neurochem. 44, 1716-1723.

Cooper A.J.L. and Plum F. (1987) Biochemistry and Physiology of Brain Ammonia. Physiol. Rev. 67, 440-519.

Cremer J.E., Heath D.F., Patel A.J., Balaz R. and Cavanagh J.B. (1975) An experimental model of CNS changes associated with chronic liver disease: portocaval anastomosis in the rat., in Metabolic Compartmentation and Neurotransmission: Relation to Brain Structure and Function (Berl S., Clarke D.D. and Schneider,D. eds.) pp 461-478, Plenum Press, New York.

Duffy T.E. and Plum F. (1982) Hepatic Encephalopathy. in The Liver: Biology and Pathobiology (Arias I., Popper H., Schachter D. and Shafritz D.A. eds.,) pp 693-715, Raven Press. New York.

Farooq M. and Norton W.T. (1978) A modified procedure for isolation of astrocyte and neuron enriched fractions from rat brain. J.Neurochem. 31, 887-894.

Hawkins R.A., Miller A.L., Nielsen R.C. and Veech A.L. (1973) The acute action of ammonia on rat brain metabolism in vivo. Biochem.J. 134, 1001-1008.

Jessy J. and Murthy Ch.R.K. (1985) Elevation of transamination of branched chain amino acids in brain in acute ammonia toxicity. Neurochem. Int. 7, 1027-1031.

Jessy J. and Murthy Ch.R.K. (1988) Cerebral cellular branched chain amino acid transaminases in hyperammonemia, in The Biochemical Pathology of Astrocytes .(Norenberg M.D., Hertz L. and Schousboe A. eds.,) pp 449-450. Alan R. Liss, Inc., New York.

Lowry O.H., Rosebrough N.J., Farr A.L. and Randall R.L. (1951) Protein measurement with the folin phenol reagent. J.Biol.Chem 193:265-275.

Mans A.M., Biebuyck J.K., Shelley K. and Hawkins R.A. (1982) Regional blood brain barrier permeability to aminoacids after portacaval anastamosis. J.Neurochem. 38,705-717.

Murthy Ch.R.K. and Hertz Z. (1987) Acute effect of ammonia on branched chain amino acid oxidation and incorporation into proteins in astrocytes and in neurons in primary cultures. J.Neurochem. 49,735-741.

Murthy Ch.R.K. and Hertz L (1987) Comparison between acute and chronic effects of ammonia on BCAA oxidation and incorporation into proteins in primary cultures of astrocytes and of neurons. J.Neurosci. Res. 17,271-276.

Murthy C.R.K. and Hertz L. (1988) Regulation of pyruvate decarboxylation in astrocytes and in neurons in primary cultures in the presence and absence of ammonia. Neurochem. Res. 13,57-61.

O'Connor J.-E., Guerri C., Jorda A. and Grisolia S. (1982) On the mechanisms of the protective effect of ethanol against ammonia intoxication in mice.Biochem. Biophys.Res.Commun. 107,1508-1516.

Patel A.J. and Balaz R. (1970) Manifestation of metabolic compartmentation during the maturation of the rat brain. J.Neurochem. 17,955-971.

RatnaKumari L., Subbalakshmi G.Y.C.V. and Murthy Ch. R.K. (1986) Acute effects of ammonia on the enzymes of citric acid cycle in rat brain. Neurochem.Int.8,115-120.

Snyder S.D., Raine C.S., Farooq M. and Norton W.T. (1980) The bulk isolation of oligodendroglia from whole rat forebrain: A new procedure using physiologic media. J.Neurochem. 34,1614-1621.

Subbalakshmi G.Y.C.V. and Murthy Ch.R.K. (1983) Acute metabolic effects of ammonia on the enzymes of glutamate metabolism in isolated astroglial cells. Neurochem. Int. 5,593-597.

Subbalakshmi G.Y.C.V. and Murthy Ch.R.K. (1985) Isolation of astrocytes, neurons and synaptosomes of rat brain cortex: Distribution of enzymes of glutamate metabolism. Neurochem. Res. 10,239-250.

Sullivan S.G., Dancis J. and Cox R.P. (1976) Modulation of branched chain -keto acid decarboxylase activity in rat liver mitochondria by hypophysectomy. Arch. Biochem. & Biophy. 176,225-234.

Yu A.C.H., Schousboe A. and Hertz L. (1984) Influence of pathological concentrations of ammonia on metabolic fate of ^{14}C labelled glutamate in astrocytes in primary cultures. J.Neurochem.42,594-597.

Yudkoff M., Nissim I., Pleasure D., Kim S., Hummeler K. and Segal S. (1983) Ammonia and amino acid interaction in cultured brain cells: Studies with $^{15}NH_4$, in Glutamine, Glutamate and GABA in the Central Nervous System.(Hertz L., Kvamme E., McGeer E.G. and Schousboe A. eds.,) pp 389-398. Alan R. Liss, Inc., New York.

PROGRESSIVE HYPERAMMONEMIA AND INSULIN RESISTANCE AFTER PORTACAVAL SHUNT IN HOMOZYGOUS FAMILIAL HYPERCHOLESTEROLEMIA

Ijaz A. Qureshi, Marie Lambert, Emile Levy, Muriel Godard and Jacques Letarte
Pediatric Research Center, Sainte-Justine Hospital and Departments of Pediatrics and Nutrition, University of Montreal, Montreal, Quebec, Canada H3T 1C5

ABSTRACT

We have followed-up four homozygous hypercholesterolemic children who had undergone an end-to-side portacaval shunt. Periodic evaluations of protein and carbohydrate tolerance after surgery have been performed. The surgery was done at ages ranging between 5 to 12 years. They have been followed-up for up to nine years. Before surgery, their plasma cholesterol levels ranged between 437 mg/dL and 1300 mg/dL. These have been reduced to about 50% with a gradual regression of xanthomas. The plasma triglycerides have been restored to normal. Protein loading tests were performed regularly. After surgery, fasting and post-load ammonia levels are abnormally high, indicating protein intolerance which becomes progressively more severe with time. A clinical trial with benzoate has shown that the excretion of hippurate nitrogen increases at the expense of urinary urea nitrogen. In one of the patients, the fasting plasma ammonia became normal during benzoate therapy with a reduction in plasma glutamine nitrogen. A standard oral glucose tolerance test was conducted in each of these children before and after portacaval shunt. Whereas no differences are discernible between the glucose curves for the pre-shunt and post-shunt periods, there is a tendency for insulin to increase progressively after shunt with the highest values being recorded after 8 years in one patient. At the same time, an increase in plasma growth hormone level was noted.

In conclusion, patients with familial homozygous hypercholesterolemia treated with portacaval shunt develop progressive protein intolerance and secondary hyperammonemia. This may be caused by the diversion of ammonia-rich portal blood and a progressive decrease in the hepatic functional capacity. It is thus important that protein intake in these children be controlled. Periodic sodium benzoate therapy could help to eliminate the nitrogen overload. Hyperinsulinemia develops progressively after surgery indicating an insulin resistance. The increase in growth hormone after shunt may be one of the factors responsible for insulin resistance and it is probable that a carbohydrate intolerance may eventually develop.

INTRODUCTION

Familial hypercholesterolemia is characterized clinically by: a) a selective elevation in the plasma levels of low density lipoprotein (LDL), b) deposition of LDL derived cholesterol as xanthomas and atheromas, and c) inheritance as an autosomal recessive dominant trait with a gene dosage effect, which means that homozygotes are more severely affected than heterozygotes (Frederickson et al., 1978). The homozygotes, numbering about 1 in a million persons in North America, are resistent to conventional drug therapy. Before the advent of liver transplantation recourse, their plasma cholesterol levels could be lowered only by performing repeated plasmapheresis or the surgical creation of a portacaval shunt (Starzl et al., 1973).

During the first 10 years of portacaval shunt surgery, at least 45 homozygous patients had undergone the operation in North America (Mitchell, 1983). Their ages at the time of operation ranged from 2½ to 24 years. Only six of these patients had died, including three within 30 days of shunt. Most of these children have been followed up post-shunt particularly in respect of their lipid metabolism and cardiovascular status, while information relating to modifications in the protein or carbohydrate metabolism is quite scanty. Only one child has been reported to have developed acute hepatic encephalopathy after portal diversion for homozygous hypercholesterolemia (Hassal et al., 1984). A brief reference to the same child was made earlier by Starzl et al. (1983) in their review of 12 patients.

Our team at the Sainte-Justine Hospital have followed up on four homozygous hypercholesterolemic children, who had undergone an end-to-side portacaval shunt. We have conducted periodic evaluations of protein and carbohydrate tolerance with hormonal status after shunt surgery. The results have indicated that all these children developed a progressive chronic hyperammonemia with insulin resistance. One of these children was also treated with acylating compound sodium benzoate, to relieve an episode of hyperammonemia, by mobilizing alternate pathways of nitrogen excretion (Brusilow et al., 1979).

MATERIALS AND METHODS

Subjects

The particulars of the four patients of Type II homozygous familial hypercholesterolemia are given in Table 1. Case histories of two of these patients, N.B. and S.B. have already been published in detail (Blanchard et al., 1979). The shunt surgery was performed at ages ranging from 5-12 years. These patients have now been followed up post-shunt for 4 to 9 years. One of these patients (S.B.) died due to cardiac complications. In another patient (F.J.) a reconstruction of portacaval shunt had to be performed. The plasma lipoprotein profiles of the four patients are of Types IIa and IIb. Before shunt surgery, their plasma cholesterol levels ranged between 437 and 1300 mg/dL (normal values: 110-230). After portacaval shunt, the cholesterol levels have generally been reduced significantly, with a gradual regression of xanthomas. The plasma triglycerides, which ranged between 10-299 mg/dL before shunt surgery have been reduced to 34-103 mg/dL after shunt. High triglyceride levels are not characteristic of hypercholesterolemia, and may have been seen in these patients before shunt due to unrestricted caloric intake.

Table 1. Homozygous familial hypercholesterolemia: plasma lipid profile.

Patient	Lipoprotein type	Age at shunt (years)	Before shunt		After shunt	
			Cholesterol	Triglycerides	Cholesterol	Triglycerides
N.B.	IIB	9.9	437-828[a]	10-278	305-524	34-82
S.B.	IIA	5.0	888-1300	64-200	492-645	50-81
P.M.	IIA	10.0	465-810	32-299	390-575	46-88
F.J.	IIA	11.8*	602-725	62-100	429-596	30-103

[a] mg/dL; normal adult values: cholesterol, 110-230; triglycerides, 10-140.

* Reconstruction of a portacaval shunt. After the original portacaval shunt at 8.0 years, the cholesterol levels came down from 640-1118 to 657-700 mg/dL.

Methodology

A protein tolerance test was conducted in these children periodically, by measuring plasma ammonia levels before and after a protein load. In two of these children, the tolerance test was also done before shunt surgery, for comparison with post-shunt values. This test was originally designed by our team to diagnose chronic hyperammonemias of hereditary etiology, and has been successfully used for this puprose (Qureshi et al., 1981). The patient is kept on stable dietary protein intake for three days prior to the test. On the day of the test, a fasting blood sample is drawn at 8:30 AM. The child is then given a load of 1.5 g protein/kg body weight. A post-load blood sample is drawn at 10:00 h. The blood samples are centrifuged within 15 minutes of drawing and kept frozen until analysed by the ion-exchange resin technique (Hutchinson et al., 1962). Plasma glutamine is measured by the specific glutaminase technique of Welbourne et al. (1972). The glutaminase releases ammonia, which is also quantitated by the ion exchange method and corrected for base values.

For the oral glucose tolerance test (OGTT), a load of 1.75 g/kg (maximum: 100 g) was taken by mouth in 250 mL of lemon juice flavoured water. All doses were calculated on the basis of ideal weight for height. Patients remained seated in their bed throughout the test. Venous blood was drawn at specific intervals, and glucose was measured as already described (Polychronakos et al., 1980). Human growth hormone (HGH) and serum immunoreactive insulin were measured by a double antibody technique (Schalch and Parker, 1964; Hales and Randle, 1963). The $\sum$ glucose, $\sum$ insulin and $\sum$ HGH was calculated by adding respective values of concentrations measured at 0, 60, 120 and 180 minutes of the test.

The integrated insulinogenic index was calculated as the ratio of the integrated rise of insulin over the integrated rise of glucose for the 180 minute period following glucose injection. This index was used as a measure of overall insulin secretory response to blood glucose rise (Polychronakos et al., 1980).

Measurements of total cholesterol and triglycerides were done by methods based on the Technicon autoanalyser as detailed in an earlier publication (Blanchard et al., 1979).

RESULTS AND DISCUSSION

The graph represented in figure 1 demonstrates fasting and post-load levels of plasma NH_3, in each of the three children in which a protein tolerance test was conducted by measuring plasma NH_3 before and after a protein load. In all cases, post-shunt fasting and post-load NH_3 levels are abnormally high indicating protein

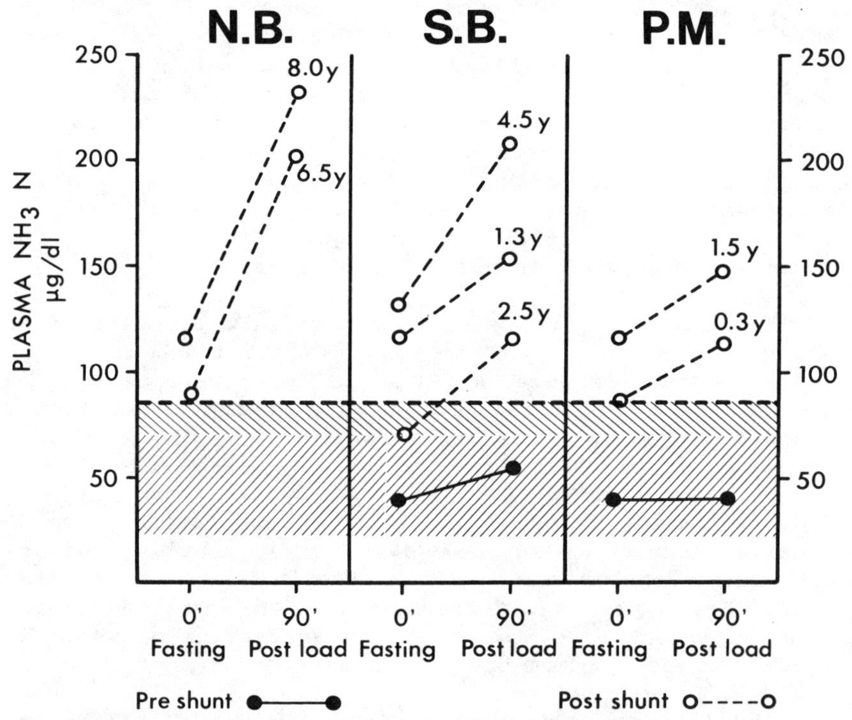

Figure 1. Fasting and post-load levels of plasma NH_3 nitrogen in three homozygous hypercholesterolemic children given a protein tolerance test at different times before and after a portacaval shunt. The closed circles indicate pre-shunt testing and open circles post-shunt testing with the post-shunt time given in years (y). The shaded area represents normal ranges of plasma NH_3 N concentration before and after protein loading in normal subjects.

intolerance, which becomes progressively worse with time elapsed after shunt surgery. The protein intolerance seen in these children at the present time, is as severe as many of the chronic cases of hyperammonemia tested by the same protocol by our team (Qureshi et al., 1981).

In our experience with hereditary hyperammonemic syndromes, sodium benzoate therapy is of considerable help as it eliminates excess nitrogen as hippurate, due to increased synthesis of glycine necessary for benzoate conjugation (Qureshi et al., 1984). Table 2 shows the results of a one week trial of sodium benzoate therapy (250 mg/kg/d) in one of the shunted hypercholesterolemic children. The figures indicate that the excretion of hippurate nitrogen increases at the expense of urinary urea N, as compared to a one-week control period. The fasting plasma NH_3 N became normal during benzoate therapy, with a reduction in plasma glutamine N.

Figure 2 represents sums of plasma glucose and of insulin concentrations measured at 0, 60, 120 and 180 minutes of the OGTT-180 test, in each patient separately. Whereas no differences are discernible between the $\sum$ glucose values calculated for pre-shunt and post-shunt periods, there is a tendency for the $\sum$ insulin to increase progressively after shunt, with the highest values recorded in N.B. at 8 years post-shunt. Similar results were obtained when the integrated surface area of the OGTT curves were compared instead of the sum of the concentrations.

Figure 3 shows similar graphs representing the sums of HGH concentrations and the insulinogenic index of various points of the OGTT curve. An increase in plasma HGH levels post-shunt was seen in two patients out of three, and was progressive in one. However, the insulinogenic index has a tendency to increase progressively, similar to the increase in the $\sum$ insulin seen in the previous graph (Figure 2).

Table 3 shows similar hyperammonemic and hyperinsulinemic indices in respect of a child (F.J.) who had a portacaval shunt done at a Quebec University Hospital (CHUL) at age 8. This shunt had to be reconstructed at Sainte-Justine Hospital at age 12.

Table 2. Sodium Benzoate Therapy in an Hypercholesterolemic Patient (S.B.). Post-shunt Plasma and Urinary Metabolites.

Sodium benzoate (mg/kg/d)	Total urinary N (g/d)	Urinary urea N (%)	Urinary hippurate N (%)	Plasma NH_3 N (μg/dL)	Plasma glutamine N (μg/dL)
-	5.8 ± 0.6[a]	73 ± 1	0.7 ± 0.1	93-132[b]	525-863[b]
250	7.8 ± 1.7	65 ± 2	9.1 ± 1.2	58-76	459-719

[a] Mean ± SEM; N = 7.

[b] Fasting values.

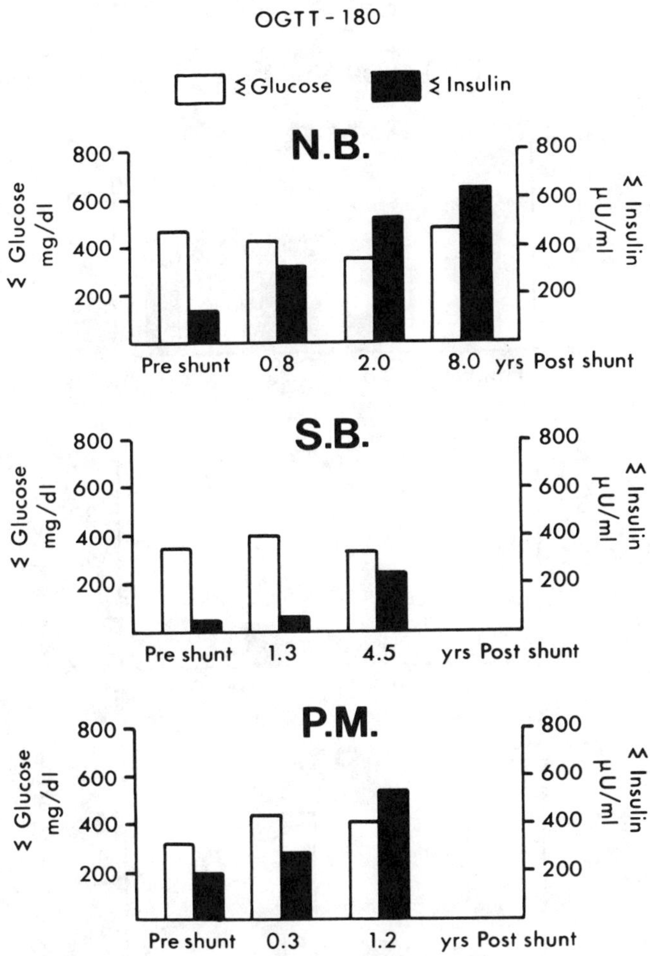

Figure 2. Results of the standard 180 minute oral glucose tolerance test (OGTT) in three homozygous hypercholesterolemic children represent sums (∑) of plasma glucose and of insulin concentrations measured at 0, 60, 120 and 180 minutes of the test, conducted at different times pre- and post-shunt.

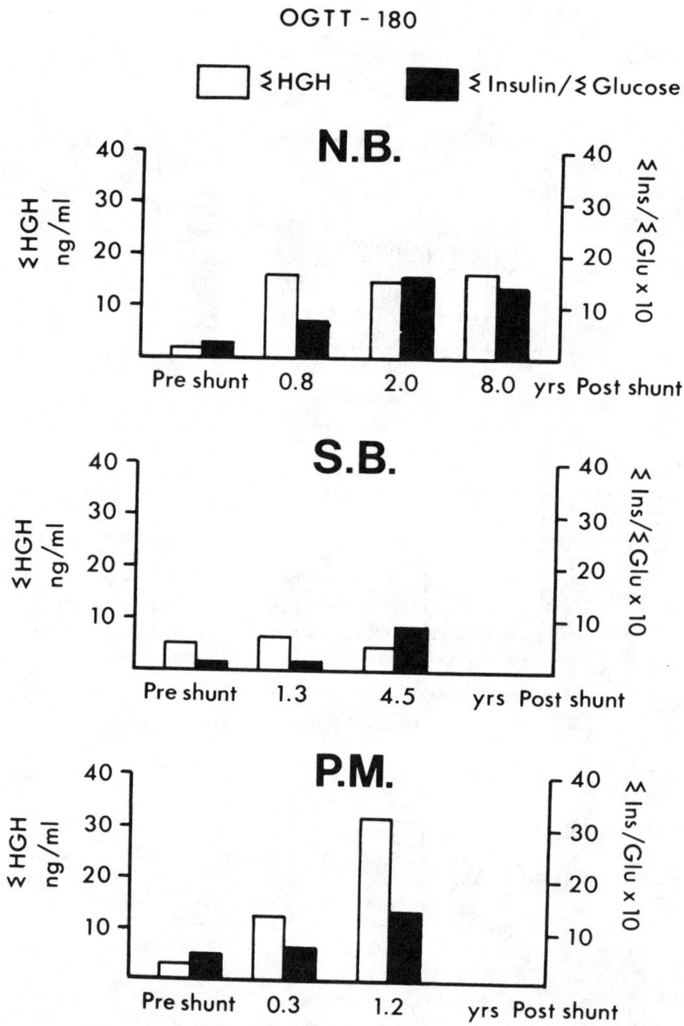

Figure 3. Sum (∑) of growth hormone (HGH) concentrations and the insulinogenic index (∑ Ins/∑ glucose) at various points of the OGTT curve. The OGTT-180 was conducted in three homozygous hypercholesterolemic children at different times pre- and post-shunt.

Table 3. Hyperammonemic and Hyperinsulinemic Parameters in a Hypercholesterolemic Child (F.J.)* Before and After Reconstruction of a Portacaval Shunt.

Years post-shunt	Fasting Plasma NH_3[a] (μg N/dL)	Post-load Plasma NH_3 (μg N/dL)	$\sum$ glucose (mg/dL)	$\sum$ insulin (μU/mL)	$\frac{\sum \text{insulin}}{\sum \text{glucose}}$
Pre-reconstruction	78.8	109.1	360	-	-
0.8	104.6	127.4	368	339	0.92
3.3	124.7	148.3	409	-	-
3.8	128.2	166.7	448	755	1.69

* LDL receptor binding in this child was measured at 2.63 ng of protein/mg cellular protein. Normal values: 75-100.

[a] Normal fasting NH_3 N levels: 24-80 μg/dL.

It can be said that patients with familial homozygous hypercholesterolemia treated with portacaval shunt develop hyperammonemia and progressive protein intolerance. This may be caused by the diversion of ammonia rich portal blood and a progressive decrease in the functional capacity of the liver. The gluconeogenesis caused by an increased level of fasting glucagon after shunt surgery in hypercholesterolemia, as reported by some authors, may also be a contributing factor. It is important that protein intake in these children be controlled. Sodium benzoate therapy could help to eliminate nitrogen overload. Hyperinsulinemia is seen to develop progressively, with indication of insulin resistance. However, fasting plasma insulin levels remain generally normal after shunt surgery. The increase in HGH after shunt surgery may only be one of the factors responsible for insulin resistance, since the increase in HGH is not always seen. It is probable that carbohydrate intolerance may eventually appear.

ACKNOWLEDGEMENTS

Work reported in this article was supported by the Medical Research Council of Canada grant # MA-9124.

REFERENCES

Blanchard, H., Bensousan, A.L., Collin, P.P., Desjardins, J.L., Ouimet, A. and Letarte, J. L'anastomose porto-cave terminolatérale dans l'hypercholestérolémie familiale de type II homozygote. Chir. Pediatr. 20: 221-226, 1979.

Brusilow, S.W., Valle, D.L. and Batshaw, M.L. New pathways of nitrogen excretion in inborn errors of urea synthesis. Lancet ii: 452-454, 1979.

Frederickson, D.S., Goldstein, J.L. and Brown, M.S. The familial hyperlipoproteinemias. In: Stanbury, J.B., Wyngaarden, J.B. and Frederickson, D.S. (Eds). The Metabolic Basis of Inherited Disease. McGraw Hill, New York, pp. 604-655, 1978.

Hales, C.N. and Randle, P.S. Immunoassay of insulin with insulin antibody precipitate. Biochem. J. 88: 127-130, 1963.

Hassal, E., Benson, L., Hart, M. and Krieger, D.E. Hepatic encephalopathy after porta-caval shunt in a non-cirrhotic child. J. Pediatr. 105: 439-441, 1984.

Hutchinson, J.H. and Labby, D.H. New method for the micro-determination of blood ammnonia by use of cation exchange resin. J. Lab. Clin. Med. 60: 170-178, 1962.

Mitchell, S.C. Porta-caval shunt in familial hyper-cholesterolemia. Lancet i: 193, 1983.

Polychronakos, C., Letarte, J., Collu, R. and Ducharme, J.R. Carbohydrate intolerance in children and adolescents with Turner syndrome. J. Pediatr. 6: 1009-1014, 1980.

Qureshi, I.A., Letarte, J., Ouellet, R., Lelièvre, M. and Laberge, C. Ammonia metabolism in a family affected by hyperargininemia. Diabete Metabol. 7: 5-11, 1981.

Qureshi, I.A., Letarte, J., Ouellet, R., Batshaw, M.L. and Brusilow, S. Treatment of hyperargininemia with sodium benzoate and arginine-restricted diet. J. Pediatr. 104: 473-476, 1984.

Schalch, D.S. and Parker, M.L. A sensitive double antibody immunoassay for human growth hormone in plasma. Nature 203: 1141-1142, 1964.

Starzl, T.E.., Chase, H.P., Putnam, C.W. and Porter, K.A. Porta-caval shunt in hyperlipoproteinemia. Lancet ii: 940-944, 1973.

Starzl, T.E., Chase, H.P., Ahrens, E.H., McNamara, D.J., Bilheimer, D.W., Schaeffer, E.J., Rey, J., Porter, K.A., Stein, E., Francovilla, A. and Benson, L.N. Portacaval shunt in patients with familial hypercholesterolemia. Ann. Surg. 198: 273-283, 1983.

Welbourne, T., Weber, M. and Bank, N. The effect of glutamine on ammonium excretion in normal subjects and in patients with renal disease. J. Clin. Invest. 56: 1852-1858, 1972.

MESENTERIC VENOUS STENOSIS REDUCES HYPERAMMONEMIA IN THE PORTACAVAL SHUNTED RAT

Kaj Johansen, M.D., Ph.D., Catherine Girod, Samuel S. Lee, M.D., Didier Lebrec, M.D.

From the Unité de Recherches de Physiopathologie Hepatique (INSERM U-24), Hopital Beaujon, 92118 Clichy, France

ABSTRACT

Hyperammonemia is a constant finding following portacaval anastomosis (PCA), and has been incriminated in the genesis of post-shunt encephalopathy (PSE) in the human. We developed a model for mesenteric venous hypertension in the rat by modification of a commonly-used technique for studying extrahepatic portal hypertension. We then examined serum ammonia levels in rats undergoing sham operation, mesenteric vein stenosis (MVS) alone, PCA alone, and MVS plus PCA.

MVS animals had significant ($p < 0.05$) elevation in mesenteric venous pressures two to three weeks after operation. Serum ammonia levels were normal in rats undergoing sham operation and MVS, and were significantly elevated ($p < 0.001$) in rats with PCA. However, combination of MVS and PCA resulted in a significant ($p < 0.01$) reduction in serum ammonia levels.

These data suggest that ammonia absorption from the intestine is a function of splanchnic venous pressure. This finding may be relevant to the management of the neuropsychiatric deterioration seen following PCA in the human.

INTRODUCTION

A progressive and debilitating neuropsychiatric deterioration has long been recognized in experimental animals and in humans undergoing portacaval anastomosis (PCA). This syndrome, termed portal-systemic encephalopathy (PSE), remains poorly understood despite extensive investigations in recent years. Most of these studies have focused upon alterations in hepatic function, changes in concentrations of various materials in serum or cerebrospinal fluid (CSF), and appearance of certain abnormal cerebral neuropeptides thought to be operative in PSE.

Ammonia has been incriminated in the pathogenesis of post-PCA neurologic dysfunction since the 1890s, when Pavlov's laboratory pointed out the deleterious impact of oral protein in shunted dogs. Serum ammonia levels have been found to be neither completely sensitive nor specific for PSE in humans (1). However, levels of CSF glutamine, the product of ammonia metabolism in the brain, are accurate predictors of PSE. Further, methods which reduce serum ammonia levels (oral lactulose or antibiotics, reduced oral protein intake, colectomy or ileostomy) clearly improve or resolve PSE in the shunted human. Recent studies (2) conclude that ammonia is, at the least, an important cofactor in the genesis of PSE.

Studies in animals (3-7) and humans (5,8,9) have suggested that the hyperammonemia commonly noted after PCA is a consequence of the degree to which the the portal venous system is decompressed. We designed the following experiment to study the effects of elevated splanchnic venous pressure on serum ammonia levels in the portacaval-shunted rat.

MATERIALS AND METHODS

Thirty-two Sprague-Dawley rats (Charles River, Saint-Aubin-les-Elbeuf, France) weighing 280-430 gm were utilized in this experiment. They were fed tap water and rat chow *ad libitum*, and were kept at a maximum of six rats to a cage in a vivarium with a 12-hour light-dark cycle.

The rats were divided into four groups. The first group (sham) underwent laparotomy, dissection of the portal vein and the vena cava, and clamping of the portal vein for ten minutes. The second group (mesenteric vein stenosis, MVS) underwent the same procedures as sham animals, but in addition underwent graded stenosis of the mesenteric vein by a modification of a technique commonly used in this laboratory (10) and elsewhere (11) to produce permanent extrahepatic portal hypertension. In brief, the hepatic end of the superior mesenteric vein just caudal to the splenic vein was dissected and, in combination with a 0.96 mm (outside diameter) polyethylene catheter passed alongside the mesenteric vein, ligated firmly with a 3-0 silk tie. Withdrawal of the catheter from the ligature results in a stenosis of known calibre in the vein, producing a consistent and permanent elevation of portal pressure.

The third group of rats (PCA) underwent end-to-side portacaval anastomosis by standard techniques (12), including 6X optical magnification and a continuous 8-0 or 9-0 anastomotic silk suture. The fourth group (PCA+MVS) underwent both PCA and MVS.

Two to three weeks following operation the rats underwent measurement of body weight, determination of mesenteric venous pressures, and collection of blood samples for ammonia determination.

Data were expressed as mean ± 1 SD, and evaluated statistically by the Wilcoxon rank test for unpaired data, with significance assumed for p values less than 0.05.

RESULTS

Serum ammonia results are displayed in Figure 1.

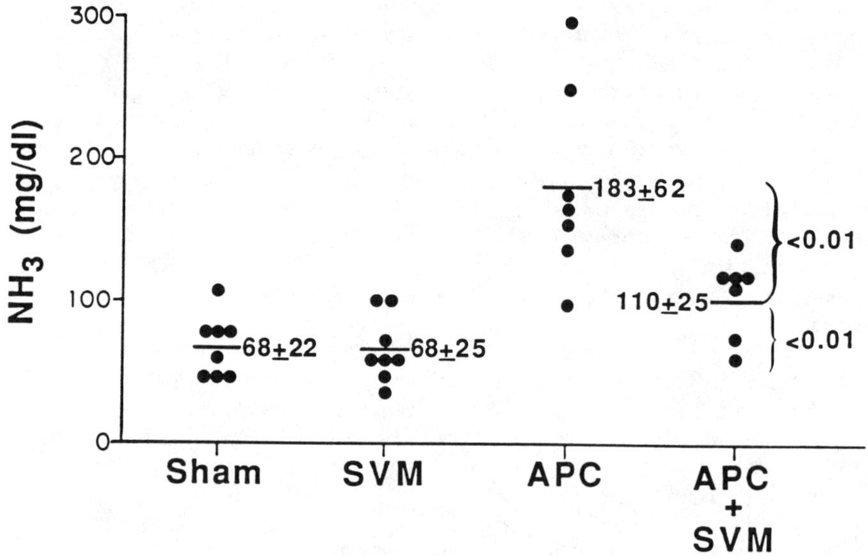

Sham and MVS animals had serum ammonia levels of 67.8 ± 22.1 and 68.3 ± 25.3 mg/dl (difference not statistically significant, NSS). PCA animals had mean ammonia levels of 183.3 ± 62.5 mg/dl ($p < 0.001$ compared with both sham and MVS animals). PCA+MVS animals had mean ammonia levels of 109.9 ± 25.4 mg/dl, significantly ($p < 0.01$) less than PCA animals [although still significantly ($p < 0.01$) greater than values for sham and MVS animals].

Maintenance of mesenteric venous hypertension was confirmed; at the time of measurement mesenteric venous pressure averaged 7.5 ± 1.6 mmHg in sham rats, and 13.0 ± 0.9 mm Hg in PCA + MVS animals ($p < 0.05$).

DISCUSSION

The pathophysiology of the neurologic dysfunction which accompanies portasystemic anastomosis - portal-systemic encephalopathy (PSE) - remains obscure. The work of Bismuth and colleagues (13) and Hess et al (14) has suggested that both hepatic insufficiency and portasystemic shunting are necessary for the development of PSE. While the success of the distal splenorenal shunt in diminishing the risk of PSE was initially ascribed to this procedure's preservation of portal perfusion of the liver (15,16), other data suggest instead that the maintenance of mesenteric venous hypertension may play a role in the favorable outcomes associated with that operation (17).

The failure of hepatic detoxification of gut-derived neurotoxins, by virtue of their bypassing of the liver, has been proposed as a major cause of PSE following portasystemic anastomosis. Though not always elevated in patients with PSE, and occasionally abnormally high in neurologically normal individuals, ammonia (and CSF glutamine, the cerebral metabolite of ammonia) continues to be implicated in the pathogenesis of PSE (1-3,6-9). Accordingly, this trial's demonstration that the hyperammonemia characteristically noted following PCA in rats can be reduced by a technique (graded mesenteric vein stenosis) which results in moderate persistent mesenteric venous hypertension is of both clinical and physiologic interest.

For example, these data appear to explain the demonstration of Orloff et al that tolerance of oral protein in shunted portacaval dogs was greater when the shunt was diminutive in calibre (7). Other studies have addressed this issue as well. For example, Price et al (5) showed that absorption of urea, xylose, and ammonia from the jejunum was augmented after a traditional large-stoma PCS. Similarly, Enquist and colleagues (4) found that D-xylose absorption from the small intestine diminished as the portal vein was gradually occluded. Harrison et al showed significantly increased ammonia absorption from Thiry-Vella colon loops in dogs following PCS (3); the same investigators also showed that grossly elevated fecal ammonia levels fell dramatically following PCA in cirrhotic patients, with a coincident rise in serum ammonia concentrations (9). The current study shows that post-PCS ammonia levels can be significantly diminished by maintaining moderate portal hypertension.

What is the mechanism? Ammonia is known to be passively absorbed from the gut by non-ionic diffusion. Lebrec's group (10) has demonstrated markedly increased portal flow following PCA, and it seems likely that factors which increase mesenteric blood flow would increase passive diffusion across the gut. The converse - that diminishing portal flow (e.g. by mesenteric vein stenosis) reduces absorption from the gut - seems plausible and is the topic of the current study. Interestingly, Hess and Jerusalem demonstrated substantial structural changes in intestinal mucosa of rats undergoing completely-decompressing PCA (14). Their findings suggest that morphologic changes accompanying PCA may be operative in the physiologic alterations accompanying portal diversion.

Prior experiments and the current study potentially clarify several clinically important points. If portal pressure is an important variable in the development of post-shunt neurologic dysfunction, the high rate of PSE following large-bore (total) portasystemic shunts is explained, as is the lower incidence of PSE following partial portal decompression [distal splenorenal, coronary-caval, prosthetic H-graft (18)], even when prograde portal flow is lost (19). This analysis permits the conclusion that, for all their beneficial effects against recurrent variceal hemorrhage, large-bore shunts may simultaneously cause excessive neurotoxin absorption from the gut; maintenance of moderately-elevated portal pressures during portal decompressive operations may be protective against post-shunt neurologic dysfunction.

REFERENCES

1. Walker CO, Schenker S. Pathogenesis of hepatic encephalopathy with special reference to the role of ammonia. Am J Clin Nutr 1977; 23:619-632.

2. Zieve L. The mechanism of hepatic coma. Hepatology 1981; 1:360-368.

3. Harrison ID, Smith AH, Shields R. Ammonia absorption from the canine colon after portacaval shunt. Brit J Surg 1977; 64:851-856.

4. Enquist IF, Golding MR, Iello RG, et al. The effect of portal hypertension on intestinal absorption. Surg Gynecol Obstet 1965; 120:87-91.

5. Price JB, McCullough W, Peterson L, Britton RC, Voorhees AB. Effects of portalsystemic shunting on intestinal absorption in the dog and man. Surg Gynecol Obstet 1967; 125:305-310.

6. Windsor CW, Goodhead B, Orloff MJ. Effects of portacaval shunts in experimental liver disease on intestinal absorption of ammonia. Surg Forum 1968; 19:333-335.

7. Orloff MJ, Wall MH, Hickman EB, Neesley T. Influence of stomal size of portacaval shunts on peripheral blood ammonia levels. Ann Surg 1963; 158:172-175.

8. Castell DO, Moore EW. Ammonia absorption from the human colon: the role of non-ionic diffusion. Gastroenterology 1971; 60:33-42.

9. Sein MM, Hardy-Smith A, Shields R. Effect of portacaval shunt on colon transfer of ammonia in cirrhosis of the liver. Gut 1974; 15:837-840.

10. Lebrec D, Blanchet L. Effect of two models of portal hypertension on splanchnic organ blood flow in the rat. Clin Sci 1985; 68:23-28.

11. Halvorsen JF, Myking AO. Prehepatic portal hypertension in the rat - immediate and long-term effects on portal vein and aortic pressure of a graded portal vein stenosis, followed by occlusion of the portal vein and splenorenal collaterals. Eur Surg Res 1979; 11:89-98.

12. Lee S, Fisher B. Portacaval shunt in the rat. Surgery 1961; 14:668-672.

13. Castaing D, Bismuth H. L'Anastomose porto-cave chez le rat: 20 ans de modele experimental. Gastroenterol Clin Biol 1984; 8:469-479.

14. Hess F, Jerusalem C, Willemen A. Altered portal pressure secondary to portacaval and portasystemic shunts in the rat: the effect on liver function and intestinal integrity. Eur J Surg Res 1982; 14:286-297.

15. Starzl TE, Porter KA, Francavilla JA. A hundred years of the hepatotrophic controversy. Hepatotrophic factors. Ciba Symposium, Amsterdam, Elsevier (1978), 111-138.

16. Warren WD. Control of variceal bleeding; reassessment of rationale. Am J Surg 1983; 145:8-16.

17. Rikkers LF. Portal hemodynamics, intestinal absorption, and post-shunt encephalopathy. Surgery 1983; 94:126-133.

18. Sarfeh IJ, Rypins EB, Fardi M, Conroy RM, Mason GR, Lyons KP. Clinical implications of portal hemodynamics after small-diameter portacaval H-graft. Surgery 1984; 96:223-229.

19. Johansen K. Partial portal decompression: initial clinical experience with small-stoma portacaval shunt. Am J Surg, in press.

HYPERAMMONEMIC ENCEPHALOPATHY SYNDROME DUE TO URINARY BLADDER DISTENTION AND INFECTION

R.R. Varma, J.W. Schenck, A. Goswami

Medical College of Wisconsin

Milwaukee, Wisconsin, USA

The role of ammonia in the pathogenesis of hepatic encephalopathy has been widely investigated. Hyperammonemia and cerebral dysfunction without liver disease or portal hypertension, or both can be due to a variety of causes which include acquired or congenital deficiency of Kreb's urea cycle enzymes such as ornithine transcarbamylase. One of the well established acquired forms is that of Kapilla-Reye's syndrome (1). Other causes include ureterosigmoidostomy performed to divert urinary flow to the colon and valproic acid toxicity. We have previously reported an interesting syndrome in a patient with hyperammonemic encephalopathy caused by infection without liver disease in a neurogenic bladder. The urinary tract infection (2) was caused by a urease producing organism. Veins from the urinary bladder form plexus on the urinary bladder surface and drain via internal iliac vein into the inferior vena cava, thus bypassing the portal circulation.

The purpose of this presentation is to provide an update of the hyperammonemic encephalopathy, urinary distention and infection, and to briefly discuss the relevance of these findings.

A 49 year old white female with history of chronic, mild, urinary tract infection was seen because of recurrent lethargy, asterixis, elevated venous blood ammonia, urinary retention and enlarged bladder. Insertion of Foley catheter led to marked improvement in her mental status and decrease in venous blood ammonia levels. Extensive workup showed no evidence of liver disease or drug abuse.

Three males and three females ages 49-71 presented with the above features. Urinary distension was due to urethral strictures in three patients and due to neurogenic bladder in the others. Temperature range was 37.5°C to 38.5°C. Blood alcohol and blood studies were essentially unremarkable. Cutaneous stigmata of liver disease were absent. All patients were well developed and well nourished and muscle wasting was not seen. A lumbar puncture was performed and showed normal spinal fluid cell count, protein, glucose and pressures. The spinal fluid cultures were unremarkable. Spinal fluid glutamine was measured in only two patients and was elevated to 23 and 40 mg/dl (normal 6-15). Blood urea nitrogen and serum creatinine values were in the range of 34-64 and 1.2-2.8 mg/dl respectively. The initial venous blood ammonia (Table I) ranged between 335-412 micro mols per liter (Normal: 11-35 micro mols per liter). The urinary pH at those times were between 8.8-9.1. During these times mental obtundation, lethargy and asterixis were present. Following Foley catheter insertion for a period of 12-48 hours the venous blood ammonia values was in the range of 32-155 micro mols/L and the pH values decreased between 7.8 and 8.1. The insertion of Foley catheter was followed by marked improvement in their mental status. The following liver biochemistry was done: Serum AST, ALT, alkaline phosphatase, bilirubin level, total protein, albumin. They were all found to be within normal limits. Dynamic upper abdominal CT and ultrasound showed normal liver and spleen. There was no evidence of portal hypertension. A percutaneous liver biopsy was performed in two patients and showed a normal histology and electron microscopy. In these two patients liver specimens were examined for Kreb's urea cycle enzymes and all five of the enzymes were also within normal limits. Kreb's cycle enzymes assayed were: Carbamylphosphate synthetase, ornithine transcarbamaylase, arginosuccinate synthetase, arginosuccinate lyase, arginase. The urinary bladder was massively dilated in five patients and modestly distended in the other. Urinary bladder calculi were seen in one patient. Culture of the surface of one of the bladder stones demonstrated the urease producing diphtheroids (2). Culture of the central part of the stone soaked in 95% ethanol for 2 hours to sterilize the surface layers and then crushed in a sterile mortar showed Serratia marcescens that did not produce urease. Urine cultures (Table II) isolated greater than 100,000 colonies of organisms which were diptheroids only in three patients, Proteous mirabilis in another. The remaining two patients had a combination of P. mirabilis, E. coli and Diptheroid infections. Urease activity was confirmed in all 5 of the 6 patients tested (3).

Table I. EFFECT OF DRAINAGE OF URINARY BLADDER

PATIENT	1	2	3	4	5	6
INITIAL VALUES						
VENOUS BLOOD AMMONIA (normal 11-35 μM/L) / URINARY pH	370 / 9.0	360 / 9.0	319 / 9.2	335 / 8.8	368 / 9.1	412 / 9.1
AFTER FOLEY CATHETER INSERTION (12-48 HRS LATER)						
VENOUS BLOOD AMMONIA (normal 11-35 μM/L) / URINARY pH	32 / 7.9	124 / 8.2	58 / 7.8	150 / 8.0	130 / 8.1	155 / 7.9

Table II. RESULTS OF UROLOGIC AND BACTERIOLOGIC STUDIES

PATIENT	1*	2*	3*	4*	5	6
URINARY BLADDER:						
DISTENTION	Severe	Severe	Severe	Severe	Modest	Severe
CALCULI	Present	Absent	Absent	Absent	Absent	Absent
URINE CULTURE						
COLONY COUNT/ML	>100,000	>100,000	>100,000	>100,000	>100,000	>100,000
ORGANISM	Diphth+	Diphth+	P mirabilis	P mirabilis & E coli	Diphth+	Diphth+ P mirabilis
UREASE ACTIVITY	Yes	Yes	Yes	Not Done	Yes	Yes

* Urine Blood Tinged

+ Diphtheroids

The effects of discontinuing antibiotics and of clamping bladder catheter on urinary urea and ammonia was evaluated in one patient. This patient had received three days of ampicillin therapy which was discontinued and the catheter was clamped intermittently, in an attempt to determine whether the bladder was the source of elevated venous blood ammonia. Urine was collected under mineral oil, with use of phenylmercuric nitrate as preservative. The effect of clamping of the Foley catheter on urinary urea, ammonia excretion, urinary pH, and urinary ammonia was studied (2,4). A progressive rise in urinary pH and in urinary ammonia concentration and excretion occurred despite stable urea secretion and normal venous blood ammonia. On the first day of clamping, the urine output for the day was 2,960 ml with urinary urea of 533 mmol, urinary ammonium 13.4 mmol/L, urinary pH 6.7, and urine ammonia 53 micro mol/L. On the second day, urine output was 2,920 ml. Urinary urea was 483 mmol/L, urinary ammonium 21.2 mmol/L, urinary pH 7.25 and urine ammonia 295 micro mol/L. On the third day, urine output was 2,390 ml, urinary urea was 461 mmol, urinary ammonium 42.0 mmol/L, urinary pH 7.5 and urine ammonia 1,030 micro mol/L. During that time, the venous blood ammonia remained normal. After discontinuing the clamping, these values were: Urine volume 3,135 ml, urinary urea 609 mmol, urinary ammonium 21.7 mmol/L, urinary pH 6.92 and urinary ammonia 142 micro mol/L. Once these changes were observed, further clamping of the catheters could not be justified, in view of the risk of recurrence of hyperammonemic stupor.

The recurrent episodes of stupor can be explained by the following mechanism in these patients. The bladder was chronically distended either from the neurogenic bladder or due to urethral stricture (5). The increased bladder surface provided a large absorptive surface. With continuous entry of urea into the bladder, the urease producing bacteria formed ammonia and alkanized urine, resulting in high partial pressure of ammonia. The elevated pH of the urine relative to the pH of venous blood in the bladder favored nonionic diffusion of ammonia directly into the inferior vena cava via the internal iliac veins (6). Thus, the ammonia from the urinary bladder bypassed the liver and reached the brain in sufficient concentration to produce encephalopathy. This view is supported by the elevated spinal fluid glutamine levels observed in two of our patients. Patients with urethral stricture underwent dilitation. One patient with neurogenic bladder and calculi underwent cystoscopy for cystolithotomy. Despite extensive workup the cause of neurogenic bladder could not be established. Our hypothesis is also supported by the fact that when the bladder was drained by Foley catheter, the blood ammonia levels fell and the stupor and asterixis resolved. With the establishment of adequate urinary flow, the patients remain symptom free 1-6 years later.

In conclusion, dilated urinary bladder complicated by retention of urine and infection caused by urease producing organism may lead to hyperammonemic encephalopathy in the absense of liver dysfunction or portal hypertension and provides a unique opportunity to study and evaluate ammonia metabolism. These data provide a strong evidence that elevated ammonia alone may lead to encephalopathy and asterixis.

REFERENCES

1. Tang TT, Siegsmund KA, Sedmak GV, Casper JT, Varma RR, McCredie SR: Reye's syndrome: A correlated electron-microscopic, viral and biochemical observation. JAMA 1975: 1339-1346.

2. Drayna CJ, Titcomb CP, Varma RR, Soergel KH: Hyperammonemic encephalopathy caused by infection in a neurogenic bladder. N Engl J Med 1981;304: 766-768.

3. Samtoy B, DeBeukelaer MM: Ammonia encephalopathy secondary to urinary tract infection with Proteus mirabilis. Pediatrics 1980;65: 294-297.

4. McDermott WV, Adams RD: Episodic stupor associated with an Eck fistula in the human with particular reference to the metabolism of ammonia. J Clin Invest 1954;33: 1-9.

5. Ullman MA, Trude AH, Medani CR: Hyperammonemic encephalopathy and urinary obstruction (Letter to Editor). N Engl J Med 1981: 1546.

6. Goss CM, ed. Gray's anatomy of the human body. 29th American ed. Philadelphia: Lea & Febiger, 1973: 673-729.

ROLE OF TOXINS AND SYNERGISM IN HEPATIC ENCEPHALOPATHY

Leslie Zieve

University of Minnesota

Minneapolis, Minnesota

The foremost hypothesis of pathogenesis of hepatic encephalopathy recognizes that in hepatic failure, toxins with coma-inducing potential accumulate and depress neuronal function by affecting fundamental neurophysiologic processes such as postsynaptic inhibition and excitation, and fundamental metabolic processes such as mitochondrial electron transfer and microsomal Na^+, K^+-ATPase activity. While accumulating, the toxins interact synergistically with each other and with various augmenting endogenous metabolic abnormalities to intensify their cellular effects. Specific toxins that both accumulate with hepatic failure and induce encephalopathy and coma in experimental animals are ammonia, methanethiol, fatty acids, and phenols (Figure 1). By interacting synergistically, they have neurological effects that are out of proportion to their individual abnormalities. Of these toxins, ammonia is the most important and has the most cumulative evidence for its role as an etiologic factor. However, the relative importance of each toxin may vary with the pathologic process. Thus, in fulminant hepatic failure the predominant toxins are mercaptans and ammonia, whereas in Reye's syndrome they are fatty acids and ammonia (Zieve, 1987a).

The most important evidence of the importance of ammonia is derived from clinical observations. Cirrhotics ingesting ammonia-generating substances such as protein in excess, amino acids, urea, ammonium-releasing resins, etc., commonly develop behavioral and neuromuscular abnormalities

Synergistic Variables	
Primary Toxins	Augmenting Abnormalities
Ammonia	Hypoxia
Mercaptans	Hypovolemia
Fatty Acids	Hypotension
Phenols	Hypoglycemia
Possibly AA abn.	Decreased albumin, Na, K, Mg, Zn

Figure 1. Primary toxins and endogenous metabolic abnormalities.

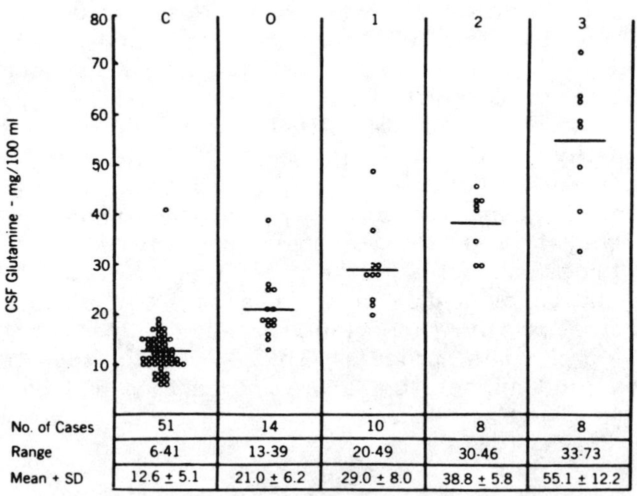

	C	0	1	2	3
No. of Cases	51	14	10	8	8
Range	6-41	13-39	20-49	30-46	33-73
Mean + SD	12.6 ± 5.1	21.0 ± 6.2	29.0 ± 8.0	38.8 ± 5.8	55.1 ± 12.2

Figure 2. Relationship of CSF glutamine to clinical severity of hepatic encephalopathy. (From Hourani et al. Arch. Int. Med. 127:1033, 1971).

such as those seen in spontaneous hepatic encephalopathy (HE). Likewise, removal of the ammonia producers often results in improvement or disappearance of the encephalopathy. The mainstay of therapy of HE today is cleansing of the bowel of ammonia-generating substances and bacteria or feeding sugars that reduce the formation of ammonia in the colon and enhance its elimination.

Among all causes of HE in patients, the severity of the encephalopathy has correlated best with the CSF glutamine, which closely reflects the brain ammonia. In the rare patient with portacaval shunt in the absence of liver disease, encephalopathy has been precipitated by the ingestion of ammonia-generating substances, and the encephalopathy worsens or improves in relation to the rise and fall in blood ammonia. Encephalopathy and coma have been induced repeatedly in normal dogs with a portacaval shunt upon feeding diets with a high ratio of protein to carbohydrate or by including other ammonia-generating substances. A large body of experimental literature on the effects of ammonium salts exists which emphasizes similarities between ammonia intoxication and HE. Thus repeated infusion of ammonium salts into normal animals results in Alzheimer Type II astrocytes in the brain that are indistinguishable from those seen in patients with hepatic coma. Similar astrocytic changes occur with time after portacaval shunts in animals (Conn and Lieberthal, 1979).

The major arguments against ammonia as an etiologic agent have been that acute ammonia intoxication is a seizure disorder, unlike the usual case of HE, and that a good correlation between the blood ammonia concentration and the severity of HE is not observed. With respect to the first argument, patients with acute fulminant hepatic failure or Reye's syndrome who may have rapid onset of hyperammonemia often have seizures, however, patients with chronic liver disease that develop HE usually do not. They have a low-grade state of chronic ammonia excess and a gradual buildup of tissue ammonia. Less additional ammonia is required to induce overt HE and the neuronal disinhibition that is triggered by ammonia is less precipitous.

The poor correlation at any given moment between blood ammonia and severity of HE is to be expected considering that most of the ammonia in the blood goes to tissues other than the brain, particularly muscle. In contrast,

CSF glutamine that reflects the buildup of ammonia in the brain shows a good correlation with the severity of HE, both clinically (Figure 2) and by the EEG (Figure 3) (Hourani et al, 1971, Oei et al, 1979).

As with ammonia, clinical observations provide the most convincing evidence of a possible etiologic role of <u>mercaptans</u>, once one recognizes that very small amounts of mercaptans induce readily reversible coma in animals. The first relevant observation was the crystallization of methanethiol (CH_3SH) and dimethyl disulfide (CH_3S-SCH_3) from the urine of a woman with fulminant hepatic failure and an intense breath odor that smelled like her urine (Challenger and Walshe, 1955). Later, encephalopathy and coma were induced in cirrhotics fed methionine, and no relation could be found between the clinical consequences of ingesting methionine and the blood ammonia or methionine levels. The encephalopathy could be prevented by prefeeding a broad-spectrum antibiotic. A fetor was noted on the breath that was somewhat but not exactly like classical fetor hepaticus (Phear et al, 1956). Fourteen years later, the breath odor was found to be due to dimethylsulfide (CH_3SCH_3). Its concentration in the breath correlated closely with the intensity of the odor (Figure 4) (Chen et al, 1970).

Dimethylsulfide is the immediate metabolic product of methanethiol which is derived largely from bacterial action on methionine. Bowel cleansing with enemas, feeding of an antibiotic such as neomycin, and ingestion of lactulose or sorbitol reduce the colonic generation of methanethiol as well as ammonia, because the colonic pH falls, so credit for the improvement of HE commonly seen after such treatments in cirrhotics must be shared by the reduction in formation of methanethiol as well as ammonia. When measurements of blood methanethiol became feasible, <u>serial</u> changes in its blood concentration were shown to correlate well with changes in the clinical severity of HE in patients with cirrhosis (Figure 5) (McClain et al, 1980). Finally, methanethiol interacts synergistically with ammonia, increasing the tissue levels of both and enhancing the action of each.

The roles of <u>phenols</u> and of fatty acids may be primarily as augmentors or modulators of the encephalopathic effects of ammonia and mercaptans. Each alone induces coma in experimental animals and interacts synergistically with

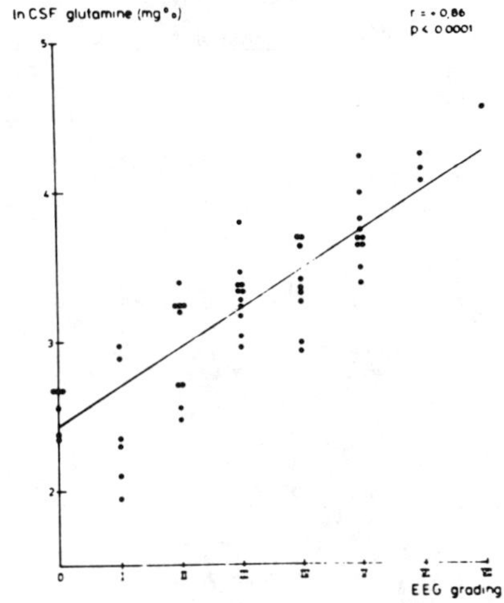

Figure 3. Relationship of CSF glutamine to severity of EEG abnormality in hepatic encephalopathy. (From Oei et al. Clin. Neurol. Neurosurg. 81:59, 1979).

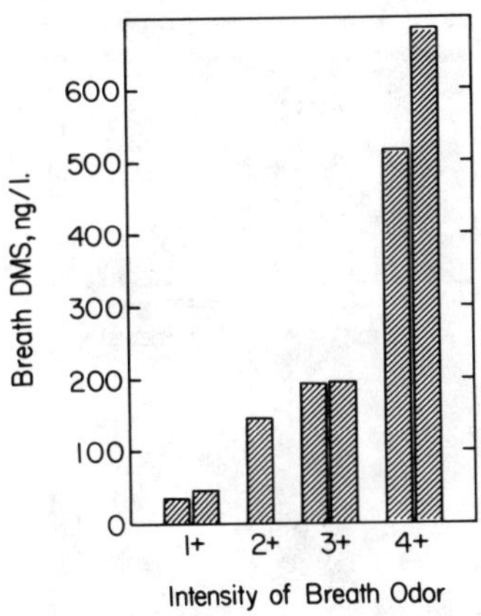

Figure 4. Relationship of breath DMS concentration to intensity of breath odor after methionine ingestion by cirrhotics. (From Chen et al, J. Lab. Clin. Med. 75:628, 1970.)

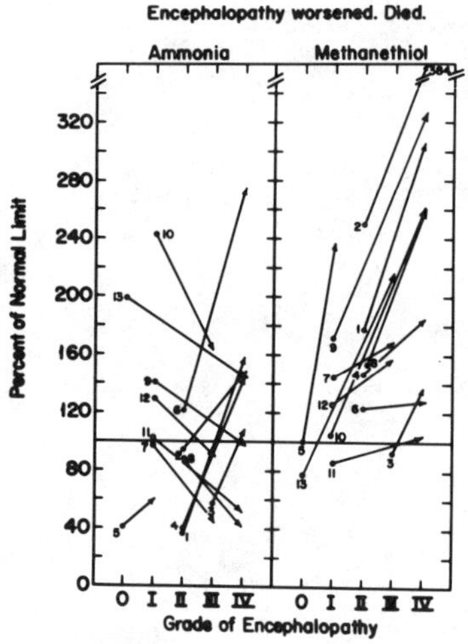

Figure 5. Serial changes in blood ammonia and methanethiol with worsening of hepatic encephalopathy in cirrhotics. (From McClain et al. Gut 21:318, 1980.)

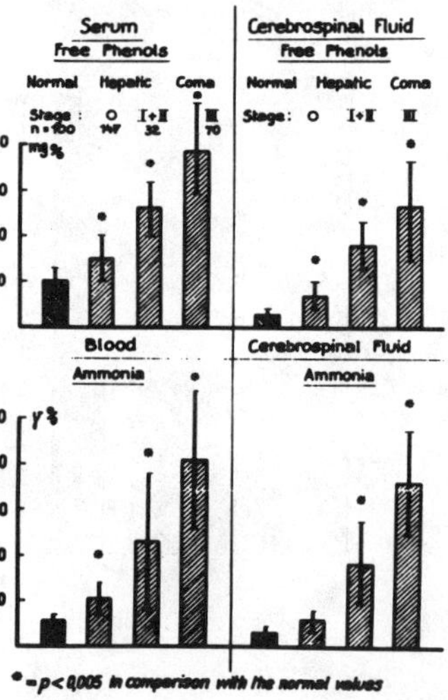

Figure 6. Relationship of serum or blood and CSF phenol and ammonia concentrations to clinical severity of hepatic encephalopathy. (From Muting. Liver Research 39:38, 1968.)

ammonia and methanethiol to enhance their effects (Zieve, 1985; Zieve and Brunner, 1985). In human hepatic failure each accumulates in blood and tissues. The correlation between serum or CSF phenol concentration and severity of HE is at least as good as that between blood ammonia and severity of HE (Figure 6) (Muting, 1968).

Fatty acids interfere with the disposition of ammonia, resulting in its disproportionate accumulation in tissues. While interacting synergistically with ammonia they reduce ammonia's propensity for eliciting seizures. In Reye's syndrome, abnormalities of short-and medium-chain fatty acids appear to be critical. At times, plasma levels are extremely elevated. More often they are not, but the fatty acids interact with significant hyperammonemia and hypoglycemia while encephalopathy develops. The precise role of fatty acids in encephalopathy associated with decompensated cirrhosis or fulminant hepatic failure has been difficult to define because measurements have not been made utilizing gas chromatography combined with mass spectrometry, which is needed to avoid the influence of extraneous factors such as phthalates that mimic fatty acids on the gas chromatograph.

Synergism among toxins and various augmenting metabolic abnormalities occurring during hepatic failure provides the key to understanding the many appearances of HE. The importance and significance of synergism per se became apparent from the results of several experimental studies, some of which are listed as follows. 1. A subcoma dose of octanoic acid reduced the coma-inducing dose of NH_4 by 52%, and increased the blood ammonia by as much as 140% (Figures 7 and 8). 2. A subcoma dose of methanethiol reduced the coma-inducing dose of NH_4 by 68%, and in the presence of a subcoma dose of NH_4 raised the incidence of coma from 0 to 100% (Figure 9). 3. No such synergistic effect was seen between pentobarbital and NH_4. However the presence of hypoxia or hypoglycemia reduced the dose of NH_4 inducing coma in rats and mice, providing experimental evidence of synergism between a hepatic failure toxin and these common metabolic abnormalities. 4. In an isolated perfused rat liver, both octanoic acid and methanethiol depressed urea synthesis by 50% to 60%. This may in part account for the increase in blood ammonia. 5. A combination of doses of NH_4Cl, methanethiol, and octanoic acid was found which caused coma in normal rats with brain levels of

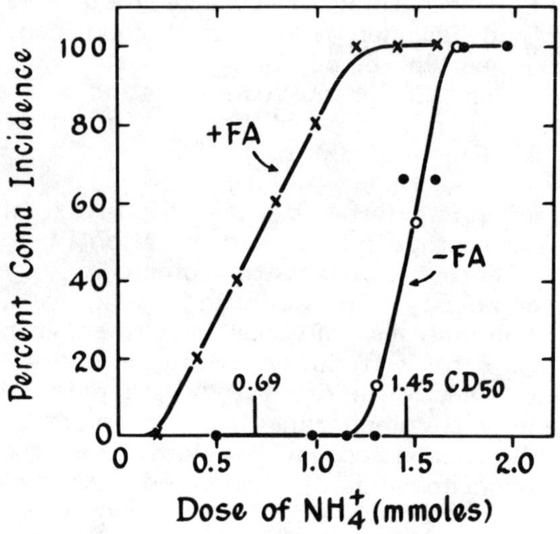

Figure 7. Dose-response curves for coma-induction with NH_4 in the presence and absence of a subcoma dose (0.4 mmol) of octanoic acid (FA). (From Zieve FJ et al. J. Pharmacol. Exp. Therap. 191:10, 1974.)

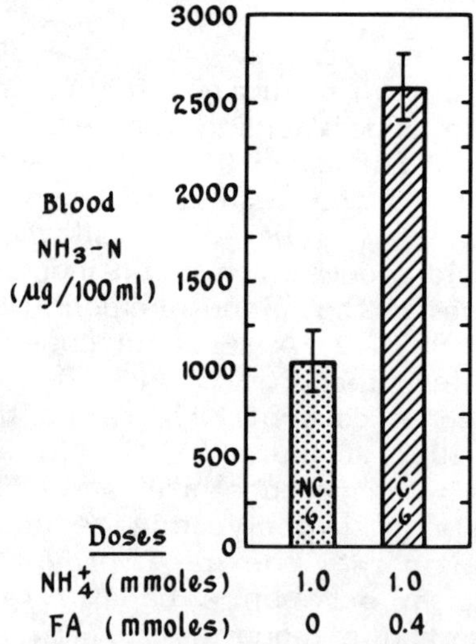

Figure 8. Blood ammonia after a subcoma dose of NH_4 in the presence and absence of a subcoma dose (0.4 mmol) of octanoic acid (FA). C= coma. NC = no coma. (from Zieve FJ et al. J. Pharmacol. Exp. Ther. 191:10, 1974.)

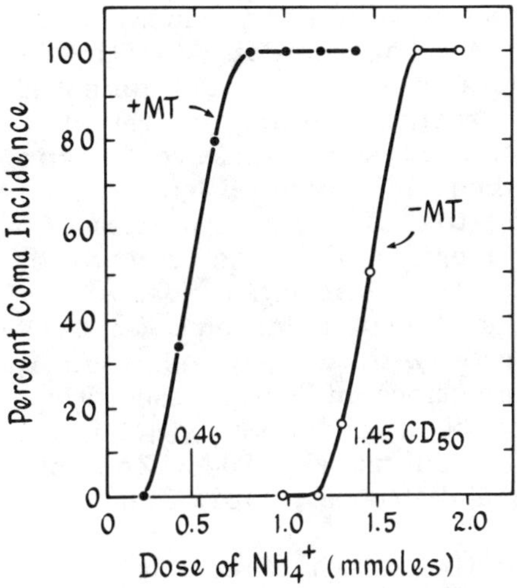

Figure 9. Dose-response curves for coma-induction with NH_4 in the presence and absence of a subcoma dose (0.12% by inhalation) of methanethiol (MT). (From Zieve et al. J. Lab. Clin. Med. 83:16, 1974.)

Brain NH_3 and MT Associated with Coma

	NH_3 nmol/g	MT pmol/g
Combination* in normal rats (11)	**1303**	**1079**
Hepatic Coma (15)	**1370**	**1135**

*** .75 mmol NH_4Cl + .15 mmol OA + 18 µmols MT**

Figure 10. Brain ammonia and methanethiol concentrations in experimental hepatic coma and in normal rats with coma-induced by a specific combination of NH_4CL, OA, and MT. (From Zieve et al. J. Lab. Clin. Med. 104:655, 1984.)

ammonia and methanethiol that were similar to those observed in experimentally induced acute hepatic coma due to ischemic necrosis of the liver (Figure 10). 6. A similar combination of doses of NH_4Cl, dimethyl disulfide and octanoic acid generated visual evoked potentials (VEP) and changes in GABA receptors of normal rats that were much like those that followed acute massive hepatic injury with galactosamine ending in coma (Figure 11). From these results, we inferred that if these quantities of ammonia and methanethiol in the brains of normal rats were sufficient to induce coma with this distinctive pattern of the VEP, perhaps they were sufficient to account for the hepatic coma observed in the rats with acutely necrosed livers due to ischemia. The implications of these observations for clinical cases of fulminant hepatic failure with hepatic coma are readily apparent (Baraldi et al, 1984a; Zeneroli et al, 1982; Zieve, 1985; Zieve and Brunner, 1985; Zieve 1987b).

Together with the known clinical information regarding toxin abnormality during hepatic failure, the above experimental data suggest that hepatic encephalopathy is most likely a consequence of the synergistic interactions among accumulated toxic factors and various augmenting metabolic abnormalities. The arguments for and against this role of the toxins are listed in the table.

Recognition that the cause of HE is multifactorial and that synergistic interactions may produce effects when the apparent abnormality of any given etiologic factor is not great, helps one to understand cases of encephalopathy that lack a dominant or obvious abnormality. Thus given an underlying state of chronic or subacute intoxication with ammonia or methanethiol in a cirrhotic, a comparatively small increase in one of the toxins or an intensification of one or more of the coexistent metabolic abnormalities (hypoxia, hypovolemia, electrolyte depletion, or acid-base imbalance) may be enough to precipitate overt encephalopathy.

The relative importance and degree of abnormality of the biological factors acting synergistically in a particular liver disease, determine the apparent pattern of abnormality clinically. The effects of a given toxin will be a consequence of the abruptness of onset and the severity and extent of the

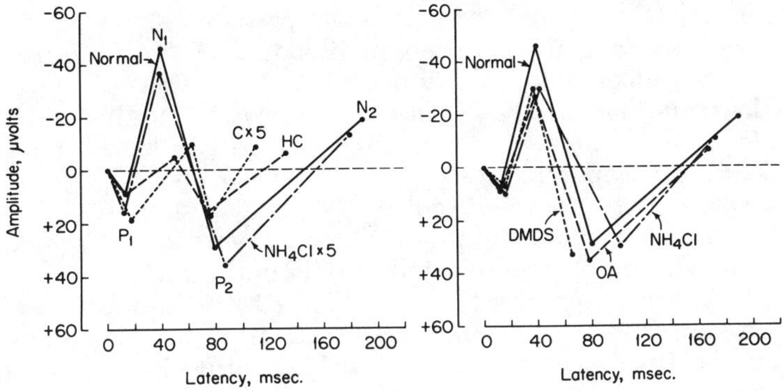

Figure 11. Visual evoked potentials in galactosamine-induced hepatic coma and in normal rats with coma induced by a specific combination of NH_4CL, OA and DMDS (dimethyl disulfide). (From Zeneroli et al. Hepatology 2:532, 1982.)

Arguments For And Against A Role For Hepatic Failure Toxins In The Pathogenesis Of Hepatic Encephalopathy (HE).

FOR

1. The toxins are increased in blood, CSF, and tissues.
2. They induce coma in animals.
3. In combination they induce coma with smaller doses.
4. In combination with NH_4, they raise the blood and brain NH_3 disproportionately.
5. A specific combination of NH_4, MT and OA induced coma in normal rats with brain levels of NH_3 and MT like those in ischemia-induced hepatic coma.
6. A similar combination of NH_4, DMDS and OA resulted in changes in GABA receptors and in visual evoked potentials like those in galactosamine-induced hepatic coma.
7. CSF glutamine has the best correlation with clinical and EEG severity of HE.
8. Ammoniagenic and mercaptogenic foods cause HE in patients with cirrhosis.
9. Elimination of these foods improve patients with HE.
10. Cleansing of bowel often good treatment of HE.
11. Lactulose or neomycin often effective treatment of HE.
12. Arginine or ornithine often improve HE in patients without severe liver disease.

AGAINST

1. The correlation of blood NH_3 and severity of HE is often poor. (To be expected. Only a small fraction of blood NH_3 goes to the brain.)
2. Not all patients with HE respond to removal of the toxins. (Failures occur in those with severe liver disease or serious complications.)
3. Mechanism of action of toxins on neurons is inferred. (State of the Art limitations.)

liver damage, the quantity of each type of toxin generated, the degree of abnormality of predisposing factors, and the severity of some precipitating factor. What the effect of these toxic factors and metabolic abnormalities is on the neuron in vivo is largely unexplored. From biochemical studies in vitro we know that mercaptans and fatty acids inhibit brain mitochondrial electron transfer, and brain microsomal Na^+, K^+-ATPase, among other effects.

Neurophysiologic studies of individual neurons have shown that ammonia affects the resting membrane potential, neuron input resistance, action potential, postsynaptic excitation, postsynaptic inhibition, and the late hyperpolarizing potential in the hippocampus. Of these, postsynaptic inhibition is most sensitive to ammonia. Disinhibition, the loss of postsynaptic inhibition or the inactivation of extrusion of Cl ions from the neuron, occurs at levels of brain NH_3 associated with the onset of encephalopathy. The other neurophysiologic changes follow as the brain ammonia concentration increases further. However most of the clinical manifestations of ammonia-induced encephalopathy can be accounted for by the disturbance in neuronal postsynaptic inhibition (Raabe and Lin, 1985).

But HE is clearly something more than pure ammonia intoxication. Similar neurophysiologic studies of combinations of toxins are unavailable. However what we do know supports the idea that HE is primarily due to a direct effect of toxins on neuronal cell function. Whether or not this leads to denervation supersensitivity of GABA or benzodiazepine receptors as suggested by Baraldi et al (1984b) is unknown at this time. Neuronal deterioration may increase the neuron's susceptibility to inhibitory neurotransmitters such as GABA, and decrease its responsiveness to excitatory neurotransmitters such as glutamate and aspartate. In the final analysis, the possible role of the GABA-ergic mechanism in HE must be evaluated through in vivo studies of the effects of antagonists. The first studies of this type have been contradictory depending on the model of experimental HE used and the severity of the HE when the antagonist was administered (Zieve, 1987b). However, observations on the arousal effects of benzodiazepine antagonists in patients with hepatic coma are of great interest, and we look forward to reports of controlled studies.

To sum up, HE is in fact multifactorial in origin. Synergistic interaction among the various factors affecting neurons is the focal point of pathogenesis, explaining many heretofore puzzling clinical observations. In particular, it may explain instances in which the abnormality in ammonia is not impressive. We are also beginning to understand how the action of these toxins on the neuron depresses its function. In conclusion, a reasonable formula for pathogenesis of HE is given in Figure 12, where T stands for toxic products, AMA for augmenting metabolic abnormalities, S for synergism, ND for neuronal deterioration, ENT for excitatory neurotransmitters, and INT for inhibitory neurotransmitters.

< Effect of ENT
|
T + AMA + S ⟶ ND ⟶ HE
|
> Effect of INT

Figure 12. Formula for pathogenesis of hepatic encephalopathy.

REFERENCES

Baraldi, M, Pinelli, G, Ricci, P, Zeneroli ML: Toxins in hepatic encephalopathy: the role of the synergistic effect of ammonia, mercaptans and short chain fatty acids. Arch. Toxicol., Suppl. 7:103-105, 1984a.

Baraldi, M, Zeneroli, ML, Ventura, E, et al: Supersensitivity of benzodiazepine receptors in hepatic encephalopathy due to fulminant hepatic failure in the rat: reversal by a benzodiazepine antagonist. Clinical Science 67: 167-175, 1984b.

Challenger, F, and Walshe, JM: Methyl mercaptan in relation to foetor hepaticus. Bioch. J. 59:372-375, 1955.

Chen, S, Zieve, L, Mahadevan V: Mercaptans and dimethyl sulfide in the breath of patients with cirrhosis of the liver. J. Lab. Clin. Med. 75:628-635, 1970.

Conn, HO and Lieberthal, MM: The Hepatic Coma Syndromes and Lactulose. Williams and Wilkins Co. Baltimore. 1979.

Hourani, BT, Hamlin EM, Reynolds, TB: Cerebrospinal fluid glutamine as a measure of hepatic encephalopathy. Arch. Int. Med. 127:1033-1036, 1971.

McClain, CJ, Zieve, L, Doizaki, WM, et al: Blood methanethiol in alcoholic liver disease with and without hepatic encephalopathy. Gut 21:318-323, 1980.

Muting, D: New aspects on the pathogenesis of hepatic encephalopathy. Liver Research 39:38-41, 1968.

Oei, LT, Kuys, J, Lombarts, AJP, et al: Cerebrospinal fluid glutamine levels and EEG findings in patients with hepatic encephalopathy. Clin. Neurol. Neurosurg. 81:59-63, 1979.

Phear, EA, Reubner, B, Sherlock, S, et al: Methionine toxicity in liver disease and its prevention by chlortetracycline. Clin. Sci. 15:93-117, 1956.

Raabe, W and Lin, S: Pathophysiology of ammonia intoxication. Experimental Neurology 87:519-532, 1985.

Zeneroli, ML, Ventura, E, Baraldi, M, et al: Visual evoked potentials in encephalopathy induced by galactosamine, ammonia, dimethyldisulfide, and octanoic acid. Hepatology 2:532-538, 1982.

Zieve, FJ, Zieve, L, Doizaki WM, Gilsdorf, RB: Synergism between ammonia and fatty acids in the production of coma: Implications for hepatic coma. J. Pharmacol. Exp. Ther. 191:10-16, 1974.

Zieve, L, Doizaki, WM, Zieve, FJ: Synergism between mercaptans and ammonia or fatty acids in the production of coma: A possible role for mercaptans in the pathogenesis of hepatic coma. J. Lab. Clin. Med. 83:16-28, 1974.

Zieve, L, Doizaki, WM, Lyftogt, C: Brain methanethiol and ammonia concentrations in experimental hepatic coma and coma-induced by injections of various combinations of these substances. J. Lab. Clin. Med. 104:655-664, 1984.

Zieve, L: Encephalopathy due to short- and medium-chain fatty acids. In DW McCandless, Ed. Cerebral Energy Metabolism and Metabolic Encephalopathy. Plenum Press. New York. 1985. pp 163-177.

Zieve, L and Brunner, G: Encephalopathy due to mercaptans and Phenols. In DW McCandless, Ed. Cerebral Energy Metabolism and Metabolic Encephalopathy. Plenum Press. New York. 1985. pp 179-201.

Zieve, L. Hepatic encephalopathy. In L. Schiff and ER Schiff, Eds. Diseases of the Liver, 6th edition. JB Lippincott Co. Philadelphia. 1987a. pp 925-948.

Zieve, L. Pathogenesis of hepatic encephalopathy. Metabolic Brain Disease 2:147-165, 1987b.

Cerebral Metabolism and Function in Hepatic Failure

BRAIN ENERGY METABOLISM IN HEPATIC ENCEPHALOPATHY

Richard A. Hawkins and Anke M. Mans

Department of Physiology and Biophysics, University of Health Sciences/The Chicago Medical School, 3333 Green Bay Road, North Chicago, IL 60064

The symptoms of hepatic encephalopathy caused by portacaval shunting (pcs) or chronic liver dysfunction in man range from very subtle subclinical abnormalities of intellectual and motor function to coma. Even in the absence of serious symptoms brain function is compromised and is more sensitive than normal to metabolic stresses and sedating pharmacological agents (Hoyumpa *et al.*, 1979; Zieve, 1979).

There are sufficient studies of cerebral energy metabolism, measured by either oxygen (CMR_{O_2}) or glucose (CMR_{Glc}) consumption, in various stages of hepatic encephalopathy to draw the conclusion that energy consumption is reduced in man and experimental animals in all stages of the disease (Tables 1 and 2). Interestingly the reductions occur even before clinical symptoms are severe enough to require professional attention, and the reductions become progressively greater as cerebral function worsens. The disease appears to represent a continuum of states where energy metabolism is reduced in a manner commensurate with cerebral function. Less attention has been focussed on the stage of latent hepatic encephalopathy, but it is clear from studies of both man and animals that brain function is compromised. In man, for instance, psychometric studies demonstrated abnormalities in a variety of functions (Elsass

TABLE 1. Cerebral energy metabolism in hepatic encephalopathy in man

Percentage difference as compared to normal

	CMR_{O2}	CMR_{Glc}
alert[1]	-30	
moderate dysfunction	-48	
coma	-52	
cirrhosis[2]	ns	ns
precoma	-42	ns
coma	-67	ns
cirrhosis[3]	ns	ns
portacaval shunting[4]		
<24h	ns	ns
>24h	-51	-64
chronic	-49	-51
encephalopathy[5]		
grade 0/1	ns	
grade 2	-18	
grade 3	-41	
grade 4	-53	
cirrhosis[6]	ns	-23
encephalopathic	-27	-32
coma	-32	-39
severe chronic encephalopathy[7]	-33	-56

1. Fazekas *et al.*(1956), 2. Erbslöh *et al.* (1958), 3. Bianchi-Porro *et al.* (1969), 4. Polli *et al.* (1969) , 5. Posner & Plum (1960), 6. Maiolo *et al.* (1971), 7. Morgan *et al.* (1980).
ns, no significant difference

TABLE 2. Cerebral energy metabolism in hepatic encephalopathy in rat

Percentage difference as compared to control

	CMR_{O2}	CMR_{Glc}
shunted 1 week[1]	-20%	
shunted 1 week[2]	-25	-55
shunted 4 week[3]	ns	
shunted 8 week	ns	
shunted 4-8 week[4]		-20
shunted 5 week[5]		-25
shunted 4 week[6]		ns
shunted 8 week		+17
shunted 12 week		+33
shunted 8 week[7]		+44

1. Eklöf *et al.* (1973), 2. Nieto *et al.* (1980), 3. Gjedde *et al.* (1978), 4. Mans *et al.* (1983), 5. Mans *et al.* (1986), 6. Cruz & Duffy (1983), 7. Lockwood *et al.* (1986). ns, no significant difference.

Two groups have described increased rates of glucose use in studies using rats (Cruz and Duffy, 1983; Lockwood *et al.*, 1986). In these studies the tracer used was deoxyglucose, an analog of glucose with similar but not identical kinetics. The interpretation of these experiments is critically dependent on the value of a kinetic constant, often called the "lumped constant", that relates glucose and deoxyglucose phosphorylation. This constant can change in pathological conditions and there is reason to believe that this is the explanation for these discrepant results (Mans *et al.*, 1984a).

et al., 1978; Rehnström *et al.*, 1977).

Experimental animals have given further insight into the disorder. Rats with pcs develop latent encephalopathy with most of the biochemical abnormalities characteristic of human liver dysfunction including elevated plasma ammonia concentrations, an altered plasma amino acid spectrum, liver and gonadal atrophy etc. (Bircher, 1979; Flynn and Kennan, 1968; Herz *et al.*, 1972a, 1972b). To casual observation they may not appear abnormal. They care for themselves, feed, groom, and show no obvious signs of cerebral dysfunction. However, closer study reveals various behavioral and electroencephalographic changes (Tricklebank *et al.*, 1978; Warbritton *et al.*, 1980), and they have a higher than normal sensitivity to medications such as barbiturates (Küpfer & Bircher, 1977).

Rats with latent encephalopathy due to pcs have a substantial decrease in brain energy metabolism (Mans *et al.*, 1983; 1986, Table 3). Glucose use is depressed by 20-30% throughout the brain. The diffuse nature of the decrease is in accordance with the concept that encephalopathy owing to liver failure or pcs is a consequence of a widespread impairment of the central nervous system. No structural or chemical lesion in a specific structure has been found, but this possibility is not yet entirely ruled out. The magnitude of the decrease in glucose use can be appreciated by comparison with the effect of barbiturates which cause a diffuse, dose-dependent depression in brain energy metabolism reaching a maximum of about 40-50% (Siesjö 1978; Davis 1987). In man and rats the dose of barbiturate necessary to induce unconsciousness reduces energy consumption by 20-30% (McCall and Taylor, 1952; Davis 1987). It appears, therefore, that cerebral function is very severely compromised by pcs and that only a modest further depression is necessary to cause unconsciousness. The onset of the depression of cerebral energy metabolism, and by inference function, is very rapid. In a carefully controlled study of brain glucose use the decrease was fully established in rats by two days (Mans *et al.*, 1988). Once established, the decrease was fairly constant for at least eight weeks. These studies demonstrated that cerebral function is rapidly brought down to a new and stable level. Whether it is

TABLE 3. Regional brain glucose use after portacaval shunting

	$\mu mol \cdot min^{-1} \cdot 100g^{-1}$ Control	Shunt	% difference
Telencephalon			
Frontal cortex	102	80	-22
Cingulate gyrus	110	84	-24
Parietal cortex	119	92	-23
Occipital cortex	98	77	-21
Pyriform cortex	82	62	-24
Insular cortex	101	83	-18
Globus pallidus	50	39	-22
Caudate nucleus	93	75	-19
Amygdala	73	55	-25
Hippocampus	76	52	-32
Septal nucleus-lateral	70	50	-29
Septal nucleus-medial	84	57	-32
Diencephalon			
Hypothalamus	79	54	-32
Habenula	111	98	-20
Thalamus	103	77	-25
Mesencephalon			
Superior colliculus	113	89	-21
Inferior colliculus	169	121	-28
Oculomotor complex	109	76	-30
Substantia nigra	75	51	-32
Red nucleus	91	65	-29
Interpeduncular nucleus	125	77	-38
Reticular formation	68	54	-21
Metencephalon			
Pons	87	62	-29
Cerebellar gray-molecular	87	64	-26
Cerebellar gray-vermis	108	82	-24
Dentate nucleus	110	88	-20
Myelencephalon			
Vestibular nucleus	130	103	-21
Cochlear nucleus	117	90	-23
Superior olive	132	100	-24
Inferior olive	93	75	-19
Reticular formation	77	61	-21

Values are the means of 8-9 rats in each group. All differences were significant at $p<0.05$. Data from Mans *et al.*, 1986.

possible to restore perfectly normal function has, however, not been established.

It is unlikely that the brain is using alternate sources of fuel after portacaval shunting. Ketone bodies are present in low concentrations and their transport is decreased after shunting (Hawkins *et al.*, 1981). The use of long-chain fatty acids by brain probably does not exceed 10%, and the neutral amino acids, in spite of their increased transport across the blood-brain barrier (Mans *et al.*, 1982) could not provide more than a few percent of the total energy requirement even upon complete oxidation in the brain.

The question arises whether the decreased rate of energy metabolism is a cause or consequence of the disease. Is there direct interference with brain energy metabolism thereby causing a decrease in the production of high energy phosphates and interference with normal neuronal function, or is the reduced rate of fuel consumption simply a reflection of a lower demand for energy because of reduced brain cell activity? There are little or no data to support the former hypothesis, but it is interesting to consider how portacaval shunting may interfere with brain energy metabolism. The most likely candidate to interfere directly with brain metabolism is ammonia.

For the present discussion ammonia is considered to include both ammonia, NH_3, and ammonium ion, NH_4^+.

Ammonia, which exists in considerably higher than normal concentrations in patients with liver dysfunction and in animals with pcs, passes fairly easily across the blood-brain barrier. Within brain ammonia is incorporated into glutamine by the following reactions.

1) 2-Oxoglutarate + NH_3 + NADH ---> Glutamate + NAD
2) Glutamate + NH_3 + ATP---> Glutamine + ADP
2-Oxoglutarate + $2NH_3$ + NADH + ATP---> Glutamine + NAD + ADP.

Reaction 1 is catalyzed by glutamate dehydrogenase and reaction 2 by glutamine synthetase. Although glutamate dehydrogenase is almost ubiquitous, glutamine synthetase is located almost exclusively in astrocytes (Martinez-

Hernandez *et al.*, 1977). Furthermore physiological studies with $^{13}NH_3$ have shown that ammonia is incorporated almost exclusively into the amide group of glutamine (Cooper *et al.*, 1979). It may be deduced from these observations that ammonia removal occurs predominately in the astrocyte layer that surrounds cerebral capillaries almost completely. Whether the glutamate needed for glutamine synthesis originates from a different astrocyte compartment or from neurons is not yet established.

When all reactions are considered, beginning with glucose, the overall production of glutamine from ammonia can be summarized as follows (Hawkins, 1973):

$$2NH_3 + 3\,NAD + Glucose \rightarrow Glutamine + CO_2 + 3\,NADH_2$$

Thus it may be seen that synthesis of glutamine increases glucose consumption. When the oxidation of $NADH_2$ by the cytochrome system to produce ATP is taken into account each mole of glutamine produced requires about 0.8 mole of glucose.

An early hypothesis, put forward by Bessman and Bessman (1954), was that when abnormally high concentrations of circulating ammonia were present the synthesis of glutamine occurred at such a rate that 2-oxoglutarate, and hence other Krebs cycle intermediates, were depleted. This would in turn cause the Krebs cycle to malfunction and in this way interfere directly with brain energy metabolism. However, this hypothesis is unlikely to be valid. The brain has a relatively high capacity to resynthesize Krebs cycle intermediates through anaplerotic reactions (Hawkins and Mans, 1983). These reactions (eg. those catalyzed by pyruvate carboxylase and malic enzyme) replenish the Krebs cycle pool by carboxylating pyruvate to either oxaloacetate or malate and thereby provide a direct link between the Krebs cycle, the glycolytic pathway and glucose.

Physiological experiments have demonstrated that rapid synthesis of glutamine does take place with the expected increase in glucose consumption, but with no sign of Krebs cycle depletion or of energy failure. When experimental rats were given a large acute ammonia load - much larger than would be likely to be encountered *in vivo*, even in a disease state - glutamine synthesis was very brisk (33 $\mu mol \cdot min^{-1} 100g^{-1}$) but there was no change in the brain content of 2-oxoglutarate or other key Krebs cycle intermediates measured in whole brain (Hawkins *et al.*, 1973, Table 4). Furthermore there was no major decrease in

TABLE 4. Key intermediates in brain after a large ammonia load

	μmol/100g	
	Control	5 Min After an Ammonia Load
Ammonia	18	153*
Citrate	32	30
2-Oxoglutarate	21	20
Malate	25	29*
Glutamate	1,160	1,130
Glutamine	671	836*

Values are the means of 6-8 rats. Data from Hawkins *et al.*, 1973.
*, p<0.0

the brain content of any of the high energy phosphates (Table 5). These experiments, as well of those of Hindfelt and Siesjö (1971) on ammonia, demonstrated that the synthesis of 2-oxoglutarate could keep pace even in an extreme circumstance. There was no sign of interference with the Krebs cycle or the production of high-energy phosphates.

TABLE 5. Energy balance after a large ammonia load

	μmol/100g	
	Control	5 Min After Ammonia
ATP	249	248
ADP	56	54
AMP	5	6
Creatine phosphate	412	386*
Phosphate	203	213

Values are the means of 6-8 rats. (The small change in creatine phosphate was probably caused by a change in intracellular pH.) Data from Hawkins *et al.*, 1973. *, $p<0.05$

The studies do, of course, leave open the possibility that certain brain structures are more succeptible to ammonia. McCandless and Schenker (1981) found that the reticular activating system may respond differently to ammonia than other selected brain structures. After giving an extremely large dose of ammonia, an order of magnitude larger than is ever likely to be encountered *in vivo*, there was a transient drop in ATP and creatine phosphate, but the values returned to near normal when the animals were in coma. On the other hand, studies of a lower primate showed no disturbance of ATP levels and only a mild, possibly pH-related, change in creatine phosphate (McCandless *et al.*, 1985). Although these studies raised the possiblity that the reticular formation may be more sensitive to an ammonia load before the onset of coma they also showed that energy production keeps up with demand during coma.

In chronic hepatic dysfunction higher than normal circulating ammonia concentrations are present for long periods. The question arises whether there might be some effect on Krebs cycle intermediates and energy production after weeks or months of liver dysfunction and exposure to ammonia. There is, however, no support for this concept either. After four or seven weeks of portacaval shunting in rats there was no major disturbance of glycolytic or Krebs cycle intermediates (Table 6) or in the brain content of high-energy phosphates (Table 7). Thus there was no sign of energy failure at a time when these rats have a marked decrease in the rate of glucose use.

TABLE 6. Key intermediary metabolites in brain after portacaval shunting

	µmol/100g			
	4 Weeks		7 Weeks	
	Control	Shunted	Control	Shunted
Glucose	192	110*	174	78*
Glucose-6-Phosphate	17	16	15	17
Pyruvate	10	10	10	9
Lactate	175	204	170	212*
2-Oxoglutarate	18	18	23	21
Ammonia	19	133*	23	110*
Glutamate	1210	1090	1290	1250
Glutamine	597	2140*	608	2240*
Malate	40	25*	23	20

Values are the means of 6-8 rats in each group. Data from Mans *et al.*, 1984.*, $p < 0.05$

TABLE 7. Adenosine nucleotides and creatine phosphate after portacaval shunting

	µmol/100g			
	4 Weeks		7 Weeks	
	Control	Shunted	Control	Shunted
ATP	293	308	296	299
ADP	60	62	63	62
AMP	2	4	4	4
Creatine Phosphate	416	389	398	431

Values are the means of 6-8 rats in each group. There were no significant differences between shunted and control groups. Data from Mans *et al.*, 1984.

Various other suggestions, based on experiments *in vitro* on how ammonia may interfere with energy metabolism, have been made (Cooper and Plum, 1987). For example, ammonia may stimulate glycolysis, inhibit the 2-oxoglutarate dehydrogenase complex or inhibit the mitochondrial malate-aspartate shuttle. Such interference would cause predictable and substantial alterations in the concentrations of several intermediary metabolites; these changes have not been observed *in vivo*. Furthermore, major disturbances in such basic processes over prolonged periods of time must eventually lead to cell death, which has not been found during chronic liver disease.

As mentioned above the continued synthesis of glutamine and concomitant replenishment of 2-oxoglutarate by anaplerotic reactions should result in a commensurate increase the rate of glucose utilization. In rats given a large acute ammonia load it was found that net glutamine synthesis occurred at a rate of 33 $\mu mol \cdot min^{-1} \cdot 100_g^{-1}$ over a 5 min period and glucose use increased by 20 μmol·min^{-1}·100·g^{-1} (Hawkins *et al.*, 1973). However, this situation is in marked contrast to that after portacaval shunting or chronic liver disease, where there is prolonged exposure to much lower levels of ammonia. In the chronic condition in man and rat there are no data obtained directly concerning glutamine synthesis, but it is possible to make calculations from experiments in which arteriovenous differences of ammonia or glutamine were measured. Table 8 shows that

TABLE 8. Glutamine production in shunted rats

	Blood Ammonia μM	Glutamine Synthesis μmol·min^{-1}·100g^{-1}	% of Glucose Use
Control	80	2	2%
Portacaval Shunt	280	4-7	5-9%

Data calculated from Gjedde *et al.*, 1978.

glutamine synthesis occurs normally at a rate of about 2 μmol·min^{-1}·100g^{-1} and rises to between 4 and 7 μmol·min^{-1}·100g^{-1} after pcs. This is a relatively low rate of glutamine synthesis. It represents only a small fraction of brain glucose consumption and is well below the maximum capacity of anaplerotic reactions to synthesize 2-oxoglutarate. In man the rate of ammonia uptake and glutamine formation is even less than in rats and represents only a small fraction of the total rate of glucose consumption (Table 9). In any event the amount of additional glucose that would have to be used for glutamine synthesis in portacaval shunted humans or experimental animals is relatively small and well within the capability of the central nervous system.

Once the decrease in glucose use is established in latent encephalopathy it is unclear whether it can be reversed by lowering ammonia. We studied the effects of a high carbohydrate diet with and without branched chain amino acids in rats with pcs (Mans *et al.*, 1984b, 1986). In one

TABLE 9. Glutamine production in man

	Blood Ammonia μM	Brain Glutamine $\mu mol \cdot min^{-1} \cdot 100g^{-1}$	% of Glucose Use
Control	100	0.3	< 1%
Portacaval Shunt	160	0.7	< 3%

Data from Posner and Plum, 1960.

experiment, the diets were fed orally over a period of two weeks, and in another, the mixtures were infused intravenously for a period of 4 days. Brain glutamine content was decreased by both treatments, probably because of the reduced production of ammonia owing to the low protein content of the diet. However, glucose use remained depressed and was not affected by any of the diets. Thus in this instance reduction of ammonia and brain glutamine did not result in the normalization of cerebral function.

To sum up, it must be concluded that portacaval shunting with consequent increases in the concentration of plasma ammonia does not impose much of a metabolic burden on brain. There is no reason to believe that the diminution of cerebral function or the lowering of the rate of energy metabolism is a consequence of enhanced glutamine synthesis interfering directly with energy metabolism. In nearly all circumstances studied high-energy phosphates were found at normal levels. Glutamine synthesis in portacaval shunted animals or humans could cause a slight rise in the rate of glucose use, but certainly not the decrease in both glucose use and oxygen consumption that has been observed so frequently (Tables 1 and 2). Thus the question whether the decreased glucose use, or energy consumption, is the cause or consequence of metabolic encephalopathy can be definitively answered. Decreased glucose use is simply the result of a decreased need for energy. In other words it is most likely that there is a reduction in nerve cell activity throughout the brain which results in a decreased need for energy and is consequently reflected in a decrease in fuel consumption. This decrease far overwhelms the small increase in glucose use necessary for ammonia detoxification. The significant reduction in the rate of energy metabolism observed in chronic portacaval shunted experimental animals and patients, who are still conscious, is very severe; it is comparable to that produced by sedatives or subanesthetic doses of barbiturates. Perhaps it should not be surprising that anything that causes a further depression of cerebral function will tip the balance and cause unconsciousness.

Acknowledgement: The authors' work was supported by NIH grants NS-16389 and NS-16737.

References

Bessman, S. and Bessman, A. (1954) The cerebral and peripheral uptake of ammonia in liver disease with an hypothesis for the mechanism of hepatic coma. *J. Clin. Invest.* 34, 622-628.

Bianchi-Porro, G., Maiolo, A.T., and Della Porta, P. (1969) Cerebral blood flow and metabolism in hepatic cirrhosis before and after portacaval shunt operation. *Gut* 10, 894-897.

Bircher, J. (1979) The rat with portacaval shunt: an animal model with chronic hepatic failure. *Pharmac. Ther.* 5, 219-222.

Cooper, A and Plum, F (1987) Biochemistry and physiology of brain ammonia. *Physiol. Reviews* 67, 440-519.

Cooper, A., McDonald, J., Gelbard, A., Gledhill., and Duffy., T (1979) The metabolic fate of ^{13}N-labeled ammonia in rat brain. *J. Biol. Chem.* 254, 4982-4992.

Cruz, N.F. and Duffy, T.E. (1983) Local cerebral glucose metabolism in rats with chronic portacaval shunts. *J. Cereb. Blood Flow Metab.* 3:311-320.

Davis, D.W. (1987) Regional cerebral energy metabolism during intravenous anesthesia with etomidate, ketamine or thiopental. Ph.D. Thesis, The Pennsylvania State University.

Eklöf, B., Holmin, T., Johannsson, H. and Siesjö, B.K. (1974) Cerebral blood flow and cerebral metabolic rate for oxygen in rats with portacaval anastomosis. *Acta. Physiol. Scand.* 90, 337-344.

Elsass, P., Lund, Y., and Ranek, L. (1978) Encephalography in patients with cirrhosis of the liver. A neuropsychological study. *Scand. J. Gastroenterol.* 13, 241-247.

Erbslöh, F., Bernsmeier, A., and Hillesheim, H.R. (1958) Der Glucoseverbrauch des Gehirns und seine Abhängigkeit von der Leber. *Arch. Psychiatr. Nervenkr.* 196, 611-626.

Fazekas, J.F., Ticktin, H.E., Ehrmantraut, W. R., and Alman, R.W. (1956) Cerebral metabolism in hepatic insufficiency. *Am. J. Med.* 21, 843-849.

Flynn, P.J. and Kennan, A.L. (1968) The rat with a portacaval anastomosis. *Arch. Path.* 85, 138-148.

Gjedde, A., Lockwood, A.M., Duffy, T.E., and Plum, F. (1978) Cerebral blood flow and metabolism in chronically hyperammonemic rats: Effect of an acute ammonia challenge. *Ann. Neurol.* 3, 325-330.

Hawkins, R.A. and Mans, A.M. (1983) Intermediary metabolism of carbohydrates and other fuels. *Handbook of Neurochemistry* 3, 259-294.

Hawkins, R.A., Miller, A.L., Nielsen, R.C. and Veech, R.L. (1973) The acute action of ammonia on rat brain metabolism *in vivo*. *Biochem. J.* 134, 1001-1008.

Hawkins, R.A., Mans, A.M. and Biebuyck, J.F. (1981) Regional blood-brain barrier permeability in hepatic encephalopathy. *J. Cereb. Blood Flow Metab.* 1, 5385-5386.

Herz, R., Sautter, V., and Bircher, J. (1972a) Fortuitous discovery of urate nephrolithiasis in rats subjected to portacaval anastomosis. *Experientia* 27-28.

Herz, R., Sautter, V., Robert, F. and Bircher, J. (1972b) The Eck fistula rat: definition of an experimental model. *Eur. J. Clin. Invest.* 2, 390-397.

Hindfelt, B. and Siesjö, B.K. (1971) Cerebral effects of acute ammonia intoxication. *Scand. J. Clin. Lab. Invest.* 28, 365-374.

Hoyumpa, A.M., Desmond, P.V., Avant, G.R., Roberts, R.K. and Schenker, S. (1979) Hepatic encephalopathy. *Gastroenterology* 76, 185-195.

Küpfer, A. and Bircher, J. (1977) Mechanisms for the exagerated sedative effect of pentobarbital in rats with experimental hepatic failure. *Experientia* 33, 807.

Lockwood, A. H., Ginsberg, M. D., Rhoades, H. M., and Gutierrez, M. T. (1986) Cerebral glucose metabolism after portacaval shunting in the rat. *J. Clin. Invest.* 78, 86-95.

Maiolo, A.T., Porro, G.B., Galli, C., Sessa, M., and Polli, E.E. (1971) Brain energy metabolism in hepatic coma. *Exp. Biol. Med.*, 4, 52-70.

Mans, A.M., Biebuyck, J.F., Shelly, K. and Hawkins, R.A. (1982) Regional blood-brain barrier permeability to amino acids after portacaval anastomosis. *J. Neurochem.* 38, 705-717.

Mans, A.M., Biebuyck, J.F., Davis, D.W., Bryan R.M., and Hawkins, R.A. (1983) Regional cerebral glucose utilization in rats with portacaval anastomosis. *J. Neurochem.* 40, 986-991.

Mans, A., Hawkins, R. and Biebuyck, J. (1984a) Regional cerebral glucose utilization after portacaval shunting. *J. Cereb. Blood Flow Metab.* 4, 123-126 (Letter).

Mans, A.M., Biebuyck, J.F., Davis, D.W. and Hawkins, R.A. (1984) Portacaval anastomosis: brain and plasma metabolite abnormalities and the effect of nutritional therapy. *J. Neurochem.* 43, 697-705.

Mans, A.M., Davis, D.W., Biebuyck, J.F. and Hawkins, R.A. (1986) Failure of glucose and branched-chain amino acids to normalize brain glucose use in portacaval shunted rats. *J. Neurochem.* 47, 1434-1443.

Mans, A.M., DeJoseph, M.R., Davis, D.W., Viña, J.R., and Hawkins, R.A. (1988) Early establishment of characteristic metabolic abnormalities after portacaval shunting. *Soc. Neuroscience Abstr.* 14, 1218.

Martin, J.R., Baettig, K., Bircher, J. (1980) Maze patrolling, open-field behavior and runway activity following experimental portacaval anastomosis in rats. *Physiol. Behav.* 25, 713-719.

Martinez-Hernandez, A., Bell, K.P., and Norenberg, M.D. (1977) Glutamine synthetase: glial localization in brain. *Science* 195, 1356-1358.
McCall, M.L. and Taylor, H.W. (1952) Effects of barbiturate sedation on the brain in toxemia of pregnancy. *JAMA* 149, 51-54.
McCandless, D.W. (1981) Metabolic turnover in the reticular activating system in ammonia induced coma. *Neurobehav. Toxicol. and Teratol.* 3, 257-260.
McCandless, D.W. and Schenker, S. (1981) Effect of acute ammonia intoxication on energy stores in the cerebral reticular activating system. *Exp. Brain Res.* 44, 325-330.
McCandless, D.W., Looney, G.A., Modak, A.T., and Stavinoha, W.B., (1985) Cerebral acetylcholine and energy metabolism changes in acute ammonia intoxication in the lower primate Tupaia glis. *J. Lab. Clin. Med.* 106, 183-186.
Morgan, M.Y., Jakobovits, A.W., James, I.M., and Sherlock, S. (1980) Successful use of bromocriptine in the treatment of chronic hepatic encephalopathy. *Gastroenterology* 78, 663-670.
Nieto, C., Arias, J., Alsasua, A., and Garia de Jalon, P.D. (1980) Changes in brain oxidative metabolism in rats with portacaval shunt. *Experientia* 36, 1403-1403.
Polli, E., Bianchi-Porro, G., and Maiolo, A.T. (1969) Cerebral metabolism after portacaval shunt. *The Lancet* 1, 153.
Posner, J.P. and Plum, F. (1960) The toxic effects of carbon dioxide and acetazolamide in hepatic encephalopathy. *J. Clin. Invest.* 39, 1246-1258.
Rehnström, S., Simert, G., Hansson, J.A., Johnson, G., and Vang, J. (1977) Chronic hepatic encephalopathy. A psychometrical study. *Scand. J. Gastroenterol.* 12, 305-311.
Siesjö, B.K. (1978) Anaesthesia, analgesia and sedation. In: *Brain Energy Metabolism,* John Wiley & Sons, New York, pp. 233-265.
Tricklebank, M.D., Smart, J.L., Bloxam, D.L. and Curzon, G. (1978) Effects of chronic experimental liver dysfunction and L-tryptophan on behavior in the rat. *Pharmacol. Biochem. Behav.* 9, 181-189.

Warbritton, J.D. III., Geyer, M.A., Jeppsson, B. and Fischer, J.E. (1980) Decreased startle reactivity in the end-to-side portacaval shunted rat. *Pharmacol. Biochem. Behav.* 12, 739-742.

Zieve, L. (1979) Hepatic encephalopathy (summary of present knowledge with an elaboration on recent developments), in *Progress in Liver Diseases,* Grune and Stratton, New York, pp. 327-341.

IN VIVO NMR SPECTROSCOPY STUDIES OF CEREBRAL METABOLISM IN RATS AFTER PORTAL-CAVAL SHUNTING.

Susan M. Fitzpatrick, Kevin L. Behar and Robert G. Shulman

Dept. of Molecular Biophysics and Biochemistry,
Yale University, New Haven, CT 06511 USA

Introduction

Hepatic encephalopathy (HE) is a clinical condition resulting from either insufficient liver function or the construction of a shunt from the portal to the systemic circulation. As a result, several compounds known to be neurotoxic are no longer removed from the blood by the liver. Hepatic encephalopathy is characterized by increasing neurological impairment leading to coma (for review see Duffy and Plum, 1982). Although information is available from both clinical studies and from research using experimental animal modes of HE, there is still much that is unknown about the biochemical mechanisms involved in the pathogenesis of this disorder. Of the several toxic compounds increased in the blood and in brain tissue during the course of HE, the finding of an increased ammonia concentration has received the most attention (Butterworth et al., 1987). Administration of ammonium salts to normal animals or to animals with portacaval shunts reproduces experimentally many of the neurochemical abnormalities that accompany decreased liver failure including changes in the concentrations of the amino acids glutamate and glutamine (Hindfelt et al., 1977). Despite the case made for other suspected toxins (see Zieve, 1981) treatments designed to lower blood and brain ammonia have met with the most success in the management of HE (see Cooper and Plum, 1987).

The techniques of <u>in vivo</u> NMR spectroscopy have been used successfully to investigate cerebral metabolism during control conditions and during the conditions of hypoxia (Behar et al., 1983), seizures (Petroff et al., 1984), acute hyperammonemia (Fitzpatrick et al., 1989) and hypoglycemia (Behar et al., 1985). We are investigating the effects of portacaval shunting on cerebral energy and intermediary metabolism with ^1H and ^{31}P NMR spectroscopy. Particular emphasis has been placed, in this report, on the metabolism of glutamate and glutamine, two amino acids directly involved in cerebral ammonia metabolism (see above).

Methods:

Animal Preparation

Animals were prepared as previously described (Fitzpatrick <u>et al</u>., 1989). Porta-caval shunted rats were purchased from Hilltop Lab Animals, Inc., Scottdale, PA. and used 4 weeks after shunting. Male Sprague-Dawley rats (180-210g), fasted 24h, were tracheostomized, paralyzed with d-Tubocurarine-Cl (3mg/Kg) and ventilated with 70% N_2O/30% O_2. Surgical procedures were performed under enflurane anesthesia; its use was discontinued prior to placing the animals in the magnet. Body temperature was maintained near 37°C with a Tygon jacket connected to a temperature-regulated circulating water bath. The tail artery (PE-50) and vein (PE-10) were cannulated to permit arterial blood sampling and the i.v. infusion of ammonium acetate. Arterial blood pH, pCO_2, pO_2 and glucose were measured from 100 μl samples usually taken 4 or 5 times during the experimental period. Animals maintained a pCO_2 between 35-42 mmHg, a $pO_2 > 100$ mmHg, a blood pH of 7.38 ± 0.04, and a fasting arterial blood glucose concentration of ≈4.5mM (range of 3.9-5.5 mM). The mean arterial blood pressure was recorded continuously throughout the time that the animal was in the bore of the magnet. Animals that did not maintain a mean arterial blood pressure greater that 90 mmHg (after correction for the additional hydrostatic pressure resulting from the vertical orientation of the animal in the magnet) were discarded from this study. The scalp was incised along the midline and the skin retracted. The skull was cleaned of all tissues between the temporal ridges. The animal was placed vertically in a plexiglass holder and attached to the surface coil probe. For additional details about the placement of the animal within the bore of the magnet see Behar and Fitzpatrick (this volume).

Ammonia infusion:

Portacaval shunted rats challenged with ammonia received an intravenous bolus injection of 50 mM ammonium acetate (200 μl per 200g body weight) followed by a progressively decreasing rate of infusion of 250 mM ammonium acetate. The rate of infusion was adjusted every 60 sec for the first 18 minutes and then maintained at a constant rate for the duration of the experiment. The total volume administered during a one-hour experiment was approximately 4 ml. In a separate control group of animals, the infusion of ammonium acetate increased the blood NH_3 concentration approximately 10-fold (for further details see Fitzpatrick et al., 1989).

Spectroscopy in vivo:

^{1}H and ^{31}P spectra were obtained at 8.4 Tesla with a Bruker AM-360 wide-bore spectrometer operating at 360.13 and 145.78 MHz, respectively. Spectra were obtained for 20 min prior to and 60 min during the infusion of ammonium acetate. The surface coil consisted of a 8x12-mm single-turn ellipse. An estimate of the depth and lateral extent of the volume of tissue detected by the surface coil for the pulse sequence and acquisition parameters used to obtain ^{1}H and ^{31}P spectra in this study has been reported elsewhere (see Fitzpatrick et al., 1989). Details of the spectroscopy used in acquiring spectra during the experiments have been previously described (Behar et al., 1985; 1987). Briefly, ^{1}H spectra were acquired using semi-selective pulses (Hore 1983; Plateau and Gueron, 1982) in a phase-cycled spin-echo sequençe (Bendall and Gordon, 1983; Hetherington and Rothman, 1985). The spin-echo evolution time was 4-msec. Additional suppression of the water signal was obtained with a period of presaturation, followed by a dephasing delay (Hetherington, 1988). The intensity maximum of the excitation profile was positioned on the proton resonance of the methyl group of N-acetylaspartate (2.023 ppm) yielding a spectral intensity profile for resonances from about 3.3 to 0.8 ppm. Relevant peak assignments are shown in Fig. 2 and are based on previous studies from this laboratory with extracts and brain tissue (Behar et al., 1983; 1985). Sixty-four transients were acquired with a 1.5 sec recycle time; 1.6 min per spectrum. Spectra were processed by the convolution difference technique followed by Gaussian multiplication of the free-induction decay for resolution enhancement. ^{31}P spectra were obtained with a pulse-acquire sequence and collected as 256 transients using a 0.6 sec recycle time (2.56 min per spectrum). Spectra were processed with convolution difference followed by exponential multiplication of the free-induction decay for optimization of the signal-to-noise ratio.

Determination of intracellular pH

The intracellular pH (pH_i) was calculated from the chemical shift difference between inorganic phospate (Pi) and PCr in the ^{31}P spectra as specified in Petroff et al., (1985), according to the equation: $pH_i = 6.77 + \log (\delta - 0.94) / (3.33 - \delta)$, where δ is the chemical shift of P_i relative to PCr using a titration solution containing 1.36 mM Mg^{2+}. Phosphocreatine was assigned a chemical shift of −2.35 ppm. The intracellular pH of portacaval shunted rats, 7.13 ± 0.02 (n=3), was not significantly different than that of non-shunted weight-matched controls (see Fitzpatrick et al., 1989).

RESULTS AND DISCUSSION

Effects of ammonia infusions

The amount of ammonia infused into portacaval shunted rats in this study (see Methods) did not affect the intracellular pH of the brain or the level of cerebral high energy phosphate compounds as measured by ^{31}P-NMR (see fig 1.). A previous study from this labroatory, in which the same ammonia infusion was given to non-shunted rats, reported similar findings (Fitzpatrick et al., 1989).

Portacaval shunted rats had higher concentrations of glutamine than non-shunted control rats evidenced by the ^1H NMR spectra in fig 2. There was, however, some variability in the concentration of glutamine in portacaval shunted rat brain (compare figs 2,3). Glutamate levels appear to be similiar in both shunted and non-shunted rats (fig 2 A,B) After about 20 min of the ammonia infusion, cerebral glutamine levels were measurably increased and continued to rise throughout the infusion period (see fig 3). Cerebral glutamate levels fell to a certain extent, but it is unlikely that the increase in glutamine can be accounted for solely by the decrease in glutamate. It has been reported (Butterworth et al., 1988) that cortical glutamine levels did not rise in portacaval shunted rats challenged with an ammonia load. One explanation for the apparent discrepancy between our study and that of Butterworth et al. could be the time at which cortical glutamine levels were measured. We did not observe a measurable increase in the cerebral glutamine concentration until at least 20 minutes after the start of the ammonia infusion. Butterworth's group sacrificed their animals about 20 minutes after an i.p. injection of ammonia acetate and it is possible that their measurements of cortical glutamine were made before the concentration had significantly increased. It is difficult to explain why it would take 20 minutes for the cerebral glutamine concentration to increase in response to an increased brain ammonia concentration, however, we observed a similar time course for changes in cerebral glutamine concentrations in a study of acute hyperammonemia in control rats (Fitzpatrick et al., 1989).

We are presently continuing our investigation of the effects of portacaval shunting on cerebral metabolism and measuring the turnover of glutamate and glutamine in vivo using the technique of proton-observed, carbon-decoupled NMR spectroscopy.

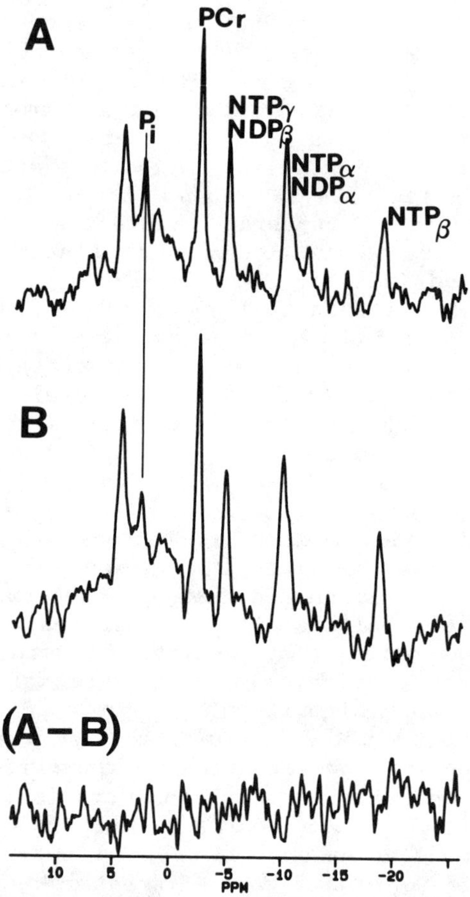

Figure 1. ^{31}P spectra acquired in vivo from the brain of a porta-caval shunted rat prior to (A) and during (B) an infusion of ammonium acetate (see Methods).
(A-B) represents the difference spectrum obtained by subtracting the spectrum acquired prior to the start of the ammonium acetate infusion from the spectrum acquired during the ammonium acetate infusion. Note that the chemical shift of the P_i resonance relative to the chemical shift of PCr (an estimate of intracellular pH, see Methods) is unchanged by the ammonium acetate infusion. Pi, inorganic phosphate; NTPα,β,γ, nucleoside triphosphate α,β,γ; NDPα,β, nucleoside diphosphate α,β. Phosphocreatine (PCr) was assigned a chemical shift of -2.35 ppm.

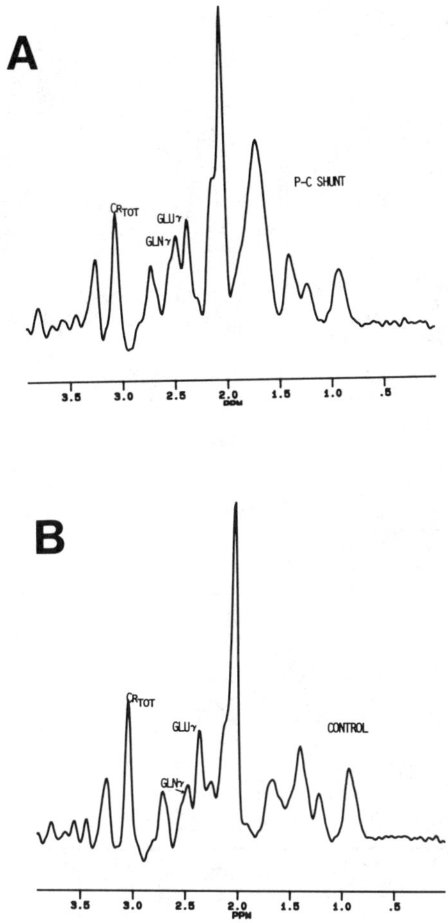

Figure 2. (A) ^1H spectra acquired in vivo from the brain of a portacaval shunted rat prior to an infusion of ammonium acetate. (B) ^1H spectra acquired in vivo from the brain of a weight matched control rat. Note the increased amplitude of the GLN' γCH_2 resonance in the spectrum from the brain of a portacaval shunted rat (A) relative to the spectrum acquired from the brain of a non-shunted rat (B). Cr_{tot}, total creatine (creatine + phosphocreatine); GLN, glutamine γCH_2; GLU, glutamate γCH_2.

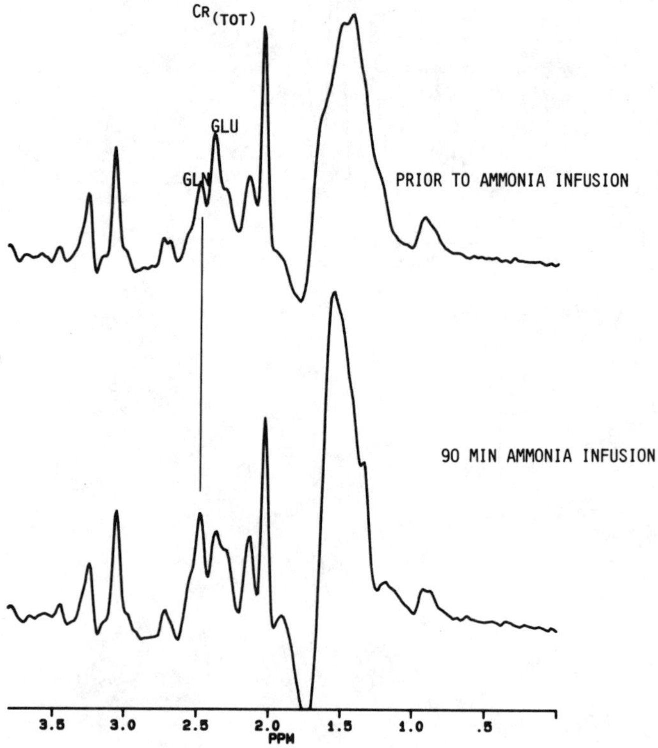

Figure 3. ^{1}H Spectra acquired <u>in vivo</u> from the brain of a porta-caval shunted rat prior to and during an infusion of ammonium acetate as described in methods. Note the increased amplitude of the GLN γCH_2 resonance ($\approx$2.4 ppm) after 90 minutes of ammonia infusion. Abbreviations are explained in the legend to figure 2.

Acknowledgements: This work was supported by USPHS DK-27121.

REFERENCES

Behar, K.L., Rothman, D.L., Fitzpatrick, S.M., Hetherington, H.P., and Shulman, R.G. (1987). Combined ^{1}H and ^{31}P NMR studies of the rat brain *in vivo*: Effects of altered intracellular pH on metabolism. *In: Physiological NMR Spectroscopy: From Isolated Cells to Man* (S. Cohen, ed.), pp. 81-88. New York Academy of Science, New York.

Behar, K.L., den Hollander, J.A., Stromski, M.E., Ogino, T., Shulman, R.G., Petroff, O.A.C. and Prichard, J.W. (1983). High resolution ^{1}H NMR study of cerebral hypoxia *in vivo*. *Proc. Natl. Acad. Sci. USA* *81*, 4945-4948.

Behar, K.L., den Hollander, J.A., Petroff, O.A.C., Hetherington, H.P., Prichard, J.W. and Shulman, R.G. (1985). Effect of hypoglycemic encephalopathy upon amino acids, high energy phosphates, and pH in the rat brain *in vivo*: detection by sequential ^{1}H and ^{31}P NMR spectroscopy. *J. Neurochem*. *44*, 1045-1055.

Bendall, M.R., and Gordon, R.D. (1983). Depth and refocusing pulses designed for multipulse NMR with surface coils. *J. Magn. Reson*. *53*, 365385.

Butterworth, R.F., Giguere, J.F., Michaud, J., Lavoie, J. and Pomier Layrargues, G. (1987). Ammonia: key factor in the pathogenesis of hepatic encephalopathy. *Neurochem. Pathol*. *6*, 1-12.

Butterworth, R.F., Girard, G., and Giguere, J.F. (1988). Regional differences in the capacity for ammonia removal by brain following portocaval anastomosis. *J. Neurochem*. *51*, 486-490.

Cooper, A.J.L. and Plum, F. (1987). Biochemistry and physiology of brain ammonia. *Physiol. Rev*. *67*, 440-519.

Duffy, T.E. and Plum, F. (1982). Hepatic encephalopathy. In: The Liver: Biology and Pathobiology (I. Arias, H. Popper, D. Schacter and D.A. Shafritz, eds.), pp. 693-751. Raven Press, New York.

Fitzpatrick, S.M., Hetherington, H.P., Behar, K.L. and Shulman, R.G. (1989). Effects of acute hyperammonemia on cerebral amino acid metabolism and pH in vitro, measured by ^{1}H and ^{31}P Nuclear Magnetic Resonance. J. Neurochem. 52 (in press).

Hetherington, H.P. (1988). The design and application of multipulse NMR techniques to in vivo ^{1}H NMR. Ph.D. thesis, Yale Univ., New Haven, CT.

Hetherington, H.P. and Rothman, D.L. (1985). Phase cycling of composite refocusing pulses to eliminate dispersive refocusing magnetization. J. Magn. Reson. 65, 348-354.

Hindfelt, B., Plum, F. and Duffy, T.E. (1977). Effect of acute ammonia intoxication on cerebral metabolism in rats with portocaval shunts. J. Clin. Invest. 59, 386-396.

Hore, P.J. (1983). Solvent suppression in Fourier transform nuclear magnetic resonance. J. Magn. Reson. 55, 386-396.

McKhann, G.M. and Tower, D.B. (1961). Ammonia toxicity and cerebral oxidative metabolism. Am. J. Physiol. 200, 420-424.

Patlack, C.S. and Pettigrew, K.D. (1976). A method to obtain infusion schedules for prescribed blood concentration time courses. J. Appl. Physiol. 40, 458-463.

Petroff, O.A.C., Prichard, J.W., Behar, K.L., Alger, J.R., den Hollander, J.A. and Shulman, R.G. (1985). Cerebral intracellular pH by ^{31}P nuclear magnetic resonance spectroscopy. Neurology 35, 781-788.

Petroff, O.A.C., Prichard, J.W., Behar, K.L., Alger, J.R. and Shulman, R.G. (1984). In vivo phosphorus nuclear magnetic resonance spectroscopy in status epilecticus. Ann. Neurol. 16, 169-177.

Plateau, P. and Gueron, M. (1982). Exchangeable proton NMR without base-line distortion, using new strong-pulse sequences. J. Am. Chem. Soc. 104, 7310-7311.

Zieve, L. (1981). The mechanism of hepatic coma. Hepatology 1, 360-365.

Effects of Hypercarbia and Porta-Caval Shunting on Amino Acids and High Energy Phosphates of the Rat Brain: a ^1H and ^{31}P NMR Study

K.L. Behar* and S.M. Fitzpatrick

Departments of Neurology and Molecular Biophysics and Biochemistry, Yale University New Haven, CT 06510.*

INTRODUCTION

Intracellular acidosis and hyperammonemia are associated with numerous metabolic encephalopathies. Conditions such as cerebral hypoxia, ischemia, seizures, and hypercarbia are accompanied by marked increases in intracellular $[H^+]$ and $[NH_4^+]$. Changes in either $[H^+]$ or $[NH_4^+]$ lead to changes in the concentrations of cerebral amino acids and metabolic intermediates of both the glycolytic and TCA cycle (Miller *et al*, 1975; Cooper and Lai, 1987). The mechanism(s) involved and the site(s) of regulation of substrate flux by changes in the concentrations of these ions is not known with certainty. It is known that nearly all encephalopathies that involve an increase in ammonium ion concentration are accompanied by a decrease in the concentration of cerebral glutamate and an increase in glutamine. Hypercarbic acidosis also results in a decrease in the concentration of glutamate in rat brain and an increase of glutamine and ammonium ion (Folbergrová *et al*, 1972; Miller *et al*, 1975). It is important to note that neither acute hyperammonemia (Fitzpatrick *et al*, 1989) nor porta-caval shunting affects intracellular pH (Fitzpatrick *et al*, 1988); therefore, changes in pH_i cannot be responsible for the decrease in glutamate, observed in response to elevated ammonium ion. While the decrease in cerebral glutamate is modest during acute hyperammonemia (Fitzpatrick *et al*, 1989), the decreases observed during intracellular acidosis are large (Behar *et al*, 1987).

We have investigated the effects of sustained hypercarbia of 60-min duration on the levels of the amino acids glutamate and glutamine in the rat brain *in vivo* by ^1H NMR spectroscopy. The levels of the cerebral high energy phosphates, phosphocreatine (PCr) and nucleoside triphosphate

(NTP), inorganic phosphate (P_i) and pH_i were investigated by ^{31}P NMR. Both ^1H and ^{31}P NMR spectra were acquired alternately on the same animal using a double-tuned surface coil. The effects of chronic hyperammonemia on the metabolic response of the brain to intracellular acidosis was investigated in rats at 4-weeks after portacaval shunting. This study was aimed at better understanding changes in metabolism occuring during acidosis in the normal and chronically hyperammonemic brain.

METHODS

Animal Preparation: Male Sprague-Dawley rats weighing 180-240 gm were used for all experiments. End-to-side porta-caval shunted rats were obtained from Hilltop Lab Animals, Inc., Scottdale, PA., and used 4-weeks after shunting. Food was withheld for 24 hours prior to the NMR experiments. The animals were tracheotomized under 1.5% enflurane anesthesia, ventilated with 30% O_2 and 70% N_2O, and immobilized with d-tubocurarine-Cl (0.1 mg, s.c.). Enflurane administration was discontinued prior to placement of the animal in the magnet. Hypercarbia was produced by replacing N_2O with CO_2 in the gas mixture. The tail artery and vein were cannulated for sampling of arterial blood and for the infusion of glucose, respectively. Arterial blood samples were collected periodically before and during hypercarbia for determination of the pCO_2, pO_2, pH, and glucose concentration. Arterial blood pressure was recorded continuously during the NMR experiment using a Statham pressure transducer connected to the arterial line via a polyethylene T-junction at the base of the magnet. Prior to the induction of hypercarbia, D-glucose was infused (i.v.) at a rate that resulted in a rapid and sustained increase in arterial blood glucose. Because a progressive decrease in arterial glucose levels were observed during hypercarbia in a previous study (Behar *et al*, 1987), glucose was infused to avoid hypoglycemia. Core body temperature was maintained near 37° by circulating warm water through a jacket constructed of Tygon tubing. A scalp incision was made along the midline and the periosteum was removed. The skin and fascia were retracted and sutured laterally so as to provide a large tissue-free region of skull. The wound was covered with Saran Wrap® to prevent desication. The animal was placed in a Plexiglas® cradle that was attached to the surface-coil probe as shown in Fig. 1. The surface coil was placed on the skull and centered approximately 2-mm posterior to the bregma.

Spectroscopy: NMR spectra were obtained at a magnetic field strength of 8.4-Tesla using an AM-360 wide-bore spectrometer (Bruker Instruments) interfaced to an ASPECT-3000 computer. The single-turn

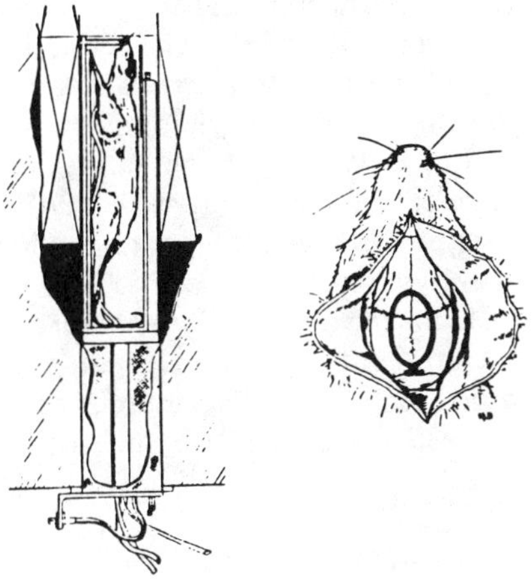

FIG. 1 Animal preparation for NMR spectroscopy. A. View of rat in NMR spectroscopy probe after insertion into the magnet. B. Placement of surface coil on rat skull after retraction of the scalp.

8x14-mm elliptical surface-coil, which was used to transmit rf pulses and detect the NMR signal, was double-tuned to the ^{1}H and ^{31}P resonance frequencies of 360.13 and 145.78 MHz, respectively. Magnetic field homogeneity was optimized by shimming on the ^{1}H free-induction-decay of tissue water yielding a linewidth of 20-25 Hz. ^{1}H spectra were collected with a phase-cycled $1-\bar{1}$ semi-selective spin-echo pulse sequence using presaturation, as described previously (Behar *et al*, 1987; and references cited therein). The interpulse delay was 1.5-sec and each spectrum represents the sum of 32 transients (48-sec). The spin-echo delay time was 4-msec. The maximum intensity of the semi-selective pulse was positioned at the chemical shift of N-acetylaspartate (N-AcAsp) at 2.02 ppm; resonances appearing at higher or lower frequencies from N-AcAsp are reduced in intensity in response to the excitation profile of the semi-selective pulses. The peak amplitudes of the γ-CH_2 of glutamate and glutamine were measured in each ^{1}H spectrum and expressed as a ratio to the amplitude of the total creatine CH_3 peak at 3.03 ppm, which represents the unresolved resonances of creatine and phosphocreatine. The values of Glu/Cr_{tot} and Gln/Cr_{tot} were determined and each was subtracted from the average value calculated over the pre-hypercarbic control period for each animal. The amplitude of the creatine resonance was not corrected for the intensity reduction due to the semi-selective pulses or for relaxation time. Hence, the ratios presented in Fig. 6 and Fig. 7 cannot be used directly to compute concentrations.

^{31}P NMR spectra were collected alternately with ^{1}H spectra using a conventional pulse-acquire sequence. The interpulse delay was 0.6-sec resulting in the partial saturation of phosphorus resonances due to incomplete spin-lattice (T1) relaxation. Spectra were collected as the sum of either 256 (2.56-min) or 512 transients (5.12-min). The reduction in signal intensity due to saturation will represent a constant fraction independent of the concentration of the metabolite yielding the NMR signal. Changes in the peak amplitudes of phosphocreatine (PCr), inorganic orthophosphate (P_i), and nucleoside triphosphate (NTPβ) were measured in the ^{31}P spectrum and expressed as a percentage of their respective values measured in the pre-hypercarbic period.

Intracellular pH was determined from the pH-dependent shift of P_i in the ^{31}P spectrum based on a titration of P_i measured at a Mg^{2+} concentration of 1.36 mM and an ionic composition that mimics cerebral intracellular cytosol (Petroff *et al*, 1985), according to the following equation:

$$pH = 6.77 + \log[(\delta - 0.94)/(0.33 - \delta)],$$

where δ is the chemical shift of P_i relative to PCr. Phosphocreatine was assigned a chemical shift of -2.35 ppm.

RESULTS

Effects of hypercarbia on physiological parameters.

No significant differences were observed in blood gases or pH between non-shunted and portacaval (PC)-shunted rats before hypercarbia. Mechanical ventilation resulted in mean P_aCO_2 values of 36 ± 6 and 37 ± 3 mmHg, respectively (Table 1). Carbon dioxide administration led to similar increases in the P_aCO_2 (203-214 mmHg) in both non-shunted and PC-shunted rats for all times measured; values were not significantly different between the two groups at 30 or 60-min. The increased P_aCO_2 was maintained near 200 mmHg over the 60-min timecouse of hypercarbia. Arterial pH decreased during hypercarbia, reaching minimum values between 6.62 and 6.72 after 10-min in both groups of animals. An increase in P_aO_2 occured following hypercarbia in both non-shunted and PC-shunted rats and was due to the effect of acidosis on the dissociation of oxygen from hemoglobin. The concentration of arterial glucose, which was increased prior to hypercarbia by i.v. infusion, increased further during hypercarbia. The arterial blood pressure increased transiently after CO_2 administration but stablized shortly after to values near those measured during the pre-hypercarbic period.

Changes in intracellular pH and high energy phosphates: the ^{31}P spectrum.

The ^{31}P NMR spectrum of a rat brain before and during hypercarbia is shown in Fig.2. The resonances of phosphocreatine (PCr), inorganic orthophosphate (P_i), and the three phosphates of nucleoside triphosphate (NTP) are labeled in the figure. The assignments of these resonances have been described previously (Glonek *et al*, 1982). Intracellular pH was measured from the chemical shift of P_i as described in Methods. The values of pH_i before hypercarbia were not different between non-shunted (7.18 ± 0.08) and PC-shunted (7.14 ± 0.03) rats. Hypercarbia led to rapid intracellular acidification in the brains of both non-shunted and PC-shunted rats, as shown in the timecourses obtained for both groups of animals (Fig. 3). The values of pH_i were not different between the two groups at 10-min (non-shunted, 6.68 ± 0.06 (n=4)); PC-shunted, 6.70 ± 0.03 (n=4)) or at 1-hour (non-shunted, 6.76 ± 0.05; PC-shunted, 6.73 ± 0.01).

Hypercarbia resulted in a decrease in the level of PCr and an increase in P_i (Fig.4). The level of PCr decreased to between 60-75% of control at 10-min in both non-shunted and PC-shunted rats, and PCr was maintained near this level throughout the 60-min period of hypercarbia. The level of P_i increased to about twice the normocapic value, corresponding to the decrease in PCr. The level of NTP was unchanged during hypercarbia in both non-shunted and PC-shunted rats.

Changes in cerebral amino acids: the ^1H spectrum.

^1H spectra of the brains' of non-shunted and PC-shunted rats obtained

TABLE I. Effects of hypercarbia on the physiological parameters of normal and portacaval shunted rats.

P-C shunt (N=4)	pH	P_aCO_2 mmHg	P_aO_2 mmHg	glucose mg/dL	MABP mmHg
Normocarbia	7.36±0.02	37.0±3	139±32	204±74	154±7
11 min hypercarbia	6.72±0.02	179±24	202±26	276±46	151±15
30 min hypercarbia	6.68±0.05	203±12	203±30	296±87	152±14
60 min hypercarbia	6.66±0.05	208±20	195±54	253±56	160±3
Non-shunted (N=4)					
Normocarbia	7.31±0.10	36.0±6	157±23	272±30	164±14
30-min hypercarbia	6.64±0.05	206±13	220±10	297±35	178±11
60-min hypercarbia	6.62±0.06	214±12	230±20	310±37	162±8

The normocarbic values of portacaval shunted and non-shunted rats were obtained prior to the administration of CO_2 to each animal. As explained in Methods, glucose was administered intraveneously by an infusion protocol designed to maintain a constant level of glucose in the blood prior to hypercarbia. Values are presented as the mean ± s.d. of four rats for each group. MABP, mean arterial blood pressure.

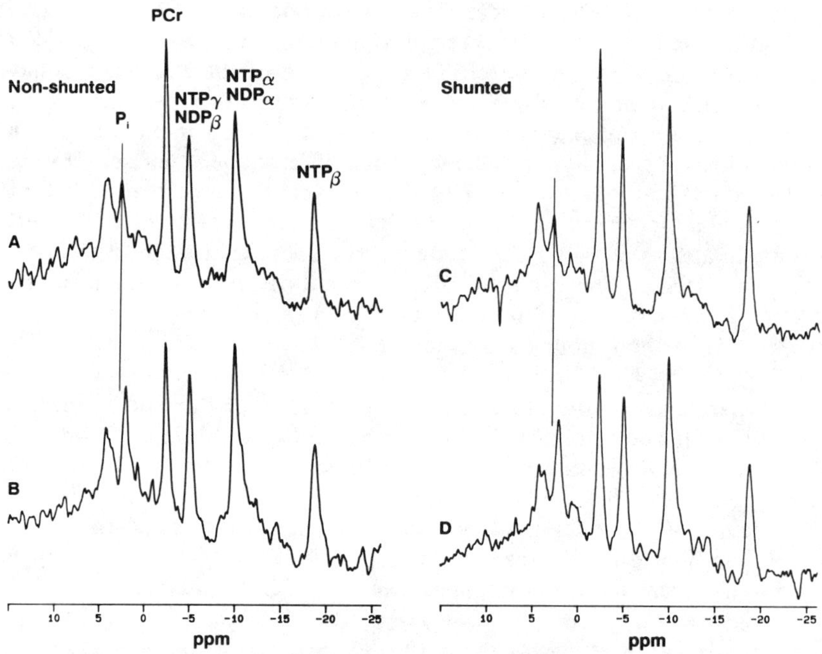

FIG. 2 ^{31}P NMR spectra of non-shunted and portacaval shunted rat brain before and during hypercarbia. Left column (non-shunted): Spectrum A was acquired before hypercarbia. Spectrum B was acquired after 60-min hypercarbia. Right column (portacaval shunted): Spectrum C was acquired before hypercarbia. Spectrum D was acquired after 60-min hypercarbia. The vertical line is centered on the P_i resonance of the spectra obtained during the pre-hypercarbic period (A and C). During hypercarbia (B and D) the P_i resonance is shifted to the right (upfield shift), indicative of acidosis.

in vivo before and during hypercarbia are shown in Fig. 5 and Fig. 6, respectively. The assignment of resonances are based on previous studies of acid extracts and tissue of rat brain (Behar *et al*, 1983; 1985; Ogino *et al*, 1986). As shown in Fig. 6, cerebral glutamine content was greatly increased in PC-shunted rats over that of non-shunted rats, which is consistent with previous studies (Butterworth and Giguere, 1986). Hypercarbia of 60-min duration led to a large decrease in the intensity of glutamate (Glu γ-CH_2, 2.36 ppm) and a lesser increase in glutamine (Gln γCH_2, 2.45 ppm) in both non-shunted (Fig. 5B) and shunted (Fig. 6B) rat brain. In order to quantitate the changes in glutamate and glutamine, their peak amplitudes were normalized to the amplitude of total creatine (Cr_{tot}), which remains constant under most conditions. The glutamate-to-creatine ratio decreased to 55 ± 6% of the pre-hypercarbic value in PC-shunted rats (n=3) compared to 53 ± 4% in non-shunted controls (n=4). Because the γ-methylene resonances of glutamate and glutamine overlap the α- and β-methylenes of GABA and N-acetylaspartate, respectively (Behar *et al*, 1985), the actual percentage change will be underestimated. Therefore, changes in the peak amplitude ratios were expressed as the difference between the average value of the ratio determined before hypercarbia and each value following hypercarbia. The change in the glutamate-to-creatine ratio ($\Delta Glu/Cr_{tot}$; Fig. 7) was ≈0.3 after 1-hr in both shunted and non-shunted rats. The glutamine-to-creatine ratio was increased slightly after 1-hour of hypercarbia (PC-shunts, 22 ± 9% [n=3]; controls, 26 ± 7% [n=4]) or a *change* ($\Delta Gln/Cr_{tot}$) of about 0.1 (Fig. 7). The changes in the levels of cerebral glutamate and glutamine in shunted and non-shunted rats followed similar timecourses. The decrease in glutamate appeared to follow a biphasic timecourse--an initial rapid decrease during the first 10-min was followed by a more gradual decline.

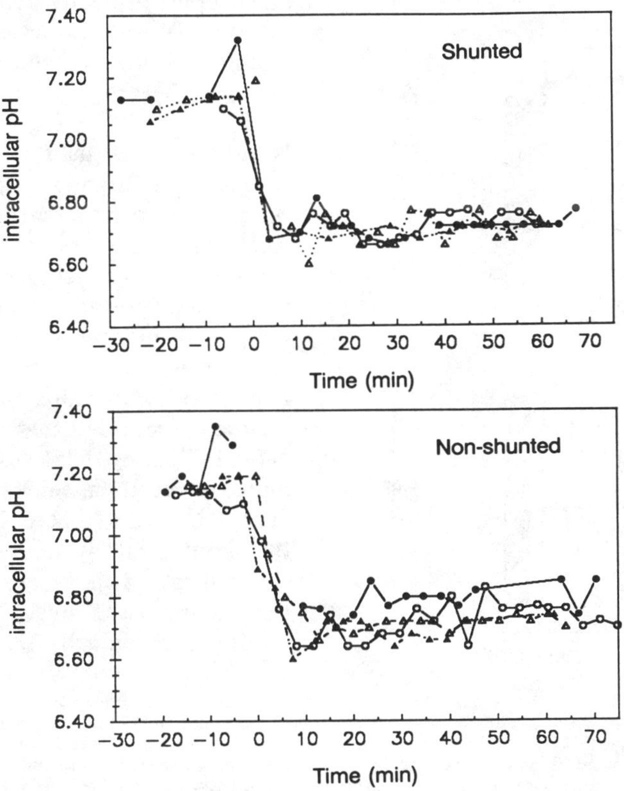

FIG. 3 Effect of hypercarbia on the cerebral intracellular pH of non shunted and porta caval shunted rats. Carbon dioxide was administered at time zero. The pH_i was determined for each animal by measuring the chemical shift of P_i, as described in Methods. Each graph displays the results of four animals.

DISCUSSION

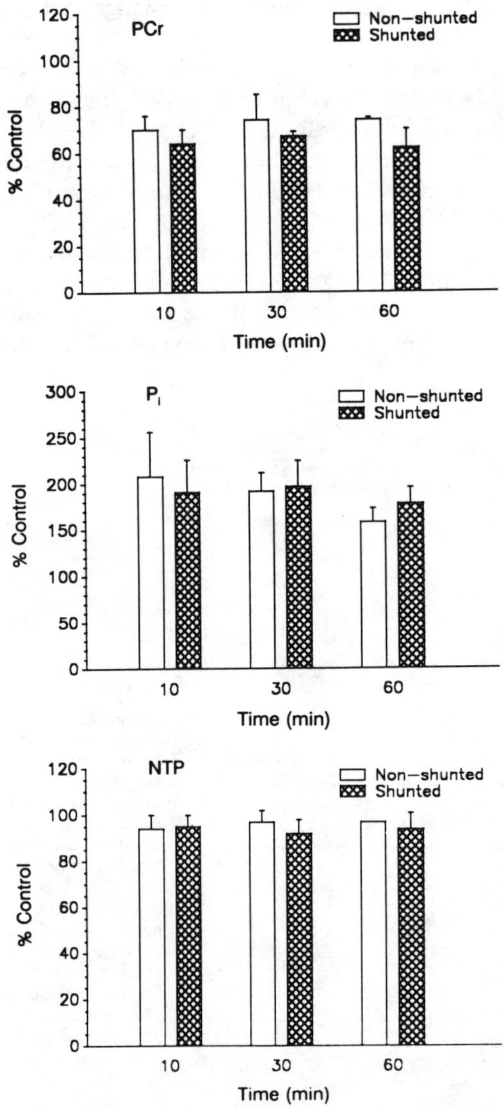

High energy phosphate metabolism and intracellular acidosis

Hypercarbia led to rapid intracellular acidification and a decrease in PCr with no change in NTP. These results are consistent with other studies of hypercarbia on brain energy metabolism by ^{31}P NMR (Behar *et al*, 1987; Cady *et al*, 1987; Litt *et al*, 1985; Petroff *et al*, 1985), and support previous studies using freeze-extraction methods (Miller *et al*, 1975; Folbergrová *et* al, 1972; Siesjö *et al*, 1972). No differences in the response of high energy phosphates to hypercarbia was observed between shunted and non-shunted rats. The magnitude of the reciprocal changes in PCr and Pi were equivalent between the two groups and no change was detected in NTP, indicating that a sustained intracellular acidosis of pH 6.7 is tolerated in PC-shunted rats as in non-shunted rats. This suggests that the increased sensitivity of the chronically

FIG. 4 Effects of hypercarbia on cerebral high energy phosphates and inorganic phosphate of non-shunted and portacaval shunted rats. The amplitudes of phosphocreatine (PCr), nucleoside triphosphate (NTP), and inorganic orthophosphate (P_i) were measured from the ^{31}P spectrum and plotted as a percentage of the average amplitude measured during the pre-hypercarbic period for the same animal. The time axis indicates the time, in min, after CO_2 administration. Error bars represent ± s.d. for four rats in each group.

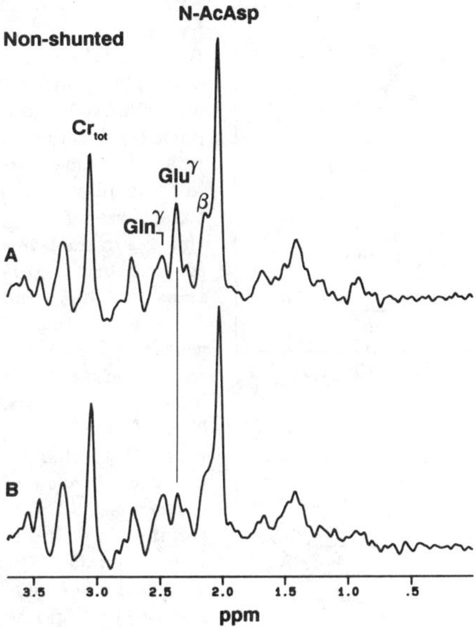

FIG. 5 ^{1}H NMR spectra of a non-shunted (control) rat brain before and during hypercarbia. Spectrum A was acquired before hypercarbia. Spectrum B was acquired after 60-min of hypercarbia. Resonances are labeled as follows: Glu, glutamate γ-CH_2; Gln, glutamine γ-CH_2; Cr_{tot}, creatine + phosphocreatine -CH_3; N-AcAsp, N-acetylaspartate -CH_3.

hyperammonemic brain to hypercarbia (Posner and Plum, 1960) is not a result of energy depletion, at least in the cerebral cortex.

The fact that portacaval shunting did not affect the metabolic response of the rat brain to intracellular acidification was surprising. Regulation of cerebral pH is believed to be under metabolic control during hypercarbia (Siesjö, 1973) yet no difference was found between porta-caval shunted and non-shunted rats in their response to CO_2; the pHi in both groups fell by 0.4 units in response to an equivalent increase in P_aCO_2. The astrocyte is believed to have an important role in the regulation of cerebral pH_i and histochemical studies have shown glial cells to possess high activities of carbonic anhydrase (Giacobini, 1962). Although astrocytes are particularly susceptible to the toxic effects of ammonia and characteristic morphological changes occur in

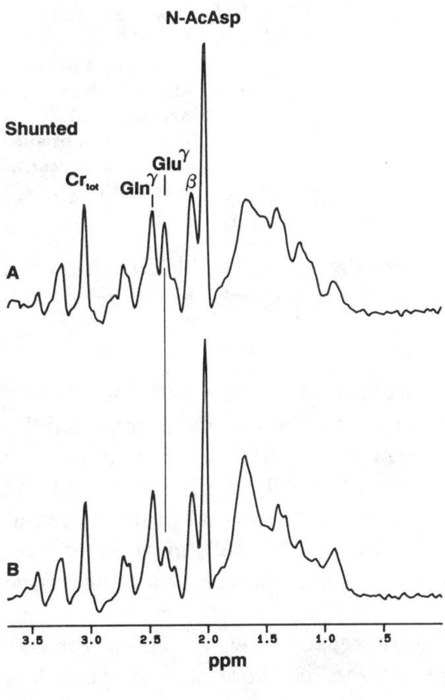

FIG. 6 ^{1}H NMR spectra of a porta- caval shunted rat brain before and during hypercarbia. Spectrum A was acquired before hypercarbia. Spectrum B was acquired after 60-min hypercarbia. The high intensity peaks between 0.8 and 1.9 ppm arrise from lipids. Portacaval shunted rats have significantly narrower skulls than non-shunted rats resulting in some pick-up of signal from poorly shimmed tissue near the periphery of the surface-coil.

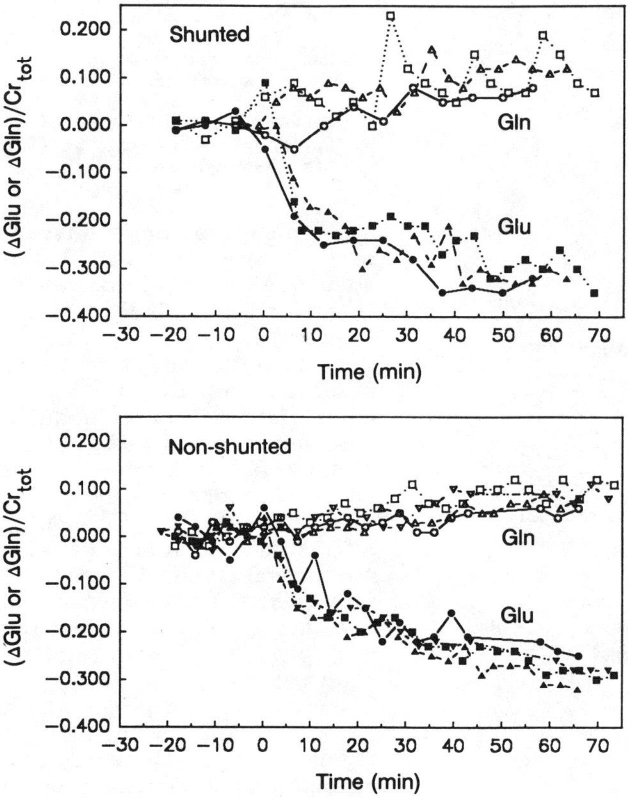

FIG. 7 Effects of hypercarbia on cerebral glutamate and glutamine in non-shunted and portacaval shunted rats. The upper figure displays the timecourse of changes in the levels of cerebral glutamate (Glu) and glutamine (Gln) in portacaval shunted rats before and during hypercarbia. The lower figure represents changes in the levels of glutamate and glutamine in non-shunted rats. The amplitudes of glutamate and glutamine were measured from ^{1}H spectra, expressed as a ratio to the amplitude of total creatine (Cr_{tot}), and subtracted from the average values of the ratios obtained before hypercarbia (see Methods).

astrocytes after portacaval shunting (Norenberg, 1987), this condition did not affect the magnitude or timecourse of the response to hypercarbia.

Amino acid metabolism

The decrease in the content of cerebral glutamate and increase in glutamine that we observed during hypercarbia has been reported by others (Miller *et al*, 1975; Folbergrová *et al*, 1972b). However, the biphasic pattern for the disappearance of cerebral glutamate during hypercarbia has not been reported previously. The initial rapid decrease, which after 20-min amounted to more than 60 to 80% of the decrease observed after 1-hr, was followed by a much slower but progressive decrease.

The content of cerebral glutamine increased linearly during the 60-min period of hypercarbia, although the rate of increase was much less and matched more closely the slower component of glutamate disappearance observed between 20 and 60-min of hypercarbia. These results are in disagreement with those obtained by Van Leuven *et al* (1972) and Weyne *et al* (1978), where hypercarbia in rats led to a decrease in the concentration of glutamate that could be accounted for quantitatively by the rise in glutamine. Their conclusions were modified, however, on the basis of results obtained from rats treated with methionine sulfoximine (MSO), an inhibitor of glutamine synthetase (Weyne *et al*, 1977). The increase in glutamine was reduced markedly after MSO treatment; however, a proportionate reduction in the decrease of glutamate was not observed, leading them to conclude that other pathways in addition to glutamine synthetase are important for glutamate removal during hypercarbia.

Glutamine synthetase, which is primarly responsible for ammonia detoxification in the brain (Cooper and Lai, 1987), is localized predominately in astrocytes (Norenberg and Martinez-Hernandez, 1979). Little information is available on acid/base regulation by astrocytes *in vivo*. Conditions that affect astrocyte metabolism might be expected to have a significant effect on pHi regulation that involves net changes in the concentrations of organic acids (Siesjö, 1973). The fact that the disappearance of glutamate was also biphasic during hypercarbia in portacaval shunted rats, and the increase in glutamine did not differ from that in non-shunted rats, indicates that pre-existing hyperammonemia, *per se*, does not significantly alter the metabolic response of the brain to acidosis. This appears to be true for high energy phosphates as well.

Acknowledgments: This work was supported by USPHS DK-27121.

REFERENCES

Behar KL, Rothman DL, Fitzpatrick SM, Hetherington HP, Shulman RG (1987) Physiological NMR Spevctroscopy: From Isolated Cells to Man. *Annals N.Y. Acad. Sci.* Vol **508**, pp. 81-88.

Behar KL, den Hollander JA, Stromski ME, Ogino T, Shulman RG, Petroff OAC, Prichard JW (1983) *Proc. Natl. Acad. Sci. USA* **80**:4945 -4948.

Behar KL, den Hollander JA, Petroff OAC, Hetherington HP, Prichard JW, Shulman RG (1985) *J. Neurochem.* **44**:1045-1055.

Butterworth RF, Giguere J-F (1986) *Metab. Brain Disease* **1**:221-228.

Cady EB, Chu A, De L Costello AM, Delpy DT, Gardiner RM, Hope PL, Reynolds EOR (1987) *J. Physiol.* **382**:1-14.

Cooper AJL, Lai JCK (1987) *Neurochem. Path.* **6**:67-95.

Fitzpatrick SM, Behar KL, Hetherington HP, Shulman RG (1988) *Abstr.*

Soc. Magn. Reson. Med.
Fitzpatrick SM, Hetherington HP, Behar KL, Shulman RG (1989) *J. Neurochem.* (in press).
Folbergrová J, MacMillan V, Siesjö BK (1972a) *J. Neurochem.* **19**:2497-2505.
Folbergrová J, MacMillan V, Siesjö BK (1972b) *J. Neurochem.* **19**:2507-2517.
Giacobini E (1962) *J. Neurochem.* **9**:169-177.
Glonek T, Kopp SJ, Kot E, Pettegrew JW, Harrison WH, Cohen MM (1982) *J. Neurochem.* **39**:1210-1219.
Cooper AJL, Lai JCK (1987) *Neurochem. Path.* **6**:67-95.
Litt L, Gonzalez-Mendez R, Severinghaus JW, Hamilton WK, Shuleshko J, Murphy-Boesch J, James TL (1985) *J. Cereb. Blood Flow Metabol.* **5**:537-544.
Miller AL, Hawkins RA, Veech RL (1975) *J. Neurochem.* **25**:553-558.
Norenberg MD (1987) *Neurochem. Path.* **6**:13-33.
Norenberg MD, Martinez-Hernandez A (1979) *Brain Res.* **161**:303-310.
Ogino T, Behar KL, Shulman RG (1986) *Abstr. Soc. Magn. Reson. Med.* **3**:985-986.
Petroff OAC, Prichard JW, Behar KL, Rothman DL, Alger JR, Shulman RG (1985) *Neurol.* **35**:1681-1688.
Petroff OAC, Prichard JW, Behar KL, Alger JR, den Hollander JA, Shulman RG (1985) *Neurol.* **35**:781-788.
Posner JB, Plum F (1960) *J. Clin. Invest.* **39**:1246-1258.
Siesjö BK (1973) *Scand. J. Clin. Lab. Invest.* **32**:97-104.
Siesjö BK, Folbergrová J, MacMillan V (1972) *J. Neurochem.* **19**:2483-2495.

POSITRON-EMISSION TOMOGRAPHY IN THE STUDY OF HEPATIC ENCEPHALOPATHY

Alan H. Lockwood, M.D.

Associate Professor of Neurology
Positron Diagnostic and Research Center
University of Texas Health Science Center

6431 Fannin, Houston TX 77030, U.S.A

Positron-emission tomography (PET) is a relatively new tool for the *in vivo* investigation of the biochemistry and physiology of the brain. The availability of compounds for the measurement of cerebral blood flow, cerebral glucose metabolism, ammonia metabolism, and ligands for the study of neurotransmission in healthy control subjects and patients with symptomatic and asymptomatic liver disease make this technique well suited for the investigation of mechanisms leading to the development of hepatic encephalopathy.

Although many of the details that must be considered in the design of PET studies are technically complex, the basic elements can be understood easily and consist of the PET camera itself, a radiopharmaceutical source, usually a cyclotron, and appropriate mathematical models of the process under investigation (Phelps, et al., 1982).

After an unstable radioactive nucleus decays by emitting a positron, this sub-atomic particle travels a short distance through the tissue and is annihilated in a matter-antimatter interaction with an electron that converts the mass of the two particles into two gamma photons that move in almost exactly opposite directions from the annihilation point. PET cameras consist of arrays of detectors that register the presence of these paired gamma photons. The electronics of the camera are designed so that it registers the presence of only those pairs of photons that are produced by the annihilation of a single positron and detected simultaneously by two of the detectors in the array. The straight line defined by the two crystals must, by definition, pass through the site of the annihilation of the positron. By registering the straight-line paths through literally millions of decay sites, it is possible to create a quantitative computerized reconstruction of the spatial distribution of the isotope during the imaging session.

A number of corrections to the raw data generated by the camera must be made for the quantitation to be accurate. These include subtraction of events that randomly activate two detectors, corrections for photons absorbed by the tissue under investigation that should have activated a detector pair, corrections for scattered radiation that activates detector pairs, and corrections for electronic and design factors that affect the sensitivity of the camera in different parts of the field of view. When these corrections are made correctly, an easily-defined relationship exists between the number of events measured in each picture element or pixel of the image and the amount of radioactivity in the tissue represented by the pixel. The result is a quantitative digital image of a tomographic slice of the brain that depicts the spatial distribution of the tracer.

The commonly-used positron-emitting isotopes include C-11, N-13, O-15, and F-18 that have half-lives of 20.4 min, 10.0 min, 123 sec, and 110 min respectively. The importance of these atoms in organic molecules and their analogs makes it theoretically possible to produce an enormous variety of tracers. The short half lives, however, mandate easy access to a source of radionuclides. In practice, a cyclotron must be a part of the PET facility, along with the sophisticated radiochemistry laboratories that are required for the synthesis and quality control procedures that are necessary for safe, routine production of radiopharmaceuticals. The synthesis of

O-15 water, N-13 ammonia, and F-18 fluorodeoxyglucose are all reasonably standard, and these compounds are routinely used for studies of blood flow, ammonia metabolism, and cerebral glucose metabolism, respectively. C-11 and F-18, because of their longer half-lives, provide the chemist with the time needed to synthesize more complex molecules, such as those required for studies of neurotransmission.

In order to interpret PET images and to extract quantitative data from PET data, it is necessary to use a mathematical model that relates tracer detected in each pixel of the brain image to the process being studied. Invariably this requires quantitative measurement of the specific activity of the traced substance in the radial artery, or arterialized venous blood, which serves to depict the arterial supply to each pixel of the brain image. In the simplest case, the uptake rate of a compound can be equated with the instantaneous measure of the quantity of tracer at the time of measurement divided by the specific activity of the tracer precursor pool integrated from the beginning of the study to the time of the measurement. In practice, things are rarely this simple, since PET cameras require relatively long periods of time to observe a sufficiently large number of events to create a statistically valid image as opposed to the desired instantaneous measure (mathematically the image is the integral of the activity over the duration of the scan, and is analogous to a time exposure compared to a snap shot). In addition, corrections for the decay of the isotope during the study, the presence of unreacted or unbound tracer in the image, values of the rate constants defining movement of tracer between compartments of the model, and, in the case of analogs, an analog correction factor (called the lumped constant in the case of fluorodeoxyglucose) must be known.

Although not mandatory, facilities for quantitative tissue autoradiography are an important desirable part of a PET facility. There are many important hypotheses that are more appropriately tested using animal models rather than patients.

CEREBRAL BLOOD FLOW AND GLUCOSE METABOLISM

The ability to measure cerebral blood flow (CBF) and the local cerebral metabolic rate for glucose (lCMRG) are critical in any PET facility. CBF and lCMRG measurements allow inferences to be made concerning the amount of work being done by a brain region, since increases in brain work are paralleled by increases in these variables. In addition, knowledge of CBF makes it possible to compute values for additional variables when blood flow studies are paired with other complementary investigations.

CBF measurements are based on the Fick principle of conservation of matter which states that the rate of change of activity in brain with respect to time (B′) equals the rate of tracer delivery to the brain minus the rate of loss by all routes:

$$B' = F(a - v) - \beta B$$

where F is CBF, a and v are the arterial and venous concentrations of tracer and β the isotope decay constant. Since v is not measurable, the expression B/λ is equated with v and substituted (λ is the blood-brain partition coefficient), which with the addition of the variable m, which describes the completeness of equilibration of the tracer between blood and brain yields:

$$B' = mFa - Bk$$

where $k = \beta + mF/\lambda$. Integrating this equation twice over the times 0 to the end of the scan yields an expression equating the cumulative tracer activity in brain during the scan, the variable measured by the PET camera, with an expression containing B, F, a , and k, all of which are measured or have assumed values, except for F. Since the equation can't be solved explicitly for F in terms of the other parameters, small incremental values of F are inserted into the equation along with measured values for a, and the resulting pixel value for that F value is computed. When a series of F vs pixel values is known, the PET image can be transformed into an image of flow by substituting F values for each pixel value. This approach to measuring CBF is perhaps the most widely used, but not the exclusive method for measuring blood flow in the brain (Ginsberg, et al. 1982; Herskovitch, et al. 1983; Raichle, et al. 1983).

Methods for measuring glucose metabolism, perhaps the method most commonly used to study brain function, are even more complex (Sokoloff, et al. 1977; Reivich, et al., 1979). Since glucose is so rapidly metabolized by the brain, relative to the length of time required to obtain an adequate image, the glucose analog fluorodeoxyglucose, labeled with F-18 is used (FDG). FDG crosses the blood-brain barrier via the same carrier protein that transports glucose and is phosphorylated to form FDG-6-PO_4 which is for practical purposes trapped in the brain forever. The model equation used to compute the metabolic rate requires measurements of arterial plasma glucose and FDG levels during the period of tracer uptake (usually 45 minutes) an image of the brain (usually obtained 45 minutes after the injection of the tracer) and assumptions about the values of the rate constants that describe forward and reverse flux of glucose across the blood-brain barrier, the rate constant for the phosphorylation, and a final constant, usually termed the lumped constant, that corrects for differences in the affinity for glucose and FDG exhibited by the carriers. Since the values of these constants may change during disease, and routine measurement of the values, especially the lumped constant, is not a simple task, there may be questions about the validity of the method when applied to the study of certain conditions, especially those in which delivery of glucose by blood flow is an important rate-limiting step. However, for many applications, the FDG method is believed to be sufficiently accurate to warrant wide-spread use. These are conditions likely to be met during the study of hepatic encephalopathy (Cruz and Duffy, 1983).

Extensive lCMRG data from rats with portacaval shunts (Cruz and Duffy, 1983; Lockwood, et al, 1986) and preliminary data from cerebral blood flow and glucose metabolism measurements in humans (Lockwood, et al., 1986b) support the hypothesis that not all brain centers are equally sensitive to the factors that produce hepatic encephalopathy. Patients with encephalopathy have increased glucose metabolism in the caudate-putamen and cerebellum, relative to whole-brain metabolism. This finding was accentuated by comparing caudate-putamen with frontal metabolism, where metabolism fell relative to the global glucose metabolic rate. CBF:metabolism ratios remained constant, suggesting that the major pathway of glucose metabolism via glycolysis and the Krebs cycle is unaffected by encephalopathy-producing toxins.

In a larger series of patients in which flow only was measured, a one-way analysis of variance of regional CBF data from 9 normal subjects showed highly significant variations between regions. A similar analysis of data from identical regions from the brains of 6 patients failed to show significant between-region flow variation. Although region-to-region tests for significant flow differences all failed, flow was lower in each region in the minimal encephalopathy group, an event with a probability of chance occurrence = 0.0157. When flow data were normalized by dividing the regional flow value by the whole slice flow value, normalized flow in visual cortex and cerebellum was significantly lower in the encephalopathic subjects than in the normal controls.

These preliminary data strengthen the hypothesis that the toxins that produce hepatic encephalopathy affect different brain regions to different degrees. These differences in sensitivity lead to differences in the functional interaction between various brain regions and result in a functional reorganization of the brain that is expressed as the alterations of consciousness (both the on-off and cognitive aspects of consciousness are abnormal), visual-spatial disorganization, increases in muscle tone, increased deep tendon reflexes, extensor plantar signs, asterixis, and the other neurological abnormalities associated with hepatic encephalopathy (Lockwood, et al., 1986a).

AMMONIA METABOLISM

The cyclotron-produced isotope N-13 is ideally suited for the study of ammonia metabolism. The chemical reactions for the production of labeled ammonia are straightforward, and easily implemented in any cyclotron facility. The ammonia produced has a very high specific activity and can be injected without affecting the arterial ammonia concentration in the patient, and disturbing the steady state for ammonia metabolism, which is assumed to hold during a study.

Studies from a number of laboratories have shown that ammonia crosses the blood brain barrier rapidly, and is quickly converted to the amide group of glutamine (Lockwood et al 1979, Cooper et al 1979). Since the retention time for the labeled glutamine formed in the brain is very long, the brain acts as an infinite sink for ammonia for studies of short (25 minutes) duration, since the metabolized ammonia remains in the brain for the duration of the study. Prior studies from our laboratory have also shown that when ammonia is injected as a bolus, all labeled ammonia is cleared from the arterial pool within 10 minutes. These facts make it possible to measure the rate of uptake and metabolism of ammonia by the brain (cerebral metabolic rate for ammonia, CMRA) very simply:

CMRA = pixel tracer content ten minutes after tracer injection/integral of arterial ammonia specific activity over 10 minutes.

In a previous study in which we measured the whole-brain ammonia metabolic rate using ^{13}N-ammonia, we found a strong linear relationship between the arterial ammonia concentration and the cerebral metabolic rate for ammonia (Lockwood et al, 1979). The study group consisted of 5 normal individuals and 17 patients with liver disease and hyperammonemia. When the study population was segregated into normals and patients with normal mentation and compared with those who had encephalopathy, the encephalopathic group had higher blood ammonia concentrations and higher brain ammonia metabolic rates than the non-encephalopathic subjects. This is strong evidence in support of the ammonia hypothesis.

When a CBF study is coupled with a ammonia metabolism measurement, and the subject's head is held without motion between the studies, flow and metabolism measures for each pixel are obtained, and additional conclusions concerning the behavior of ammonia can be made (Lockwood et al, 1984). Again, the Fick principle, is pertinent to ammonia and:

$$CMRA = F(a - v)$$

where a and v in this case refer to ammonia in arterial and venous blood. Solving for (a - v) and dividing by a yields E, the extraction fraction:

$$E = (a - v)/a$$

When E and F are known, the Renkin - Crone model of capillary permeability can be applied to ammonia to compute PS, the permeability - surface area product:

$$-PS = \ln(1 - E)F$$

Thus from the images of CBF and CMRA it is possible to generate images depicting E and PS with the same anatomical-spatial relationships that pertain to the original functional images of CBF and CMRA.

An analysis of these images makes it possible to examine the relationships between measured variables. For example, a large region of interest drawn on a CBF image and superimposed extraction image shows that extraction is inversely proportional to flow. In a graph containing about 1000 points, each representing a discrete brain region, flow = -0.13(extraction) + 0.32, $r = -0.64$, with a p value for the $r \ll 0.001$, a highly significant relationship.

The assumption that permits the application of the Renkin-Crone model, notably that there is no significant back-diffusion of the tracer from brain to blood, is supported by biochemical data that show rapid metabolic trapping of labeled ammonia as the amide nitrogen group of glutamine, and time-activity curves generated by radiation detectors over the brain that show a steep rise and plateauing of brain activity after ammonia is injected as a bolus.

The validity of this assumption also makes additional modeling strategies applicable to the study of cerebral ammonia metabolism (Patlack et al, 1983). In a very general model that assumes that there may be many brain compartments that permit ingress and egress of ammonia and at least one compartment in which ammonia is trapped, it can be shown that if the amount of trapped ammonia at time t divided by the plasma concentration of tracer at t plotted as a function of the integral of blood tracer content from time zero to t divided by the plasma tracer content at t yields a straight line whose slope is Ki, the rate constant that describes the movement of tracer from blood to brain. The y intercept of this function is a volume that is less than or equal to the vascular volume plus the volume of the compartments permitting ingress and egress. From the Ki, the total influx can be computed by multiplying Ki by the concentration of tracer in the vascular pool, and the turnover time in the brain ammonia pool can be calculated by dividing the product of the plasma tracer concentration, the blood:brain ammonia ratio, and the intercept by the flux.

NEUROTRANSMITTERS

Although PET has been used to study neurotransmitters, these studies are among the most difficult and controversial areas in the field. Progress in understanding neurotransmission in patients with hepatic encephalopathy is likely to be slow using this technique.

A portion of this difficulty stems from the nature of the receptor complexes themselves. The γ-aminobutyric acid (GABA) receptor, currently of great interest in the proposed mechanisms and therapy of hepatic encephalopathy, may be the most difficult of all receptors to investigate using PET. The receptor itself consists of a number of elements, including the GABA binding site, a benzodiazepine binding site, a GABA modulin, a chloride ion channel, and a coupler linking the GABA recognition site and ion channels (Mazzari, et al, 1981). In addition to this specific binding site, there are also non-specific sites. There are also sites in the nervous system that are not believed to have GABA-ergic input that exhibit GABA sensitivity and GABA auto-receptors that respond to GABA but, when activated, inhibit GABA release. The complexity of the GABA system and its capacity for multiple actions present formidable problems in the evaluation of in vivo GABA actions (Enna, 1981).

In order to study neurotransmitter function in the brain, an appropriate ligand, labeled with a positron emitting nuclide must be synthesized. In general, nuclides with long half-lives are desirable so that the chemist, who generally likes to complete all synthesis and purification within 3 half-lives, has an adequate amount of time to produce and purify the compound. This leaves a sufficient period of time for serial imaging of the brain after the ligand is administered. These images are used in the model to deduce the kinetics of uptake and distribution. In practice, C-11 and F-18 are the most useful nuclides, and a variety of compounds labeled with them have been produced.

In general, imaging studies are conducted without saturating the receptor site under study. The ligands must therefore have a very high specific activity. If saturation occurs, then a pharmacological effect of the ligand may be induced that alters the desired steady state condition in the subject. In addition, as the specific activity of the ligand increases, the amount of tracer bound to nonspecific binding sites increases, relative to the amount bound to the higher-affinity specific binding site, which is usually the focus of the study. An exception to this latter rule occurs when a non-radioactive ligand is administered after the initial imaging study in an attempt to displace the tracer from high affinity binding sites. These studies still require a high specific activity tracer for the initial portion of the investigation.

To study receptor systems by PET, a mathematical model of the behavior of the labeled ligand is required: it is here that perhaps the most difficult problems emerge. In addition to the considerations listed above relative to the study of the GABA receptor complex, models of ligand behavior must also account for the characteristics describing the delivery of ligand to the locus of the receptor. These variables include knowledge of blood flow and the mechanism(s) by which the ligand moves across the blood-brain barrier such as carrier-mediated transport (describable in usual enzyme kinetics terms), diffusion (describable in terms of permeability and surface area), and pinocytosis.

In spite of technical and cost considerations, it is likely that well-designed PET studies will produce answers to critical questions regarding the mechanisms and therapy of hepatic encephalopathy. It is one of the few techniques available for the study of these problems in human subjects, and enables the investigator to avoid legitimate questions that might be raised when animal models are used.

This work was supported in part by NIH grant DK 38307.

REFERENCES

Cooper AJL, McDonald JM, Gelbard AS, et al. The Metabolic fate of ^{13}N-labeled ammonia in rat brain. J Biol Chem 254:4982, 1979.

Cruz NF, Duffy TE. Local cerebral glucose metabolism in rats with portacaval shunts. J Cereb Blood Flow Metabol 3:86, 1983.

Enna SJ. Commentary: GABA receptor pharmacology: functional considerations. Biochem Pharmacol 30:907, 1981.

Ginsberg MD, Lockwood AH, Busto RG, et al. A simplified in vivo autoradiographic strategy for the determination of regional cerebral blood flow by positron-emission tomography: theoretical considerations and validation studies in the rat. J Cereb Blood Flow Metabol 2:89, 1982.

Herskovitch P. Markham J, Raichle ME. Brain blood flow measured with $H_2{}^{15}O$. I Theory and error analysis. J Nucl Med 24:782, 1983.

Lockwood AH, Bolomey L, Napoleon F. Blood-brain barrier to ammonia in man. J Cereb Blood Flow Metabol 4:516, 1984.

Lockwood AH, Ginsberg, MD, Rhoades HM, Gutierrez MT. Cerebral glucose metabolism after portacaval shunting in the rat: patterns of metabolism and implications for the pathogenesis of hepatic encephalopathy. J Clin Invest 78:86, 1986a.

Lockwood AH, Marani S, Yap E, et al., Cerebral blood flow and glucose metabolism in hepatic encephalopathy. Ann Neurol 22:167, 1986b.

Lockwood AH, McDonald JM, Reiman, RE, et al. The dynamics of ammonia metabolism in man: effects of liver disease and hyperammonemia. J Clin Invest 63:449, 1979.

Mazzari S, Bassotti M, Guiditi A, Costa E. Gaba receptors as supramolecular units. In GABA and Benzodiazepine Receptors, Advances in Biochemical Pharmacology, Vol, 26 Costa E, DiChihara G Gessa GL. Eds. Raven Press, New York, p. 1, 1981.

Mintun MA, Raichle ME, Killbourn MR, et al. A quantitative model for the in vivo assessment of drug binding sites with positron emission tomography. Ann Neurol 15:217, 1984.

Patlak CS, Blasberg RG, Fenstermacher JD. Graphical evaluation of blood-to-brain transfer constants from multiple-time uptake data. J Cereb Blood Flow Metabol 3:1, 1983.

Phelps ME, Mazziotta JC, Huang S-C. Study of cerebral function with positron computed tomography. J Cereb Blood Flow Metabol 2:113, 1982.

Raichle ME, Martin WRW, Herschkovitz P, et al. Brain blood flow measured with $H_2{}^{15}O$ II. Implementation and validation. J Nucl Med 24:790, 1983.

Reivich M, Kuhl D, Wolf A, et al. The $[^{18}F]$-fluorodeoxyglucose method for the measurement of local cerebral glucose utilization in man. Circ Res 44:127, 1979

Sokoloff L, Reivich M, Kennedy, et al. The [^{14}C]-deoxyglucose method for the measurement of local cerebral glucose utilization: theory, procedure and normal values in the normal and anesthetized albino rat. J Neurochem 28:897, 1977.

THE USE OF CULTURED ASTROCYTES IN THE STUDY OF HEPATIC ENCEPHALOPATHY

Michael D. Norenberg

Veterans Administration Medical Center,
Department of Pathology, University of Miami
School of Medicine-Jackson Memorial
Hospital, Miami, FL 33101

INTRODUCTION

Considerable data implicates astrocytes in the pathogenesis of hepatic encephalopathy (HE). The Alzheimer type II astrocyte change constitutes the histopathologic hallmark in man and experimental animals with portal systemic encephalopathy (for reviews, see Norenberg, 1981, 1987). Ammonia, the chief toxin implicated in the pathogenesis of HE also causes similar astrocytic changes. Additionally, the detoxification of ammonia in brain occurs through the action of glutamine synthetase, an enzyme which in brain is exclusively localized in astrocytes (Norenberg and Martinez-Hernandez, 1979). Other implicated toxins, such as short-chain fatty acids also appear to be predominantly metabolized in astrocytes (Cremer, 1975). Finally, fulminant hepatic failure leads to the development of cerebral edema, a condition caused by water accumulation in astrocytes (Traber et al., 1987).

The Alzheimer type II astrocyte change (Fig. 1) consists of a cell with an enlarged vacuolated nucleus, marginated chromatin and often a prominent nucleolus. In contrast to reactive fibrous gliosis as that seen in glial scar tissue, there is no cytoplasmic hypertrophy or increase in cytoplasmic processes. An increased accumulation of lipofuscin pigment may be present.

The significance of the astroglial change seen in HE has been elusive. Based on morphologic studies, it was concluded that initially the astrocytes were

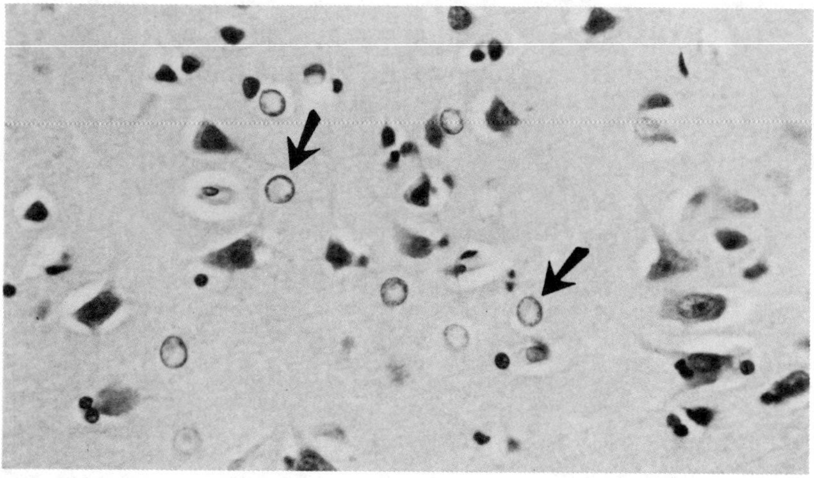

Fig. 1. Cerebral cortex from a patient who died with hepatic encephalopathy showing Alzheimer type II astrocytes (arrows). Note the characteristically enlarged, pale nuclei with marginated chromatin. x300.

metabolically active, perhaps for the purpose of detoxifying ammonia (and perhaps other toxins as well) (Norenberg, 1977). Eventually, degenerative changes ensue, possibly indicative of metabolic exhaustion. It is this later degenerative phase that corresponds to the Alzheimer type II change. We suggested that toxins produce astroglial injury (a gliopathy) thereby impairing critical astroglial activities resulting in CNS dysfunction.

For many years astrocytes have been simply viewed as passive supportive elements of the CNS. Recently, there has been a profusion of knowledge about these cells. Chiefly through the use of cell culture and bulk-isolated methods, astrocytes have been shown to be active in the uptake of K^+, glutamate, GABA and other neurotransmitters, and in the uptake and release of neuromodulators such as taurine and adenosine. These cells also appear to act as immunomodulators and have the capacity to phagocytize foreign substances and myelin debris. Astrocytes are additionally involved in the metabolism/detoxification of ammonia, CO_2, short-chain fatty acids, and also participate in water and pH regulation in brain. They can generate neurotrophic substances and are capable of providing neuronal guidance during development and regeneration (for comprehensive reviews on glial functions, see Hertz, 1982; Fedoroff and Vernadakis, 1986; Norenberg et al., 1988a).

In order to carry out these critical functions, astrocytes are endowed with a number of membrane receptors (to most of the known neurotransmitters in addition to benzodiazepines) and their corresponding second messenger systems (Murphy and Pearce, 1987; Kimelberg, 1988), various ion channels (Bevan et al, 1985), carrier systems and specific enzymes (Hertz, 1982). These systems largely allow astrocytes to "sense" the extracellular environment and thus generate appropriate responses to changes in the extracellular milieu.

In order to explore the significance of the astrocyte change in HE, some investigators have used bulk-isolated astrocytes derived from animals in HE (Albrecht, 1988). While there are some drawbacks associated with this approach (chiefly concerned with the purity and integrity of these fractions), so far it remains the only method for evaluating glial cells derived from an animal model of HE. Useful data have been obtained with this approach and in many instances, similar findings to those described in cell cultures have been reported.

Although cell culture represents an in vitro system with certain inherent artificial aspects, it nonetheless obviates the critical concern of bulk isolation methods and for this reason, our research group has chosen to employ this method. Astrocytes were derived from cerebral cortices of neonatal rats. In some instances, cultures were also given dibutyryl cyclic AMP (dBcAMP), an agent that induces differentiation in these cells. At 3-6 weeks of age, cells were exposed to various toxins (chiefly ammonia) and a variety of studies carried out.

EXPERIMENTAL FINDINGS

Depending on dose and time of exposure, ammonia (2-10 mM) had cytopathic effects including cytoplasmic basophilia, vacuole and dense body formation (Gregorios et al., 1985a) (Figs. 2.B,C). Prolonged exposure to high doses resulted in changes similar to the Alzheimer type II change as well as in cell death (Fig. 2.D).

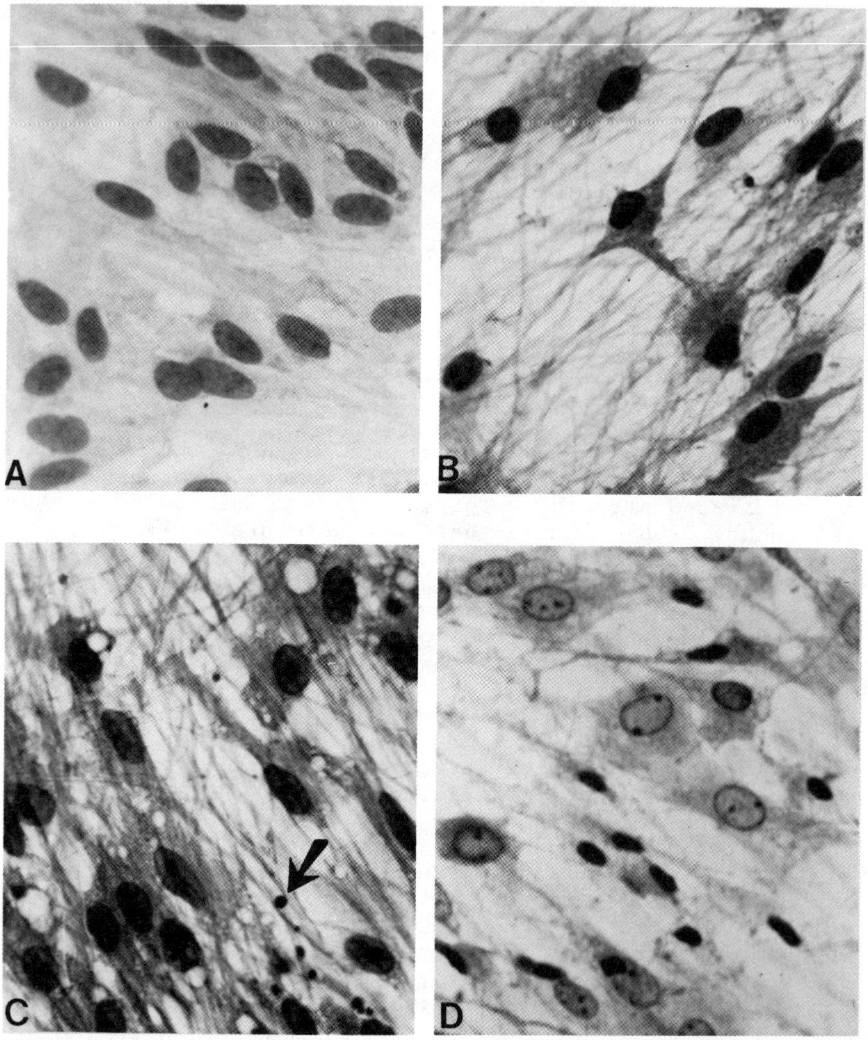

Fig. 2. A. Control astrocytes in culture. Cells have a pale cytoplasm with ill-defined borders. B. Treatment with 10 mM ammonium chloride for 1 day produces cells having a stellate appearance and dark-staining cytoplasm. C. After 4 days of ammonia exposure, the cells show prominent cytoplasmic vacuoles and dense bodies (arrow). D. Astrocytes adjacent to an area of necrosis show enlarged pale nuclei resembling Alzheimer type II cells. All x900.

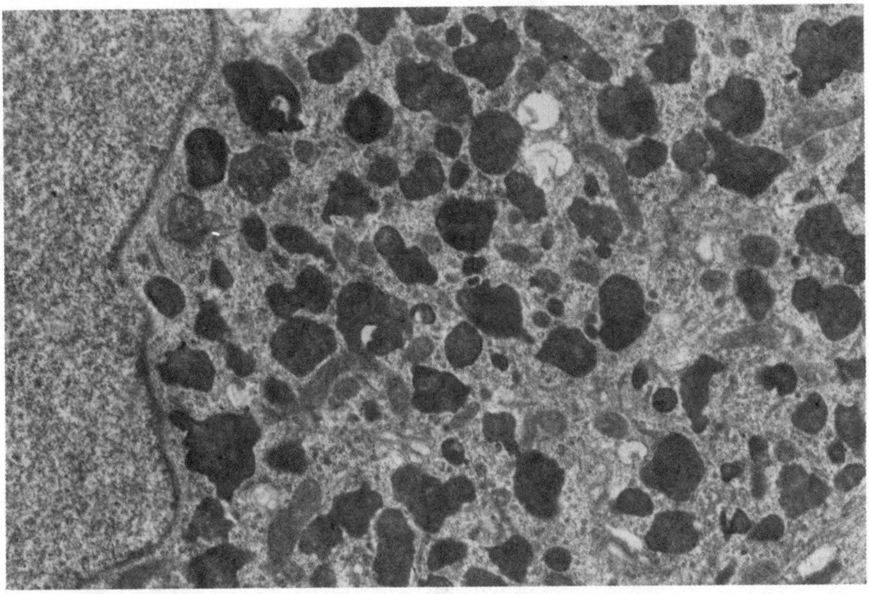

Fig. 3. Electron micrograph of an astrocyte treated with 10 mM ammonium chloride for 4 days showing numerous dense bodies. x9800.

Ultrastructural findings included a variety of degenerative alterations in organelles (Gregorios et al., 1985b). The most striking of these changes was the appearance of numerous dense bodies resembling lipofuscin pigment (Fig. 3) in keeping with findings in man and experimental animals (Voorhies et al., 1983). In addition, a loss of intermediate filaments was noted.

We subsequently showed that cells that had been maintained in dBcAMP showed significantly reduced cytotoxic effects of ammonia. Treating the cells with cyclic AMP analogues, phosphodiesterase inhibitors and beta-adrenergic agonists, all of which increase intracellular levels of cyclic AMP, also protected the cells from the ammonia-induced effects (Gregorios et al., 1986).

The effects of ammonia on the uptake of glutamate, GABA and K^+ have likewise been investigated (Norenberg et al., 1986a). Cultures treated with ammonia for 1 day showed a slight increase in the uptake of glutamate and

GABA. However, with 4 days of treatment, a decrease occurred in GABA (about 25%) and glutamate (about 13%) uptake. A decrease in the uptake of potassium, however, occurred after only one day of treatment (about a 50% reduction with 10 mM ammonia). These effects on the glial uptake of neuroactive agents could contribute to the excitability state of the CNS.

The protective effect of cyclic AMP suggested the possibility that ammonia might prevent the generation of cyclic AMP, perhaps by interfering with beta-adrenergic receptors. Baseline levels of cyclic AMP were not affected by ammonia treatment. However, beta-adrenergic-stimulated increase in cyclic AMP was reduced by 60% after 4 days of 10 mM ammonia treatment (Liskowsky et al., 1986). No changes were found in number or affinity of beta-adrenergic receptors. This suggests the possibility of a derangement in G protein or may reflect internalization of these receptors. Regardless, these findings indicate an aberration in a second messenger system. It is noteworthy that beta-adrenergic receptor blockers have been reported to precipitate HE (Tarver et al, 1983; Wiesner, 1986). While it is likely that such agents may act "peripherally", i.e., on the portal vein, one has to consider the possibility that beta-adrenergic blockers may act "centrally", i.e., on astrocytes.

A prevalent hypothesis regarding HE has been that of a derangement in monoamines, possibly secondary to an increase in the ratio of aromatic (AAA) to branched-chain (BCAA) amino acids (Fischer, 1982) or due to an alteration in amino acid transport across the blood-brain barrier (Hawkins et al., 1984). Although controversial (Conn, this volume), some studies seem to indicate that BCAAs may be beneficial in HE (Bernardini and Fischer, 1982). In our studies (Norenberg and Norenberg, 1986a), astrocytes were given amino acids (0.4-1.6 mM) for 1 day; on the second day the cells were given 10 mM ammonium chloride in addition to the amino acids. As can be seen in TABLE 1, pretreatment with BCAAs reduced the ammonia-induced cytotoxicity. Concurrent administration of BCAAs with ammonia had no protective effect. Aromatic amino acids had variable results with tryptophan exacerbating the ammonia-induced lesions and phenylalanine providing considerable protection.

While the mechanism of protection provided by BCAAs is not known, the implication of these findings is that they have a direct effect on neural cells, in addition to possibly altering the transport of amino acids into

TABLE 1

EFFECT OF AMINO ACIDS ON AMMONIA TOXICITY

	% necrosis
None (ammonia only)	80
Isoleucine	15-20
Valine	45
Leucine	50
Phenylalanine	30-35
Tyrosine	65-70
Tryptophan	90
Alanine	70

brain. It is possible that BCAAs improve brain bioenergetics by restoring the malate shuttle (Jessy and Murthy, 1985) or perhaps by activating glutamic dehydrogenase (McGiven et al., 1973). Likewise, the mechanism of exacerbation by tryptophan is not clear. It is of interest that of all amino acids measured in the CSF of patients with HE, tryptophan shows the best correlation with the level of encephalopathy (Ono et al., 1978). Tryptophan is a precursor of quinolinic acid, an excitotoxic agent that has been implicated in HE (Moroni et al., 1986). It is possible that quinolinic acid exerts a toxic effect on astrocytes.

Consistent with a disturbance in monoamines, we have recently shown a decrease in monoamine oxidase (MAO) (approximately 23% in both types A and B) (Norenberg et al., 1988b). Noteworthy, monoamine oxidase type B may be chiefly an astrocytic enzyme (Levitt et al., 1982). Since MAO is the principal mechanism for metabolizing neurotransmitter monoamines, a deficiency in astrocytic MAO may contribute to monoamine derangements in HE. A defect in MAO has been found in dogs with portacaval shunts (Faraj et al., 1980).

The main effect of cyclic AMP is to stimulate protein phosphorylation. In view of the protective effect of cyclic AMP on ammonia-induced changes in astrocytes, we investigated how protein phosphorylation might be affected by ammonia treatment. A 3-fold reduction in phosphate incorporation into a 66-kdalton protein was observed with increasing concentration of ammonia (Neary et al., 1987). Such an effect was not seen in cells that had been previously treated with dBcAMP. The nature of the protein so affected by ammonia is not known. However, since protein phosphorylation plays a key role in many brain functions, e.g. status of ion channels, receptors, enzymes (for review see Nestler and Greengard, 1984), altered protein phosphorylation may contribute to the pathogenesis of ammonia-induced encephalopathy.

An important current view implicates excessive benzodiazepine (BZD) receptor activity (Schafer et al., 1983; Baraldi et al., 1984, 1987) or the presence of endogenous BZD ligands (Mullen et al., 1988). While the focus of activity has been on the "central" BZD receptor, i.e., the one on neurons and related to the GABA-chloride ionophore complex, it should be noted that astrocytes possess BZD receptors (Bender and Hertz, 1985a), although they are of the "peripheral" type, i.e., like those seen in the heart, kidney and adrenal gland. The function of the peripheral receptor in brain is unknown but may be related to a calcium channel (Bender and Hertz, 1985b). In recent studies we have shown a 25-30% increase in receptor binding affinity without a change in receptor number (Ducis et al., 1988). Interestingly, the changes were observed only in cells which were not given any dBcAMP. The mechanism of this enhanced affinity is not known. However, ammonia has been shown to alter membrane fluidity (O'Connor et al., 1984) and such fluidity changes may affect receptor function. Preliminary investigations in our laboratory have shown increased membrane fluidity after ammonia treatment and as in the BZD receptor studies, these changes were also observed only in the cells not treated with dBcAMP.

The possibility that the astrocytic BZD receptor may be linked to a calcium channel led us to investigate the effect of ammonia on calcium flux. The uptake of calcium was measured over 1.5 minutes using 45-calcium. One day treatment with 10 mM ammonium chloride had no effect; however, after three days of treatment, a 35% reduction in calcium influx occurred. Since calcium regulates many cellular processes (Alkon and Rasmussen, 1988), such a defect may have wide-ranging effects on glial functions. This defect in calcium transport furthermore represents another second messenger abnormality occurring in ammonia-treated astrocytes.

As previously noted, fulminant hepatic failure is associated with astrocyte swelling. To determine whether ammonia causes swelling in cultured astrocytes, we employed the 3-O-methyl-glucose method of Kletzien et al. (1975) which measures the intracellular water space. One-day treatment with 10 mM ammonia caused an 11%

increase in water space, while a 4-day treatment brought about a 29% increase. How ammonia causes cell swelling is not fully understood. It may be due to an ammonia-induced inhibition in Na, K-ATPase (Norenberg et al., 1986a) or it may reflect a derangement in ion exchange mechanisms (e.g. Na^+/H^+ exchanger, anion exchanger), co-transporters or ion channels. Experiments currently in progress are aimed at exploring the latter possibilities.

One of the characteristic features of astrocytes is the presence of glial fibrillary acidic protein (GFAP), an intermediate filament found only in astrocytes. Work on human HE has shown a decrease in GFAP (Sobel et al., 1981; Kretzschmar et al., 1985; Kimura and Budka, 1986) and as noted above, ammonia caused a decrease in glial filaments in encephalopathic animals (Norenberg, 1977) and in culture (Gregorios et al., 1985b). We have quantitated GFAP by ELISA and have shown that a 1-day treatment with 10 mM of ammonium chloride results in a 25% decrease while a 4-day treatment results in a 33% loss (Norenberg et al., 1988c). As remarked by Sobel et al. (1981), HE may reflect the only human disease associated with a loss of GFAP. While the function of intermediate filaments is not known, HE may provide a probe with which to explore the functions of this protein in astrocytes.

Carnitine has been shown to exert some protective effect against ammonia toxicity (O'Connor et al., 1982).

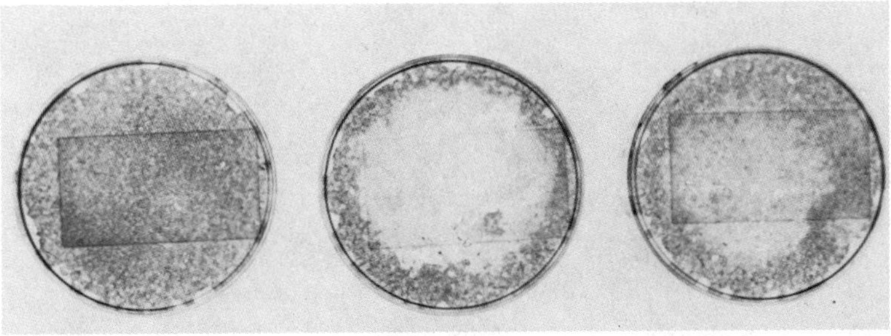

Fig. 4. Plate at left shows the control culture. The center plate has been treated with 10 mM ammonium chloride. The lucency is an indication of massive cell necrosis as most of the cells have sloughed off. The plate on the right was treated concurrently with ammonium chloride and 8 mM carnitine. While some necrosis has occurred, most of the cells remained viable.

The administration of carnitine to ammonia-treated astrocytes largely prevented the ammonia-induced cytotoxicity (Fig. 4) (Norenberg and Norenberg, 1988). Carnitine is involved in the transport of long-chain fatty acids into mitochondria (Fritz, 1959) and this action may improve the bioenergetic state of astrocytes. Additionally, carnitine can buffer potentially harmful acyl coA compounds (Roe et al., 1984). However, we have no data that ammonia treatment generates acyl coAs.

We have also investigated the effect of other toxins that have been implicated in HE. Octanoic acid (0.05-10 mM) gave rise to mitochondrial and cytoplasmic swelling and pallor, while phenol (0.1-10 mM) caused a marked loss of ribosomes (Norenberg et al., 1986b,c) These changes were different from that produced by ammonia. The effect of these agents on the BZD receptor was also different in that octanoic acid had no effect but phenol increased binding affinity while at the same time decreasing receptor number (unpublished observations).

An important concept relative to toxins is that of

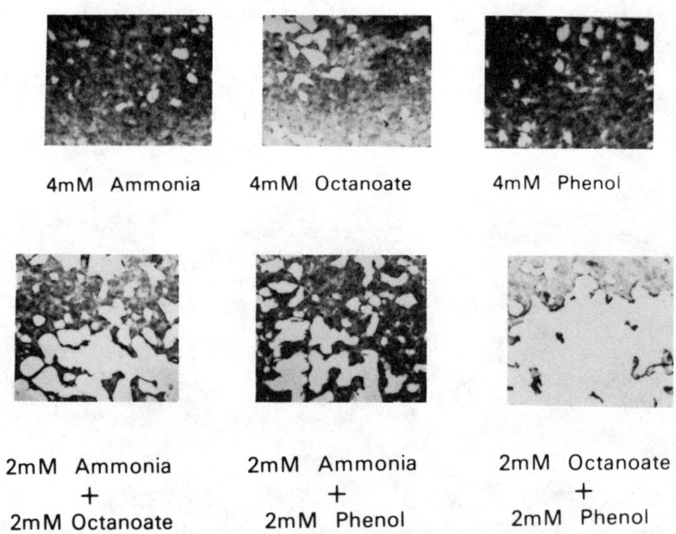

Fig. 5. Astrocyte cultures were exposed either to a single toxin or combination of toxins. Note the greater extent of necrosis when combinations of toxins even at low concentrations were added. Gaps in the plates are indicative of cell necrosis.

synergism (Zieve, 1981). To investigate whether such a phenomenon could be observed in culture, we exposed astrocytes to varying concentrations of toxins (Norenberg and Norenberg, 1986b). As shown in Fig. 5, synergism could be demonstrated in this system.

SUMMARY

The use of primary astrocyte cultures has been extremely useful in exploring the effects of various putative toxins. This system closely reproduces the morphologic changes seen in vivo and thus we feel that the biochemical and functional derangements observed in these cells in culture reflect events in vivo. Such in vivo phenomena as synergism, response to BCAAs and carnitine can be observed in this system. Moreover, many of the presently invoked pathogenetic mechanisms can potentially be explained by astroglial involvement: the effects of ammonia and other toxins, the role of monoamines, and BZD receptor abnormalities. We believe that HE represents a gliopathy and that further progress in the characterization of functional deficits in astrocytes should yield a better understanding of the pathogenesis of HE and perhaps provide novel therapeutic approaches.

ACKNOWLEDGMENTS

I wish to thank my colleagues whose contributions made this article possible: Lee W. Mozes, Jocelyn B. Gregorios, David R. Liskowsky, Joseph T. Neary, Ilze Ducis and Luz-Oliva B. Norenberg. I am most grateful to Celeste Bowen for her patience, persistence and careful preparation of the manuscript.

REFERENCES

Albrecht J., Hilgier W., Lazarewicz J.W., Rafawolska U., and Wysmyk-Cybula U. (1988) Astrocytes in acute hepatic encephalopathy: metabolic properties and transport functions, in The Biochemical Pathology of Astrocytes (Norenberg M.D., Hertz L., and Schousboe A., eds.) pp. 465-476, Alan R. Liss, New York.

Alkon D.L., and Rasmussen H. (1988) A spatial-temporal model of cell activation. Science 239, 998-1005.

Baraldi M., Zeneroli M.L., Ventura E., Penne A., Pinelli G., Ricci P., and Santi M. (1984) Supersensitivity of benzodiazepine receptors in hepatic encephalopathy due to fulminant hepatic failure in the rat: reversal by benzodiazepine antagonist. Clin. Sci. 67, 167-175.

Baraldi M., Zeneroli M.L., Ventura E., and Vezzelli C. (1987) An increase in cerebral benzodiazepine receptors induced by a subacute administration of ammonia, mercaptans and short-chain fatty acids in rats. Clin. Sci. 73, 669-671.

Bender A.S., and Hertz L. (1985a) Binding of (3H) RO5-4864 in primary cultures of astrocytes. Brain Res. 341, 41-49.

Bender A.S., and Hertz L. (1985b) Pharmacological evidence that the non-neuronal diazepam binding sites in primary cultures of glial cells is associated with a calcium channel.. Europ. J. Pharm. 110, 287-288.

Bernardini P., and Fischer J.E. (1982) Amino acid imbalance and hepatic encephalopathy. Annu. Rev. Nutr. 2, 419-454.

Bevan S., Chiu S.Y., Gray P.T.A., and Ritchie J.M. (1985) The presence of voltage-gated sodium, potassium and chloride channels in rat cultured astrocytes. Proc. R. Soc. London 225, 299-313.

Cremer J.E., Heath D.F., Teal H.M., Woods M.S., and Cavanagh J.B. (1975) Some dynamic aspects of brain metabolism in rats given a portocaval anastomosis. Neuropathol. Appl. Neurobiol. 1, 293-311.

Ducis I., Norenberg L.O.B., and Norenberg M.D. (1988) Effect of ammonia on the benzodiazepine receptor in cultured astrocytes. Soc. Neurosci. Abst. 14, 1082.

Faraj B.A., Camp V.M., Ansley J., Ali F.M., and Malveaux E.J. (1980) Impaired monoamine oxidase activity in dogs with portacaval shunt. Biochem. Pharm. 29, 28-31.

Fedoroff S., and Vernadakis A. (eds.) (1986) Astrocytes: Cell Biology and Pathology of Astrocytes, Academic Press,Orlando.

Fischer J.E. (1982) Amino acids in hepatic coma. Digest. Dis. Sci. 27, 97-102.

Fritz I.B. (1959) Action of carnitine on long-chain fatty oxidation by liver. Am. J.Physiol. 197, 297-304.

Gregorios J.B., Mozes L.W., Norenberg L.O.B., and Norenberg M.D. (1985a) Morphologic effects of ammonia on primary astrocyte cultures. 1. Light microscopic studies. J. Neuropathol. Exp. Neurol. 44, 397-403.

Gregorios J.B., Mozes L.W., and Norenberg M.D. (1985b) Morphologic effects of ammonia on primary astrocyte cultures. II. Electron microscopic studies. J. Neuropathol. Exp. Neurol. 44, 404-414.

Gregorios J.B., Mozes L.W., Norenberg L.O.B., and Norenberg M.D. (1986) Effect of cyclic AMP on ammonia-induced alterations in primary astrocyte cultures. J. Neuropath. Exp. Neurol. 45, 396-407.

Hawkins R.A., Mans A.M., and Biebuyck J.F. (1984) Blood-brain barrier dysfunction in hepatic encephalopathy, in Advances in Hepatic Encephalopathy and Urea Cycle Diseases pp. 251-260, Karger, Basle.

Hertz L. (1982) Astrocytes, in Handbook of Neurochemistry, 2nd ed. (Lajtha A., eds.) vol.1, pp. 319-355, Plenum Press, New York.

Jessy J., and Murthy C.R.K. (1985) Elevation of transamination of branched chain amino acids in brain in acute ammonia toxicity. Neurochem. Int. 7, 1027-1031.

Kimmelberg H.K.(1988) Glia Cell Receptors, Raven Press, New York.

Kimura T., and Budka H. (1986) Glial fibrillary acidic protein and S-100 protein in human hepatic encephalopathy; immunocytochemical demonstration of dissociation of two glia-associated proteins. Acta Neuropathol. (Berl) 70, 17-21.

Kletzien R.F., Pariza M.W., Becker J.E., and Potter V.R. (1975) A method using 3-0-methyl-D-glucose and phloretin for the determination of intracellular water space of cells in monolayer culture. Anal. Biochem. 68, 537-544.

Kretzschmar H.A., DeArmond S.J., and Forno L.S. (1985) Measurement of GFAP in hepatic encephalopathy by ELISA and transblots. J. Neuropath. Exp. Neurol. 44, 459-471.

Levitt P., Pintar J.E., and Breakfield X.O. (1982) Immunocytochemical demonstration of monoamine oxidase B in brain astrocytes and serotonergic neurons. Proc. Natl. Acad. Sci. USA 79, 6385-6389.

Liskowsky D.R., Norenberg L.O.B., and Norenberg M.D. (1986) Effect of ammonia on cyclic AMP production in primary astrocyte cultures. Brain Res. 386, 386-388.

McGiven J.D., Bradford N.M., Crompton M., and Chappel J.B. (1973) Effect of L-leucine on the nitrogen metabolism of isolated rat liver mitochondria. Biochem. J. 134, 209-215.

Moroni F., Lombardi G., Carl V., Lal S., Etienne P., and Nair N.P. (1986) Increase in the content of quinolinic acid in cerebrospinal fluid and frontal cortex of patients with hepatic failure. J. Neurochem. 47, 1667-71.

Mullen K.D., Martin J.V., Mendelson W.B., Bassett M.L., and Jones E.A. (1988) Could an endogenous benzodiazepine ligand contribute to hepatic encephalopathy? Lancet i, 457-459.

Murphy S., and Pearce B. (1987) Functional receptors for neurotransmitters on astroglial cells. Neuroscience 22, 381-394.

Neary J.T., Norenberg L.O.B., Gutierrez M.P., and Norenberg M.D. (1987) Hyperammonemia causes altered protein phosphorylation in astrocytes. Brain Res. 437, 161-164.

Nestler E., and Greengard P.(1984) Protein phosphorylation in the nervous system, John Wiley, New York.

Norenberg M.D. (1977) A light and electron microscopic study of experimental portal systemic (ammonia) encephalopathy. Lab. Invest. 36, 618-627.

Norenberg M.D. (1981) The astrocyte in liver disease, in Advances in Cellular Neurobiology (Fedoroff S., and Hertz L., eds.) vol.2, pp. 303-352, Academic Press, New York.

Norenberg M.D. (1987) The role of astrocytes in hepatic encephalopathy. Neurochem. Pathol. 6, 13-33.

Norenberg M.D., and Martinez-Hernandez A. (1979) Fine structural localization of glutamine synthetase in astrocytes of rat brain. Brain Res. 161, 303-310.

Norenberg L.O.B., and Norenberg M.D. (1986a) Effect of amino acids on ammonia toxicity in primary astrocyte cultures. Soc. Neurosci. Abst. 12, 269.

Norenberg L.O.B., and Norenberg M.D. (1986b) Synergism of hepatic coma toxins in primary astrocyte cultures. Proc. Internat. Cong. Neuropathol. 10, 48.

Norenberg M.D., and Norenberg L.O.B. (1988) Protective effects of carnitine on ammonia-treated astrocytes in primary culture. J. Neuropathol. Exp. Neurol. 47, 325.

Norenberg M.D., Mozes L.W., Norenberg L.O.B., and Gregorios J.B. (1986a) Effects of ammonia on primary astrocyte cultures: morphological and biochemical considerations, in Dynamic Properties of Glial Cells - Cellular and Molecular Aspects (Grisar T., Franck G., and Hertz L., Norton W.T., Sensenbrenner M., and Woodbury D., eds.) pp.353-362, Pergamon Press, Oxford.

Norenberg M.D., Norenberg L.O.B., and Gregorios J.B. (1986b) The effect of octanoic acid on astrocytes in primary cultures. J. Neuropath. Exp. Neurol. 45, 365.

Norenberg M.D., Norenberg L.O.B., and Gregorios J.B. (1986c) The effect of phenol on astrocytes in primary culture. Proc. Internat. Cong. Neuropathol. 10, 48.

Norenberg M.D., Hertz L., and Schousboe A. (eds.) (1988a) The Biochemical Pathology of Astrocytes, Alan R Liss, New York.

Norenberg M.D., Forster E.H., Norenberg L.O.B., and Neary J.T. (1988b) Monoamine oxidase activity in cultured astrocytes treated with ammonia. Trans. Amer. Soc. Neurochem. 19, 237.

Norenberg M.D., Norenberg L.O.B., and Neary J.T. (1988c) Ammonia induced decrease in glial fibrillary acidic protein (GFAP) in cultured astrocytes. Soc. Neurosci. Abst. 14, 1082.

O'Connor J.E., Costell M., and Grisolia S. (1982) Prevention of ammonia toxicity by L-carnitine: metabolic changes in brain. Neurochem. Res. 9, 563-570.

O'Connor J.E., Guerri C., and Grisolia S. (1984) Effects of ammonia on synaptosomal membranes. Biochem. Biophys. Res. Comm. 119, 516-523.

Ono J., Hutson D.G., Dombro R.S., Levi J.U., Livingstone A., and Zeppa R. (1978) Tryptophan and hepatic coma. Gastroenterology 74, 196-200.

Roe C.R., Millington C.S., Maltby D.A., Bohan T.P., and Hoppel C.L. (1984) L-carnitine enhances excretion of propionyl coenzyme A as propionyl carnitine in propionic acidemia. J. Clin. Invest. 73, 1785-1788.

Schafer D.F., Fowler J.M., Munson P.J., Thakur A.K., Waggoner J.G., and Jones E.A. (1983) Gamma-aminobutyric acid and benzodiazepine receptors in an animal model of fulminant hepatic failure. J. Lab. Clin. Med. 102, 870-880.

Sobel R.A., DeArmond S.J., Forno L.S., and Eng L.F. (1981) Glial fibrillary acidic protein in hepatic encephalopathy. An immunohistochemical study. J. Neuropathol. Exp. Neurol. 40, 625-632.

Tarver D., Walt R.P., Dunts A.A., Jenkins W.J., and Sherlock S. (1983) Precipitation of hepatic encephalopathy by propranolol in cirrhosis. Brit. Med. J. 287, 585.

Traber P.G., DalCanto M., Ganger D., and Blei A.T. (1987) Electron microscopic evaluation of brain edema in rabbits with galactosamine-induced fulminant hepatic failure: Ultrastructure and integrity of the blood-brain barrier. Hepatology 7, 1272-1277.

Voorhies T.M., Ehrlich M.E., Duffy T.E., Petito C.K., and Plum F. (1983) Acute hyperammonemia in the young primate. Physiologic and neuropathological correlates. Pediatric. Res. 17, 970-975.

Wiesner R.H. (1986) Does propranolol precipitate hepatic encephalopathy? J. Clin. Gastroenterol. 8, 74-76.

Zieve L. (1981) The mechanism of hepatic coma. Hepatology 1, 360-365.

BRAIN EDEMA
IN EXPERIMENTAL FULMINANT HEPATIC FAILURE

Andres T. Blei MD and Peter G. Traber MD. Lakeside VAMC & Northwestern Univ, Chicago, Il 60611 and Ann Arbor VAMC & Univ. of Michigan, Ann Arbor, MI 48105

Brain edema and intracranial hypertension are fatal complications of fulminant hepatic failure (FHF) (1). Demonstration of temporal lobe or uncal herniation at autopsy indicate that they represent a leading cause of death in this disease (2). Catalogued as "complications" of acute liver failure (3), few studies have addressed their pathogenesis. Studies of the mechanisms that bring about both edema and intracranial hypertension are difficult to carry out in these critically ill patients. In this regard, hepatologists tend to equate brain edema with intracranial hypertension; while edema relates to an increase in brain water, intracranial hypertension to a rise in pressure within the skull. In this presentation, we will review our work in the experimental animal, an approach that allowed us to separately explore each aspect of this disturbance.

A. <u>Background</u>: Conceptually (4), brain edema can arise from two mechanisms. In a cytotoxic injury , changes in osmolarity or toxic effects on membrane pumps result in cellular swelling; this type of edema will initially predominate in the cellular gray matter. Vasogenic edema will result from a disruption of the blood-brain barrier and will initially be seen in the less densely packed white matter. Although in clinical medicine this classification may be somewhat arbitrary, it emphasizes the site of initial injury, an important consideration in FHF.

The permeability of the barrier in human acute liver

failure has not been studied. In the experimental animal, conflicting results have been obtained. While gross abnormalities have been detected in the hepatectomized rat in a premoribund stage (5), more subtle changes have been shown in the rabbit with galactosamine-induced FHF using autoradiography of alphaaminoisobutyric acid (6). Such changes have not been reproduced in rats with galactosamine-induced FHF using both the previous marker (7) and GABA (8). On the other hand, the same laboratory that reported changes in the permeability of the barrier in rats (9) also noted a decrease in Na+-K+ ATPase activity of rat brain membranes after the addition of sera from FHF patients (10). Elements of both mechanisms could be operative; even in cytotoxic injury a net increase in water must occur for brain edema to develop.

While few studies have measured brain water content in experimental models of FHF, their results are not homogenous. In the hepatectomized rat, Livingstone et al showed a 5.4% increase in the water content of cortical gray matter at a preterminal stage (11). In the rat with galactosamine-induced FHF, Grofflin et al had only noticed an increase in water of the cerebellum (12), while Dixit et al observed a more diffuse increase in the same model (13). Zieve reported no changes of brain water in the rat after hepatic devascularization (14).

B. <u>The measurement of brain water</u>: How to acount for these differences ? Variability of experimental models and/or intervals between study and demise of the animal could be influencing the reviewed results. All these studies employed the dry-weight technique, where water content is measured from the difference of tissue wet and dry weights. In addition to being insensitive, this technique cannot precisely determine regional variations in water content. Small tissue samples are difficult to process; discrepancies in the previous reports could have originated in part from the use of this method.

The gravimetry technique was developed by Marmarou and colleagues to overcome these difficulties (15). In brief, a density gradient column is constructed with two solutions of different specific gravity. The column is calibrated to assure a linear relation between depth and specific gravity. Specimens of brain tissue as small as 10 mg can then be placed on the column; its equilibration point after 2 minutes reflects the specific gravity of the tissue, which is directly related to its water content (16).

The gravimetry technique requires tissue sampling; noninvasive methods would be ideal to monitor changes in water content. It is possible that Magnetic Resonance may provide such information in the future (17). Feasibility of this technique is limited at the present time to anesthetized preparations; unfortunately, animals with FHF tolerate poorly any kind of anesthesia.

C. Studies in the rabbit with galactosamine-induced FHF: We used the gravimetry technique to determine brain water content in animals with galactosamine-induced FHF at different stages of encephalopathy (17a). Changes in objective signs developed a mode of 20 hours after the injection of 4.25 mM/kg of galactosamine to white New Zealand rabbits. Although five stages have been described in the progression of encephalopathy (18), we chose to study animals at two defined points in their course: during mild encephalopathy (n=6), when rabbits were unable to maintain head posture and severe encephalopathy (n=6), when they were unable to right themselves up.

Controls (n=6) included animals injected with dextrose vehicle. An additional group of galactosamine-injected animals (n=6) had evidence of liver injury (with similar elevation of transaminases to the encephalopathic groups) but no signs of encephalopathy. In all groups, an arterial catheter placed in an ear vessel allowed measurements of arterial pressure and heart rate, together with blood sampling. At the time of study, animals were sacrificed with a pentobarbital overdose and the brains quickly removed; 10 mg samples from different regions were systemically obtained from sliced brains and applied to a bromobenzene-kerosene density gradient.

Water content in gray matter of cerebral cortex (figure 1) rose in parallel with the development of encephalopathy; neither hypoglycemia, hypoxia or hypotension accounted for this finding. Unchanged water values were observed in rabbits with hepatic injury but without encephalopathy, a "de facto" second control group. The hippocampus, another area rich in gray matter, also exhibited a significant rise. The noted increase in brain water could be estimated to result in up to a 30% rise in brain volume of the affected area. No significant changes were observed in the water content of white matter of all 4 groups.

To further explore the permeability of the blood-brain barrier, rabbits were injected with horseradish

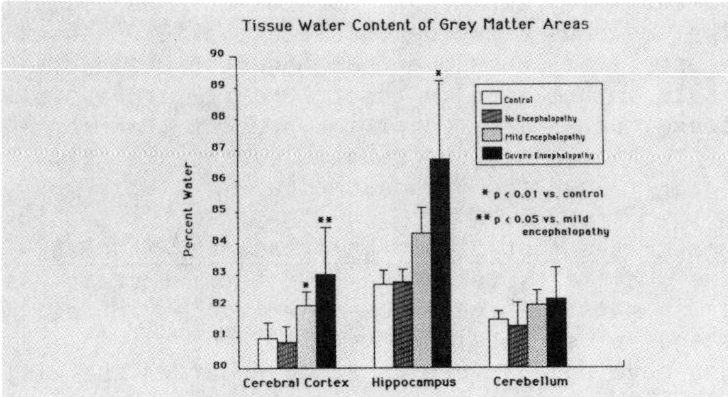

Fig 1. Tissue H20 in gray matter areas. Controls received D5W; the 2nd group developed liver injury but no encephalopathy. $\bar{x} \pm$ SD. Reproduced from ref 17a with permission from the publishers.

peroxidase, an electron-dense protein of MW 44000, to which the cerebral endothelial cell is normally impermeable (19). After perfusion-fixation of the brain, no evidence of horseradish peroxidase could be observed within endothelial cells or brain parenchyma (19a)

Qualitative morphological studies revealed striking changes in the astrocytes of the cerebral cortex (figure 2). These were characterized by marked hydropic changes

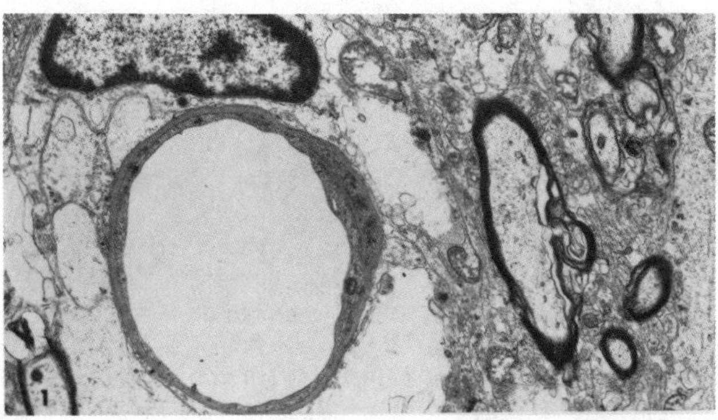

EM showing a capillary vessel surrounded by greatly dilated astrocytic foot processes (x18,000). Reproduced from ref 19a with permission from the publishers.

with vacuolation of the perinuclear cytoplasm, pericapillary and periaxonal foot-processes. Neurons appeared normal; no gross changes could be detected within white matter.

Taken as a whole, these studies describe the development of brain edema in a well characterized model of FHF, its progression with the development of encephalopathy, and suggest that cortical astrocytes may be the preferential site for cytotoxic injury and resultant water accumulation; the blood-brain barrier appeared grossly intact.

D. Studies in the rat after hepatic devascularization: We decided to study a conceptually different model of experimental FHF, the rat after hepatic devascularization. Although this model can be criticized for its lack of potential recovery, it does overcome the possibility of a direct effect of the injected drug or toxin to the brain.

This model has been extensively used for the study of hepatic encephalopathy (20). A two-stage devascularization is acomplished by a portacaval shunt followed in 24-48 hs by a hepatic artery ligation. In our preparation, at the time of the second surgery a chronic catheter was placed in a femoral artery and a thermistor wire placed between intestinal loops to measure core body temperature; both exited from the animal's back. The need for accurate temperature measurements was prompted by initial observations that indicated marked hypothermia in devascularized animals; this had been noted in a conceptually similar preparation, the hepatectomized rat (11).

Adult male S/D rats were divided into 3 groups (n=5 per group): portacaval shunt (PCS) + sham hepatic artery ligation (HAL), PCS + HAL with animals at room temperature (also labeled IFHF-A) and PCS + HAL with animals at controlled body temperature (IFHF-B); in the latter group, animals were kept in an incubator. Encephalopathic rats (the latter two groups) were studied at the time of stage IV encephalopathy, defined by Zieve (20) as the inability of animals to right themselves up; animals of the PCS group were studied at a time equidistant from both devascularized groups.

Table 1 shows changes in several variables at the time of determination of brain water. Marked hypothermia was seen in the devascularized group kept at room temperature, with a corresponding adjustment of arterial pressure and heart rate to this change. Time to stage IV was significantly prolongued in the "cold" animals.

Table 1 - Studies in rats after hepatic devascularization

	(C) PCS	(IFHF-A) PCS + HAL	(IFHF-B) PCS + HAL
Body temp(C)	37.5±0.4	26.9±2.8*	38.4±1.6
MAP (mmHg)	89±12	60±25**	140±18
HR(beats/min)	430±48	213±102*	412±55
Hours to s.IV	11.9±4.2	17.0±3.1**	9.4±1.0

mean ± SD. *p<0.01 vs. both groups. **p<0.01 vs IFHF-B.

Due to the small size of the rat's white matter, only cortical gray matter was evaluated with the gravimetry technique. The water content of cortical gray matter was significantly increased in the devascularized rats with normal body temperature (figure 3); the small size of white matter precluded sampling from this area. Animals with portacaval shunt and sham HAL (group C) did not differ in their water content from 6 normal rats (80.1 ± 0.3%).

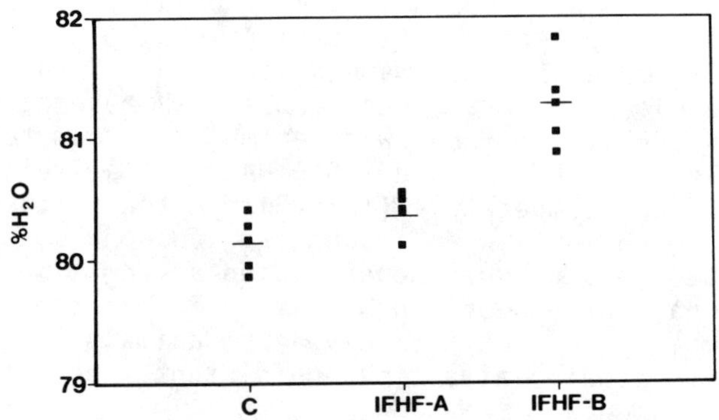

Fig 3. Individual values show a clear increase in "warm" devascularized rats (IFHF-B) vs "cold" (IFHF-A) or PCA alone (group C). Reproduced from ref 20a with permission from the publishers.

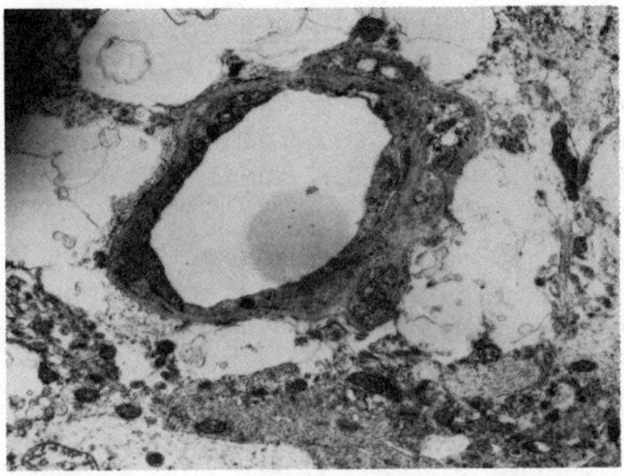

Fig 4. Marked swelling of pericapillary astrocyte foot processes in a warm" devascularized rat. Dark vesicles containing horseradish peroxidase are seen in the capillary endothelial cell. (x 18,000)

Astrocyte swelling was again observed in the cortical gray matter. Different from the observation in the rabbit, vesicles containing horseradish peroxidase were now observed in the cerebral endothelial cell in all three groups (fig 4). This had been previously seen after portacaval anastomosis (21) and interpreted as evidence of a breakdown of the blood-brain barrier. Recent studies, however, have raised the possibility that these vesicles are part of a system of tubular invaginations of the cell membrane that in a two-dimension projection may appear as an intracytoplasmic vesicle (22). Thus, it is unclear whether the barrier is abnormally permeable in this model; reported changes in hepatectomized rats (11) occurred in hypothermic, premoribund animals.

Our studies in two experimental models of FHF suggest that cytotoxic injury to astrocytes may be a key factor in the pathogenesis of brain edema. The involvment of this cell in neuropathological studies of both experimental (23) and human (24) chronic hepatic encephalopathy raise the possibility that the unique development of massive astrocyte swelling in FHF may be related to the rapid rate of rise of cytotoxin(s) in the brain, which result in water accumulation in acute but not chronic liver failure. We propose that brain edema may be part of the "spectrum" of hepatic encephalopathy rather than a distinct "complication" of FHF.

E. <u>Pathogenesis of the cytotoxic injury</u>: If one accepts the observation that cortical astrocytes are a site for water accumulation, unique features of the metabolic activities of this cell (24a) could lead to plausible hypotheses on the mechanisms for brain edema. Amongst them, ammonia metabolism should be reexamined.

Glutamine synthetase, enzyme which catalyzes the incorporation of ammonia into glutamate to form glutamine, is present in the central nervous system only within astrocytes; carbonic anhydrase is also present in this cell, catalyzing the formation of HCO3- and H+ from CO2 (of neuronal origin) and H2O. VanGelden has proposed a cycle of entry and exit of water from the astrocyte based on these two enzymes. In this scheme (figure 5), Cl-/HCO3- exchange results in entry of a molecule of chloride (and thus Na+) into the cell, which is accompanied by a molecule of water; the presence of Na+-K+ ATPase assures the exchange of Na+ for K+. On the other hand, exit of glutamine from the cell into the interstitial space is osmotically accompanied by a molecule of water, a process which may be important in the formation of CSF.

Ammonia will result in astrocyte swelling when injected into young primates (26). Although it has not been shown to increase water content of the brain in short-term

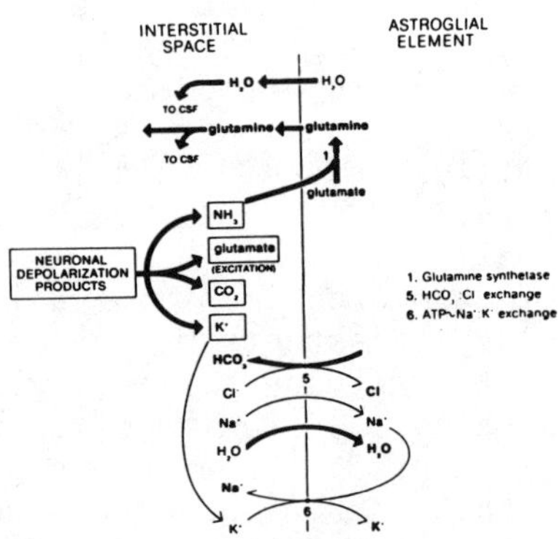

Fig 5 - Proposed cycle of water entry and exit from astrocytes. Reproduced from ref. 25 with permission from the author and publishers.

infusions in rabbits (27), it does cause swelling of brain slices (28) at concentrations that approximate those seen in experimental FHF (29). It will result in hydropic changes of isolated astrocyte preparations (30). Although the precise site where ammonia could exert its edematogenic effect has not been delineated, an increase in glutamine has been proposed as a cause of brain edema in conditions associated with hyperammonemia (31).

Alternative views emphasize inhibition of Na+-K+ ATPase as the likely site for a cytotoxic effect (1). This membrane pump is directly involved in maintenance of cellular osmolarity; a reduced brain activity of this enzyme has been shown in rats with galactosamine-induced FHF (32). Digoxin-like immunoreactivity has been shown in the plasma of patients with FHF (33), an activity that may be responsible for the inhibition of rat brain membrane Na+-K+ ATPase seen in advanced coma (10). If cytotoxic injury is mainly of one cell, the astrocyte, it would be of great interest to determine this enzymatic activity in this structure. In a study in thioacetamide-induced FHF in rats, Na+-K+ ATPase activity in astrocyte-enriched fractions was actually increased (34). Further work in this area is urgently needed.

The development of brain edema may well be multifactorial and the assumption that one cytotoxin plays a pathogenic role does not rule out the effects of others. Additional studies using cortical brain slices, primary astrocyte cultures, cocultures of astrocyte and glial cells or glioma cell lines may provide additional insight into this problem.

F. Intracranial hypertension in experimental FHF: Few studies have measured intracranial pressure in experimental FHF. In the devascularized pig, Hanid et al (35) showed a progressive rise in ICP until death; the animals were kept anesthetized after the devascularization procedure and results are thus difficult to interpret. In our study in the rabbit with galactosamine-induced FHF, a cannula was inserted into the lateral ventricle using stereotaxic coordinates and animals allowed to recover for 7 days. Once galactosamine was injected and encephalopathy developed, intracranial pressure was continously monitored in the abscence of anesthesia.

In early stages of encephalopathy in the rabbit, intracranial pressure does not rise although brain edema is already detectable (figure 6). Pressure rises at later

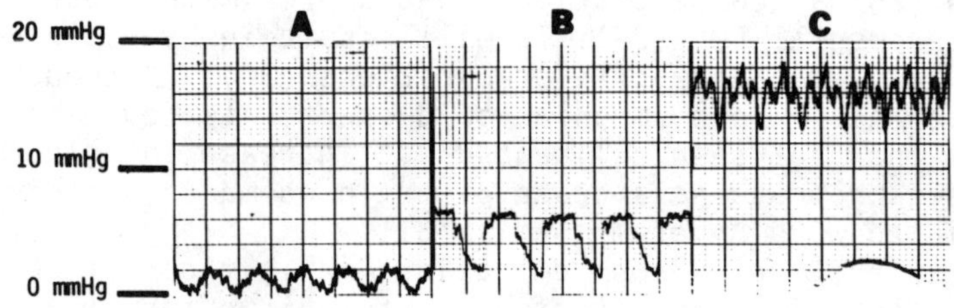

Fig 6. Tracing of panel A is of a normal rabbit; B: mild encephalopathy, with greater respiratory excursions; C: severe encephalopathy, with intracranial hypertension (>9 mmHg). Reproduced from ref 17a with permission.

stages, when brain edema is marked. As observed in humans with FHF (36), the elevations may occur intermittently, the so called "A waves" (37). When intracranial pressure approaches mean arterial pressure, cerebral perfusion pressure (arterial - intracranial pressure) falls to a critical level, resulting in additional ischemia. Death may occur at this stage.

An elevation of intracranial pressure may occur as a result of an increase in the volume of cerebral tissue, brain blood volume, the volume of cerebrospinal fluid, or a combination of these factors (38). A role for changes in blood volume are of particular interest, as an increase in cerebral blood flow has been noted in humans with FHF (39). This cerebral hyperemia may be part of a generalized hemodynamic picture seen in FHF, consistent with the so called "hyperdynamic circulation": elevated cardiac output, increased regional blood flows and decreased peripheral resistance (40). Hemodynamic studies in experimental FHF are difficult to perform but this is an area of importance for future studies.

G. Future directions: Ellucidation of the pathogenesis of brain edema and intracranial hypertension, leading causes of death in human FHF, will continue to require studies in the experimental animal in order to control the multiple variables present in this disease. Every animal model has shortcomings; thus, an observation in one preparation should be strengthened by findings in another. Magnetic Resonance may be a non-invasive method to study water accumulation. In vitro studies in brain slices or isolated cells will provide insight into pathogenic mechanisms. Together with in vivo monitoring of intracranial pressure, this strategy should allow testing of new approaches to treat this lethal problem.

REFERENCES

1. Ede RS, Williams RJ. Hepatic encephalopathy and cerebral edema. Sem Liv Dis 6:352-79, 1986.
2. Gazzard BG, Portmann B, Murray-Lyon IM, et al. Causes of death in fulminant hepatic failure and relationship to quantitative histological assessment of parenchymal damage. Q J Med 64: 615-26, 1975.
3. Jones EA, Schafer DF. Fulminant hepatic failure in Zakim D, Boyer TA, eds. Hepatology. Philadephia; WB Saunders 1981: 415-445.
4. Klatzo I. Neuropathological aspects of brain edema. J Neuropathol Exp Neurol 26: 1-14, 1967.
5. Potvin M, Morrison HF, Hinchey EJ, et al. Cerebral abnormalities in hepatectomized rats with acute hepatic coma. Lab Invest 50: 560-4, 1984.
6. Horowitz ME, Schafer DF, Molnar P, et al. Increased blood-brain transfer in a rabbit model of fulminant hepatic failure. Gastroenterology 84:1001-11, 1983.
7. Lo WD, Ennis SR, Goldstein GW, Betz AL. The effects of galactosamine-induced hepatic failure upon blood-brain barrier permeability. Hepatology 7:452-6, 1987.
8. Knudsen GM, Poulsen HE. Blood-brain barrier permeability in galactosamine-induced hepatic encephalopathy. No evidence for increased GABA transport. J Hepatol 6: 187-92, 1988.
9. Zaki AEO, Wardle EN, Canalese J, et al. Potential toxins in acute liver failure and their effects on blood-brain barrier permeability. Experientia 39: 988-91, 1983
10. Seda HWM, Hughes RD, Gove CD, et al. Inhibition of rat brain Na+-K+ ATPase activity by serum from patients with fulminant hepatic failure. Hepatology 4: 74-9. 1984.
11. Livingstone AS, Potvin M, Goresky CA, et al. Changes in the blood-brain barrier in hepatic coma after hepatectomy in the rat. Gastroenterology 74: 664-71, 1977.
12. Grofflin UB, Tholen V. Brain edema in the rat with

galactosamine-induced severe hepatitis. Experientia 34:1501-3, 1978.

13. Dixit V, Chang TMS. Effects of prostaglandin E2 on brain edema and liver histopathology in a galactosamine-induced fulminant hepatic failure rat model. Biomat., Art. Cells, Art. Org, 15:559-73, 1987.
14. Zieve L. Reply to Mullen, KD. Hepatology 7:618, 1987.
15. Marmarou A, Poll W, Shulman K, et al. A simple gravimetric technique for measurement of cerebral edema. J Neurosurg 49:5307, 1978.
16. Nelson SR, Mantz ML, Maxwell JA. Use of specific gravity in the measurement of cerebral edema. J Appl Physiol 30:268-71, 1971.
17. Haida M, Yamamoto M, Matsumura H, et al. Intracellular and extracellular spaces of normal adult rat brain determined from the proton nuclear magnetic relaxation times. J Cerebr Blood Flow Metab 7:552-6, 1987.

17a. Traber PG, Ganger DR, Blei AT. Brain edema in rabbits with galactosamine-induced hepatic failure. Regional differences and effects on intracranial pressure. Gastroenterology 91:1347-56, 1986

18. Schafer DF, Pappas SC, Brody LE, et al. Visual evoked potentials in a rabbit model of hepatic encephalopathy: sequential changes and comparisons with drug-induced comas. Gastroenterology 86:540-5. 1984.
19. Bradbury M. The blood-brain barrier: Transport across the cerebral endothelium. Circ Res 57: 213-22, 1985.

19a. Traber PG, DalCanto M, Ganger DR, Blei AT. Electron microscopic evaluation of brain edema in rabbits with galactosamine-induced fulminant hepatic failure: Ultrastructure and integrity of the blood-brain barrier. Hepatology 7: 1272-77, 1987

20. Zieve L, Doizaki WM, Zieve FJ. Synergism between mercaptans and ammonia or fatty acids in the production of coma: A possible role for mercaptans in the pathogenesis of hepatic coma. J Lab Clin Med 83:16-28, 1974.

20a. Traber PG, DalCanto M, Ganger DR, Blei AT. Effect of body temperature on brain edema and encephalopathy in the rat after hepatic devascularization. Gastroenterology 96:885-81, 1989

21. Laursen H, Westergaard E. Enhanced permeability to horseradish peroxidase across cerebral blood vessels in the rat after portacaval anastomosis. Neuropathol Appl Neurobiol 3: 29-43, 1977.
22. Bundgaard M. Pathways across the vertebrate blood-

brain barrier: morphological viewpoints. Ann NY Acad Sci 481: 7-19, 1986.

23. Norenberg MD, Lapham LW. The astrocyte response in experimental portal-systemic encephalopathy: an electron microscopic study. J Neurop Exp Neurol 33:422-35,1974.
24. Martin H, Voss K, Hufnagl P, et al. Morphometric and densitometric investigations of protoplasmic astrocytes and neurons in human hepatic encephalopathy. Exp Pathol 32:241-50, 1987.

24a. Norenberg MD. The astrocyte in liver disease in Fedoroff S, Hertz L, eds. Adv in cellular neurobiology. New York, Academic Press, pp 303-52, 1981.

25. Van Gelden NM. Metabolic interactions between neurons and astroglia: Glutamine synthetase, carbonic anhydrase and water balance in Basic Mechanisms of Neuronal Hyperexcitability. Alan Liss 1983, New York. pp 5-29.
26. Voorhies TM, Erlich ME, Duffy TE, et al. Acute hyperammmoniemia in the young primate: physiologic and neuropathologic correlates. Pediatric Res 17: 970-5, 1983.
27. Kindt GW, Brock M, Altenau LL, Poll E. Blood brain barrier and brain oedema in ammonia intoxication. Lancet 1:201, 1977.
28. Benjamin AM, Okamoto K, Quastel JH. Effects of ammonia ions on spontaneous action potentials and on contents of sodium, potassium, ammonium and chloride ions in vitro. J Neurochem 30: 131-143, 1978.
29. Mans AM, Saunders SJ, Kirsch RE, Biebuyck JF. Correlation of plasma and brain amino acid and putative neurotransmitter alterations during acute hepatic coma in the rat. J Neurochem 32: 285-92, 1979.
30. Gregorios JB, Mozes LW, Norenberg MD. Morphologic effects of ammonia on primary astrocyte cultures. II. Electron microscopic studies. J Neuropathol Exp Neurol 44:404-14, 1985.
31. Watson AJ, Chambers T, Karp JE, et al. Transient idiopathic hyperammoniemia in adults. Lancet 2: 1771-74, 1985.
32. Ede RJ, Gove CD, Hughes RD, et al. Reduced brain Na+-K+ ATPase activity in rats with galactosamine-induced hepatic failure: relationship to encephalopathy and brain edema. Clin Science 72: 365-71, 1987.
33. Yang SS, Gughes RD, Williams R. Digoxin-like immunoreactive substances in severe acute liver disease due to viral hepatitis and paracetamol overdosage. Hepatology

8: 93-7, 1988.

34. Albrecht J, Wysmyk-Cybula U, Rafalowska U. Na+/K+ ATPase activity and GABA uptake in astroglial cell enriched fractions and synaptosomes derived from rats in the early stage of experimental hepatic encephalopathy. Acta Neurol Scand 72:317-20, 1985.
35. Hanid MA, Mackenzie RL, Jenner RE, et al. Intracranial pressure in pigs with surgically induced liver failure. Gastroenterology 76:123-31, 1979.
36. Hanid MA, Davis M, Mellon PJ, et al. Clinical monitoring of intracranial pressure in fulminant hepatic failure. Gut 21: 866-9, 1980.
37. Rosner MJ, Becker DP. Origin and evolution of plateau waves. Experimental observation and a theoretical model. J Neurosurg 60:312-24, 1984.
38. Venes J. Intracranial pressure monitoring in perspective. Childs Brain 7:236-51, 1980.
39. Ede R, Gove CD, Williams R. Increased cerebral blood flow in fulminant hepatic failure due to paracetamol overdose. in Advances in ammonia metabolism and hepatic encephalopathy. Soeters PB, Wilson JHP, Meijer AJ, Holm E, eds. Excerpta Medica, Amsterdam 1987; pp 567-70. Gut 28:A1330, 1987.
40. Bihari D, Alexander ES, Waterson M, Williams R. Tissue hupoxia during fulminant hepatic failure. Critical Care Med 13: 1034-6, 1985.

ENCEPHALOPATHY OF REYE'S SYNDROME: STUDIES OF THE PATHOGENIC PROCESS IN ANIMAL MODELS

M.G. Murphy, S.C. Digout, and J.F.S. Crocker

Depts. of Pharmacology and Pediatrics, Dalhousie University and the Izaak Walton Killam Hospital for Children, Halifax, Nova Scotia, CANADA B3H 4H7

Reye's Syndrome (RS) is a frequently fatal disease that occurs in children as a rare sequelae to a relatively benign viral illness such as influenza or varicella (1). Its hallmark features include fatty infiltration, particularly of the liver, mitochondrial degeneration, and severe cerebral encephalopathy. The latter is the usual cause of death. The sequence of events that occurs during the early stages of RS are not known; however, this information is of great importance, since unlike true viral encephalopathies, tissue destruction in this disease can be minimal and reversible, and the disease can be recoverable (2). If appreciable brain edema does develop, however, approximately 20% of the patients sustain residual CNS damage.

The purposes of this article are to review current knowledge concerning the encephalopathy of RS (and the features that distinguish it from other non-RS encephalopathies), and to summarize information from studies with animal models that have attempted to resolve questions regarding its pathogenesis. The latter will include results of work with our own chemical/viral mouse model.

The Encephalopathy of Reye's Syndrome

The encephalopathy of RS is non-inflammatory and stereotypic in its progression, and in advanced cases presents as severe edema with abnormalities in cerebral circulation (3) and potentially lethal increases in intracranial pressure (ICP)(4,5).

The elevated intracranial pressure seen in RS is associated with cerebral edema that has been proposed by some to be cytotoxic (6), and by others to be vasogenic (7) in nature. In a recent review of data from 150 patients, Partin summarized the abnormal morphological characteristics of RS brain as (i) swollen astrocytes with expanded cytoplasm that was often devoid of glycogen granules, but with relatively normal mitochondria, (ii) neurons that contained pale, enlarged and pleomorphic mitochondria, and (iii) large watery blebs within myelin sheaths (8). Capillary endothelial cells were normal, with normal mitochondria, which suggested that mitochondrial failure was not responsible for the cerebral edema.

Elucidation of the relationship between hepatic dysfunction and central nervous system disease presents one of the greatest challenges in RS research. The liver disease does not appear to accurately mirror the process in the brain with respect to either morphological changes (8) or alterations in enzyme activities (9). This has raised the question of whether the CNS injury is a response to the same toxic stimulus as that acting at the liver, or whether the CNS injury is a secondary effect of liver damage. In reviewing the mechanisms that have been proposed to account for RS and non RS encephalopathy, we have taken the approach that the initial injury involves a toxic insult.

Reye's Syndrome and Other Hepatic Encephalopathies

(i) The Toxic Insult

In their initial documentation of Reye's Syndrome, Reye et al. suggested that the disease was the result of an interaction between one or more toxic agents and an infecting virus (10). This concept has received considerable support (11-13); however, the search for a single causative agent has been unsuccessful (12). Our own feeling is that a wide variety of chemicals have the potential to generate metabolic changes that enhance viral virulence. Even structurally-unrelated agents may trigger an RS response if they are able to change the body's production or utilization of antiviral substances, alter intermediary (e.g.,lipid) metabolism, and/or alter membrane structure. Amoung the many agents that have been implicated are aflatoxin B_1, 4-pentenoic acid (a derivative of hypoglycin A, the toxin responsible for `Jamaican Vomiting Sickness'), salicylates, acetaminophen, pesticides, and industrial surfactants.(12).

In recent years, there has been renewed interest in the role of chemicals and drugs as toxins that are capable of potentiating viral virulence, particularly in light of the apparent decrease in RS concomitant with reduced consumption of salicylates (14,15). In sufficiently high doses, salicylates are capable of producing severe encephalopathy, either early or late in the toxic event (16). However, there is some uncertainty as to whether salicylate toxicity is always associated with rises in intracranial pressure. Two additional features that clearly distinguish salicylism from RS are the lack, in the former, of striking hyperammonemia (17), and the absence of RS-specific mitochondrial degeneration (18). However, the current feeling is that rather than being a sole causative agent, salicylates exacerbate an RS condition by imposing an added metabolic burden on an organ that has been already compromised by a primary mitochondrial injury.

(b) Metabolic Bases of Encephalopathy

Several circulatory substances have been frequently proposed to be either the direct or indirect `toxins' responsible for Reye's encephalopathy in humans. These include high concentrations of ammonia (19), short-, medium- and long-chain fatty acids (20-22) and as described above, salicylates. It is not known whether each constitutes part of a common pathophysiological mechanism of cerebral pathology, but several observations suggest that individually, each is not the sole factor involved in RS.

Hyperammonemia, whether due to inborn errors of metabolism or an acute hepatic disease, can lead to large increases in intracranial pressure and cerebral edema (23,24). RS patients have elevated levels of serum ammonia which is effectively taken up and metabolized by the brain. Many investigators support the concept that the brain ammonia burden in RS children correlates well with the severity of the resulting encephalopathy (25-27). The major questions regarding the relationship between ammonia toxicity and RS concern differences in brain pathology, including the added presence of myelin blebs in RS.

Inborn errors of metabolism, including medium-chain acyl-CoA dehydrogenase deficiency (MCAD), are often clinically indistinguishable from RS, and, in patients younger than 2 years, may frequently be misdiagnosed as RS (28-30). Expression of these inherited metabolic disorders, which can also be triggered by viral infections, may involve fatty infiltration of the liver (usually macrovesicular in nature) and production of elevated levels of medium-chain and dicarboxylic fatty acids and/or ammonia. If ammonia levels are elevated, they are either very high relative to those seen in RS, or return to normal much more slowly than do those in RS patients (29). Behaviorally, these patients do not go through the characteristic agitation/disorientation stage before the onset of coma, and in most cases, they lack abnormal posturing or signs of increased intracranial pressure (30).

Elevations in circulating free fatty acids have been suggested to be important factors in the development of encephalopathy, largely on the basis of their effects in experimental animals. Several observations in humans would tend to argue against this. First, there is considerable discrepancy as to whether fatty acid levels are elevated (31), and even if they are, there appears to be little correlation between freefattyacidemia and the clinical course of the disease (26). Furthermore, levels of serum fatty acids that are as high as those in RS can occur in response even to fasting, which is not associated with encephalopathy (32). Thus, they are generally considered relatively minor factors in the pathogenesis of cerebral dysfunction. One specific abnormality in serum fatty-acid profiles that has been implicated in RS is a significant increase in circulating polyunsaturates (PUFA). Holman and coworkers (22) reported that serum from 17 RS patients contained high levels of most PUFA, and as such, proposed that increased prostaglandin production (which has the ability to produce cerebral swelling) was involved in the pathogenesis of RS encephalopathy.

The factor(s) that initiate and perpetuate RS, and the sequence of events that lead to loss of functional integrity of body organs remain obscure, since most of the initial metabolic events associated with the disease have occurred before the patient can be diagnosed as having RS. Furthermore, the mitochondrial injury associated with metabolic abnormalities appears to be transient in nature (33). Accordingly, the only promising means of elucidating the early events in the pathogensis of RS, and thereby provide information necessary for successful therapeutic intervention, is to develop an animal model in which the prominent features of the syndrome are duplicated.

Studies of RS Encephalopathy using Animal Models

In general, there is relatively little data available regarding the CNS characteristics (either pathophysiological or morphological) of animal models that have been developed to study the pathogenesis of RS.

Studies with the ferret model have provided perhaps the best example of the effects of hyperammonemia (34-36). Deshmukh and co-workers (34,35) developed their ferret model based on varying combinations of influenza B (FluB), salicylates and the hyperammonemia-producing arginine-deficient (AD) diet. The animals required the AD diet to develop hyperammonemia and the neurological symptoms of cerebral encephalopathy. Neither of these two events took place if FluB or salicylates were given alone or in combination. A synergistic effect could be demonstrated, however, since the ammonia levels could be further elevated, and mortality rate increased, if salicylates and FluB were added to the experimental paradigm. These authors did not carry out extensive morphological studies, but they have recently reported that there is intracellular edema and vacuolization of the epithelial cells of the choroid plexus, similar to that reported in patients with RS (37). In earlier studies, Linnemann et al.(36) demonstrated that ferrets infected with FluA or FluB, either with or without salicylates, developed selective neuronal necrosis compatible with anoxic injury.

Several models have evolved from the theory that RS involves an environmental toxin. Bourgeois et al. (38) examined the effects of feeding young cynomolgus monkeys aflatoxin B_1 produced from fungi that contaminated food eaten by a young child who died of RS. They found that the marked encephalopathy with coning that developed was associated with neuronal degeneration, increased perivascular space, swelling and separation of the myelin sheath - all consistent with RS. Others have examined the effects of feeding animals margosa oil (a long-used medicinal oil in several Asian countries), based on reports of deaths in children who had been given the oil for viral infections (39). The active component of the oil has been proposed to be aflatoxin (39) in combination with another, unidentified compound (40). However, rats fed this oil showed no evidence of either brain morphological changes or cerebral edema (which is seen in humans), despite a florid encephalopathy. This raises the question of species specificity with respect to precisely duplicating a pathological response.

Several studies of the CNS component of RS have evolved from the observation that children with RS have elevated levels of serum medium and long-chain free fatty acids. Trauner and coworkers found that continuous infusion of octanoic acid into rabbits resulted in swelling of cerebral astrocytic foot processes (41) and early damage of paraventricular brainstem neurons (42). They interpreted these findings as suggestive of early edema, although no gross edema was evident. At least one investigator has suggested that the octanoate model better represents MCAD deficiency than RS (17). Kim et al.(43) extended the fatty acid hypothesis when they demonstrated that both medium-chain monocarboxylic (e.g., octanoate) and dicarboxylic acids (e.g., suberic acid) blocked organic acid transport in the choroid plexus _in vitro_. Accordingly, they suggested that the increased levels of these components in RS patients could contribute to increased intracranial pressure via this mechanism. In another series of studies, Kang et al. (44) examined the effects of long-chain fatty acids on development of neuropathy, since Ogburn and coworkers had reported that levels of polyunsaturated fatty acids (PUFA) were high in RS sera (22). Injection of PUFA intra-cisternally into rabbits resulted in a progressive, dose-dependent deepening coma state. To our knowledge, this is the only study in which progressive encephalopathy has been documented. The effect was specific for unsaturated fatty acids, with the most toxic effects observed with the naturally-occuring arachidonic ($C_{20:4w6}$) and docosatetraenoic ($C_{22:4w6}$) acids. With respect to a possible source of PUFA, these authors demonstrated that under hyperosmolar conditions (such as those that exist in RS patients), these acids could penetrate the blood-brain barrier from the circulation. High levels of free PUFA have the potential to produce cerebral edema (45); it is not known whether these acyl constituents are elevated in RS brain. Interestingly, the phenomenon could be reproduced by injection of melettin, an activator of phospholipase A_2 that catalyzes hydrolysis of fatty acids from membrane phospholipids. One mammalian virus, dengue virus, is known to have phospolipase activity (46).

At least three models for RS have involved the use of mice (47-52). Brownstein et al.(47) described a RS-like illness that arose spontaneously in 6- to 9-week-old BALB/cByJ mice after they had been housed for 2 to 3 weeks in rooms that contained indigenous mouse viruses. In addition to neurological abnormalities, the mice had astrocytic neuropathy that was consistent with Alzheimer's Type II astrocytes. Astrocyte numbers were increased, they had swollen, pallorous nuclei with compressed chromatin, and prominent nucleoli. Electron microscopy revealed deposition of lipid droplets in the astrocytes, reduced glycogen, and no change in mitochondrial structure. By contrast, neuronal mitochondria were swollen, and had disrupted internal structure consistent with the damage seen in RS. The major difficulty with this model is that it could not be reproduced more than five times. The mouse model of Davis and coworkers (48-50) involves intravenous injection of juvenile mice with very high doses of mouse-adapted human influenza B (FluB). In earlier studies, this group stated that this procedure, which resulted in a high rate of mortality, produced a non-inflammatory mild cerebral edema (48) and a disrupted blood-brain barrier (50). Unfortunately they did not provide any data. Recently, the authors suggested that this experimental paradigm (and by association, RS) involves a nonpermissive infection of liver and brain, the latter of which is based on the presence of viral antigen reactivity within brain capillary endothelial cells (49). One of the major problems with this model is that it fails to reproduce the hepatic mitochondrial characteristics of RS. The mouse model that has been developed in our laboratories is described in detail below.

There have been several studies in which animal models have been reported to produce encephalopathy, which have either not included more detailed morphological or functional studies, or in which the findings have been negative. These include models based on feeding rats

4-pentenoic acid (the derivative of hypoglycin that is the active toxin in `Jamaican Vomiting Sickness') (53), injecting rats with endotoxin from E.coli (54), and treating mice with encephalomyocarditis (EMC) virus in combination with various chemical agents, including salicylates and sodium pentachlorophenol (51). In another model that involved chemicals without an associated virus infection, Zieve and coworkers demonstrated that the combination of ammonia and free fatty acids acted synergistically in producing hepatic coma in cats and rats (55). Based on the knowledge that complex metabolic abnormalities are prerequisite to the final toxic insult in RS, they proposed that the human disease involves a lack of one or more essential substances, the identity of which is/are unknown at present.

The Chemical/Viral Mouse Model for RS

During the past few years, we have developed and have been characterizing an animal model for RS based on the hypothesis that it is a consequence of an altered reaction to a relatively benign virus due to a previous chemical/drug exposure. This theory evolved as a result of our observations that most of the children that we encountered with RS came from rural areas in which an active forestry spray program was being carried out (56). We have established that neonatal mice exposed to nontoxic amounts of an industrial emulsifier and subsequently infected with sublethal doses of either encephalomyocarditis (EMC) or mouse-adapted influenza B (FluB) virus develop many of the features of human RS, and that these are accompanied by a high rate of mortality (52). In the initial studies of the hepatic aspects of the mouse model, we demonstrated liver pathology consistent with that seen in RS, abnormal liver enzyme activities, microvesicular fat deposition (with EMC virus) and abnormal lipid profiles (with FluB) (57). As well, serum lipid content was low (58), serum ammonia levels were high (52), and interferon response was reduced (59).

Recently, we have begun to verify that the mice undergoing their version of RS also have a number of characteristics that mimic the encephalopathic aspects of RS. We now describe our recent findings concerning the neurological features of our model, including the significant elevations in intracranial pressure associated with chemical plus virus treatment and the increased levels of lipid peroxides related to dermal exposure of the mice to nontoxic doses of industrial surfactant.

The Mouse Model

The chemical/viral mouse model for RS has been described in detail elsewhere (52). Male and female Swiss white mice (Charles River Canada, Inc.) were mated, and the day after the largest number of mice were born, they were pooled and randomly devided into groups of eight pups per nursing mother. The same day, the pups were assigned arbitrarily to one of two equal groups, one of which was exposed dermally by abdominal painting to minimal essential medium (MEM), while the other was painted with commercial industrial surfactant, Toximul MP8 (Tox)(Charles Tenant Co., Toronto, Canada). This treatment was repeated daily for 11 days. On day 12, half of each of the MEM and Tox groups received 10 ul of mouse-adapted influenza B (Lee) virus (FluB)(640 HA units/0.5 ml) intranasally under light ether anesthesia; the other half of each received anesthetic alone. Deaths were recorded daily.

Experimental Results

(a) Mortality Rates

Mortality rates using the chemical/viral paradigm varied from one experiment to another, depending on the relative strength of the infecting virus. Deaths in the group that received virus alone (FluB) ranged from very few to 50-60% of the animals. Mice given the combined treatment (Tox + FluB) consistently had a higher mortality rate, whether it was 30% when there were few FluB-related deaths, or 90% when 50% of the animals in the virus group died. Deaths were very rare in those groups given either MEM or Tox alone.

(b) Intracranial Pressure Studies

Between 48 and 72 hr after virus exposure, intracranial pressure (ICP) was measured in each of the four groups of mice (10 per group) using a modification of the procedure described by Love and Leslie (60). Under halothane anesthetic and using sterile conditions, a 25 gauge needle, attached to a pressure transducer and microinfusion pump, was inserted into the subarachnoid area. After the mean baseline value was obtained, the pressure was balanced at zero using a manometer, and calibrated for pressures ranging from 0 - 30 mm H_2O.

In all experiments, the mice survived the procedure very well (the mortality rate was < 2%), and were awake and active within 5 min of anaesthetic withdrawal. In five separate experiments, we found significantly high values for ICP in the groups of mice treated with either FluB or Tox plus FluB. In one representative experiment, the mean value for ICP in the virus group was 1.8 mm H_2O, while the corresponding value for the mice that received combined treatment was 5 mm H_2O. Occasionally there was a random high value in the Tox or MEM groups; this was invariably associated with a seizure. These studies have been particularly significant in that no one has previously reported measuring intracranial pressure in an animal model for Reye's Syndrome.

(c) Lipid Peroxidation Studies

Several xenobiotics that have been implicated in the etiology of RS, including pentenoic acid, aflatoxin and pesticides and industrial solvents, have been shown to induce lipid peroxidation (LPO) in liver and brain (61). We have measured ambient levels of LPO at various times during the mouse experiments, essentially as described by Wong et al. (62). Results from preliminary experiments have shown that abdominal painting of the mice with Toximul MP8 results in dramatic increases in LPO products in both liver and brain. LPO levels, measured as the formation of malondialdehyde (MDA), increased in control (MEM) brains between postnatal days 2 and 6, and remained stable at approximately 4 nmol/mg protein for the duration of the experiment. Tox exposure increased these ambient levels, with a maximum effect at the end of the painting stage (day 11). Within 2 days of cessation of surfactant exposure, the Tox-dependent effect was no longer apparent. Further infection with FluB did not appear to retard the return to normal values. The results of these studies suggest that although cerebral LPO values increase with extended exposure to surfactant, they are not by themselves a direct cause of death in this mouse model.

SUMMARY

Reye's Syndrome is a disease(s) that involves very complex changes in the integrity and function of different body organs. We do not yet know whether this specific syndrome shares one or more common pathophysiologic mechanisms with other forms of hepatic encephalopathy, since we do not know precise cause-effect relationships, or the sequence of events, that occurs in any of them. Animal models that duplicate the the cardinal features of these conditions have provided essential information regarding the potential role of different `toxins' or other insults (e.g.,infections) in pathogenesis of these multifactorial diseases. As such, they offer the only means whereby in future these life-threatening disorders may be prevented, or at least successfully treated.

References

1. Crocker JFS, Bagnell PC (1981) Reye's Syndrome--A clinical review. Can. Med. Assoc. J. 124, 375.
2. Brown T, Hug G et al. (1976) Transiently reduced activity of Carbamyl phosphate synthetase and ornithine transcarbamylase in liver of children with Reye's Syndrome. N. Eng. J. Med. 294, 861.
3. Frewen TC, Kissoon N (1987) Cerebral blood flow and brain oxygen extraction in Reye Syndrome. J. Pediatr. 110, 903.
4. Venes JL, Shaywitz BA, Spencer DD (1978) Management of severe cerebral edema in the metabolic encephalopathy of Reye-Johnson Syndrome. J. Neurosurg. 48, 903.
5. McWilliam RC, Stephenson JPB (1985) Life threatening intracranial hypertension in Reye's Syndrome treated with intravenous thiopentone. Eur. J. Pediatr. 144, 383.
6. DeVivo DC, Keating JP, Haymond MW (1975) Intensive supportive approach to the management of Reye's Syndrome, in Reye's Syndrome (JD Pollack, ed). Grune & Stratton, New York, pp.315-327.
7. Hopkins JW, Giannotta SL, Kindt GW (1979) Pathologic vasodilatation as a cause of increased intracranial pressure in Reye's Syndrome, in Reye's Syndrome II (JFS Crocker, ed). Grune & Stratton, New York, 486.
8. Partin JS, McAdams AJ et al.(1978) Brain ultrastructure

in Reye's Syndrome. Acute injury and recovery processes in three children. J. Neuropathol. Exp. Neurol. 377, 96.

9. Robinson BH, Taylor J et al. (1978) Reye's Syndrome: Preservation of mitochondrial enzymes in brain and muscle compared with the liver. Pediatr. Res. 12, 1045.

10. Reye RDK, Morgan G, Baral J (1963) Encephalopathy and fatty degenerations of the viscera: A disease entity in childhood. Lancet 2, 749.

11. Pollack JD (1979) Models of chemical and virus interaction and their relation to a multiple etiology Reye's syndrome. in Reye's Syndrome II (JFS Crocker,ed) pp 341-360.

12. Crocker JFS (1988) The role of exogenous agents in Reye's Syndrome. in Reye's Syndrome. Br.Med.Soc.Round Table Discussions 8, 45.

13. Colon AR, Ledesman F, et al. (1974) Viral potentiation of chemical toxins in the experimental syndrome of hypoglycemia, encephalopathy and visceral fatty degeneration. Am. J. Digest. Dis. 19, 1091.

14. Remington PL, Rowley D et al. (1986) Decreasing trends in Reye Syndrome and aspirin use in Michigan. Pediatrics 77, 93.

15. Hurwitz ES, Barrett MJ et al. (1987) Public health service study of Reye's Syndrome and medications. JAMA 257, 1905.

16. Quint PA, Allman FD (1984) Pediatrics 74, 1117.

17. De Vivo DC (1984) Editorial: Do animals develop Reye Syndrome. Lab. Invest. 51, 367.

18. Daugherty CC, Gartside PS et al.(1987) A morphometric study of Reye's Syndrome. Am. J. Pathol. 129, 313.

19. Fitzgerald JF, Clark JH et al. (1982) The prognostic significance of of peak ammonia levels in Reye Syndrome Pediatrics 70, 997.

20. Trauner DA, Nyhan WL, Sweetman L (1975) Short chain organic acidemia and Reye's Syndrome. Neurology 25, 296.

21. Ogburn PL Jr (1976) Medium and long chain serum free fatty acids in a case of Reye's Syndrome. South. Med. J. 69, 632.

22. Ogburn PL Jr, Sharp H et al.(1982) Abnormal polyunsaturated fatty acid patterns of serum lipids in Reye's Syndrome. Proc. Natl. Acad. Sci. USA 79, 708.

23. Shih VE (1976) Congenital hyperammonemic syndromes. Clin. Perinatol. 3, 3.

24. Ware AJ, D'Agostino AN, Combes B (1971) Viral hepatitis complicating the Dubin-Johnson Syndrome.

Gastroenterology 61, 877.

25. Heubi JE, Partin JC, et al. (1987) Reye's Syndrome: Current concepts. Hepatology 7, 155.
26. DeLong GR, Glick TH (1982) Encephalopathy of Reye's Syndrome: A review of pathogenetic hypotheses. Pediatrics 69, 53.
27. Goldberg DM, Brown D (1987) Clin. Biochem. 20, 127.
28. Roe CR (1988) Metabolic disorders producing a Reye-like Syndrome. in Reye's Syndrome. Br. Med. Soc. Round Table Discussions 8, 85.
29. Greene CL, Blitzer MG & Shapiro E (1988) Inborn errors of metabolism and Reye's Syndrome: A differential diagnosis. J. Pediatr. 113, 156.
30. Taubman B, Hale DE & Kelley RI (1987) Familial Reye-Like Syndrome: A presentation of medium-chain Acyl- Coenzyme A dehydrogenase deficiency. Pediatrics 79, 382.
31. McArthur B, Ashok P, et al. (1984) Short-chain fatty acids and encephalopathy of Reye's Syndrome. Neurology 34, 831.
32. Corvilain J, Loeb H, et al. (1961) Effect of fasting on levels of plasma nonesterified fatty acids in normal children, normal adults and obese adults. Lancet 1, 534.
33. La Montagne JR (1983) Summary of a workshop on influenza B viruses and Reye's Syndrome. J. Infect. Dis. 148, 943.
34. Deshmukh DR, Maassab HF & Mason M (1982) Interactions of aspirin and other potential etiologic factors in an animal model of Reye's Syndrome. Proc. Natl. Acad. Sci. USA 79, 7557.
35. Rarey KE, Davis JA & Deshmukh DR (1987) Effect of upper respiratory infection on hearing in the ferret model. Lab.Invest. 56, 249.
36. Linnemann CC, Ueda K, et al. (1979) Salicylate intoxication and influenza in ferrets. Pediatr. Res. 13, 44.
37. Brown RE, Madge GE (1972) Lipid and lipoprotein studies in Reye's Syndrome. Virginia Med. Monthly 99, 271.
38. Bourgeois CH, Shank RC, et al. (1971) Acute aflatoxin B, toxicity in the macaque and its similarities to Reye's Syndrome. Lab Invest. 24, 206.
39. Sinniah D, Baskaran G, et al. (1982) Reye-like Syndrome due to Margosa oil poisoning. Am. J. Gastroent. 77, 158.
40. Sinniah D, Schwartz PH, et al. (1985) Investigation of an animal model of a Reye-like Syndrome caused by Margosa oil. Pediatr. Res. 19, 1346.
41. Trauner DA (1982) Pathologic changes in a rabbit model of

Reye's Syndrome. Pediatr. Res. 16, 950.

42. Trauner DA, Huttenlocher PR (1978) Short chain fatty acid induced central hyperventilation in rabbits. Neurology 28, 940.

43. Kim CS, O'Tuama LA, et al. (1983) Effect of increasing carbon chain length on organic acid transport by the choroid plexus: a potential factor in Reye's Syndrome. Brain Res. 259, 340.

44. Kang ES, Olson G, et al. (1984) Development of encephalopathic features similar to Reye Syndrome in rabbits. Proc. Natl. Acad. Sci USA 81, 6169.

45. Chan PH, Fishman RA, et al. (1983) Induction of brain edema following intracerebral injection of arachidonic acid. Ann. Neurol. 13, 625.

46. Malewicz B, Parthasarathy S, et al. (1981) Rapid phospholipase A_2 stimulation and diacylglycerol cholinephosphotransferase inhibition in baby hamster kidney cells during initiation of dengue virus infection. Biochem. Biophys. Res. Comm. 101, 404.

47. Brownstein DG, Johnson EA, Smith AL (1984) Spontaneous Reye's-like Syndrome in BALb/cB4J mice. Lab. Invest. 51, 386.

48. Davis LE, Cole LL, et al. (1983) Experimental influenza B toxicity in mice--a possible model for Reye's Syndrome. Lab. Invest. 48, 140.

49. Davis LE (1987) Influenza B virus model of Reye's Syndrome. Lab. Invest. 56, 32.

50. Davis LE, Kornfeld M (1986) Mouse influenza B virus model of Reye's Syndrome. Amer. J. Pathol. 122, 190.

51. Hug G, Bosken J, et al. (1981) Reye's Syndrome simulacra in liver of mice after treatment with chemical agents and encephalomyocarditis virus. Lab. Invest. 45, 89.

52. Crocker JFS, Renton KW, et al. (1986) Biochemical and morphological characteristics of a mouse model of Reye's Syndrome induced by the interaction of influenza B virus and a chemical emulsifier. Lab. Invest. 54, 32.

53. Glasgow AM & Chase HP (1975) Production of the features of Reye's Syndrome in rats with 4-pentenoic acid. Pediatr. Res. 9, 133.

54. Yoder MC, Egler JM, et al. (1985) Metabolic and mitochondrial morphological changes that mimic Reye Syndrome after endotoxin administration to rats. Infect. Immun. 47, 329.

55. Zieve FJ, Zieve L, et al. (1974) Synergism between ammonia and fatty acids in the production of coma:

Implications for hepatic coma. J Pharmacol Exp Ther 191, 10.

56. Crocker JFS, Rozee KR, et al. (1974) Insecticide and viral interaction as a cause of fatty visceral changes and encephalopathy in the mouse. Lancet 2, 22.
57. Murphy MG, Archambault-Schertzer L, et al. (1987) Hepatic lipid abnormalities in a chemical/viral mouse model for Reye's Syndrome. Lipids 22, 217.
58. Murphy MG, Archambault-Schertzer L, et al. (1986) Serum lipid abnormalities in a chemical/viral mouse model for Reye's Syndrome. Lipids 21, 378.
59. Rozee KR, Lee SHS, et al. (1982) Is a compromised interferon response an etiologic factor in Reye's Syndrome. Can. Med. Assoc. J. 126, 798.
60. Love JA & Leslie RA (1984) The effects of raised ICP on lymph flow in the cervical lymphalic trunks in cats. J Neurosurg. 60, 577.
61. Murphy MG, Carroll C, et al. In Preparation
62. Wong SHY, Knight JA, et al. (1987) Lipoperoxides in plasma as measured by liquid-chromatographic separation of malondialehyde--Thiobarbituric acid adduct. Clin. Chem. 33, 214.

THE RAT WITH CARBON TETRACHLORIDE-INDUCED CIRRHOSIS AS A MODEL OF HEPATIC ENCEPHALOPATHY.

Paul Martin, Maria R. Ferreira, Sergio H. Gammal, H. Chris Fromm, E. Anthony Jones.

Liver Diseases Section, NIDDK, National Institutes of Health.
Bethesda, Md. U.S.A.

Animal models of liver disease have greatly facilitated studies of the manifestations of hepatocellular failure (1), including hepatic encephalopathy (HE) (2). Certain models of fulminant hepatic failure (FHF) (3,4) appear to be particularly useful tools for studying the neural mechanisms which mediate HE in this syndrome. Studies of these models have provided evidence for an involvement of the gamma-aminobutyric acid/benzodiazepine receptor complex in the pathogenesis of HE due to FHF(5-9). However, the most common clinical scenario for HE in humans is not FHF but chronic liver disease, especially cirrhosis (10). HE in patients is often precipitated or exacerbated by an easily recognizable precipitating factor such as a gastrointestinal hemorrhage or the overzealous administration of diuretics (11). The study of HE in this clinical setting would be facilitated by the availability of a suitable animal model. Research on such a model would obviate the ethical and logistic constraints inherent in human studies. Unfortunately, no entirely satisfactory model of HE complicating chronic liver disease appears to exist (12). To address this issue we have produced cirrhosis in rats using the protocol described by Proctor and Chatamra (13) and have assessed whether an HE-like syndrome, which would be suitable for experimental studies of HE, can be readily induced in this model. Our findings suggest that in the rat with CCl_4-induced cirrhosis and progressive hepatic decompensation, an easily recognizable syndrome of HE does not occur.

METHODS

Animals:

Male Wistar rats (150-200g) (Taconic Farms, Germantown, NY) were housed in standard animal cages and fed standard chow ad libitum.

Induction of cirrhosis:

Phenobarbital powder (Sigma, St. Louis, Mo) was added to the drinking water (350 mg/L tap water) to induce cytochrome P_{450}. Starting one week after the initiation of phenobarbital administration, at weekly intervals 60 animals were weighed and gavaged with CCl_4(Fisher, Fair Lawn, NJ) using steel feeding tubes (Popper and Sons, New Hyde Park, NY) and no sedation. The initial dose was 75 ug per rat. If an animal gained at least 5g weight during the previous week, the dose of CCl_4 administered was increased by an increment of 25 ug. Conversely, if there had been a loss of at least 5g weight during the previous week, the dose was reduced by 25 ug. If the weight changed by less than 5g during the previous week, the dose was not changed. Once an animal exhibited fluid retention with ascites, detectable by marked bulging of the flanks, CCl_4 administration was discontinued. Each animal was observed over the subsequent two weeks to assess whether the ascites was persistent. Occasionally the ascites would resolve, in which case further doses of CCl_4 were administered according to the schedule above. Once ascites was judged to be stable, phenobarbital was withdrawn from the animal's drinking water. Behavioral and electrophysiological studies were conducted at least one week after the last dose of CCl_4.

Control groups:

One control group (n=10) received neither phenobarbital nor CCl_4. Another control group (n=10) received phenobarbital but no CCl_4.

Maneuvers:

If no features of encephalopathy were noted in CCl_4-treated rats, one of the following additional maneuvers was undertaken in an effort to precipitate encephalopathy:

1. A brisk diuresis was induced by the intramuscular injection of furosemide (6 mg/kg body weight) (n=6). The diuresis was associated with a 6% $\pm$ 1(SD)% weight loss, hypokalemia (3.9 $\pm$ 0.2 vs. 4.2 $\pm$ 0.2 mEq/L in controls, $p<0.02$) and increased BUN (33 $\pm$ 4 vs. 21.6 $\pm$ 3.7 mg/dl in controls, $p<0.01$).

2. Human blood (kindly supplied by the Blood Bank, Clinical Center, NIH) was administered by gavage in quantities sufficient to produce melena (n=5) (i.e. up to 60 ml p.o.)

3. Urease (Sigma, St. Louis, Mo.) 110 ug was administered intamuscularly to induce or augment hyperammonemia (6) in both cirrhotic (n=5) and control (n=6) animals. Mean plasma ammonia concentrations were 90 ug/dl in controls and 330 ug/dl in cirrhotic animals, and after urease treatment were 2963 ug/dl in controls and 2762 ug/dl in cirrhotic animals.

Behavioral studies:

Animals were carefully observed for signs of encephalopathy, such as decreased motor activity, incoordination, ataxia or decreased level of consciousness (4,6).

Electrophysiologic studies:

Visual evoked responses (VERs) were recorded via chronically implanted epidural electrodes as previously described (6). After obtaining a baseline VER, serial tracings were obtained from rats as they received weekly doses of CCl_4 and developed cirrhosis.

Biochemical and pathological studies:

Animals were sacrificed by exsanguination under ether anesthesia. Blood was obtained for routine hematology and serum biochemistry, which included estimations of prothrombin time, plasma ammonia, serum aspartate and alanine aminotransferases and serum albumin. Liver tissue was retrieved for histological examination; sections were stained with hemotoxylin and eosin, and Masson's trichrome.

RESULTS

The model of cirrhosis:

Of the 60 rats that received CCl_4, 23 died during the period of treatment due to either CCl_4 toxicity (renal or hepatic) or to gavage-associated aspiration-induced respiratory failure. Thus 37 rats or 62% of the original total treated were available for study. The number of treatments with CCl_4 that were necessary to induce cirrhosis was 13 ± 5. At the time rats were considered to be cirrhotic, they weighed significantly less than control rats (454 ± 48.5 vs. 519 ± 31g in controls, $p<0.01$) in spite of the fact that cirrhosis was complicated by fluid retention. The mean serum bilirubin was 0.58 mg/dl in cirrhotic animals and 0.1 mg/dl in controls ($p<0.01$). The mean serum albumin was 1.9g/dl in the cirrhotic group and 2.8 g/dl in controls ($p<0.01$). Prothrombin time was 13.7 ± 1.2 seconds in CCl_4-treated group compared to 12.6 0.69 in controls, ($p<0.05$).

Behavioral findings:

No control or cirrhotic animal exhibited any unequivocal behavioral changes of encephalopathy. Furthermore none of the provocative maneuvers applied to CCl_4-treated rats induced any signs of overt encephalopathy. The only behavioral changes that were observed were seizures, which occurred in two of the CCl_4-treated rats with urease-induced hyperammonemia.

Electrophysiological findings:

Epidural electrodes remained functional throughout the entire period of the study in the majority of animals in whom they had been implanted. There was no significant difference between corresponding latencies and amplitudes of peaks in the VER for the control group that received phenobarbital and the control group that did not receive this drug. Changes in latencies and amplitudes of peaks of the VER in CCl_4-treated animals were small and did not differ significantly from those of baseline recordings. Furthermore, none of the provocative maneuvers in CCl_4-treated animals resulted in any significant changes in latencies or amplitudes of peaks of the VER.

Hepatic pathology:

At the time of sacrifice, all animals with obvious ascites were found to have firm nodular livers. Histology of these livers confirmed the presence of well-established micronodular cirrhosis.

DISCUSSION.

Though chronic administration of CCl_4 has been known for many years to induce cirrhosis in laboratory animals, only in recent years has a regimen for giving the toxin been available that results in a high yield of cirrhotic animals without excessive mortality (13). Using this method, we obtained rats with histologically confirmed cirrhosis and many of the features of chronic decompensated liver disease seen in man (1). These features included obvious ascites, hypoalbuminemia , hyperbilirubinemia and prolonged prothrombin time. Others have demonstrated the presence of portal hypertension in this model (14). Chronic administration of CCl_4 allows the gradual evolution of parenchymal liver disease comparable to that which occurs in humans who develop cirrhosis following repetitive exposure to a hepatotoxin such as alcohol (1). Another model, which has been extensively studied in relation to chronic HE, is the rat with a surgically constructed porto-caval shunt (15). While this model is associated with dramatic changes in liver blood flow, it probably reproduces the complex derangements of hepatic functions characteristic of cirrhosis less well than does the rat chronically treated with CCl_4. Indeed, the encephalopathy that developes in dogs with surgically constructed porto-caval shunts can be prevented by adequate nutrition (16).

The quantities of phenobarbital administered to rats in their drinking water were insufficient to cause phenobarbital-induced changes in either behavior or the VER (5, 6). In contrast to the behavioral and VER changes associated with HE in models of FHF (4-7), and in contrast to the alleged occurrence of HE in rats chronically treated with CCl_4 reported by others (16), rats with decompensated cirrhosis in this study did not exhibit overt behavioral or significant VER changes consistent with HE. Furthermore, attempts to precipitate HE or an HE-like state in these cirrhotic rats by additional maneuvers were unsuccessful. Two of these maneuvers, induction of a brisk diuresis and introduction of blood into the gastrointestinal tract, are known to precipitate or exacerbate HE in humans with cirrhosis (11). However, neither of these two maneuvers induced behavioral or VER changes consistent with HE in the decompensated cirrhotic rats. If a brisk diuresis proportionally equivalent to that induced in cirrhotic rats by furosemide in this study were to be induced in a human with cirrhosis and ascites the precipitation of overt signs of encephalopathy would seem to be likely (10). In the gavage experiments, the volume of blood administered to each rat was greater than the animal's total circulating blood volume. If a volume of blood proportionally equivalent to that administered to cirrhotic rats in this study were to be delivered into the intestine of a human with cirrhosis and ascites overt signs of encephalopathy might well be precipitated (10). At least three potential factors should be considered as potential explanations for the failure of gavaging with blood to induce overt encephalopathy in the cirrhotic rats. First, it is possible that maintenance of hepatocellular function was sufficient for adequate metabolism of the increased load of nitrogenous substances delivered to the liver as a consequence of the administered blood. However, the degree of hypoalbuminemia present in the cirrhotic animals suggests that hepatocellular function was significantly impaired. Second, it is possible that the degree of portal-systemic shunting of blood was insufficient for HE to develop.

However, the presence of florid ascites, portal hypertension (14) and established micronodular cirrhosis seems to make this possibility unlikely. Third, in the typical clinical situation gastrointestinal hemorrhage is often associated with hypovolemia or hypoxia or both, whereas gavaging with blood would not be expected to induce hypovolemia or hypoxia. However, neither of these two factors alone or in combination, induce HE in cirrhotic patients (11).

It is less certain whether the third maneuver, the selective induction of hyperammonemia using urease, satisfactorily mimics a precipitating factor of HE in man. Many factors which precipitate or exacerbate HE and increase plasma ammonia levels in patients with cirrhosis also induce increased plasma concentrations of many other potentially neuroactive substances (1). Induction of hyperammonemia alone in experimental animals does not appear to reproduce the behavioral and VER manifestations of HE (6,18), and the induction of a selective increase in plasma ammonia concentrations in our model of chronic liver disease did not precipitate overt behavioral or VER changes consistent with HE. Indeed the occurrence of seizures in two of the cirrhotic rats with urease-induced hyperammonemia suggests that the known neuroexcitatory and seizure-inducing properties of ammonia (6,18-21) were not significantly modified by the presence of decompensated chronic liver disease.

CONCLUSION

The rat with CCl_4-induced decompensated cirrhosis, while apparently being suitable for studying some manifestations of chronic hepatocellular failure, does not readily develop HE and consequently does not appear to be a satisfactory model for studying HE complicating chronic liver disease.

SPECULATIONS

1. HE may not be a consistent feature of decompensated chronic hepatocellular disease in all species.
2. Human beings may belong to an atypical subgroup of mammalian species which have a predilection to develop HE as a complication of cirrhosis.

REFERENCES

1. Arias I.M., Jaboky W.B., Popper H. et al. (1988) The Liver: Biology and Pathobiology. Raven Press. New York.

2. Soeters P.B., Wilson J.H.P., Meijer A.J., Holm E. eds. (1988) Advances in Ammonia Metabolism and Hepatic Encephalopathy. Excerpta Medica, Amsterdam.

3. Blitzer B.L., Waggoner J.G., Jones E.A. et al (1978) A model of fulminant hepatic failure in the rabbit. Gastroenerology 74, 664-671.

4. Geller D., Gammal S.H., Mullen K.D. et al. (1988) An improved rat model of hepatic encephalopathy due to fulminant hepatic failure. The importance of supportive therapy. In: Advances in Ammonia Metabolism and Hepatic Encephalopathy. (Soeters P.B., Wilson J.H.P., Meijer A.J., and Holm E. eds), pp. 213-217. Excerpta Medica, Amsterdam.

5. Schafer D.F., Pappas S.C., Brody L.E., et al. (1984) Visual evoked potentials in a rabbit model of hepatic encephalopathy. 1. Sequential changes and comparisions with drug-induced comas. Gastroenterology 86, 540-545.

6. Jones D.B., Mullen K.D., Roessle M. et al. (1987) Hepatic encephalopathy: application of visual evoked responses to test hypotheses of its pathogenesis in rats. J. Hepatol. 4, 118-126.

7. Bassett M. L., Mullen K.D., Skolnick P. et al. (1987) Amelioration of hepatic encephalopathy by pharmacologic antagonism of the $GABA_a$/benzodiazepine receptor complex in a rabbit model of fulminant hepatic failure. Gastroenterology 93, 1069-1077.

8. Basile A.S., Gammal S.H., Mullen K.D. et al. (1988) Differential responsiveness of cerebellar Purkinje neurons to GABA and benzodiazepine receptor ligands in an animal model of hepatic encephalopathy. J. Neuroscience 8, 2414-2421.

9. Gammal S.H., Jones E.A., Bassett M.L. et al. (1988) Antagonists of the GABA/benzodiazepine receptor complex ameliorate hepatic encephalopathy in two models of fulminant hepatic failure. In: Advances in Ammonia Metabolism and Hepatic Encephalopathy. (Soeters P.B., Wilson J.H.P., Meijer A.J., and Holm E. eds.), pp 325-332. Excerpta Medica, Amsterdam.

10. Conn H.O., Lieberthal M.M. (1978) The Hepatic Coma Syndromes and Lactulose. Williams and Wilkins. Baltimore.

11. Hoyumpa A.M., Desmond P.V., Avant G.R. et al. (1979) Hepatic encephalopathy. Gastroenterology 76, 184-195.

12. Mullen K.D., McCullough A.J. (1989) Problems with animal models of chronic liver disease. Suggestions for improvement in standardization. Hepatology (in press).

13. Proctor E., Chatamra K. (1982) High yield micronodular cirrhosis in the rat. Gastroenterology 83, 1183-1190.

14. Vorobioff J., Bredfeldt J.E., Grozmann R.J. (1984) Increased blood flow through the portal system in cirrhotic rats. Gastroenterology 87, 1120-1126.

15. Bircher J. (1979) The rat with porto-caval shunt: an animal model with chronic hepatic failure. Pharmacol. Ther. 5, 219-222.

16. Thompson J.S., Schafer D.F., Haun J. and Schafer G.J. (1986) Adequate diet prevents hepatic coma in dogs with Eck fistulas. Surg. Gynecol. Obstet. 162, 126-130.

17. Yamamoto H., Sugihara N. (1987) Blood ammonia levels and hepatic encephalopathy by carbon-tetrachloride in rats. Toxicology and Applied Pharmacology 91, 279-285.

18. Pappas S.C., Ferenci P, Schafer D.F. and Jones E.A. (1984) Visual evoked potentials in a rabbit model of hepatic encephalopathy. 11. Comparision of hyperammonemic encephalopathy, postictal coma, and coma

induced by synergistic neurotoxins. Gastroenterology 86, 546-551.

19. Raabe W., Lin S. (1984) Ammonia, postsynaptic inhibition and CNS-energy state. Brain Res. 303, 67-76.

20. Torda C. (1953) Ammonium ion content and electrical activity of the brain during the preconvulsive and convulsive phases induced by various convulsants. J. Pharmacol. Exp. Ther. 107, 197-203.

21. Roberge A., Charbonneau R. (1969) Protective effects of arginine against hepatic coma, insulin shock and epilepsy. Life Sci. 8, 369-377.

GABA and Benzodiazepine Receptors in Hepatic Encephalopathy

HEPATIC ENCEPHALOPATHY AND BENZODIAZEPINE RECEPTOR LIGANDS

E. Anthony Jones, Sergio H. Gammal, Anthony S. Basile, Kevin D. Mullen, Mark L. Bassett, Daniel F. Schafer, Phil Skolnick.

Liver Diseases Section and Laboratory of Neuroscience, NIDDK, National Institutes of Health, Bethesda, MD.

Benzodiazepine (BZ) receptor ligands interact with specific binding sites on the $GABA_A$ receptor/chloride channel complex on neurons in the central nervous system (CNS) (1) (Figure 1). Such ligands could contribute to or ameliorate hepatic encephalopathy (HE) if the supramolecular complex itself is involved in this syndrome. The complex includes distinct receptors for BZ ligands and GABA, and a chloride ionophore on which there are binding sites for barbiturates and cage convulsants, such as picrotoxin. Activation of the effector component of this complex, the chloride channel, is mediated by the binding of GABA to its receptor and is potentiated by the binding of a BZ agonist or a barbiturate to discrete loci on the complex. Activation induces conformational changes in the complex which result in opening of the chloride channel and hyperpolarization of the neuron (1) (Figure 1). This GABA-gated chloride ion conductance is the basis of GABA-mediated inhibitory neurotransmission or "GABAergic tone." If the complex is involved in HE, the basic abnormality may be related to increased GABAergic tone, since this phenomenon would induce impaired motor function and decreased consciousness (2), prominent features of the syndrome of HE. Such behavioral changes are readily mimicked by agents that augment GABAergic neurotransmission, such as BZ agonists or barbiturates (2-5), which increase the frequency or open-time of chloride channel openings respectively (6). Indeed, it was the observation in rabbits

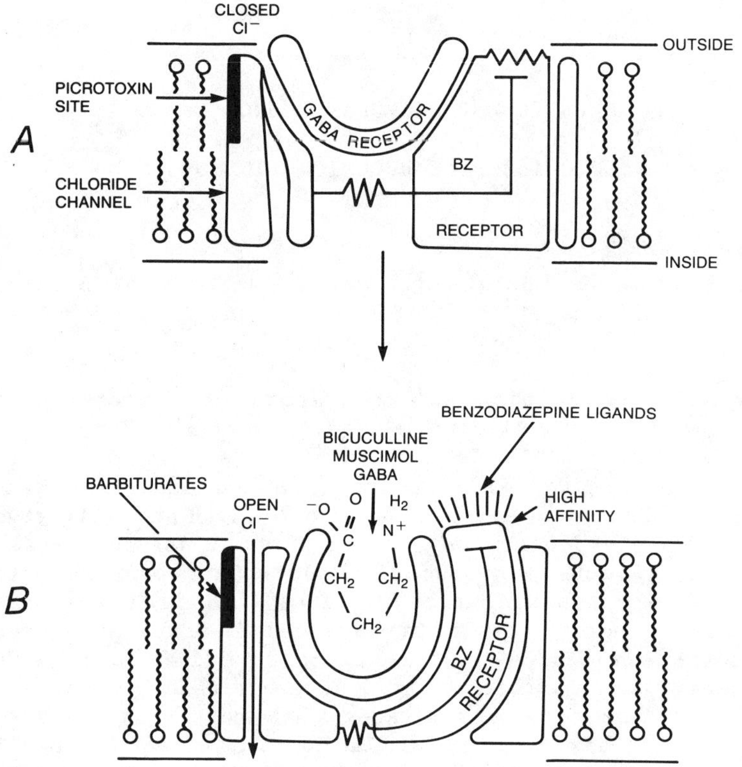

Figure 1: The GABA$_A$/benzodiazepine receptor/chloride ionophore complex. Receptors are depicted for: (a) GABA, (b) barbiturates and (c) benzodiazepine ligands. GABA, GABA receptor agonists (e.g. muscimol), and GABA receptor antagonists (e.g., bicuculline) bind to GABA receptors; barbiturates are thought to bind to specific recognition sites near the chloride ionophore; benzodiazepine receptor agonists (e.g., diazepam), benzodiazepine receptor antagonists (e.g., flumazenil) and benzodiazepine receptor inverse agonists (e.g., 6,7 dimethoxy-4-ethyl-3-carbomethoxy-β-carboline) bind to benzodiazepine receptors.

A: shows the receptor complex in an inactivated state with the Cl^- channel closed. B: shows the receptor complex in an activated state with the Cl^- channel open. Activation is induced by GABA or GABA agonists binding to GABA receptors, or by barbiturates or benzodiazepine agonists interacting with their specific receptors in the presence of GABA. Activation of the receptor complex is associated with conformational changes and opening of the Cl^- channel which will hyperpolarize the cell.

Reproduced in a slightly modified form with permission from Biological Psychiatry (1).

that the abnormalities of the waveform of the visual evoked response (VER) in experimental HE resembled those in pentobarbital-induced coma that originally led to the hypothesis that HE is associated with increased GABA-ergic tone (3,7,8).

Original findings which were considered to provide additional support for the concept of an involvement of the supramolecular complex in HE were a gut source for GABA and, in a rabbit model of fulminant hepatic failure (FHF) (9), an increase in serum levels of GABA-like activity, an increase in the permeability of the blood brain barrier to an isomer of GABA and data compatible with an increase in the densities of GABA and BZ receptors (8). However while these particular findings served a most useful role in focussing attention on the supramolecular complex in HE, it is important to emphasize that none of them is a necessary requisite for the occurrence of increased GABAergic tone (Table 1). While other factors in liver failure could well be responsible for this phenomenon (10), the initial findings have succeeded in stimulating a new wave of research activity relating to the pathogenesis of HE. Perhaps inevitably much of this activity has been devoted to vigorously attacking the potential validity of the hypothesis. This has largely taken the form of citing studies of GABA concentrations, blood brain barrier permeability and the status of GABA and BZ receptors. Unfortunately such studies do not directly address the issue of whether GABAergic tone is increased in HE (10) (Table 1).

A role for BZ receptor ligands in HE has become apparent as a consequence of attempting to answer two questions. First, is GABAergic tone increased in HE? Second, if it is, what mechanism(s) are responsible?

TABLE 1

THE HYPOTHESIS IMPLICATING THE $GABA_A$ RECEPTOR COMPLEX IN THE PATHOGENESIS OF HEPATIC ENCEPHALOPATHY

A. The hypothesis is currently not dependent on studies of-

- Serum GABA levels.
- CSF GABA levels.
- Brain GABA levels.
- The gut as a source of GABA.
- Blood brain barrier permeability.
- Changes in GABA receptor density and affinity.
- Changes in BZ receptor density and affinity.
- GABA transaminase activity.
- Glutamate decarboxylase activity.
- Direct excitatory actions of BZ receptor antagonists.
- Unsatisfactory models of HE.

B. The hypothesis is currently supported by studies of-

- Seizure induction by drugs which decrease GABA-ergic tone (2,11).
- GABA-mediated changes in VERs (2-5).
- GABA and BZ agonist-induced depression of neuronal firing rate (14)
- BZ antagonist-induced changes in neuronal firing rate (14)
- BZ antagonist-induced changes in neuronal sensitivity to a GABAmimetic (14).
- BZ and GABA antagonist and Cl^- channel blocker-induced changes in behavioral manifestations of HE (2,5,22,34,35).
- BZ and GABA antagonist and Cl^- channel blocker induced changes in electrophysiologic manifestations of HE (2,5,22,34,35).
- Neurochemical evidence for presence of endogenous BZ agonists (22,36-38).
- Apparently satisfactory models of FHF (9,27-29, 33).

Several observations in animal models of FHF suggest that HE is indeed associated with augmented GABA-ergic tone. Increased resistance to drugs, such as bicuculline (2) or mercaptoproprionic acid (11), which induce seizures by reducing GABAergic tone has been demonstrated. Bicuculline is a GABA receptor antagonist (12), whereas mercaptoproprionic acid reduces the synthesis of GABA by inhibiting glutamate decarboxylase (13). The abnormalities of the waveform of VERs resemble those induced by drugs that augment GABAergic tone, such as pentobarbital (3,4) and the GABA agonist, muscimol (3). Neurons from the CNS exhibit increased sensitivity to depression by agonists of the supramolecular complex, such as muscimol, but not to depression by an α-adrenoreceptor agonist (14). Behavioral and electrophysiologic ameliorations of the encephalopathy have been induced by pharmacologically antagonizing binding sites on the supramolecular complex (Table 1). For example, bicuculline has been used to antagonize the GABA receptor and isopropylbicyclophosphate to block the chloride channel (2). Additional evidence suggesting the presence of increased GABAergic tone in models of HE involves studies of BZ receptor ligands (Table 1).

There are several possible mechanisms that could be responsible for increased GABAergic tone in HE. One possibility would be an increased sensitivity of the supramolecular complex due to an increase in the density of receptors, or structural changes in receptors that alter the apparant affinities of their ligands. The available data relating to this possibility are inconsistent (e.g. 15-19). Interpretation of such studies are facilitated by recognizing that while an increase in the density of supramolecular complex receptors affords one potential mechanism for increased GABAergic tone (i.e., by theoretically permitting more agonist-receptor interaction), such a finding would not itself be an indication that increased GABAergic tone is necessarily present, or conversely, the lack of such a finding would not be an indication that increased GABAergic tone is not increased (10). Another possibility would be increased delivery of GABA to its receptor. This could occur from presynaptic neurons or from plasma across the blood brain barrier. There appears to be a paucity of data on the release of GABA from presynaptic neurons in HE. However, plasma GABA has been convincingly shown to cross the blood brain barrier in increased amounts in one model of FHF (20), and increased GABA in brain extracellular fluid, as occurs for example following γ-vinyl GABA-induced inhibition of GABA transaminase, is known to induce an encephalopathy that resembles HE (2). A third, and perhaps most important, possibility would be the presence of an endogenous ligand which potentiates the action of GABA. For example, if a substance with BZ agonist-like properties were present in liver failure it may contribute to the manifestations of HE by allosterically potentiating GABAergic tone (1,21). Recent evidence strongly supports this last possibility (10,22).

The first experimental findings that suggested a possible relationship between BZ receptors and HE were generated in studies of VERs (3). VERs occur in response to repeated flashes of light. They are contained within the electroencephalogram (EEG), and are derived from it by averaging signals that are time-locked to the photic stimulus. The result is an average smooth VER trace which is devoid of extraneous background EEG activity. VERs are considered to represent summations of the electrical activity from volleys of excitatory and inhibitory postsynaptic potentials through neuronal networks in the brain region being monitored (occipital cortex) (23-26). Using an extensively characterized model of HE (9,27-29), an early finding (3), which has subsequently been repeated (2), was that the abnormalities of the VER in HE closely resembled those in diazepam-induced coma (Table 1). As BZs like diazepam induce neural inhibition by interacting with BZ receptors and allosterically potentiating the action of GABA (1,21), one of the possible interpretations of these findings would be that GABAergic tone is increased in HE due to events at the BZ receptor. These VER studies (3) provided a rationale for further studies of BZ receptor ligands in relation to HE. In particular, if events at BZ receptors play a role in the neurochemical mechanisms that mediate HE, then antagonizing the BZ receptor should ameliorate the syndrome.

Flumazenil (Ro 15-1788, Hoffmann La Roche) is regarded as the prototype of a class of drugs known as BZ receptor antagonists. Structurally it is an imidazobenzodiazepine (Figure 2). Functionally, it counteracts the effects of BZ ligands by competitively displacing them from BZ receptors. It is virtually without behavioral effects in normal animals, but it abruptly reverses BZ-induced sedation (30-32). The relationship of

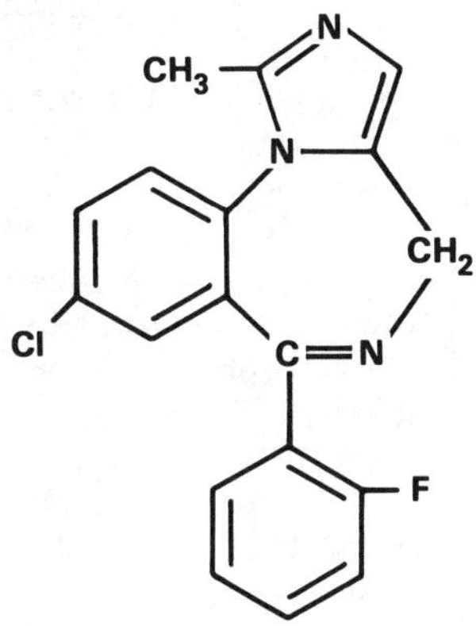

MIDAZOLAM

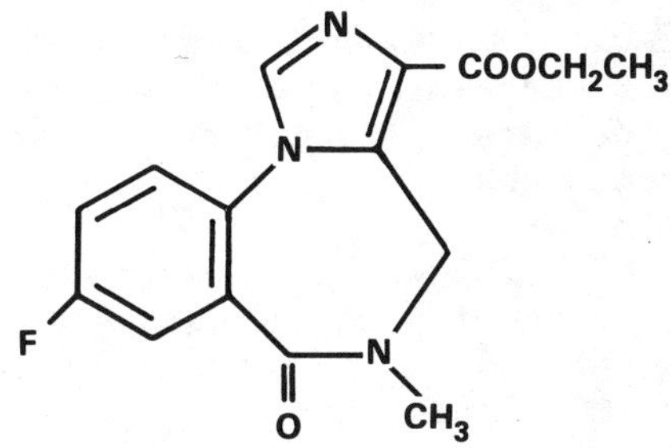

FLUMAZENIL

Figure 2: The chemical formulae of midazolam and flumazenil (Ro 15-1788). Midazolam is a 1,4-benzodiazepine which acts as an agonist at the benzodiazepine receptor. Flumazenil is an imidazobenzodiazepine which acts as an antagonist (and weak partial agonist) at the benzodiazepine receptor. While there are obvious differences between the structures of the two drugs, certain structural similarities between them are apparent.

this drug to BZ agonists like diazepam, is analogous to the relationship between naloxone and opiates, such as morphine.

Using a subjective behavioral response score based on assessments of muscle tone, responses to a painful stimulus, alertness, and spontaneous motor activity, a transient, but unequivocal behavioral amelioration of HE was consistently induced in the rabbit model (2) by the IV injection of flumazenil. These behavioral ameliorations were associated with a significant correction of the abnormal VER in this model (i.e., correction of an increased amplitude of the first positive wave). Injection of the vehicle for the drug had no detectable effect on behavior or the VER (2). These observations suggest that in this model of HE, an endogenous ligand with BZ-like properties may have been displaced from BZ receptors by the BZ receptor antagonist, thereby decreasing increased GABAergic tone by disinhibiting neurons.

It was obviously important to determine whether antagonism of the BZ receptor could also induce ameliorations of HE in another model in a different species. The alternative model chosen, which had also been extensively characterized, was an improved model of thioacetamide-induced FHF in the rat (33). To assess objectively the behavioral responses of this model to flumazenil, motor activity of rats was quantitated using a computerized open field activity meter which measures a variety of different movements over time using infrared sensors. In rats with HE, the intraperitoneal injection of flumazenil induced significantly more motor activity, including travelling a greater distance, than did injection of the vehicle for the drug. Flumazenil-induced behavioral ameliorations of HE were associated with a partial correction of the associated abnormalities of the VER in this model (notably a significant increase in the decreased amplitude of the first negative wave) (34). These findings suggest not only that an endogenous BZ-like compound can occur in HE, but also that the presence of such a compound in HE is neither species nor model specific.

To stengthen further the potential relationship between BZ receptor-mediated augmentation of GABAergic neurotransmission and HE, a technique with a higher resolution than the VER was applied. This involved recording the changes in spontaneous electrical activity or firing rate of Purkinje neurons in cerebellar slices in response to drugs that act specifically at the BZ receptor. Cerebellar slices were continuously perfused with an oxygenated medium to which a drug could be added. The spontaneous activity of single neurons was recorded using microelectrodes. This electrical activity was amplified, filtered and fed through a window discriminator to separate action potentials from background noise. When this technique was applied to control rabbit neurons, low concentrations of flumazenil were shown to have no effect on the spontaneous firing rate, whereas high concentrations depressed neuronal activity due to a weak partial agonist action of the drug. In contrast, when neurons from the rabbit model were studied application of flumazenil was associated with a qualitatively different response in that there was a robust concentration dependent increase in neuronal activity (14). Given the specificity of this drug for BZ receptors, the most plausible explanation for this phenomenon would be that flumazenil had displaced an agonist (i.e., BZ-like pharmacologic properties) ligand that was potentiating GABAergic tone. Flunitrazepam was shown to be 3-4 times more potent in depressing the activity of neurons from this model of HE than control neurons (14). This latter finding could be due to increased availability of GABA at its receptor. Alternatively the presence of increased concentrations or availability of an endogenous BZ agonist ligand could be responsible for amplifying the responsiveness to flunitrazepam.

If current concepts of the mechanism of BZ receptor-induced modulation of GABA-gated chloride channel function are correct (1,21), and if increased concentrations or availability of an endogenous BZ agonist ligand were present in HE, then a BZ receptor antagonist, such as flumazenil, should decrease the observed hypersensitivity of neurons from the rabbit model of HE to muscimol-induced depression of their spontaneous firing rate. Indeed, such pretreatment was shown to abolish completely this hypersensitivity (14). This finding provides further evidence for the presence of an endogenous BZ agonist ligand in HE.

To summarize, it has been shown in animal models of FHF that: (i) there are similarities between the VER in HE and BZ agonist-induced sedation; (ii) CNS neurons exhibit increased sensitivity to depression by a BZ agonist; and (iii) a BZ receptor antagonist induces excitation of individual CNS neurons and behavioral and electrophysiologic ameliorations of the encephalopathy. Thus, it can be inferred from studies utilizing BZ receptor ligands that in models of HE due to FHF: (i) GABAergic tone is increased; (ii) ameliorations of the encephalopathy can be induced by competitive blockade of BZ receptors, and (iii) an increased concentration or availability of an endogenous ligand with BZ agonist-like properties is probably present. The results of these animal studies also provide a rationale for trials of a BZ receptor antagonist, such as flumazenil, in the management of human beings with HE.

REFERENCES

1. Paul S.M., Marangos P.J., and Skolnick P. (1981) The benzodiazepine-GABA-chloride ionophore receptor complex. Common site of minor tranquillizer action. Biol. Psych. 16, 213-229.

2. Bassett M.L., Mullen K.D., Skolnick P. et al. (1987) Amelioration of hepatic encephalopathy by pharmacologic antagonism of the $GABA_A$/benzodiazepine receptor complex in a rabbit model of fulminant hepatic failure. Gastroenterology 93, 1069-1077.

3. Schafer D.F. Pappas S.C., Brody L.E. et al. (1984) Visual evoked potentials in a rabbit model of hepatic encephalopathy. I. Sequential changes and comparisons with drug-induced comas. Gastroenterology 86, 540-545.

4. Jones D.B., Mullen K.D., Roessle M. et al. (1987) Hepatic encephalopathy: application of visual evoked responses to test hypotheses of its pathogenesis in rats. J. Hepatol. 4, 118-126.

5. Gammal S.H., Jones E.A., Bassett M.L. et al. (1988) Antagonists of the GABA/benzodiazepine receptor complex ameliorate hepatic encephalopathy in two models of fulminant hepatic failure, in Advances in Ammonia Metabolism and Hepatic Encephalopathy. (Soeters P.B., Wilson J.H.P., Meijer A.J., and Holm E. eds), pp. 325-332. Excerpta Medica, Amsterdam.

6. Study R.E., and Barker J.L. (1981) Diazepam and (-)-pentobarbital Fluctuation analysis reveals different mechanisms for potentiation of γ-aminobutyric acid responses in cultured neurons. Proc. Natl. Acad. Sci. USA 78, 7180-7184.

7. Schafer D.F., Fowler J.M., Brody L.E. et al. (1980) Hepatic coma and inhibitory neurotransmission: the enteric bacterial flora as a source of γ-aminobutyric acid. Gastroenterology 79, 1052. (Abstract).

8. Schafer D.F. Jones E.A. (1982) Hepatic encephalopathy and the γ-aminobutyric acid neuro-transmitter system. Lancet i, 18-20.

9. Blitzer B.L., Waggoner J.G., Jones E.A. et al. (1978) A model of fulminant hepatic failure in the rabbit. Gastroenterology 74, 664-671.

10. Jones E.A., and Skolnick P. (1989) Benzodiazepine receptor ligands and the syndrome of hepatic encephalopathy, in Progress in Liver Diseases, volume IX. (Popper H., Schaffner F., eds). W.B. Saunders, Philadelphia, 'In press'.

11. Ferreira M.R., Gammal S.H., and Jones E.A. (1988) Hepatic encephalopathy (HE): Evidence of increased GABA-mediated neurotransmission in a rat model of fulminant hepatic failure (FHF). Gastroenterology 94, A616. (Abstract).

12. Olsen R.W. (1982) Drug interactions at the GABA receptor-ionophore complex. Ann. Rev. Pharmacol. Toxicol. 22, 245-277.

13. Karlsson A., Fonnum F., Malthe-Sorrenssen D. et al. (1974) Effect of the convulsive agent 3, mercaptopropionic acid on the levels of GABA, other amino acids and glutamate decarboxylase in different regions of the rat brain. Biochem. Pharmacol. 23, 3053-3061.

14. Basile A.S., Gammal S.H., Mullen K.D. et al. (1988) Differential responsiveness of cerebellar Purkinje neurons to GABA and benzodiazepine receptor ligands in an animal model of hepatic encephalopathy. J. Neurosc. 8, 2414-2421.

15. Baraldi M., and Zeneroli M.L. (1982) Experimental hepatic encephalopathy: changes in the binding of γ-aminobutyric acid. Science 216, 427-429.

16. Schafer D.F., Fowler J.M., Munson P.J. et al. (1983) Gamma-aminobutyric acid and benzodiazepine receptors in an animal model of fulminant hepatic failure. J. Lab. Clin. Med. 102, 870-880.

17. Maddison J.E., Dodd P.R., Morrison M. et al. (1987)

Plasma GABA, GABA-like activity and the brain GABA-benzodiazepine receptor complex in rats with chronic hepatic encephalopathy. Hepatology 7, 621-628.

18. Roy S., Pomier-Layrargues G., Butterworth R.F. et al. (1988) Hepatic encephalopathy in cirrhotic and portacaval shunted dogs: lack of changes in brain GABA uptake, brain GABA levels, brain glutamic acid decarboxylase activity and brain postsynaptic GABA receptors. Hepatology 8, 845-849.

19. Ferenci P., Zimmerman C., and Ebner J. (1988) Postsynaptic γ-aminobutyric acid (GABA) receptors in thioacetamide-induced acute liver failure in rats, in Advances in Ammonia Metabolism and Hepatic Encephalopathy. (Soeters P.B., Wilson J.H.P. Meijer A.J. and Holm E. eds), pp. 218-224. Excerpta Medica, Amsterdam.

20. Bassett M.L. Mullen K.D. Scholz B., et al. (1985) Increased brain GABA uptake in precoma encephalopathy in a rabbit model of fulminant hepatic failure. Hepatology 5, 1032. (Abstract).

21. Squires R.F. (1984) Benzodiazepine receptors, in Handbook of Neurochemistry, 2nd ed. Volume 6. (Lajtha A. ed), pp. 261-306. Plenum, New York.

22. Mullen K.D., Martin J.V., Mendelson W.B. et al. (1988) Could an endogenous benzodiazepine ligand contribute to hepatic encephalopathy? Lancet i, 457-459.

23. Chiappa K.H., and Ropper A.H. (1982) Evoked potentials in clinical medicine. N. Eng. J. Med. 306, 1140-1147.

24. Zemon V. Kaplan E., and Katloff F. (1980) Bicuculline enhances a negative component and diminishes a positive component of the visual evoked cortical potential in the cat. Proc. Natl. Acad. Sci. USA 77, 7476-7478.

25. Nakayama K., (1982) The relationship of visual evoked potentials to cortical physiology. Ann. N.

Y. Acad. Sci. 388, 21-36.

26. Dyer R.S., Jensen K.F. and Boyer W.K. (1987) Focal lesion of visual cortex: effects on visual evoked potentials in rats. Exp. Neurol. 95, 100-115.

27. Traber P.G. Ganger D.R., and Blei A.T. (1986) Brain edema in rabbits with galactosamine-induced fulminant hepatitis. Regional differences and effects of intracranial pressure. Gastroenterology 91, 1347-1356.

28. Traber P.G., Canto M.D., Ganger D.R. et al. (1987) Electron microscopic evaluation of brain edema in rabbits with galactosamine-induced fulminant hepatic failure. Ultrastructure and integrity of the blood brain barrier. Hepatology 7, 1272-1277.

29. Mullen K.D., Schafer D.F., Cuchi P. et al. (1988) Evaluation of the suitability of galactosamine-induced fulminant hepatic failure as a model of hepatic encephalopathy in the rat and the rabbit, in Advances in Ammonia Metabolism and Hepatic Encephalopathy. (Soeters P.B. Wilson J.H.P., Meijer A.J., and Holm E. eds), pp. 205-212. Excerpta Medica, Amsterdam.

30. Hunkeler W., Mohler H., Pieri L. et al. (1981) Selective antagonists of benzodiazepines. Nature 290, 514-516.

31. Bonetti E., Pieri L., Cumin R. et al. (1982) Benzodiazepine antagonist Ro 15-1788: Neurological and behavioral effects. Psychopharmacology 78, 8-18.

32. Haefely W. (1983) Antagonists of benzodiazepines: functional aspects, in Benzodiazepine Recognition Site Ligands: Biochemistry and Pharmacology. (Biggio G., and Costa E. eds), pp. 73-93. Raven Press, New York.

33. Geller D. Gammal S.H., Mullen K.D. et al. (1988) An improved rat model of hepatic encephalopathy due to fulminant hepatic failure. The importance of supportive therapy, in Advances in Ammonia Metabolism and Hepatic Encephalopathy. (Soeters P.B., Wilson

J.H.P., Meijer A.J., and Holm E. eds), pp. 213-217. Excerpta Medica, Amsterdam.

34. Gammal S.H., Basile A.S., Geller D. et al. Involvement of the benzodiazepine receptor in hepatic encephalopathy: Quantitative behavioral and electrophysiological evidence in a rat model. (In preparation).

35. Baraldi M., Zeneroli M.L., Ventura E. et al. (1984) Supersensitivity of benzodiazepine receptors in hepatic encephalopathy due to fulminant hepatic failure in the rat: reversal by a benzodiazepine antagonist. Clin. Sci. 67, 167-175.

36. Mullen K.D. Szanter K.M., Kaminsky K. et al. (1989) Detection and characterization of endogenous benzodiazepine activity in both animal models and humans with hepatic encephalopathy. (This volume).

37. Basile A.S., Ostrowski N.L., Gammal S.H. et al. (1989) The $GABA_A$ receptor complex in hepatic encephalopathy: Autoradiographic evidence for the presence of an endogenous benzodiazepine receptor ligand. Gastroenterology. (Abstract).

38. Basile A.S. Gammal S.H., Jones E.A. et al. (1989) The $GABA_A$ receptor complex in an experimental model of hepatic encephalopathy: Evidence for elevated levels of an endogenous benzodiazepine receptor ligand. Gastroenterology. (Abstract).

DETECTION AND CHARACTERIZATION OF ENDOGENOUS BENZODIAZEPINE ACTIVITY IN BOTH ANIMAL MODELS AND HUMANS WITH HEPATIC ENCEPHALOPATHY

KD Mullen, KM Szauter, K Kaminsky
PD Tolentino, AJ McCullough
Cleveland Metropolitan General Hospital
Case Western Reserve University
3395 Scranton Road, Cleveland, Ohio 44109

The search for endogenous ligands for central nervous system benzodiazepine(BZ) receptors and the definition of the fundamental mechanism responsible for the mediation of hepatic encephalopathy(HE) have had parallel courses in futility until very recently.[1] These apparently disparate fields of interest have converged because of experimental observations made in animal models of HE. The catalyst for this joining of forces was the development of BZ receptor antagonists. These drugs e.g. Ro 15-1788 (Flumazenil) and CGS-8216 largely devoid of effect in controls reverse the sedative effects of synthetic agonist BZ compounds. Administration of these agents to animal models of HE resulted in considerable amelioration in the manifestations of HE[2,3] which has been corroborated to some degree in isolated human reports. While the response in HE was not as complete as that seen with administration of the drug to controls treated with agonist BZ drugs it was of sufficient magnitude to suggest a significant neurological event was taking place which required further study.

Analogous to the situation of brain opiate receptors it appears unlikely that the brain would have receptors for synthetic BZ drugs by chance alone. Instead it is probable that these recognition sites are present because of exposure of brain to compounds which mediate a CNS modulatory effect by binding to these recognition sites. However the BZ receptor equivalents of endogenous ligands for opiate receptors, endorphins and encephalins have yet to be identified.[4] Despite the apparent lack of effect of BZ receptor antagonists in normal controls, enough information is available to suggest that given certain test situations opposing CNS effects can be seen (anxiogenic and anxiolytic) with administration of these drugs leading File and Pellow to suggest that there may be two endogenous ligands for the BZ receptor.[5] One with proconflict, excitatory properties and the other simulating the sedative, anticonflict effects of diazepam. Based on our current understanding of the mechanism of action of BZ receptor antagonists a plausible hypothesis to entertain for the reversal of HE is that it is due to accumulation of this hypothesized sedative endogenous ligand in a critical subregion of brain. The improvement in HE by BZ receptor antagonists would thus be due to displacement of this neurodepressant compound off the BZ receptor. Compounds which bind to central BZ receptors have been shown convincingly to mediate their CNS effects through modulation of GABAergic neural transmission. Thus the benzodiazepine hypothesis should be looked upon as a variant of the GABA hypothesis first postulated by Schafer and Jones rather than a distinct new entity.[6]

Rabbit Studies

Based on this hypothesis a study was undertaken in 1985 to determine if any evidence for the presence of this endogenous BZ ligand could be found. The rabbit model of fulminant hepatic failure induced by D-galactosamine(GalN) was first used because it had been shown to respond to BZ receptor antagonists.[3] Sera and CSF obtained from rabbits in stage III HE was assayed for endogenous BZ activity using a radioreceptor assay. Compared to controls CSF obtained from animals with HE displayed significant BZ activity using the BZ ligand (^{3}H) beta-carboline-ethyl-ester (BCEE). Subsequent studies using another (^{3}H) ligand Ro 15-1788 initially did not show any differences between HE and controls leading to a speculation of a shared selectivity of the endogenous BZ ligand for BZ_1 receptor subtypes as has been noted with BCEE. Further studies however showed that in fact no selectivity was present and Ro 15-1788 binding was also displaced significantly by HE CSF. The limited material available was also analyzed for GABA and purine compounds using HPLC. While GABA levels were modestly elevated with HE hypoxanthine concentrations were elevated 6 fold. However removal of hypoxanthine using xanthine oxidase treatment did not alter activity effectively excluding it as a contributor to the endogenous BZ activity of HE CSF. Deproteinized Sera did not display any significant BZ activity.[7]

Limited studies with extracts of brains from rabbits with HE have shown that the BZ activity is heat stable; low molecular weight (< 1000 daltons) and behaved as a competitive inhibitor of BZ binding.

Rat Studies

Gammal et al have recently reported that the thioacetamide rat model of hepatic failure is similar to the GalN rabbit model in that it showed reversal of HE with administration of BZ receptor antagonists.[8] Brain extractions in this model have yielded similar BZ activity which was heat stable and low in molecular weight (< 1000 daltons). The activity appeared to have excellent specificity for central BZ receptors and to be non-polar in nature.

Human Studies

Unlike animals the problem with humans with HE is to ensure that they have not received BZ drugs. Because of the frequency of endoscopy in patients with severe liver disease many are not suitable for study in regards to "endogenous" BZ activity because they receive BZ drugs for pre-medication. Despite this limitation of enrollment which at times precluded study of as many as 80% of hospital admissions, a relatively large number of samples have been collected from BZ drug free patients (no BZ drugs for > 2000 hours) in various stages of HE.

Initially cerebrospinal fluid but not sera collected from patients with advanced hepatic encephalopathy was shown to contain significant benzodiazepine binding activity using a radioreceptor assay. The principle of this assay is that deproteinized physiologic fluids which contain BZ compounds will displace (^{3}H) Ro 15-1788 (0.2 M) binding to fixed amounts of cortical membrane BZ receptors in proportion to their concentration of the BZ compound. The term significant BZ activity applies to the situation where significantly greater ($p < 0.05$) displacement of the (^3H) Ro 15-1788 binding occurs with groups of samples from one source versus another (e.g. HE vs control). It should be noted that in normal controls some displacement of BZ binding is seen. This activity is the source of most if not all previous studies to try and identify an endogenous BZ ligand. The BZ activity is expressed as either % inhibition of total specific BZ binding or as equivalent to a concentration of a known BZ (e.g. diazepam or oxazepam) derived from comparison to a standard curve generated for this compound. The results for human CSF were as follows: (see TABLE 1)

Subsequently it was discovered that the activity in sera/plasma was highly protein bound (≈ 90%). Using a refinement of our extraction procedures we detected significant benzodiazepine binding activity in patients with advanced HE (see Table 2). The extraction technique employed (C18 octadecylsiliate minicolumn) clearly showed that the activity was non-polar in nature. We now measure BZ activity using a modified radio immunoassay (^{125} I-oxazepam, sheep antibenzodiazepine antibody) as well as with the radioreceptor assay. The results in the first 48 human samples tested using the RIA are detailed in Table 2. Correspondingly high levels were seen with the radioreceptor assay.

Based on the available data to this point it appears that a compound with BZ immunoreactivity and receptor activity is accumulating in patients with severe hepatic encephalopathy. Since ingestion of BZ drugs can be excluded as the cause for these findings the source of this

TABLE 1

ENDOGENOUS BENZODIAZEPINE ACTIVITY IN HUMAN CSF

	n	% INHIBITION	DIAZEPAM BINDING UNITS
CONTROLS	8	6.8 ± 3.5	11.0 ± 5.6
HE	8	33.5 ± 23.7	54.3 ± 38.4
p		< 0.01	< 0.01

TABLE 2

RADIOIMMUNOASSAY OF BZ ACTIVITY IN HUMAN PLASMA

Mean + SD in ng/ml Oxazepam Equivalents

Group	Cirrhotic No HE	Cirrhotic HE I-II	Cirrhotic III-IV	Uremia	Controls
n	7	11	9	5	16
ng/ml	16.2 ± 5.8	15.1 ± 11.3	>>150	18 ± 9.1	10 ± 6.3
p vs control	NS	NS	< 0.001	NS	

activity remains to be established. The growing list of putative endogenous BZ ligands (Table 3) contains a very diverse array of compounds. Initial characterization of the BZ activity has shown it to be heat stable (100°C X 10 minutes), and to have a molecular weight less than 1000 daltons by ultrafiltration. As mentioned earlier the activity is non-polar in nature and extracts into a variety of solvents (methanol, isoamyl-alcohol/benzene, acetonitrile). Using both the RIA and radioreceptor assay we have excluded the other previously proposed HE toxins and bilirubin as contributors to the BZ activity. The available putative BZ ligands listed have also been tested and while some (especially betacarbolines) are active in the radioreceptor assay only Desmethyldiazepam and diazepam have significant activity in both assays. Preliminary high performance liquid chromatography studies suggest that the BZ activity in HE resides in at least two distinct fractions with retention times close to desmethyldiazepam. However formal identification remains to be determined.

Conclusions

These apparently exciting results have to be viewed with caution not withstanding the fact that they provide a

TABLE 3

ENDOGENOUS BENZODIAZEPINE LIGANDS

- Beta Carboline Compounds
- Diazepam Binding Inhibitor
- Purine Compounds
- Nicotinamide
- Thromboxane A_2, Prostaglandins A_1, A_2
- Unidentified High and Low MW Compounds
- Desmethyldiazepam and Diazepam

rationale basis for the apparent response of HE to benzodiazepine receptor antagonists. Wildmann and colleagues have recently reported the finding of an assortment of benzodiazepine compounds in plants.[9] It is possible that the detection of BZ activity in patients with severe liver disease reflects the ingestion of these compounds in the diet. Since the liver is involved in the biotransformation and detoxification of benzodiazepines, accumulation of these compounds may occur in the presence of liver disease. However until the BZ activity is formally identified, shown to be neurodepressant, and to be present in sufficient quantities to mediate a CNS effect it will remain just a hypothesis that a BZ like compound is mediating HE. Recently the conflicting anxiogenic-anxiolytic actions of Ro 15-1788 seen in certain test situations has been attributed to the effect of this compound on adenosine mediated neural transmission.[10] While electrophysiologic studies in cerebellar brain slices suggest strongly that GABA/BZ receptor mediated events are predominantly operative in HE[11] it remains a possibility that some non-BZ receptor type of mechanism is involved in the pathogenesis of this syndrome. However at present the concept that a BZ like compound is contributing to the neural inhibitor of HE remains an attractive hypothesis. One issue that still needs to be resolved is why BZ compounds induce such a profound "encephalopathy" in patients with liver disease compared to normal controls. The underlying mechanism which converts a sedative type compound to a virtual anesthetic in some patients is not simply explicable based on the endogenous BZ hypothesis. Perhaps ammonia or one of the other HE toxins may play a role in this "amplification" effect.

References

1. Mullen KD, Martin JV, Mendelson WB, Bassett ML, Jones EA. Could an endogenous benzodiazepine ligand contribute to hepatic encephalopathy? Lancet i 457-459,1988.
2. Baraldi M, Zeneroli ML, Ventura E, et al. Supersensitivity of benzodiazepine receptors in hepatic encephalopathy due to fulminant hepatic failure in the rat reversal by benzodiazepine antagonist. Clin Sci 67:167-175,1984.

3. Bassett ML, Mullen KD, Skolnick P, Jones EA. Amelioration of hepatic encephalopathy by pharmacologic antagonism of the $GABA_A$-benzodiazepine receptor complex in a rabbit model of fulminant hepatic failure. Gastro 93:1069-1077,1987.
4. Hamon M, Soubrie P. Searching for endogenous ligand(s) of central benzodiazepine receptors. Neurochem Int, 5:663-672,1983.
5. File S, Pellow S. Do the intrinsic actions of benzodiazepine receptor antagonists imply the existence of an endogenous ligand for benzodiazepine receptors? In: Advances in Biochemical Psychopharmacology: GABAergic Transmission and Anxiety. Raven Press, New York, 98-107,1986.
6. Schafer DF, Jones EA. Hepatic encephalopathy and the gamma-aminobutryric acid neurotransmitter system. Lancet i, 18-20,1982.
7. Mullen KD, Martin JV, Bassett ML, et al. Hepatic encephalopathy: A syndrome modulated by an endogenous benzodiazepine ligand? Hepatology 6:1221,1986.
8. Gammal SH, Geller D, Skolnick P, et al. Unequivocal amelioration of hepatic encephalopathy by benzodiazepine antagonists in a rat model of fulminant hepatic failure. J Hepatol 4. Suppl. 1 S17,1987.
9. Wildmann J, Mohler H, Vetler W, Ranalder V, Schmidt K, Maurer R. Diazepam and N-desmethyldiazepam are found in rat brain and adrenal and may be of plant origin. J Neural Transm 70:383-388,1987.
10. Phillis JW, O'Regan MH. The role of adenosine in the central actions of the benzodiazepines. Prog Neuropsychopharmacol Biol Psychiat, 12,389-404,1988.
11. Basile AS, Gammal SH, Mullen KD, Jones EA, Skolnick P. Differential responsiveness of cerebellar purkinje neurones to GABA and benzodiazepine receptor ligands in an animal model of hepatic encephalopathy. J of Neuroscience 8:2414-2421,1988.

Acknowledgments

This work was supported in part by grants from the Cuyahoga County Hospital Foundation, Case Western Reserve University Biomedical research support grant #RR-05410-26, and RO1 DK-39527-01

ISOLATED CNS NEURONS FROM A MODEL OF HEPATIC ENCEPHALOPATHY EXHIBIT INCREASED SENSITIVITY TO A BENZODIAZEPINE

Sergio H. Gammal, Anthony S. Basile, Phil Skolnick and E. Anthony Jones

Liver Diseases Section and Laboratory of Neuroscience, NIDDK, National Institutes of Health, Bethesda, M.D.

Patients with cirrhosis appear to exhibit an exaggerated response to the administration of benzodiazepines (BZs) both in terms of clinical neurologic and electroencephalographic changes (1). One potential explanation for increased sensitivity to BZs would be a decrease in their hepatic metabolism. However patients with similar degrees of hepatic dysfunction sometimes show different sensitivities to these drugs. In one recent study, after the administration of a specific dose of the BZ, triazolam, to control subjects and patients with cirrhosis the plasma levels of this drug were similar in both groups, but the sedative response was significantly greater in the cirrhotic group (2). These findings suggest that impaired hepatic metabolism of BZs may not be sufficient to account for the increased sensitivity to these drugs in liver disease and that an alternative potential explanation for this phenomenon might be an increased sensitivity of neurons in the central nervous system (CNS) to BZs.

BZs mediate their pharmacologic actions as a consequence of their interaction with specific receptors on the GABA/BZ receptor chloride ionophore (supramolecular) complex. This interaction results in allosteric potentiation of the action of GABA (i.e. GABA-gated chloride conductance and hence hyperpolarization of neurons) (3). To assess whether hepatocellular failure may be associated with increased sensitivity of CNS neurons to a BZ the responsiveness of Purkinje neurons from a rabbit model of fulminant hepatic failure (FHF) to depression by the BZ receptor agonist, flunitrazepam, was studied in vitro. As a control the sensitivity of neurons to depression by an α-adrenoreceptor agonist, phenylephrine, was also studied. In contrast to flunitrazepam, the actions of phenylephrine are mediated by its binding to receptors which do not form part of the supramolecular complex.

MATERIALS AND METHODS

Model of liver failure

FHF was induced in rabbits by the intravenous administration of galactosamine (4 mmol/kg) (4). In this model the syndrome of acute hepatocellular failure was complicated by the syndrome of hepatic encephalopathy (HE) which evolved through 4 easily recognized stages: I. lethargy; II. mild ataxia, poorly maintained head posture; III. severe ataxia, hind limb extension, loss of righting reflex; IV. coma. Both stages II and III were associated with a marked reduction in spontaneous motor activity (4).

Experimental groups:

Control rabbits (n=19) and rabbits in stage II, III HE (n=27) were studied.

Cerebellar slice preparation:

Rabbits were sacrified by decapitation and cerebella rapidly removed. Cerebellar slices were placed in a recording chamber continuously perfused with an oxygenated medium. Only one Purkinje neuron in each slice was studied. Ethanolic solutions of flunitrazepam or aqueous solutions of phenylephrine were administered through a manifold in the medium delivery tubing. Each individual Purkinje neuron was studied before, during and after administration of a drug for 5-15 minutes, the total duration of study varying between 15 and 35 minutes (5).

Data acquisition and analysis:

Action potentials were recorded extracellularly from single Purkinje neurons using glass microelectrodes. The electrical activity of single neurons was processed (to eliminate interference and hence obtain a higher resolution record) and then recorded electronically. Mean neuronal firing rates were determined before, during and after drug application. Responses from 3-5 cells were averaged and used to construct a concentration-response curve. IC50 values for these curves were determined using probit analysis and the 95% confidence intervals for these values were determined using the Litchfied - Wilcoxon test.

RESULTS

The mean firing rates of neurons from the control and FHF groups were not significantly different (22.5 ± 1.8 [SEM] discharges per sec. (dps) vs 24.91 ± 1.6 dps, respectively) and were similar to the firing rates of Purkinje neurons in rat cerebellar slices (19.4 ± 0.8 dps) (5). The extra-cellular action potentials recorded from the rabbit neurons displayed a biphasic waveform characteristic of Purkinje neurons and exhibited a stable firing pattern.

Perfusion of cerebellar slices with flunitrazepam (0.1-10 um) produced a concentration-dependent reduction in the spontaneous activity of neurons from both control and FHF groups (Figure 1). However, flunitrazepam was aproximately 3.5 times more potent in depressing the spontaneous activity of neurons from the FHF group than from the control group. The IC50 values were 1.7 uM for controls and 0.5 uM for the FHF group. The 95% confidence intervals were 1.0-2.9 uM and 0.3-1.0 uM respectively. The selectivity of this increased sensitivity of neurons from the FHF group to depression by a BZ agonist was examined by studying the effects of phenylephrine on the spontaneous activity of these neurons. Mean values for the degree of depression of spontaneous activity induced by phenylephrine (50 uM) for neurons from rabbits with FHF and from control rabbits were not significantly different (40 ± 7.9% vs 37 ± 5.6% respectively).

DISCUSSION

The cerebellar slice maintained in vitro appears to be appropriate for studies of the role of the GABA/BZ receptor complex in liver failure for several reasons. First, the cerebellum is rich in GABAergic innervation. Second, the slice preparation permits the monitoring of the responsiveness of single neurons (substantially isolated from local and intrinsic synaptic inputs) to known concentrations of drugs in a controlled environment. Third, the in vitro cerebellar slice technique minimizes effects of ion and metabolic inbalances, peripheral drug metabolites and alterations in blood brain barrier permeability that would probably be encountered when studying neurons in situ in animal models of liver failure.

In this study flunitrazepam was significantly more potent in depressing the firing rate of neurons from rabbits with FHF than from control rabbits. There are several possible explanations

FIGURE 1

The inhibitory effects of flunitrazepam on the spontaneous firing rate of Purkinje neurons in rabbit cerebellar slices are shown in the form of concentration-response curves. Closed triangles, FHF neurons; closed squares, control neurons. Each data point represents the mean ± SEM of responses from 3-5 cells.

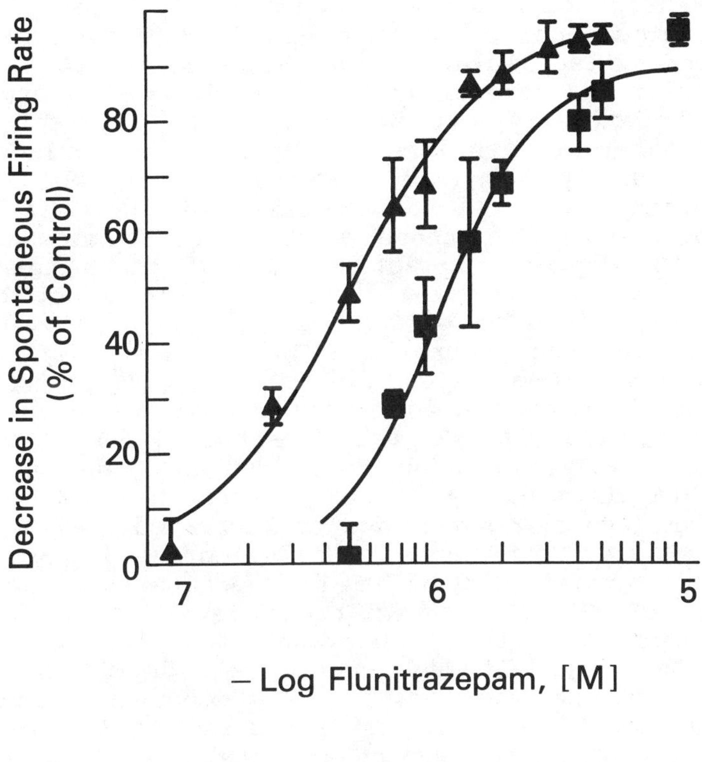

for this finding. One would be an increased responsiveness of the supramolecular complex to the drug due to an increase in the density of receptors on the complex or an increase in affinity of ligands for receptors on the complex. However, the status of BZ receptors in the CNS in this model of liver failure do not appear to be altered (6,7). Indeed, recent studies of another model do not suggest that liver failure is associated with changes in BZ receptors (8). Another possibility would be generalized changes in polarization and permeability of neurons due to changes induced by liver failure in the concentrations of metabolites or electrolytes in the extracellular space. However, the continuous perfusion of cerebellar slices with a defined medium would partially, if not completely, correct such abnormal concentrations in the slice. Furthermore, if the finding in this study were due to changes in polarization and permeability of neurons from rabbits with FHF, these cells would be more sensitive to depressants acting at loci other than the supramolecular complex. This was found not to be the case since no differences in the senstivity of neurons in the control and FHF groups to depression by the α-adrenoceptor agonists, phenylephrine, were observed. Thus the increased sensitivity of neurons from rabbits with FHF may be specific for a certain class or classes of neuronal depressants. A third possibility would be increased responsiveness of the supramolecular complex due to increased availability of GABA at its receptor. While increased plasma-to-brain transfer of GABA has been demonstrated in this model of FHF (9), whole brain levels of GABA have not been found to be increased in a model of liver failure (10), although such levels do not necessarily reflect changes in GABA concentrations in synaptic clefts. Indeed increased GABA levels in synaptic clefts might be expected to decrease the sensitivity of neurons to BZs (5). A fourth possibility would be stress-induced modulation of the function of the supramolecular complex due to the procedure used to induce FHF (11,12,13). However, changes in function of the supramolecular complex in models of liver disease appear to be related to liver failure rather than the stressfulness of methods used to induce the model (14,15,16).

A fifth possible explanation for the finding in this study would be the presence in rabbits with FHF of an endogenous substance that mimics the electrophysiological actions of flunitrazepam or increased production or increased availability of such a substance to neurons in rabbits with FHF. The presence of such a compound may increase neuronal sensitivity to the depressant actions of flunitrazepam (17). The mechanisms controlling the release, uptake, synthesis or degradation of such a substance in control animals may be sufficient to attenuate its actions. Alternatively, such a substance may be produced as a direct consequence of liver failure. Thus, if the findings in this study can be extrapolated to the syndrome of liver failure exhibited by human beings with cirrhosis, the presence of an endogenous BZ ligand with agonist properties at the BZ receptor may afford at least a partial explanation for the observed hypersensitivity of patients with chronic liver disease to BZs.

SUMMARY AND CONCLUSIONS

Purkinje neurons from a rabbit model of FHF exhibit increased sensitivity to depression by the BZ agonist, flunitrazepam, but not to the α-adrenoceptor agonists, phenylephrine. This increased sensitivity of neurons to a BZ agonist in liver failure may be attributable to increased concentrations or availability of an endogenous BZ receptor ligand with BZ agonist properties. If such a ligand is present at BZ receptors in human beings with cirrhosis, it may contribute to or account for the observed hypersensitivity of such patients to BZs.

REFERENCES

1. Branch, R.A., Morgan, M.H., James, J., Read, A.E. (1976). Intravenous administration of diazepam in patients with chronic liver disease. Gut, 17, 975-983.

2. Baktir, G., Fisch, H.U., Karlaganis, G. et al. (1987). Mechanism of the sedative response of cirrhotics to benzodiazepines: Model experiments with triazolam. Hepatology, 7, 629-638.

3. Paul, S.M., Marangos, P.J., Skolnick, P. (1981). The benzodiazepine-GABA-chloride ionophore receptor complex: common site of minor transquilizer action. Biol. Psychiatry, 16, 213-229.

4. Blitzer, B.L., Waggoner, J.G., Jones, E.A. et al. (1978). A model of fulminant hepatic failure in the rabbit. Gastroenterology, 74, 664-671.

5. Basile, A.S., Gammal, S.H., Mullen, K.D. et al. (1988). Differential responsiveness of cerebellar Purkinje neurons to GABA and benzodiazepine receptor ligands in an animal model of hepatic encephalopathy. J. Neuroscience, 8, 2414-2421.

6. Rossle, M., Deckert, J. (1988). Quantitative autoradiography of gamma-aminobutyric acid and benzodiazepine receptors in a model of acute hepatic encephalopathy. In Advances in Ammonia Metabolism and Hepatic Encephalopathy (P.B. Soeters, J.H.P. Wilson, A.J. Meijer, E. Holm, eds.), pp. 233-237, Excerpta Medica, Amsterdam.

7. Rossle, M., Mullen, K.D., Jones, E.A. (1988). Benzodiazepine receptors in a model of acute hepatic encephalopathy: effect of triton on receptor solubilization. In Advances in Ammonia Metabolism and Hepatic Encephalopathy (P.B. Soeters, J.H.P. Wilson, A.J. Meijer, E. Holm, eds.), pp. 243-247, Excerpta Medica, Amsterdam.

8. Maddison, J.E., Dodd, P.R., Morrison, M., Johnston, G.A.R., Farrel, G.C. (1987). Plasma GABA, GABA-like activity and the brain GABA-benzodiazepine receptor complex in rats with chronic hepatic encephalopathy. Hepatology, 7, 621-628.

9. Basset, M.L., Mullen, K.D., Scholz, B. et al. (1985). Increased brain GABA uptake in precoma encephalopathy in a rabbit model of fulminant hepatic failure. Hepatology, 5, 1032.

10. Roy, S., Pomier-Layrargues, G., Butterworth, R.F. et al. (1988). Hepatic encephalopathy in cirrhotic and portocaval shunted dogs: lack of changes in brain GABA uptake, brain GABA levels, brain glutamic acid decarboxylase activity and brain postsynaptic GABA receptors. Hepatology, 8, 845-849.

11. Havoudjian, H., Paul, S.M., Skolnick, P. (1986). Rapid stress-induced modification of the benzodiazepine receptor coupled chloride ionophore. Brain Res., 375, 401-406.

12. Schwartz, R.D., Wess, M.J. Labarca, R. (1987). Acute stress enhances the activity of the GABA receptor-gated chloride ion channel in brain. Brain Res., 411, 151-155.

13. Trullas, R.H., Havoundjian, H., Zamir, N. et al. (1987). Environmentally-induced modification of the benzodiazepine/GABA receptor coupled chloride ionophore. Psychopharmacology, 91, 384-390.

14. Bassett, M.L., Mullen, K.D., Skolnick, P. et al. (1987). Amelioration of hepatic encephalopathy by pharmacologic antagonism of the GABAa-benzodiazepine receptor complex in a rabbit model of fulminant hepatic failure. Gastroenterology, 93, 1069-1077.

15. Gammal, S.H., Geller, D., Skolnick, P. et al. (1987). Unequivocal amelioration of hepatic encephalopathy by benzodiazepine receptor antagonists in a rat model of fulminant hepatic failure. J. Hepatol., 4, S17.

16. Rzepczynski, D., Zieve, S., Lindblad, S. et al. (1987). In vivo studies of GABAergic effects in experimental hepatic encephalopathy. Hepatology, 6, 902-905.

17. Study, R.E., Barker, J.L. (1981). Diazepam and (-)-pentobarbital: fluctuation analysis reveals different mechanisms for potentiation of GABA responses in cultured central neurons. Proc. Natl. Acad. Sci. USA, 78, 7180-7184.

STUDIES OF CENTRAL AND "PERIPHERAL-TYPE" BENZODIAZEPINE RECEPTORS IN AUTOPSIED BRAIN TISSUE FROM CIRRHOTIC PATIENTS WITH HEPATIC ENCEPHALOPATHY

Joël Lavoie, Gilles Pomier Layrargues and Roger F. Butterworth, Lab. of Neurochem. André-Viallet Clin. Res. Centre, Hôpital St-Luc (U. of Montréal), Montréal, Québec, Canada H2X 3J4.

SUMMARY

Previous studies with galactosamine-induced liver failure in rats and rabbits suggested that benzodiazepine (BDZ) receptors were modified in brain in hepatic encephalopathy (HE). As part of a series of studies to evaluate brain BDZ receptors in human HE, the integrity of central and "peripheral-type" benzodiazepine receptors was evaluated in brain tissue obtained at autopsy from cirrhotic patients who died with hepatic encephalopathy and an equal number of appropriate controls. Affinities and densities of clonazepam-displaceable binding of [^{3}H]-Flunitrazepam as well as [^{3}H]-Ro15-1788, a specific ligand for central benzodiazepine receptors were unaltered. These findings do not support recent hypothesises that HE results from alterations of the GABA-benzodiazepine receptor complex. On the other hand, increased densities of high affinity binding sites for ^{3}H-PK 11195 were observed in frontal cortex and caudate nuclei of cirrhotic patients with HE.

INTRODUCTION

It has been suggested that benzodiazepine receptors may be modified in hepatic encephalopathy (HE). Increased densities of benzodiazepine binding sites to neural membranes from rabbits with galactosamine-induced liver failure have been reported (Schafer et al., 1983). Similar observations were reported in galactosamine-treated rats (Zeneroli et al., 1984). Furthermore, a recent communication described increased uptake of the central benzodiazepine antagonist Ro15-1788 in vivo measured by positron emission tomography in four patients with hepatic encephalopathy (Samson et al., 1987) and it was suggested that increased densities or affinities of central benzodiazepine receptors could account for this finding. In view of the postulated role for benzodiazepine receptors in the pathogenesis of hepatic encephalopathy, we have assessed the integrity of these receptors in autopsied brain tissue from nine patients who died in hepatic coma and an equal number of appropriately matched controls. Radioligands used were [^{3}H]-flunitrazepam and [^{3}H]-Ro15-1788. In addition, in order to assess the integrity of "peripheral-type" benzodiazepine receptors in these brains, use was made of the radioligand [^{3}H]-PK 11195.

METHODS

Nine brains from cirrhotic patients who died in stage IV hepatic encephalopathy (mean age 58 yr, range 48-70 yr) and nine control brains (mean age 58 yr, range 49-67 yr) were obtained at autopsy. Control patients were free from neurological, psychiatric or hepatic diseases in all cases. Mean time intervals from death to freezing of autopsied brain samples were comparable (15.9 h for control group; 17.4 h for cirrhotic group). Cirrhosis was alcoholic in etiology in all cases but in no case had alcohol been consumed for the 2 months prior to death. In no cases (controls or cirrhotics) had benzodiazepines been prescribed in the 72 h period prior to death. Brain tissue was dissected as previously described (Lavoie et al., 1987) and stored at -70°C until time of assay.

High affinity binding of [^{3}H]-flunitrazepam, [^{3}H]-Ro15-1788 or [^{3}H]-PK 11195 was measured at 0-4°C in cerebral membrane preparations as described by Kish et al., (1984). A filter assay using Whatman GF/B filters was employed. Non-specific binding was determined by the amount of bound [^{3}H]-flunitrazepam, [^{3}H]-Ro15-1788 or [^{3}H]-PK 11195 remaining after addition of 10 μM clonazepam, 100 μM diazepam or 1 μM PK 11195 respectively.

Receptor density (Bmax) and affinity (Kd) were determined by Scatchard analysis. Correlation coefficients for Scatchard plots were 0.90 or better.

RESULTS

Scatchard analyses of clonazepam-displaceable binding of [^{3}H]-flunitrazepam to membrane preparations from prefrontal cortex yielded a single high-affinity binding site (Kd = 2.45 ± 0.35 nM (controls), Kd = 2.51 ± 0.32 nM (patients)). In the case of [^{3}H]-Ro15-1788, again, a single high-affinity site was revealed (Kd = 0.71 ± 0.03 nM (controls), Kd = 0.70 ± 0.05 nM (patients)). Typical saturation curves, Scatchard plots as well as distribution "scattergrams" for Kd's and Bmax's for [^{3}H]-flunitrazepam and [^{3}H]-Ro15-1788 binding sites are presented in Figures 1 and 2. Neither affinities nor densities of [^{3}H]-flunitrazepam or [^{3}H]-Ro15-1788 binding sites on membrane preparations from cirrhotic patients with hepatic encephalopathy were significantly different from those from control patients. Similar negative results were obtained in membrane preparations from frontal cortex and caudate nuclei (data not shown).

In contrast, Bmax's for [^{3}H]-PK 11195 binding sites were significantly increased in membrane preparations from frontal cortex and caudate nuclei of cirrhotic patients (Table 1). However binding affinities (Kd's) were unchanged.

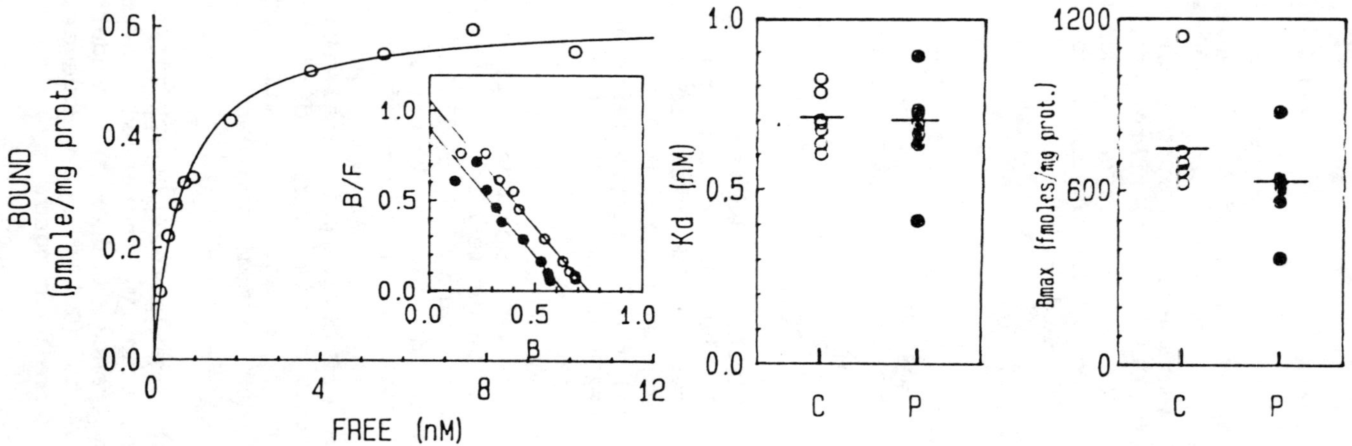

Figure 1. Saturation plots and Scatchard analyses of specific [^{3}H]-flunitrazepam binding with increasing concentrations of ligand. Data on Scatchard plots represent mean values of triplicate determinations on material from control patients o and cirrhotic patients ●. Also shown are scattergrams of Kd and Bmax values for controls (C) and patients (P).

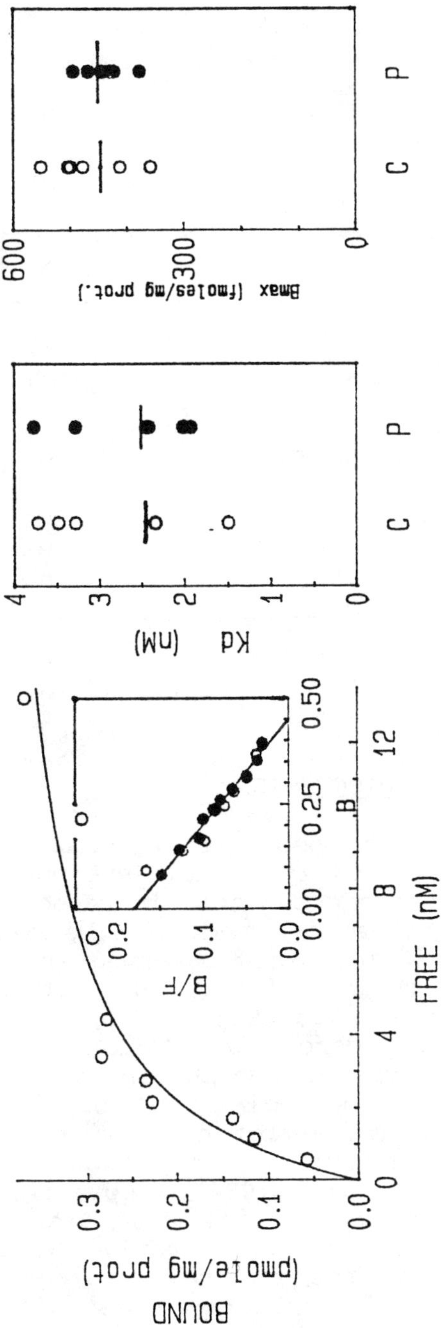

Figure 2. Saturation plots and Scatchard analyses of specific [^{3}H]-Ro15-1788 binding with increasing concentrations of ligand. Data on Scatchard plots represent mean values of triplicate determinations on material from control patients o and cirrhotic patients ●. Also shown are scattergrams of Kd and Bmax values for controls (C) and patients (P).

Table 1. Affinites and densities of [3H]-PK 11195 binding sites in autopsied brain tissue from cirrhotic patients with hepatic encephalopathy

Brain region	Binding parameter	Controls(9)	Cirrhotic patients(9)
Frontal cortex	Kd (nM)	2.4 ± 0.3	4.2 ± 0.9
	Bmax (fmole mg.prot.$^{-1}$)	669 ± 44	1096 ± 106**
Prefrontal cortex	Kd (nM)	3.2 ± 0.5	3.7 ± 0.8
	Bmax (fmole mg.prot.$^{-1}$)	704 ± 42	738 ± 56
Caudate nuclei	Kd (nM)	2.1 ± 0.2	3.5 ± 0.8
	Bmax (fmole mg.prot.$^{-1}$)	469 ± 25	618 ± 52*

*$p<0.02$, **$p<0.01$ compared to control values

DISCUSSION

It has been proposed that benzodiazepine receptors may be altered in hepatic encephalopathy. In addition, it was suggested that increased numbers of drug binding sites may mediate the increased sensitivity to benzodiazepines observed in patients with liver disease (Branch et al., 1976; Schafer and Jones, 1982; Samson et al., 1987). Support for such hypotheses originally came from studies in experimental models of hepatic encephalopathy in which increased numbers of benzodiazepine binding sites were observed on brain membrane preparations from galactosamine-treated rabbits (Schafer et al., 1983) and rats (Zeneroli et al., 1984). However, subsequent studies of benzodiazepine receptors in this experimental model using quantitative autoradiography did not confirm the original findings (Rossle and Deckert, 1988). Furthermore, studies in a different rat model of liver failure revealed normal benzodiazepine binding sites (Maddison et al., 1987). Results of the present study (Figures 1 and 2), clearly show that neither clonazepam-displaceable [^{3}H]-flunitrazepam nor [^{3}H]-Ro15-1788 binding sites are significantly altered in membrane preparations from prefrontal cortex of cirrhotic patients with hepatic encephalopathy. A review of the literature to date of brain benzodiazepine receptors in human and experimental hepatic encephalopathy is shown in Table 2.

An examination of the reported findings in Table 2 suggests that the lack of consensus regarding benzodiazepine receptors may relate to one (or more) of the following factors:

(a) the experimental model;

(b) the method used to assess binding sites (in vitro assay vs autoradiography);

(c) the identity of the radioligand and the displacing drug.

Of the three possibilities, (c) is probably the most important cause of the inconsistent findings reported in the literature. Ligands such as [^{3}H]-flunitrazepam and [^{3}H]-diazepam bind with high affinity to both central-type and peripheral-type benzodiazepine receptors. Assessment of the integrity of central benzodiazepine receptors using these ligands requires the use of a central-type displacing drug such as clonazepam. Thus, the original report by Schafer et al. (1983) showing significantly increased densities of (flunitrazepam-displaceable) [^{3}H]-flunitrazepam binding sites in the brains of galactosamine-treated rabbits refers to total (ie: central-type plus peripheral type) benzodiazepine receptors. The later studies by Rossle and Deckert (1988) employed clonazepam as displacing drug (in their [^{3}H]-flunitrazepam binding studies by quantitative autoradiography in the same animal model as Schafer et al., 1983) and thus were assessing only central-type benzodiazepine

Table 2. Benzodiazepine receptors in Hepatic Encephalopathy: a review of the literature

Finding	Radioligand	Displacing drug	Reference (material)
Increased number of receptors	[^{3}H]-flunitrazepam	Flunitrazepam	Schafer et al., 1983 (galactosamine rabbit)
Increased number of receptors	[^{3}H]-diazepam	Clonazepam	Zeneroli et al., 1984 (galactosamine rat)
Receptors unchanged	[^{3}H]-flunitrazepam	Clonazepam	Maddison et al., 1987 (ischemic rat)
Receptors unchanged	[^{3}H]-flunitrazepam [^{3}H]-Ro15-1788	Clonazepam	Rossle and Deckert, 1988 (galactosamine rabbit)
Receptors unchanged	[^{3}H]-flunitrazepam [^{3}H]-Ro 15-1788	Clonazepam Diazepam	Butterworth et al., 1988 (brain tissue from patients with HE)
Increased number of "peripheral type" receptors	[^{3}H]-PK 11195	PK 11195	Present study (brain tissue from patients with HE)

receptors. In these studies, no changes of benzodiazepine receptors were observed. Thus, apart from the one report by Zeneroli et al., 1984, a careful review of the literature reveals that central-type benzodiazepine receptors are unchanged in experimental hepatic encephalopathy. Results of the present study confirm these negative findings. Furthermore, they suggest that, if there are alterations of benzodiazepine binding sites in hepatic encephalopathy, it is most likely to be the <u>peripheral-type of receptor</u> that is modified.

In favour of such a possibility, results from our laboratory (Table 1) suggest that autopsied brain tissue from cirrhotic patients with hepatic encephalopathy contains significantly increased densities of [^{3}H]-PK 11195 binding sites. PK 11195 is a ligand with high selectivity for "peripheral-type" benzodiazepine receptor (PBR's) in brain. Currently available evidence suggests that PBR's are localized on astrocytes rather than neurons (Schoemaker et al., 1982; Benavides et al., 1987). PK 11195 displaces [^{3}H]-diazepam from cultured astrocytes with 10^4 times the potency compared to cultured neurons (Bender and Hertz, 1987). Thus, the increased densities of [^{3}H]-PK 11195 binding sites observed in postmortem brain from patients with hepatic encephalopathy probably reflects astrocytic rather than neuronal changes in these brains. Indeed astrocytic pathology (Alzheimer type II astrocytosis) with neuronal sparing is the characteristic neuropathological finding in human hepatic encephalopathy resulting from chronic liver disease. However, whether or not PBR's play a role in the pathogenesis of hepatic encephalopathy or in the hypersensitivity of cirrhotic patients to benzodiazepines awaits further studies.

Ro15-1788 (Flumazenil) is a central-type benzodiazepine antagonist. Recent studies using Positron Emission Tomography (PET) have shown increased uptake of [^{11}C]-Ro15-1788 in cerebral cortex of four patients with severe liver failure and recurrent episodes of hepatic encephalopathy (Samson et al., 1987) and it was suggested that this increased uptake may reflect increased densities or affinities of central benzodiazepine receptors in these patients. Results of the present study do not support this hypothesis. On the other hand, a recently-completed study of the pharmacokinetics of Ro15-1788 in cirrhotic patients in our laboratory clearly shows marked alterations of plasma half-life and clearance of Ro15-1788 in cirrhotic patients (Pomier Layrargues et al., this volume). Thus, altered drug pharmacokinetics, rather than changes of cerebral receptors for Ro15-1788 are probably responsible for the PET findings.

A number of uncontrolled studies suggest beneficial effects of Ro15-1788 in patients with hepatic encephalopathy (Scollo-Lavizzari and Steinmann, 1985; Grimm et al., 1987; and Meier et al. this volume). However, reports of negative findings have also started to appear (Sutherland and Minuk, 1988; Klotz and Walker, 1989). Possible mechanisms involved in these beneficial effects of Ro15-1788 (if, indeed, such effects survive the test of rigorously controlled trials) include the following:

(i) blockade of increased numbers of central benzodiazepine receptors in the brains of these patients.

(ii) antagonism of the action of an endogenous benzodiazepine-like coumpound resulting from liver failure (see Mullen et al., this volume).

(iii) a mechanism, as yet unknown, involving, for example, ammonia's toxic effects on brain.

(iv) prior administration of benzodiazepines.

Results of the present study suggest that possibility (i) is unlikely. Possibility (ii) awaits the isolation and characterization of the endogenous ligand concerned.

In animal models of hepatic encephalopathy, attempts to ameliorate neurological impairment have yielded conflicting results. Whereas administration of Ro 15-1788 to the galactosamine-treated rabbit appears to lead to amelioration of hepatic encephalopathy (Bassett et al., 1987), no such beneficial effects of benzodiazepine antagonists were noted in an ischemic model of hepatic encephalopathy in the rat (Zieve et al., 1987).

In conclusion, results of the present study reveal no changes in affinities or densities of central-type benzodiazepine receptors in brain membrane preparations from cirrhotic patients with hepatic encephalopathy. When taken in conjunction with the data from experimental models, this data is not consistent with the proposal that changes of central benzodiazepine receptors play a role in the pathogenesis of hepatic encephalopathy. However, "peripheral-type" benzodiazepine receptors appear to be modified in this condition. The role of these latter findings in the pathogenesis of hepatic encephalopathy is unclear at the present time.

ACKNOWLEDGMENTS

The authors are grateful to Dr. Jean-François Giguère for his assistance with brain dissections and to Anne-Marie Besnard for her excellent technical assistance. We appreciate the generous gifts of clonazepam and Ro15-1788 from Hoffmann-La Roche Inc. We are grateful to Dominique Roy for the preparation of the manuscript.

REFERENCES

Bassett M.L., Mullen K.D., Skolnick P. and Jones E.A. (1987). Amelioration of hepatic encephalopathy by pharmacologic antagonism of the $GABA_A$-benzodiazepine receptor complex in a rabbit model of fulminant hepatic failure. Gastroenterol. 93, 1069-1077.

Benavides J., Fage D., Carter C. and Scatton B. (1987). Peripheral type benzodiazepine binding sites are a sensitive indirect index of neuronal damage. Brain Res. 421, 167-172.

Bender A.S. and Hertz L. (1987). Pharmacological characteristics of diazepam receptors in neurons and astrocytes in primary cultures. J. Neurosci. Res., 18, 366-372.

Butterworth R.F., Lavoie J., Giguère J.F. and Pomier Layrargues G. (1988). Affinities and densities of high-affinity [^{3}H]-muscimol (GABA-A) binding sites and of central benzodiazepine receptors are unchanged in autopsied brain tissue from cirrhotic patients with hepatic encephalopathy. Hepatology 8, 1084-1088.

Grimm G., Lenz K., Kleinberger G. Laggner A., Druml W., Schneeweiss B., Gremmel F. and Holzner F. (1987). Ro15-1788 improves coma in 4 out of 5 patients with fulminant hepatic failure; verification by long latency auditory and somatosensory evoked potentials (abstract). Eur. J. Hepatol. 4, S21.

Kish S.J., Perry T.L. and Hornykiewicz O. (1984). Benzodiazepine receptor binding in cerebellar cortex: observations in olivoponto-cerebellar atrophy. J. Neurochem. 42 466-469.

Klotz U. and Walker S. (1989). Flumazenil and hepatic encephalopathy. Lancet, 155-156.

Lavoie J., Giguère J.F., Pomier Layrargues G. and Butterworth R.F. (1987). Amino acid changes in autopsied brain tissue from cirrhotic patients with hepatic encephalopathy. J. Neurochem. 49, 692-697.

Maddison J.E., Dodd P.R., Morrison M., Johnston G.A.R. and Farrell G.C. (1987). Plasma GABA, GABA-like activity and the brain GABA-benzodiazepine receptor complex in rats with chronic hepatic encephalopathy. Hepatology, 7, 621-628.

Rossle M. and Decker J. (1988). Quantitative autoradiography of gamma-aminobutyric acid and benzodiazepine receptors in a model of acute hepatic encpehalopathy. In: Advances in ammonia metabolism and hepatic encephalopathy (Soeters P.B., Wilson J.H.P., Meijer A.J. and Holm E., eds), Elsevier Science Publishers, Netherlands, pp. 233-237.

Samson Y., Bernuau J., Pappata S., Chavoix C., Bason J.C. and Maziere M.A. (1987). Cerebral uptake of benzodiazepine measured by positron emission tomography in hepatic encephalopathy. New Engl. J. Med. 316, 414-415.

Scollo-Lavizzari G. and Steinmann E. (1985). Reversal of hepatic coma by benzodiazepine antagonist Ro 15-1788. The Lancet, 1, 1324-1325.

Schafer D.F. and Jones E.A. (1982). Hepatic encephalopathy and the -aminobutyric-acid neurotransmitter system. The Lancet, 1, 18-20.

Schafer D.F., Fowler J.M., Munson P.J., Thakur A.K., Waggoner J.G. and Jones E.A. (1983). Gamma-aminobutyric acid and benzodiazepine receptors in an animal model of fulminant hepatic failure. J. Lab. Clin. Med. 102, 870-880.

Schoemaker H., Morelli M., Deshmukh P. and Yamamura H. (1982). ^{3}H-Ro5-4864 benzodiazepine binding in the kainate lesioned striatum and Huntington's diseased basal ganglia. Brain Res. 248, 396-401.

Sutherland L.R. and Minuk G.Y. (1988). Ro15-1788 and hepatic failure. Am. Int. Med. 108, 158.

Zeneroli M.L., Baraldi M. and Ventura E. (1984). -aminobutyric acid receptors in experimental hepatic encephalopathy. In "Hepatic Encephalopathy in Chronic Liver Failure" (Capocaccia L., Fischer J.E. and Rossi-Fanelli F., eds), Plenum Press, New York, pp. 25-40.

Zieve L., Ferenci P., Rzepczynski D., Ebner J. and Zimmermann Ch. (1987). A benzodiazepine antagonist does not alter the course of hepatic encephalopathy or neural -aminobutyric acid (GABA) binding. Metab. Brain Dis. 2, 201-205.

RO 15-1788 KINETICS ARE MARKEDLY IMPAIRED IN CIRRHOTIC PATIENTS.

G. Pomier-Layrargues, Jean-François Giguère, Joël Lavoie, Bernard Willems and Roger F. Butterworth. André-Viallet clinical research center, Hôpital Saint-Luc and Université de Montréal, Montreal, Quebec, Canada H2X 1P1

ABSTRACT

It has been suggested that hepatic encephalopathy (HE) might be related to changes in the GABA-benzodiazepine receptor complex. A recent PET scan study reported an increased uptake of ^{11}C RO 15-1788 (a specific ligand for brain benzodiazepine (Bz) receptors) in cirrhotic patients with HE, suggesting an increase in the density or affinity of Bz receptors. RO 15-1788, is extensively metabolized by the liver and marked changes in its kinetic parameters might best explain these findings. In the present study, we evaluated RO 15-1788 pharmacokinetics in 8 normal volunteers (group 1), 8 cirrhotic patients with moderately impaired liver function (Pugh score < 10; group 2) and 8 cirrhotic patients with severe liver dysfunction (Pugh score > 10; group 3). The subjects of each group were age and sex-matched. After an i.v. injection of 2mg RO 15-1788, blood samples were taken at fixed intervals up to 7 hours after the injection. Ro 15-1788 plasma levels were determined using a specific gas chromatographic assay and plasma clearance and half-life were calculated. Mean age was 38.5 years old in group 1, 45.1 years old in group/ 2 and 44.7 years old in group 3. There were 7 males and 1 female in each group. Cirrhosis was related to chronic alcoholism in 13 patients, to chronic B virus infection in 1 and was cryptogenic in 2. Pharmacokinetic analyses demonstrated a marked decrease in Ro 15-1788 plasma clearance (-57%) and increase in half-life (+66%) in cirrhotic patients of group 2. RO 15-1788 elimination was further impaired in cirrhotic patients of group 3: mean decrease in plasma clearance: -74%; mean increase in half-life: + 210%. These data demonstrate that Ro 15-1788 pharmacokinetics are markedly altered in cirrhosis, changes being related to the degree of liver dysfunction. These findings might best explain the reported increase in RO 15-1788 brain uptake in cirrhotic patients with HE rather than changes in brain Bz receptors.

INTRODUCTION

According to a recent hypothesis, hepatic encephalopathy (HE) might be related to disturbances in cerebral function induced by an endogenous benzodiazepine (Bz) ligand (Mullen et al, 1988). This hypothesis is based on experimental evidence showing an increased density of brain Bz receptors (Schafer et al, 1983; Baraldi et al, 1984) as well as amelioration of HE using Bz antagonists in animal models of fulminant hepatic failure (Basset et al, 1988). Futhermore, a recent PET-Scan study reported an increased brain uptake of ^{11}C RO 15-1788, a specific ligand for brain Bz receptors, in cirrhotic patients with previous HE episodes, suggesting an increase in the density or affinity of these receptors (Samson and Bernuau, 1987). These findings led to the use of the specific Bz antagonist RO 15-1788 in the treatment of HE associated with acute or chronic liver diseases (Scollo-Lavizzarri et al, 1985; Bansky et al, 1987; Burke et al, 1988; Grimm et al, 1989). However, some discrepancies were reported with regards to the status of Bz receptors in the brain of cirrhotic patients (Butterworth et al, 1988). Moreover, the therapeutic usefulness of Bz antagonists has not been established unequivocally (Sutherland and Minuk, 1988; Klotz and Walker, 1989); this is not surprising as none of the studies claiming the efficacy of RO 15-1788 were placebo-controlled trials.

RO 15-1788 is a highly specific ligand for central brain Bz receptors. It is extensively metabolized by the liver but surprisingly the influence of liver disease on its disposition has never been assessed. The aim of the present study was to measure changes in RO 15-1788 pharmacokinetics induced by cirrhosis of varying severity.

MATERIAL AND METHODS

Subjects

Three groups of subjects were studied. Group I was made up of normal volunteers. The patients with cirrhosis, as previously established by biopsy, were divided into two groups: those with compensated cirrhosis (Group II) as determined by a score of 5 to 10 according to Pugh's classification (1973) and those with decompensated cirrhosis (Group III) as shown by a score of 11 to 15 on this scale. The patients with history or presence of clinically significant cardiovascular, renal or neuropsychiatric disorders were excluded from the study, as well as women who were pregnant, lactating or who were of child-bearing potential. None of the subjects was treated with drugs known to influence the metabolizing capacity of the liver. The study was approved by our local ethics committee and all the subjects gave their informed signed consent to participate.

Pharmacokinetic Study

After an overnight fast, subjects received 2.0mg of RO 15-1788 administered by intravenous infusion over five minutes. Blood samples were taken at 2, 4, 10, 20, 30, 45, 60, 120, 180, 240, 300, 360 and 420 minutes after the start of infusion. Blood samples were collected in vacutainers containing potassium oxalate (10mg) and sodium fluoride (12.5mg) and centrifuged immediately after collection. Plasma samples were stored at $-20^{o}C$ pending assay.

Analysis of RO 15-1788

Plasma concentrations of Ro 15-1788 were analyzed by capillary gas chromatography with nitrogen-selective detection according to a previously described technique (Zell and Timm, 1986).

Calculations

Pharmacokinetic parameters (plasma clearance, half-life

and volume of distribution at steady-state) were calculated according to a non-compartmental analysis (Gibaldi and Perrier, 1982).

The data were compared between the 3 groups using a one-way variance analysis and differences were considered significant at a 0.05 p level.

RESULTS

24 patients entered the study (8 patients for each group). The clinical and biochemical characteristics of the 3 groups are summarized on table 1. As expected, serum albumin levels and prothrombin times were more altered in the patients with decompensated cirrhosis.

TABLE 1
Clinical and Biochemical Characteristics of Normal Volunteers and Cirrhotic Patients

	group I (n = 8)	group II (n = 8)	group III (n = 8)
.Age (years)	38.5±1.6	45.1±8.1	44.7±11.2
.Sex (M/F)	7/1	7/1	7/1
.Weight (kg)	76±11	73±16	70±12
Etiology of liver disease			
. Alcoholic	---	7	6
. non alcoholic	---	1	2
.Serum albumin (g/dl)	4.1±0.5	3.0±0.5*	2.4±0.5*o
.Serum bilirubin (mg/dl)	0.6±0.3	2.8±3.9	4.8±5.2*
.SGOT (UI)	17±11	44±21	85±46*o
.Alkaline phosphatase (UI)	57±15	137±51*	161±89*
.Prothrombin time (sec above control)	0.1±0.2	4.0±2.4*	6.7±2.5*o

Group I: normal volunteers; Group II: compensated cirrhosis; Group III: decompensated cirrhosis.
Values are expressed as mean ± SD.
* $p < 0.05$ as compared to group 1
o $p < 0.05$ as compared to group 2

As shown on table 2, the kinetic parameters were markedly altered in cirrhotic patients, changes being more pronounced in those with severely impaired liver function. Plasma clearance was decreased by 57% in the patients of group II and by 74% in that of group III; plasma half-life was markedly prolonged in these 2 groups by 66% in group II and by 210% in group III. The volume of distribution at steady-state was unchanged in cirrhotic patients when compared to controls.

TABLE 2
Pharmacokinetic Parameters of RO 15-1788 after a 2mg Intravenous Injection to Normal Volunteers and Patients with Cirrhosis

Parameter	Group I (n = 8)	Group II (n = 8)	Group III (n = 8)
Clp (ml/min/kg)	16.3±2.6	7.0±2.1*	4.2±1.1*o
Half-life (min)	45.7±8.5	75.7±23.3*	141.7±27.5*o
Vdss (L/kg)	0.62±0.09	0.68±0.15	0.85±0.11

Group I: normal volunteers; Group II: compensated cirrhosis; Group III: decompensated cirrhosis.
Clp = plasma clearance; Vdss = volume of distribution at steady-state.
Values are expressed as mean ± SD.
* $p < 0.05$ as compared to group I.
o $p < 0.05$ as compared to group II.

DISCUSSION

RO 15-1788 is a ligand for central Bz receptors which is highly selective as shown by in vitro studies, as well as by its ability to reverse symptoms induced by Bz overdosage (Riishede et al, 1988). Pharmacokinetic studies in normal subjects showed a short half-live and a high plasma clearance (Brogden et al, 1986). In our control group, mean plasma half-life was 46 minutes and mean plasma clearance was 16.3 ml/min/kg, these values being within the range of previously reported data. In this study, we found marked changes in the pharmacokinetics of Ro 15-1788, and we demonstrated that decreases in plasma clearance and increases in half-life were more pronounced in patients with severely impaired liver function. This is not surprising as this compound is extensively metabolized in the liver with only 0.12% excreted unchanged in the urine over a 12 hour period after intravenous injection (Brogden et al, 1988). The i.v. dose injected in this study (2mg) was chosen according to previous pharmacokinetic studies in normal subjects and also because the dosage used in most of the therapeutic trials on hepatic coma were in a 0.4mg - 2mg range.

These kinetic changes should be taken into account before using RO 15-1788 in cirrhotic patients. These findings might explain why the reported beneficial effects of this drug on the state of conciousness in cirrhotics with HE was observed after various i.v. doses and during different periods of time. Such differences could be related to differences in the degree of impairment of liver function. Moreover, if the drug is proven to be an efficient treatment of HE, it will be administered as prolonged infusion in the future, and the calculation of the infusion rate will be influenced by alterations of pharmacokinetics observed after a single i.v. dose.

The results of this study have some relevance with regards to the hypothesis suggesting a role for an endogenous Bz ligand in the pathogenesis of HE (Mullen et al, 1988). This theory relies mainly on the reported increase of brain Bz receptors in animals (Schafer et al, 1983; Baraldi et al, 1984), the enhanced brain uptake of ^{11}C Ro 15-1788 in cirrhotic patients (Samson and Bernuau, 1987) and the improvement of HE induced by Ro 15-1788 in animals (Basset et al, 1988) and humans (Ferenci et al, 1988). However this hypothesis might be criticized in the view of several observations: the brain Bz receptors are unchanged in other animal models (Maddison et al, 1988) as well as in cirrhotic patients who died with HE (Butterworth et al, 1988). Treatment of HE with Bz antagonists has been shown to be unsuccessful in a rat model of acute HE (Zieve et al, 1987) as well as in approximately 30% of patients with cirrhosis or fulminant hepatitis (Ferenci et al, 1988). Moreover, the reported increase in Ro 15-1788 brain uptake in a PET-Scan study might be best explained by an increase in the amount of drug available for uptake, due to a decrease in the systemic clearance and also an increase in the free fraction of this drug.

Finally,the absence of significant side-effects at the

dosage used in this study provides the basis for safe utilization in future controlled clinical trials in cirrhotic patients with HE.

REFERENCES

Bansky G., Meier P.J., Ziegler W.H., Walser H., Schmid M., and Huber M. (1985). Reversal of hepatic encephalopathy by benzodiazepine antagonist (Ro 15-1788). Lancet 1;2360-2361.

Baraldi M., Zeneroli M.L., Ventura E., Penne A., Pinelli G., Ricci P., and Santi M. (1984). Super-sensitivity of benzodiazepine receptors in hepatic encephalopathy due to fulminant hepatic failure in the rat: reversal by a benzodiazepine antagonist. Clin. Sci. 67;167-175.

Bassett M.L., Mullen K.D., Skolnick P., and Jones E.A. (1988). Amelioration of hepatic encephalopathy by pharmacologic antagonism of the $GABA_A$-benzodiazepine receptor complex in a rabbit model of fulminant hepatic failure. Gastroenterology 93;1069-1077.

Brogden R.N., and Goa K.L. (1988). Flumazenil, a preliminary review of its antagonist properties, intrinsic activity, and therapeutic use. Drugs 35;448-467.

Burke D.A., Mitchell K.W., Al Mardini H., and Record C.D. (1988). Reversal of hepatic coma with flumazenil with improvement in visual evoked potentials. Lancet 2;505-506.

Butterworth R.F., Lavoie J., Giguère J.F., and Pomier-Layrargues G. (1988). Affinities and densities of high-affinity [^{3}H] muscimol (GABA-A) binding sites and of central benzodiazepine receptors are unchanged in autopsied brain tissue from cirrhotic patients with hepatic encephalopathy. Hepatology 8;1079-1083.

Gibaldi M., Perrier D. (1982). Non compartmental analysis based on statistical moment theory. In: Gibaldi M., Perrier D., eds. Pharmacokinetics, New York: Decker Inc. pp. 409-418.

REFERENCES (cont'd)

Grimm g., Katzenschlager R., Schneeweiss B., Lenz K., Ferenci P., Madl C., Laggner A.N., and Gangl A. (1988). Improvement of hepatic encephalopathy treated with flumazenil. Lancet 2;1392-1394.

Klotz U and Walker S. (1989). Flumazenil and hepatic encephalopathy. Lancet 1;155-156.

Maddison J.E., Dodd P.R., Morrison M., Johnston G.A. and Farrell G.C. (1987). Plasma GABA, GABA-like activity and the brain GABA-benzodiazepine receptor complex in rats with chronic hepatic encephalopathy. 7;621-628.

Mullen K.D., Martin J.V., Mendelson W.B., Bassett M.L., and Jones E.A. (1988). Could an endogenous benzodiazepine ligand contribute to hepatic encephalopathy. Lancet 1;457-459.

Pugh R.N.H., Murray-Lyon I.M., Dawson J.L., Pietroni M.C., and Williams R. (1973). Transection of the oesophagus for bleeding oesophageal varices. Br. Med. J. 60;646-649.

Riishede L., Krogh B., Lynhe Nielsen J., Freuchen I., and Mikkelsen B.D. (1988). Reversal of flunitrazepam sedation with flumazenil. A randomized clinical trial. Acta Anaesthesiol. Scand. 32;433-436.

Samson Y., and Bernuau J. (1987). Cerebral uptake of benzodiazepine measured by positron emission tomography in hepatic encephalopathy. N. Engl. J. Med. 316,414-415.

Schafer D.F., Fowler J.M., Munson P.J., Thakur A.K., Waggoner J.G., and Jones E.A. (1983). Gamma-aminobutyric acid and benzodiazepine receptors in an animal model of fulminant hepatic failure. J. Lab. Clin. Med. 102;870-880.

Scollo-Lavizzari G., and Steinmann E (1985). Reversal of hepatic coma by benzodiazepine antagonist (Ro 15-1788). Lancet 1;1324.

Sutherland L.R., and Minuk G.Y. (1988). RO 15-1788 and hepatic failure. Ann. Intern. Med. 108;158.

<u>**REFERENCES**</u> (cont'd)

Zell M., and Timm U (1986). Highly sensitive assay of a benzodiazepine antagonist in plasma by capillary gas chromatography with nitrogen-selective detection. J. Chromatogr. 382;175-188.

Zieve L., Ferenci P., Rzepczynski D., Ebner J and Zimmermann C.H. (1987). A benzodiazepine antagonist does not alter the course of hepatic encephalopathy or neural gamma-aminobutyric acid (GABA) binding. Metab. Brain dis. 2;201-205.

PLASMA GABA-LIKE FACTOR IN HEPATIC ENCEPHALOPATHY MAY BE TAURINE.

Jill E. Maddison*, Dorothy K. Leong, Peter R. Dodd, Graham A.R. Johnston.
Dept. Veterinary Clinical Sciences* and Dept. Pharmacology, University of Sydney, N.S.W. 2006, Australia.

Plasma γ-aminobutyric acid (GABA) concentrations, determined by radioreceptor assay (RRA) and high performance liquid chromatography (HPLC), have been reported to be elevated in patients and experimental animal models of HE (1-7). However, there is a marked discrepancy between the estimation of GABA concentrations in plasma by the two different methods: values obtained by RRA are consistently higher than those obtained by HPLC (6,8). A possible explanation for this discrepancy is the presence in plasma of a substance that can displace [^{3}H]-GABA binding from its receptor but which is not GABA itself. We have previously shown that the concentration of this "GABA-like factor" is increased in animal models of acute and chronic HE (6,8).

The RRA for the determination of GABA in physiological fluids assumes that there is no substance present that can significantly interfere with GABA receptor binding (9). This has been shown to be true for cerebrospinal fluid (10,11) and in a study of plasma from a small number of normal rats (12). However, the effect of substances found in increased concentrations in the plasma of patients and animal models of HE has not been thoroughly investigated. Ferenci *et al* (13) reported that a number of amino acids known to be raised in HE, including taurine, glutamine and glutamate significantly inhibited [^{3}H]-GABA binding at 1 mM final assay concentration. However, this concentration is considerably higher than the concentration at which any of these compounds is present in normal or HE plasma. The authors concluded from fractionation studies that the inhibition of GABA binding by plasma samples in HE is not mediated by GABA alone, but can be caused by GABA and glutamine.

The aim of the present study was to investigate the effect of those amino acids (glutamine, glutamate, phenylalanine, tyrosine, citrulline and taurine) that are elevated in both acute and chronic HE, on [^{3}H]-GABA binding to rat whole brain synaptosomal membranes. The concentrations of the amino acids used in the assay were equivalent to those found in the plasma of control and HE animal models (14).

MATERIAL AND METHODS.

Effect of Amino Acids and Plasma from Rats with HE on [^{3}H]-GABA Binding to Whole Rat Brain Synaptosomal Membranes.

Triton-washed whole brain synaptic membranes from male Sprague Dawley rats were prepared as described by Maddison *et al* (5). On the day of assay, Triton-washed membranes were thawed, centrifuged at 40,000 g, washed in Tris citrate buffer and again in ice-cold water before resuspension in Tris citrate buffer. Membrane suspension aliquots (300 µl) were incubated with 2.5 nM [^{3}H]-GABA (Amersham International, Buckinghamshire, U.K.: specific activity 105 Ci/mmol) and (a) 100 µl unlabeled amino acids at 1 mM final concentration or at plasma equivalent concentrations (see Table 1); or (b)100 µl of a range of unlabeled GABA concentrations (0-1000 nM) each in the presence of 100 µl of unlabeled taurine (0-100 µM); or (c)100 µl aliquots of deproteinised plasma, to determine plasma GABA-like activity (5,6).

[^{3}H]-GABA binding assays were performed as previously described (6). Non-specific binding was determined by the amount of bound [^{3}H]-GABA remaining after the addition of 100 µM unlabeled GABA to the assay; this was not changed by the presence of any other amino acid at the concentrations used in the assay, nor by the presence of deproteinised plasma.

Data analysis. The degree of inhibition of [^{3}H]-GABA binding by the unlabeled amino acids or plasma aliquots was expressed as the percentage reduction in specific binding.

Scatchard analysis of [^{3}H]-GABA receptor binding data in the presence of unlabeled taurine was performed by nonlinear regression analysis using the computer program LIGAND (15) modified to run on the Macintosh microcomputer (Apple Computers, Inc., Cupertino, Calif.) under Microsoft BASIC version 2.0 (Microsoft Corp., Bellevue, Wash.).

Plasma GABA-like activity was calculated from the amount of bound [^{3}H]-GABA displaced by an aliquot of plasma, using a standard curve constructed from the amount of ligand displaced by unlabeled GABA. Results are expressed as mean ± SD. Values from control rats and rats with HE were compared by the Student's t test (unpaired, two-tailed).

Measurement of Taurine and GABA Concentrations in Rat Plasma by HPLC.

Plasma amino acid concentrations were determined in 18 rats with either acute HE induced by thioacetamide (5) or chronic HE due to portal vein ligation and/or bile duct ligation and 9 control rats (6). The clinical signs and biochemical parameters in the HE rats have been documented elsewhere (5,6).

Taurine and GABA concentrations were determined by cation-exchange chromatography on a Varian 5000 Liquid Chromatograph using post-column derivatisation with *ortho*-phthalaldehyde (OPA; 6). A dual buffer gradient system was used to elute taurine and GABA in two different runs. The composition of Buffer A was 0.066 M trilithium citrate in 4% methanol, adjusted to pH 2.3 (for GABA) or 2.7 (for taurine) with 6M HNO_3. Buffer B was 0.25 M lithium hydroxide, 0.065 M boric acid, 0.13 mM EDTA, adjusted to pH 10.4 with 6M HNO_3. OPA was dissolved in methanol (1g/100 ml) and 5 ml of this solution added to 95 ml borate buffer (0.16 M boric acid and 0.16 M KOH in water) and activated with 0.2 ml of mercaptoethanol.

To measure taurine, plasma was deproteinised with an equal volume of 1.2 M perchloric acid and neutralised with 0.6 M K_2CO_3. Ten µl

Table1. Concentrations of amino acids used to inhibit [^{3}H]-GABA receptor binding.

Amino acid	Concentration (µM) HE	Control
Glutamine	1000	750
Glutamate	300	150
Phenylalanine	100	50
Tyrosine	150	100
Citrulline	100	50
Taurine	300	150

Concentrations are those of the amino acid solutions added to the assay (thus final assay concentrations are 10% of these), and are approximately equivalent to plasma concentrations in rats with hepatic encephalopathy or controls.

aliquots of deproteinised plasma were injected onto the column. For GABA determination, plasma samples were deproteinised with 50 μl of 4 mM sulphosalicylic acid to each 450 μl aliquot of plasma and 100 μl aliquots injected onto the column. Initial experiments showed that the use of sulphosalicylic acid, rather than perchloric acid, to deproteinise plasma resulted in better separation and a sharper GABA peak. However, the estimation of GABA in a plasma aliquot was unaffected by the deproteinising agent used.

The mobile phase flow rate was 0.4 ml/min and OPA flow rate 0.7 ml/min. Column temperature was set at 30° C (for GABA) or 62° C (for taurine) and column pressure averaged 7000 kPa. The column was regenerated in 100% Buffer A. To elute taurine, Buffer A alone was run for 10 mins. Taurine was eluted at 3.8 min. To elute GABA, a linear gradient was started at the time of injection which reached 65% Buffer B by 1 min after injection. From 1-60 min there was a linear gradient from 65%-70% Buffer B. Under these conditions, GABA eluted at 59 min. The amount of taurine or GABA in the chromatogram was estimated from the appropriate peak height. To measure taurine, plasma was deproteinised with an equal volume of 1.2 M perchloric acid and neutralised with 0.6 M K_2CO_3. Ten μl aliquots of deproteinised plasma were injected onto the column. GABA concentrations were measured as previously described (6). Results are expressed as mean ± SD. Values from control rats and rats with HE were compared by the Student's t test (unpaired, two-tailed).

RESULTS.

Effect of Amino Acids on [^{3}H]-GABA receptor binding.

Significant inhibition of [^{3}H]-GABA binding (>20%) was noted when 10 mM (1 mM final assay concentration) glutamine (65% inhibition), glutamate (64%), tyrosine (82%) and taurine (86%) were added to the assay (Fig 1). However, when each of these amino acids was added at concentrations equivalent to those found in plasma, only taurine resulted in significant inhibition (36% at 30 μM final assay concentration; Fig 1). Moreover, for each amino acid other than taurine, there was no significant difference between the degree of inhibition caused by the concentration found in control plasma, and the degree of inhibition caused by its concentration in HE plasma. Therefore, results were combined for each amino acid other than taurine.

Further studies showed that the IC_{50} concentration for taurine in the assay was 158 μM. This corresponds to a plasma concentration of 1.58 mM. Taurine inhibition of [^{3}H]-GABA binding parameters was competitive, as the B_{max} was unaltered, but the Kd increased with increasing concentrations of taurine (Fig 2).

Plasma GABA-like Activity and Plasma GABA and Taurine Concentrations.

Results are summarised in Table 2. As reported previously (5,6), plasma GABA-like activity was significantly increased in rats with HE when compared with controls. Plasma GABA and taurine concentrations determined by HPLC were both significantly increased in HE rats compared with controls. In both control rats and HE rats plasma GABA-like activity was significantly greater than plasma GABA concentrations suggesting the presence in plasma of a substance other than GABA that can inhibit [^{3}H]-GABA binding but is not GABA itself.

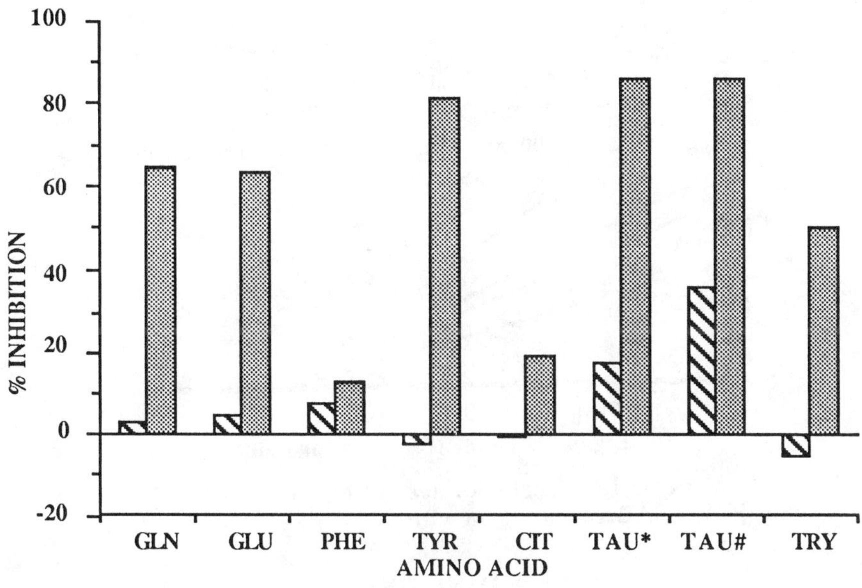

Fig. 1. Effect of amino acids on [3H]-GABA brain receptor binding in vitro. TAU* = taurine at control plasma concentration; TAU# = taurine at HE plasma concentration.
▧ Plasma equivalent concentrations
▩ 1 mM final assay concentration

To determine if the inhibition of [^{3}H]-GABA binding due to plasma aliquots (GABA-like activity) was attributable to only GABA and taurine in the plasma, the percent inhibition of [^{3}H]-GABA binding expected by the known concentrations (determined by HPLC) of taurine and GABA in each plasma sample was determined from the standard curves generated for each amino acid.

There was no significant difference between the mean inhibition of [^{3}H]-GABA binding by plasma aliquots and the sum of the inhibition predicted from the known concentrations of GABA and taurine in HE rats (63 ± 10% vs 68 ± 13%; mean ± S.D, n=18; NS) nor in control rats (38 ± 9% vs 41 ± 12%, n=9; NS). This result suggests that in both rats with HE and control rats, GABA-like activity in plasma arises from the combined inhibition of [^{3}H]-GABA binding by GABA and taurine in the plasma.

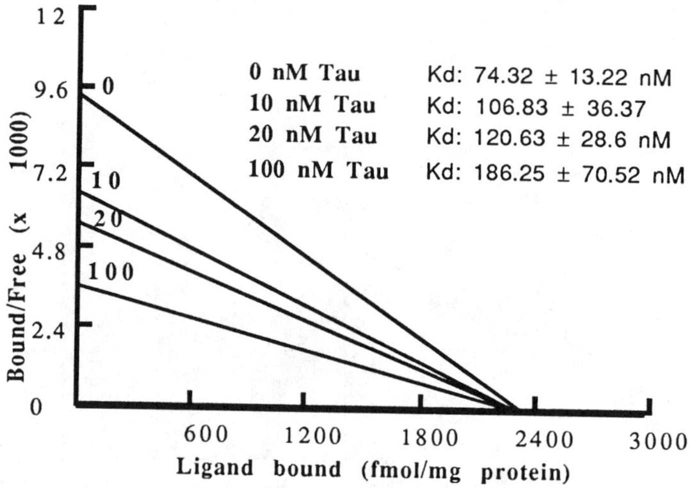

Fig. 2. Eadie-Hofsteeplot of the effect of taurine on [3H]- GABA binding in vitro.

DISCUSSION.

The presence in plasma of a substance that can inhibit [^{3}H]-GABA binding but is not GABA itself has practical implications in the use of the radioreceptor assay to estimate plasma GABA concentrations. The radioreceptor assay has been shown to overestimate the concentration of GABA in the plasma when compared with HPLC determinations of GABA concentrations (6). The difference between radioreceptor and HPLC estimates of plasma GABA has been shown to be significantly greater in rats with acute or chronic HE (6,8). This finding implies that a substance, other than GABA, that is present in the plasma of rats in increased concentrations in HE can inhibit [^{3}H]-GABA binding.

Table 2: Plasma GABA-like activity, GABA concentrations and taurine concentrations in control rats and rats with hepatic encephalopathy.

Parameter	Control (n=9)	HE (n=18)	p
GABA-like activity (nM)	411 ± 219	2051 ±1048	<0.001
GABA-like activity (% inhibition*: A)	38 ± 9	63 ± 10	<0.001
GABA (nM)	226 ± 107	884 ± 493	<0.001
GABA (predicted % inhibition*: B)	27 ± 11	51 ± 11	<0.001
Taurine (μM)	175 ± 29	234 ± 106	<0.05
Taurine (predicted % inhibition*: C)	14 ± 3	17 ± 7	NS
B + C	41 ± 12	68 ± 13	<0.001
A-(B + C)	-4 ± 9	-5 ± 10	NS

Results are expressed as mean ± SD, n = number of rats. *percentage inhibition of [^{3}H]-GABA binding to rat brain membranes by plasma aliquots (A), plasma GABA (B) and plasma taurine (C).

The present study demonstrates that of the plasma amino acids that are increased in both acute and chronic HE, only taurine can inhibit [^{3}H]-GABA binding at concentrations equivalent to those found in plasma. The IC_{50} for taurine inhibition of [^{3}H]-GABA binding is greater than the plasma concentration found in HE. However, the discrepancy between the degree of inhibition expected by the known concentration of GABA in plasma and the total inhibition caused by plasma aliquots can be attributed entirely to the predicted inhibitory effect of taurine in the plasma.

The importance of this finding in the pathogenesis of HE is difficult to ascertain. Abnormalities found in the plasma of patients or animal models of HE can only have relevance to the pathogenesis of the disorder if they are reflected in, or change, the brain milieu. In addition, putative changes must be shown to be causative rather than a consequence of the encephalopathy (16). Plasma taurine concentrations have been reported to correlate with the grade of encephalopathy in patients with chronic liver disease (17). Cascino and colleagues (18) also reported that plasma taurine concentrations were higher in patients with cirrhosis and severe HE compared with patients with cirrhosis and lesser grades of encephalopathy. However, in the same study, taurine concentrations were not altered in patients with cirrhosis and encephalopathy who had surgically created portocaval shunts.

The roles of taurine in the body have yet to be determined. Taurine is found in abundance throughout the body, particularly in excitable tissues such as heart, muscle and brain (19). The concentration of taurine in adult brain is exceeded only by glutamic acid. In mammals the highest amounts are found in cerebral and cerebellar cortex, olfactory bulbs and striatum. However, a specific function that correlates with the marked regional variation in concentration has not been found (19). It is now thought that taurine functions as an inhibitory neurotransmitter or a neuromodulator of neuronal activity, possibly by stablising excitable membranes and suppressing the release of neurotransmitters at synapses (20, 21). It has been shown that taurine can modulate GABA receptor binding and the GABA-benzodiazepine receptor complex *in vitro* (22-24). In addition, taurine has anticonvulsant properties in a variety of experimental and human epilepsies (20,25,26), although its role *in vivo* is still problematical.

There have been few studies of brain taurine levels in HE. Zanchin *et al* (27) reported that cerebral taurine uptake was unchanged in rats with portocaval shunts. Similarly, Record *et al* (28) found no difference in taurine concentrations in the frontal cortex of patients dying from fulminant hepatic failure. Taurine has limited diffusability across the blood-brain barrier (21), and taurine exchange does not appear to be important in the maintenance of brain taurine levels under normal circumstances (19).

The present study shows that the increased plasma GABA-like activity in rats with acute or chronic HE can be attributed to both GABA and taurine. However, it is clear that more extensive work needs to be performed to determine if taurine has a role in the pathogenesis of hepatic encephalopathy.

REFERENCES.

1. Schafer DF, Waggoner JC, Jones EA. Sera from rabbits in acute hepatic coma inhibit the binding of [^{3}H]γ-aminobutyric acid in neural membranes. *Gastroenterology* (1980) **78**:1320.

2. Ferenci P, Schafer DF, Kleinberger G, Hoofnagle JH, Jones EA. Serum levels of gamma-aminobutyric-acid-like activity in acute and chronic hepatocellular disease. *Lancet* (1983) **II**: 811-814.

3. Thompson JS, Schafer DF, Schafer GJ, Hodgson PE. γ-Aminobutyric acid plasma levels and brain binding in Eck fistula dogs. *J Surg Res* (1985) **38**: 143-148.

4. Minuk GY, Winder A, Burgess ED, Sarjeant EJ. Serum gamma-aminobutyric acid (GABA) levels in patients with hepatic

encephalopathy. *Hepato-gastro* (1985) **32**: 171-174.

5. Maddison JE, Dodd PR, Johnston GAR, Farrell GC. Brain γ-aminobutyric acid receptor binding is normal in rats with thioacetamide-induced hepatic encephalopathy despite elevated plasma γ-aminobutyric acid-like activity. *Gastroenterology* (1987) **93**: 1062-1068.

6. Maddison JE, Dodd PR, Morrison M, Johnston GAR, Farrell GC. Plasma GABA, GABA-like activity and the brain GABA-benzodiazepine receptor complex in rats with chronic hepatic encephalopathy. *Hepatology* (1987) **7**: 621-628.

7. Maddison JE, Dodd PR, Morrison M, Farrell GC,Johnston GAR. Plasma GABA concentrations and cerebro-cortical GABA receptor binding and function in dogs with congenital portosystemic encephalopathy. In: Soeters P, Wilson JHP, Meijer et al., eds. *Advances in ammonia metabolism and hepatic encephalopathy*. Amsterdam, The Netherlands: Excerpta Medica (1988) pp 265-274.

8. Maddison JE, Dodd PR, Johnston GAR, Farrell GC. Increase in plasma GABA-like factor that is not GABA itself in acute and chronic models of hepatic encephalopathy in rats. *Aust NZ J Med* (1986) **16** (suppl. 3): 603.

9. Enna SJ, Snyder SH. A simple, sensitive and specific radioreceptor assay for endogenous GABA in brain tissue. *J Neurochem* (1976) **26**: 221-224.

10. Enna SJ, Wood JH, Snyder SH. γ-Aminobutyric acid (GABA) in human cerebrospinal fluid: radioreceptor assay. *J Neurochem* (1977) **28**: 1121-1124.

11. Wood JH, Glaeser BS, Enna SJ, Hare TA. Verification and quantification of GABA in human cerebrospinal fluid. *J Neurochem* (1978) **30**: 219-293.

12. Ferkany JW, Smith LA, Seifert WE, Caprioli RM, Enna SJ. Measurement of gamma-aminobutyric acid (GABA) in blood. *Life Sci* (1978) **22**: 2121-2128.

13. Ferenci P, Ebner J, Zimmermann C. Overestimation of serum concentrations of γ-aminobutyric acid in patients with hepatic encephalopathy by the γ-aminobutyric acid-radioreceptor assay. *Hepatology* (1988) **8**: 69-72.

14. Maddison JE. Studies of gamma-aminobutyric acid in animal models of hepatic encephalopathy. PhD thesis, University of Sydney, N.S.W., Australia, 1987.

15. Munson PJ, Rodbard D. LIGAND: a versatile computerized approach for characterization of ligand binding systems. *Anal Biochem* (1980) **107**: 220-239.

16. Zieve L. The mechanisms of hepatic coma. *Hepatology* (1981) **1**: 360-365.

17. Cascino A, Cangiano C, Calcaterra V, Rossi-Fanelli F, Capocaccia

L. Plasma amno acids imbalance in patients with liver dieases. *Dig Dis* (1978) **23**: 591-598.
18. Rosen HM, Yoshimura N, Hodgman JM, Fischer JF. Plasma amino acid patterns in hepatic encephalopathy of differing etiology. *Gastroenterology* (1977) **72**: 483-487.
19. Mandel P, Pasantes-Morales H. Taurine in the nervous system. In: Ehrenpreis S, Kopin I., eds. *Reviews of neuroscience.* New York: Raven (1978) pp 157-193.
20. Kuriyama K. Taurine as a neuromodulator. *Fed Proc* (1980) **39**: 2680-2684.
21. Durelli L, Mutani R. The current status of taurine in epilepsy. *Clin Neuropharmacol* (1983) **6**: 37-48.
22. Medina JH, de Robertis E. Taurine modulation of the benzodiazepine-γ-aminobutyric acid receptor complex in brain membranes. *J Neurochem* (1984) **42**: 1212-1217.
23. Malminen O, Kontro P. Modulation of the GABA-benzodiazepine receptor complex by taurine in rat brain membranes. *Neurochem Res* (1986) **11**: 85-94.
24. Malminen O, Kontro P. Actions of taurine on the GABA-benzodiazepine receptor complex solubilized from rat brain. *Neurochem Int* (1987) **11**:113-117.
25. van Gelder NM. Taurine, the compartmentalized metabolism of glutamic acid, and the epilepsies. *Can J Physiol Pharmacol* (1978) **56**: 362-374.
26. Huxtable RJ. Taurine: epilepsy, inotropy and eyesight. *TIPS* (Jan 1982) pp 21-25.
27. Zanchin G, Rigotti P, Dussini N, Vassanelli P, Battistin L. Cerebral amino acid levels and uptake in rats after portocaval anastomosis: II. Regional studies in vivo. *J Neurosci Res* (1979) **4**: 291-299.
28. Record CO, Buxton B, Chase RA, Curzon G, Murray-Lyon IM, Williams R. Plasma and brain amino acids in fulminant hepatic failure and their relationship to hepatic encephalopathy. *J Clin Invest* (1976) **6**: 387-394.

Acknowledgements: This study was supported by the Australian National Health and Medical Research Council.

DIAZEPAM METABOLISM IN PERFUSED CULTURES OF ADULT RAT HEPATOCYTES

HO Jauregui, S Naik, H Santangini

Dept. of Pathology, Rhode Island Hospital

Providence, RI 02903

ABSTRACT

Recent studies implicating brain benzodiazepine receptors in the genesis of hepatic encephalopathy justify a comprehensive testing of a hollow fiber membrane liver assist device in terms of its capacity to metabolize diazepam *in vitro*. The detection of diazepam metabolites (temazepam, nordiazepam and oxazepam) indicates an active group of P-450 isoenzymes (b and e in the rat) which are responsible for many detoxification functions and are stimulated by the administration of phenobarbital. After a separation procedure following the Seglen method, 8 x 10^6 adult rat hepatocytes per ml were seeded on the surface of about 100 polysulfone hollow fibers (W.R. Grace) housed in polypropylene cartridges. The diameter of these fibers is approximately 816 μm. Prior to the seeding procedure, these fibers were coated with Vitrogen (a mixture of collagen I and III). The total capacity of the extracapillary space (the tissue culture chamber) was 5 ml and the total amount of perfused medium (extracapillary space plus reservoir) was 25 ml. The perfusate was a tissue culture medium (CEM)(GIBCO Labs) supplemented with 10% fetal bovine serum (FBS). CEM maintained good diazepam metabolic activities in these hepatocytes attached to hollow fiber membranes. For instance, at 24 hours the combined (nordiazepam, temazepam and oxazepam) activity as measured by HPLC (Waters Chromatography Div.) was 71±13 μg/25 ml media whereas at 48 hours the activity was 27±15 μg/25 ml media. At day 10, the cells still showed an activity of 7 μg/25 ml media.

A new tissue culture formulation, which avoids the incorporation of FBS, and includes ITS (insulin, selenium and transferrin) resulted in higher diazepam metabolism. For instance, during the first 24 hrs of culture, our perfused hepatocytes metabolized diazepam at a level of 117±49 μg/25 ml media and at 48 hrs 50±19 μg/25 ml of media. Control monolayer cultures were able to metabolize diazepam at 178 μg/25 ml of media at 24 hours.

These results indicate that hollow fiber membrane devices seeded with adult rat hepatocytes are able to metabolize a benzodiazepine at levels that are similar to monolayer cultures at 24 hrs. However, in hollow fiber cartridges, there is a rapid decrease in metabolic activity. The ability to maintain the integrity of P-450 functions can be improved by attachment of the cells to the surface of the fibers as well as in the design of the perfusion system, and will result in a better adaptation of the hepatocyte and facilitate the use of these devices to treat experimental liver failure in an animal model, e.g., the rabbit.

INTRODUCTION

Acute liver failure demands immediate application of a supportive hepatic system. Several technologies for liver support proceed from a focus on selected aspects of liver function such as: a)hemoperfusion for adsorption of specific toxins or chemical processing of a specific biological product (1,2), b)hemodialysis and hemofiltration for clearance of abnormal compounds within the limitations of the convective and/or diffusive permeability of the exchange membrane (3) and c)plasmapheresis, whereas the separated plasma is processed by sorbents or reagents before it is returned to the patient circulatory system (4).

From another group of interactive methodological approaches such as the parabiotic dialysis (5), we are developing a hybrid hollow fiber membrane liver support system. In this system, live isolated hepatocytes [cell clusters or tissue slides have also been used by others (6)] are either encapsulated or somehow separated, as in this case, by a semi-permeable membrane from the patient's blood (7).

There are three major options in designing this type of liver support. The first approach is to use hepatoma cells as a source of the cell component (8). Although bilirubin conjugation has been reported in this system (8), tumor cells show highly simplified functional cell activity and lack most of the hepatocyte metabolic pathways (9). A second choice, the use of cells with replicating capacity, either neonatal or embryonic hepatocytes, was reported to display important enzymatic functions (10). These results contradict extensive literature indicating that proliferating hepatocytes showed no well-developed metabolizing functions (11,12). Furthermore, successful implementation of this model for therapeutic purposes requires the acquisition of embryonal or neonatal cells in large quantities, which is logistically impractical.

The third choice, developed in our laboratory during the last 10 years, is to isolate sufficient numbers of highly purified, anchor dependent populations of adult liver cells to be attached to semipermeable artificial capillary surfaces. A good attachment to hollow fiber membranes especially designed for this task (W.R. Grace) should be followed by *in vitro* studies proving the ability of this system to display those enzymatic pathways which metabolize endogenous toxins related to the production of acute liver failure and its more life-threatening consequence: hepatic encephalopathy.

General agreement among different investigators about which metabolites are responsible for the development of hepatic encephalopathy is lacking, but there is enough scientific knowledge to prove that the major detoxification function of the liver resides in its complex oxido-reductase system, the so-called P-450 system. Our laboratory started a comprehensive study of benzodiazepines in 1982 in order to test P-450 activity. Diazepam was our choice as a model drug, mainly because the production of three extractable metabolites present in rat hepatocytes can be accurately followed (13) and permits the inference of the presence of $P\text{-}450_{b+e}$ isoenzymes in the cell in culture.

We are presenting preliminary data showing that adult rat hepatocytes seeded in Vitrogen-coated (a mixture of collagen I and III) polysulfone hollow fiber membranes metabolize diazepam for at least 10 days. Sophisticated functions of hydroxylation and N-demethylation are detected by the presence of oxazepam, temazepam and nordiazepam. As expected, temazepam was the highest metabolic product of adult rat hepatocytes seeded in our device as is the case with the whole animal (14). Initial levels of the combined activity of these three metabolites were similar to those tested in hepatocyte monolayers (13), but there was a rapid decrease in P-450 metabolic activity. The latter is probably related to our present inability to support adult rat hepatocyte attachment to these types of capillaries for more than 10 days resulting in a progressive cell loss from the device. In spite of this progressive protein decay in the tissue culture compartment of our hollow fiber membrane cartridge, we are now proceeding with animal testing because the total diazepam metabolic activity of this system suffices to create a liver support device that, in our judgement, will prolong the life of an experimental animal model of hepatic failure.

MATERIALS AND METHODS

Hepatocytes were isolated using Seglen's method (15) with minor modifications (16). Isolated hepatocytes were suspended at 8×10^6 cells/ml in CEM media supplemented with 10 mU/ml insulin, 1 μM dexamethasone and 50 μg/ml gentamicin. The perfusion or monolayer tissue culture media was also CEM supplemented as above plus 10% FBS or serum free CEM supplemented with insulin (6.25 μg/ml), transferrin (6.25 μg/ml) and selenium (6.25 ng/ml).

Tissue Culture of Hepatocytes in Capillary Chambers

Polysulfone hollow fiber membrane cartridges (W.R. Grace) were used in all the experiments. The hollow fibers were arranged in parallel bundles of 100 to 110 capillaries. Their ends were then potted in thermosetting resin (urethane or epoxy) in such a way that capillary lumens were open providing a multi-perforated "face sheet" (Fig.1). The bundle of hollow fibers was encased in a cylindrical, transparent, styrene acrylonitryl copolymer jacket featuring side ports for cell seeding. A peristaltic pump was located between the reservoir of the tissue culture medium and the chamber to provide a positive head pressure (Fig.2).

Prior to the hepatocyte seeding, the polysulfone hollow fibers were coated with Vitrogen (Collagen Corp.) by a salt precipitation procedure. These chambers were seeded with 8 x 10^6 cells/ml via the described ports. The hepatocytes were allowed to attach for 3 hrs. Perfusion was at a rate of 15 ml/min. The media in the reservoir was changed every 24 hrs and assayed for percent protein attached (17), percent of glucuronide conjugation of phenol red (18) and the amount of diazepam metabolized.

Long Term Tissue Culture of Hepatocytes and Diazepam Metabolism Evaluation

Hepatocytes were seeded in 60 mm Permanox (Miles Scientific) dishes coated with Vitrogen. The TC medium was previously described in the perfusion studies. At various time points, hepatocyte cultures were incubated with 10 μg/ml diazepam (Sigma) for 8 hrs. Aliquots (1.0 ml) of TC media were incubated with 1.0 mg β-glucuronidase (Sigma) dissolved in 1 M sodium acetate buffer (pH 5.0). After 36 hrs of incubation, 1.0 ml of saturated sodium borate buffer (pH 5.0) was added and the mixture was extracted into 10 mls of chloroform. The aqueous layer was aspirated off and 8.0 mls of the infranatant was dessicated at 55°C under a stream of air. The sample was reconstituted in 0.8 mls of a mobile phase consisting of 0.03 potassium phosphate monobasic/methanol (45-55%). Diazepam metabolites were identified by reverse phase HPLC using a Waters Assoc. instrument with

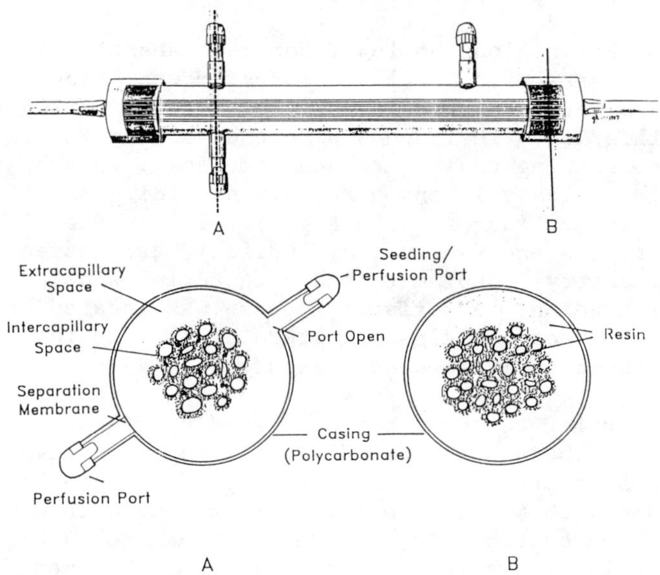

Figure 1. Hollow Fiber Membrane Chamber

CLOSED LOOP PERFUSION CIRCUIT

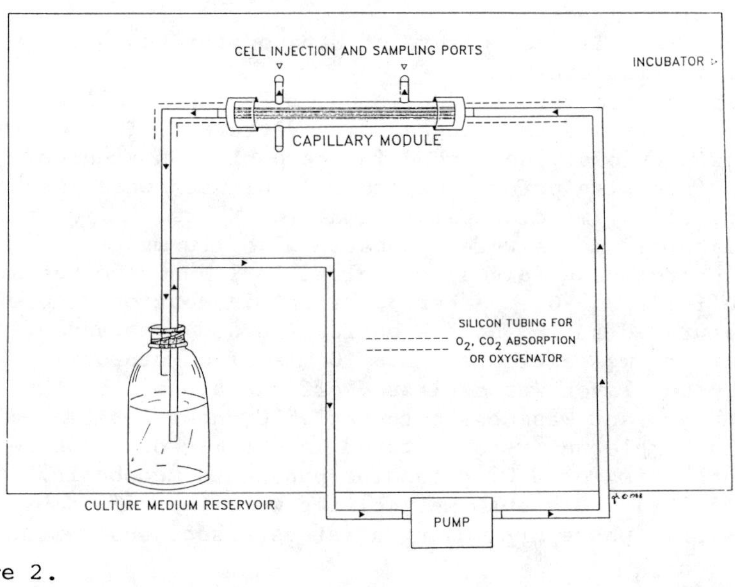

Figure 2.

a μBondadak C18 stainless steel column (3.9 min x 30 cm with detector at 254 min and 1.5 ml/min flow).

RESULTS

Diazepam metabolic activity of adult rat hepatocytes seeded in hollow membranes perfused with CEM + 10% FBS was at the level of 70±15 μg/25 ml of in the first 24 hrs of culture (Figure 3). This activity was significantly lower than the values of 117±49 μg/25 ml present in the TC media of hepatocytes growing in CEM supplemented with ITS but without FBS (Fig. 3). Neither media formulation changed the dramatic decrease of diazepam metabolic activity which was approximately 50% by 48 hrs of culture and to 75% by 72 hrs. The cells perfused with FBS-free CEM consistently showed more diazepam metabolic activity than the hepatocytes maintained with CEM + 10% FBS. Visual examinations of the chambers as well as scanning electron micrographs of the hollow fiber membrane external surfaces revealed that hepatocyte attachment and spreading was marginal in all cases and even more with chambers perfused with CEM + 10% FBS. This observation was corroborated by a net protein loss of the extracapillary chambers of about 20% at 3 hrs which correspond to hepatocytes that either did not attach or were not trapped in between the hollow capillary bundles. These cells were washed off by a gentle washing of this chamber that took place at that time and was repeated every 24 hrs afterwards. Indeed, 10% of the original hepatocytes seeded at 0 time was lost with each successive daily washing procedure (Fig.3). When replicate numbers of adult rat hepatocytes were seeded in Permanox TC dishes using CEM media (no FBS, + ITS), the cells showed better diazepam metabolic activity (178 μg/25 ml). This activity remained at high levels for 10 days (160 μg/25 ml) and decreased to 50% of initial values by day 30 (Fig. 4). Finally, Fig. 5 indicates that temazepam was the major metabolite produced by the perfused hepatocytes.

DISCUSSION

We have tested *in vitro* a hollow fiber membrane system seeded with freshly isolated rat hepatocytes and perfused with TC medium (CEM). Without the addition of

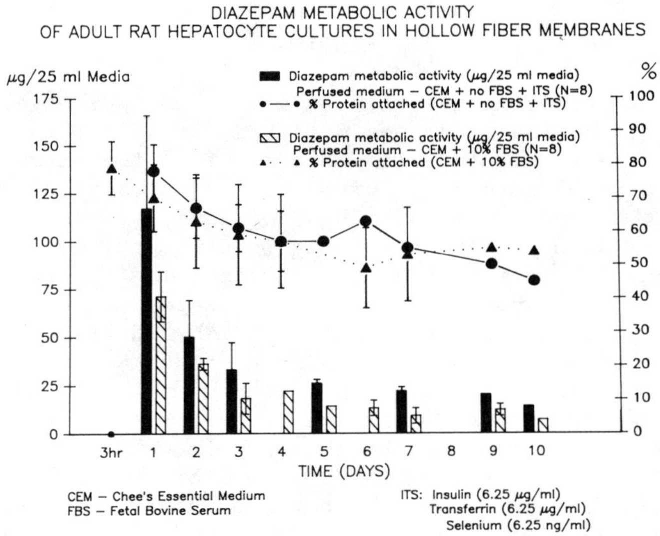

Figure 3.

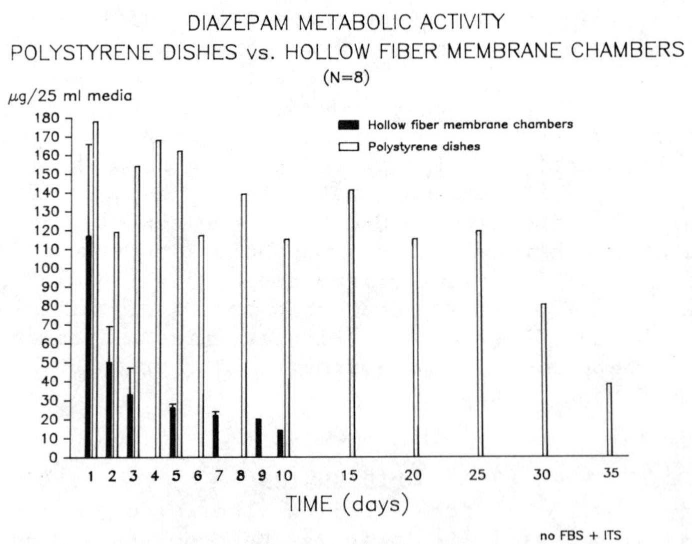

Figure 4.

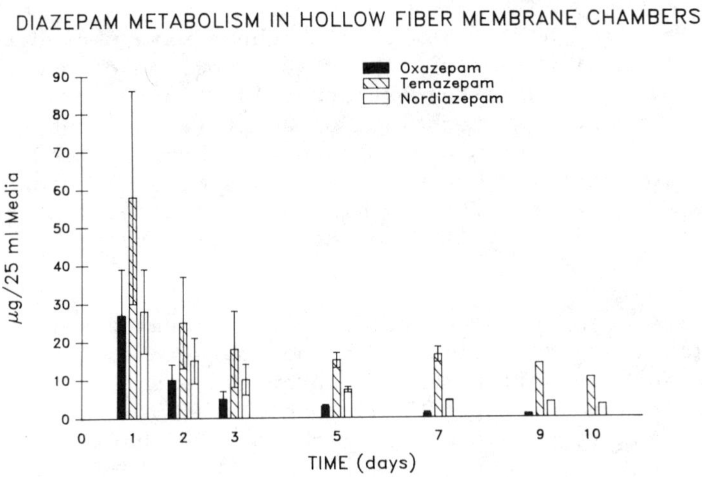

Figure 5.

FBS, but supplemented with ITS, CEM proved to be a better TC media to maintain diazepam metabolic activity. As was reported in the rat *in vivo*, (14) temazepam remained the major metabolite of diazepam in these perfused cultures. Diazepam was selected as a test drug for the hollow fiber membranes prior to recent publications incriminating diazepam-like substances as the ones that are responsible for the depolarization of the cell through an opening of the chloride channel (19).

Adult rat hepatocytes maintained in culture chambers are oxygenated by a bundle of hollow fiber membranes which have an internal "skin" that limits the molecular weights of the perfused solutes. The skin-like lining can be manufactured in such a way as to provide a constant molecular weight cutoff ranging from 40,000 to 150,000. The clinical application of this arrangement is potentially useful because the hollow fibers provide immunological isolation to the hepatocytes growing on their external surfaces. Consequently, pig or other mammalian hepatocytes may be used to support human patients in hepatic failure.

Perfused hollow membrane chambers have proved to maintain insulin levels in pancreatectomized dogs when β-cells were seeded in the extracapillary space (20). Our improvements to the technique of adult rat or rabbit hepatocyte separation and culture have been applied to the hollow fiber membrane system. A similar approach was also used by Wolf, although his utilization of hepatoma cell lines seriously limited the metabolic range of his device (8).

Two major problems remain in the development of this liver support system. The polymer composition chosen for the hollow membrane prevents platelet attachment on the internal surface, however, this composition is also preventing the desired hepatocyte attachment to the external surface. This problem will be addressed properly by the development, in progress, of a different external hollow membrane surface which will be "activated" to support a good substrate coating that should improve cell attachment. The second problem is to find a perfusion rate compatible with good oxygenation of the cells in the extracapillary chamber. Although the hepatocyte seems to thrive with low PO_2 values, it consumes high levels of O_2, as demonstrated when hepatocyte oxygen consumption was compared to pancreatic islands seeded in *ad hoc* tissue culture chambers (21). The right combination of internal and external diameters of the hollow fibers, together with proper pumping devices, will resolve this second problem. In this context, it is important to note that the perfusion circuit will be subject to different conditions when these units are put into contact with experimental animals with hepatic coma through proper AV shunts.

REFERENCES

1. Chang TMS. Hemoperfusion over microencap-sulated adsorbent in a patient with hepatic coma. Lancet 2:1371, 1972.

2. Silk DBA, Williams R. Treatment of fulminant hepatic failure by charcoal hemoperfusion in: Chang TMS, ed. Artificial Kidney, Artificial liver and Artificial Cells. New York, Plenum Press. 1978, pp 125

3. Knell AJ. Haemodialysis procedures in acute liver failure in: Brunner G and Schmitt F eds. Artificial Liver Support. Berlin, Heidelberg, Springer-Verlag: 1981, pp 147.

4. Asanuma Y, Malchesky PS, Zawicki I, et al. Clinical hepatic support by on line treatment with multiple sorbents-evaluation of system performance. Trans Am Soc Artif Intern Organs 26:400, 1980.

5. Opolon P, Huguet C, Granger A, et al. in: Williams R, Murray JM eds. Proc of Int Symp on Artif Support System for Acute Hepatic Failure. Tunbridge Wells, Kent, UK, Pitman Med Pub:1975, pp 186.

6. Koshino I, Sakamoto H, Shiada Y, et al. Use of liver slices for an artificial liver. Trans Am Soc Artif Int. Organs 25:493, 1979.

7. Knazek RA, Gullino PM, Kohler PO, et al. Cell culture on artificial capillaries: an approach to tissue growth *in vitro*. Science 178:65, 1972.

8. Wolf, CW. Liver assist by cultured cells on artificial capillaries in: Brunner G. Schmidt eds. Artificial Liver Support. Berlin, Heidelberg, New York, Springer-Verlag:1981, pp 247.

9. Bissell DM, Guzelian PS. Microsomal functions and phenotipic changes in adult rat hepatocytes in primary monolayer culture in: Gerchenson LE, Thompson EB, eds. New York, Academic Press: 1975, p 119.

10. Nakamura T, Kumegawa M. Induction of key glycolytic enzymes by insulin in mouse fetal liver cultured in a circumfusion system. Biochem Biophys Res Comm 51:474, 1973.

11. Koch K, Leffert HL. Growth control of differentiated fetal rat hepatocytes in primary monolayer culture. VI. Studies with conditioned medium and its functional interactions with serum factors. J Cell Biol 62:780, 1974.

12. Bissell DM, Hammaker K, Meyer UA. Parenchymal cells from adult rat liver in nonproliferating monolayer culture. I Functional Studies. J. Cell Biol 59:722, 1973.

13. Schwartz MA, Bommer P, Vane FM. Diazepam metabolites in the rat characterization by high mass spectrometry and nuclear magnetic resonance. Arch Biochem Biophys 121:508, 1967.

14. Schwartz MA, Koechlin BA, Postma E, et al. Metabolism of diazepam in rat, dog, and man. J Pharmacol Exp Therapeut 149(3):423, 1965.

15. Seglen PO. Preparation of isolated rat liver cells. Methods Cell Biol 13:391-398, 1973.

16. Jauregui HO, McMillan PN, Driscoll J, et al. Attachment and long term survival of adult rat hepatocytes in primary monolayer cultures; comparison of different substrata and tissue culture media formulations. In Vitro Cell & Dev Biol 22:13, 1986.

17. Hayner N, Driscoll J, Ferayorni L, et al. Ponceau S: A sensitive method for protein determination in freshly isolated and cultured cells. J. Tissue Cult Methods 7:77, 1982.

18. Driscoll JL, Hayner NT, Williams-Holland R, et al. Phenosulfonphthalein (phenol red) metabolism on primary monolayer cultures of adult rat hepatocytes. In Vitro 18:835, 1982.

19. Jones EA, Gammal SH. Hepatic encephalopathy in: Arias IM, Jakoby WB, Popper H, et al. The Liver: Biology and Pathobiology, 2nd ed. New York, Raven Press: 1988, pp 985.

20. Chick W, Like AA, Lauris V. Beta cell culture on synthetic capillaries: an artificial endocrine pancreas. Science 187:847, 1975.

21. Altman T, Panol G, Hayner N, et al. Oxygen requirements of freshly isolated cells for hybrid artificial organs. Proc Europ Soc Artif Organs 6:282, 1979.

Other Neurotransmitter Systems (amines, amino acids, peptides) in Hepatic Encephalopathy

HEPATOCEREBRAL DYSFUNCTION AND BRAIN SEROTONIN

Finn Bengtsson[1], Mogens Bugge[2] and Anders Nobin[2]

Department of Clinical Pharmacology and Laboratory for Experimental Brain Research[1] and Department of Surgery[2], Lund University Hospital, S-221 85 Lund, Sweden

SUMMARY

Alterations in brain serotonin have frequently been observed to accompany conditions of hepatocerebral dysfunction. These observations suggest that increased brain serotonin metabolism is pathogenetically related to the encephalopathy associated with liver failure. To further clarify this issue, we recently undertook a series of experimental studies in rat models of hepatocerebral dysfunction. Encephalopathy was quantitated and, by the use of enzyme inhibition techniques, the regional brain serotonin turnover was evaluated in vivo. The rat models included portacaval shunt, hepatic cirrhosis, ammonia intoxication, partial or total liver ischemia, and total hepatectomy.

Results demonstrated general but varying increases in regional brain serotonin turnover in portacaval shunted and totally hepatectomized rats. Enhanced brain serotonin formation was also found to be greater, especially in hepatectomized rats, than had previously been reported. In contrast, comatose ammonia-intoxicated and liver ischemic rats had normal or even decreased regional brain serotonin turnover. Other data render suspect the functional relevance of an altered brain serotonin turnover in hepatocerebral dysfunction.

INTRODUCTION

Hepatic dysfunction, no matter what its etiology and severity, is frequently associated with manifestations of cerebral dysfunction, i.e. encephalopathy. Our current lack of understanding of basic pathogenetic mechanisms responsible for the development of hepatocerebral dysfunction makes contemporary therapeutic interventions empirical in nature. Numerous pathogenetic mechanisms and etiological agents for hepatocerebral dysfunction have been suggested (see e.g. ref. 1). Some-brain ammonia levels, brain amino acids, e.g. tryptophan, brain serotonin and serotonin metabolite concentrations-have frequently been observed to be increased in association with hepatic disorders. These factors have therefore been suggested to be of major importance in the etiology of hepatocerebral dysfunction, although no single factor has proved to correlate clearly with the functional state. It has thus been presumed, therefore, that the encephalopathy arising in conjunction with hepatic disorders must be of multifactorial origin.

Increased brain tryptophan, serotonin and serotonin metabolite concentrations have been observed in patients and animal models with several types of hepatic disorders (2-20). To evaluate systematically the brain serotonin turnover in vivo and its possible impact on the development of hepatocerebral dysfunction, we recently undertook a series of studies using rat models representing various types of liver impairment. We determined the brain serotonin metabolism using in vivo enzyme inhibition techniques and studied the behavior and neurological outcome in these rat models. The objective of the present survey is to summarize our observations on brain serotonin metabolism seen in different states of hepatocerebral dysfunction, and to discuss the possible functional relevance of these observations.

HEPATOCEREBRAL DYSFUNCTION: NOMECLATURE

Neurologic and mental disorders associated with liver diseases conform to the group of metabolic encephalopathies. This denotes a usually fluctuating cerebral disorder related to disturbances of general metabolic activities in the body. Besides impaired liver function, metabolic encephalopathies may also accompany conditions such as hypoglycemia, uremia and septicemia. In this survey we have chosen the term hepatocerebral dysfunction only to note the prerequisite that both hepatic and cerebral dysfunction coexist. Sherlock and co-workers (21) introduced the term "portal-systemic encephalopathy" (PSE) to designate syndromes involving hepatocerebral dysfunction. This term is pathophysiologically relevant, since in many instances the portal-systemic shunting of blood around, or through (i.e. intrahepatic shunts), the liver parenchyma is more important for the development of encephalopathy than the liver impairment itself (22).

A distinction between acute and chronic (latent) PSE can be made. Acute PSE develops as a sudden episode of encephalopathy, usually precipitated by a specific event (gastrointestinal bleeding, protein/ammonia intoxication, sedative/diuretic medication, infection, etc.). Acute PSE can arise in patients with latent PSE whose liver dysfunction is not necessarily aggravated. Alternatively, chronic PSE is usually a more stable, low grade cerebral dysfunction. PSE must be distinguished from frank "hepatic coma", an entity which results from the anhepatic state and also accompanies fulminant hepatic failure. In turn, these latter two conditions are quite different; for example, the coma associated with fulminant hepatic failure comprises a complexity of potential comagenic factors such as pronounced hepatic parenchymal damage and the release of numerous toxic substances (23-26). It is probable that the coma associated with the anhepatic state or fulminant hepatic failure is more closely related to profoundly diminished hepatic function than is the case for PSE.

It is possible that all forms of hepatocerebral dysfunction have common denominators. However, it is also possible that in such conditions each one depends on a multitude of different factors to be manifested.

HEPATOCEREBRAL DYSFUNCTION: ANIMAL MODELS

To survey brain serotonin metabolism and neurologic findings in hepatocerebral dysfunction, we investigated several different rat models. These included portacaval shunts (PCS) and liver cirrosis (resulting in chronic PSE), and ammonia intoxication, partial or total liver ischemia, and total hepatectomy (reflecting acute and severe states of hepatocerebral dysfunction). Our PCS rat model was adopted from Lee and Fisher (27) with slight modifications (28). An end-to-side portacaval shunt was performed with microsurgical techniques, using a continuous 9-0 polyglycolic acid suture. Controls were either unoperated or sham-operated rats (includes all operative steps except for construction of the shunt).

Micronodular diffuse liver cirrhosis without ascites ("compensated cirrhosis") was achieved by the simultaneous administration of phenobarbitone and the liver toxin carbon tetrachloride (CCl_4) according to Proctor and Chatamra (29, 30). Cirrhotic induction was completed at 12 weeks and was verified by liver histology (31). Controls received either phenobarbitone only, or no hepatic-active agent at all. Neurochemical measurements took place one week after the last dose of CCl_4.

Ammonia (3.6 mmol/kg/h) was infused intravenously for 6 h in PCS or sham-operated rats. The ammonia solution was a buffered mixture of ammonium acetate and ammonium bicarbonate, and the pH was adjusted to 7.4 (for methodology see ref.'s 32-34).

Liver ischemia was induced in rats 48 h after the construction of a PCS. Partial liver ischemia was accomplished by division of the main hepatic artery (9,35-38). Total liver ischemia was achieved by complete liver devascularization, including division of all hepatic ligaments and all vessels to the liver (i.e. the left hepatic artery, vessels between the diaphragm and the superior surface of the liver and the vessels at the liver hilum) (4,38-40). The hepatic veins and the bile duct were left open in the total liver ischemic rats.

Total hepatectomy was performed in three stages according to Holmin et al. (41). The first stage included division of the inferior caval vein immediately below the liver. After 4 weeks an end-to-side PCS was constructed. Total hepatectomy was finally performed, one week after the PCS (16,42).

General characteristics

Each of these animal models has its own characteristic liver impairment, hemodynamic alteration and general metabolic perturbation (16,27,29,32,35,36,39,43-46). It should be noted that PCS rats (3-20 weeks after operation) exhibit a stable loss of liver mass, liver parenchymal atrophy without necrosis, moderately but invariably impaired liver function (however, normal function in the remaining liver parenchyma) and unchanged total hepatic blood flow with increased hepatic arterial and splanchnic arterial blood flow but reduced splanchnic venous pressure (e.g. 44). Rats with compensated cirrhosis (i.e. without ascites) exhibit increased liver weight, moderate to severe liver dysfunction, hyper-dynamic portal venous inflow and portal hypertension but no major portal-systemic shunting of blood (see e.g. ref. 47). Peripheral metabolic alterations in the rat models presently reported are summarized as follows.

All animals were normoxemic, normocapneic, and without plasma acid-base disturbances. All were normotensive even when severely neurologically impaired. Plasma glucose levels were slightly diminished in the PCS rats and slightly eleveted in the cirrhotic rats. Animals with severe neurologic deterioration (and their controls) were kept normoglycemic by a constant glucose infusion. All animal models exhibited elevated plasma ammonia levels. PCS rats undergoing ammonia infusion showed extremely high plasma ammonia levels, and high plasma ammonia levels were noted following total hepatectomy, PCS, and cirrhosis, whereas hyperammonemia was only moderate in liver ischemic rats. Rats subjected to total liver ischemia had extremely high plasma liver enzyme levels. Plasma liver enzyme levels were also elevated in the cirrhotic rats but not in PCS or PCS ammonia-infused rats, or in the anhepatic state.

Several of our studies referred in this survey also included measurements of plasma and brain amino acids and regional brain catecholamine turnover. However, this extensive body of information does not further clarify the present discussion and, therefore, for these topics the original articles are suggested (28,34,38,42,49,51, 52,86).

Encephalopathy

Since PCS and cirrhotic rats do not display grossly-evident encephalopathy, standardized computerized techniques to evaluate neurologic dysfunction were used. PCS rats were tested for open field behavior (spontaneous activity and exploration), somatosensory reactivity (different startle responses to an electrical foot-shock), water maze swim performance, motor coordination tasks, and step-through passive avoidance behavior (for methodology see ref. 48). In our experiments, PCS rats displayed a selective behavioral syndrome involving decreased generalized motor activity and sensory-motor integration, with no changes in motor coordination or congnitive abilities (48). It was also demonstrated that the decreased generalized motor activity in PCS rats was probably centrally mediated, since muscle and peripheral nerves were normal in these rats (49). Animals with overt encephalopathy or coma were graded according to Livingstone et al. (26) or evaluated as a score based on six different reflexes according to Bures et al. (50, se also ref. 33). In ammonia-infused PCS and sham-operated rats (34) the level of consciousness was evaluated by this latter reflex-rating score test. A normal rat has a maximal score of 15 points, which was also the score seen in both PCS and sham-operated control rats. In addition, sham-operated rats infused with ammonia had a normal score during the whole 6 hour infusion period. However, PCS rats infused with ammonia deteriorated to about 10 points at 2 hours of infusion, to about 5 points at 4 and 5 hours, and to coma (1.4 ± 0.7 points) after 6 hours of infusion.

Both partial and total liver ischemic rats were in grade 3 coma (out of 4) according to the scale of Livingstone (26) after 5 to 6 hours of ischemia (38,51), whereas totally hepatectomized rats were in grade 2 coma according to this scale after 5 hours of anhepatic state (42,52,53).

BRAIN SEROTONIN

The connection between serotonin and nerve tissue can be traced back to the 1920s (see e.g. ref. 54). Solid contemporary evidence supports serotonin as a functional neurotransmitter substance (cf. ref. 55), although this matter has not been completely resolved. Central nervous system (CNS) serotonin concentrations show substantial regional variations, with the highest concentrations found in the midbrain and hypothalamus but little in the cerebellum. CNS serotonin is, however, similarily distributed in all vertebrate classes (56). While the serotonergic projection areas are neuroanatomically outlined, the relevant functional role for each of these projections is still poorly understood. It has been suggested that serotonin is generally involved in a variety of physiological functions such as influencing sleep rhythms, sexual behavior, mood, thermoregulation, pain, etc. It may be that brain serotonergic innervation modulates general neurotransmitter activity in the CNS rather than primarily acting in specific centrally-mediated functions. In the case of the fluctuating cerebral dysfunction seen in metabolic encephalopathies, it has seemed theoretically well-founded that disturbances in such a neuromodulator function in the CNS might play an etiologic role.

Biosynthesis of serotonin in the brain

Serotonin (L-5-hydroxytryptamine; 5-HT) is formed from L-tryptophan by a two-step reaction. L-tryptophan is first converted to L-5-hydroxytryptophan (5-HTP) by the aromatic amino acid hydroxylase; tryptophan hydroxylase (L-trypto-phan-5-monooxygenase), and the 5-HTP is then decarboxylated to 5-HT by aromatic L-amino acid decarboxylase.

L-tryptophan is an essential amino acid, derived from dietary nutrients. It enters the CNS by a series of mechanisms, the description of which is outside the scope of this survey. The enzyme tryptophan hydroxylase appears to be restricted to cells that normally synthesize serotonin, because it is best located in or near the midline or raphe regions of the pons and upper brainstem; it is low in concentration in other parts of the CNS (including the cytosol). Hence, the possibility of non-specific effects of tryptophan due to formation of serotonin at non-physiological sites is usually not a consideration. However, large doses of tryptophan are required to produce small increases in the amount of serotonin formed. It should be noted that tryptophan is converted to other products than serotonin by numerous enzymatic pathways (see e.g. ref. 57) and that tryptophan is also required for protein synthesis in the brain (58). The conversion of 5-HTP to serotonin needs only aromatic L-amino acid decarboxylase, an enzyme which is not restricted to serotonin-containing neurons (it also plays a role in the biosynthesis of catecholamines and perhaps various other amines). Thus, if exogenous 5-HTP is administered it can be converted to serotonin in cells other than serotonergic ones (59), with release of serotonin from these other neurons or cells when oxidative deamination is blocked or when these neurons accept and handle serotonin as a false transmitter. Thereby serotonin may also act indirectly by promoting the release of the natural transmitter in these other neurons. Furthermore, the dopaminergic system is particularly rich in decarboxylase activity and the active accumulation of serotonin in this neuronal system of the CNS has been observed (60). With the suspicion that increased brain serotonin formation may accompany hepatocerebral dysfunction, it has therefore been of interest to survey the regional brain serotonergic distribution in experimental models of such conditions (see e.g. ref. 49).

Serotonin is stored in granules of vesicles within the serotonergic nerve terminals; it is released after nerve impulse into the synaptic cleft and is then bound to its receptor(s). In the synaptic cleft serotonin is inactivated by re-transportation into the neuron through the action of membrane carriers or re-uptake pumps on the serotonergic neuronal membrane. Inside the neuron, it may be restored in granules or degraded. The only effective route for continued metabolism of serotonin is deamination by mitochondrial monoamine oxidase (MAO), yielding 5-hydroxyacetaldehyde, which is rapidly oxidized further to 5-hydroxyindole acetic acid (5-HIAA), or in some circumstances reduced by aldehyde reductase to 5-hydroxytryptophol (depending on the ambient NAD/NADH ratio within the tissue). 5-HIAA is finally cleared from the CNS via an acid transport system. The half-life of serotonin is quite long, in the order of hours in neuronal tissue.

The initial hydroxylation of L-tryptophan involves two substrates, tryptophan and molecular oxygen, as well as a reduced pterin cofactor. This cofactor is considered to be L-erythro-5,6,7,8-tetrahydrobiopterin in vivo, as the tryptophan hydroxylase from rat brainstem has the highest affinity for this pterin (61,62). Tetrahydrobiopterin can be synthesized in brain from guanosine triphosphate (GTP) via D-erythro-7,8-dihydroneopterin triphosphate and L-erythro-7,8-dihydrobiopterin as well as other yet unidentified intermediates (see ref.'s 63,64). There is no evidence that dietary sources of biopterin can be utilized (65) and the CNS tetrahydrobiopterin is probably synthesized in situ because it penetrates the blood-brain barrier very poorly (66). During hydroxylation one atom of molecular oxygen is used in the 5-hydroxylation of tryptophan while the other is reduced to water. Electrons are donated by the pterin cofactor, and the unstable quinonoid dihydropterin that results is immediately regenerated to the tetrahydrobiopterin form by NADH(NADPH)-linked quinonoid dihydropteridine reductase, making the pterin cofactor act like a true catalyst in this reaction (61,67). Thus, for every mole of tetrahydrobiopterin that is oxidized one mole of 5-HTP is formed.

Aromatic L-alpha-amino acid decarboxylase is a pyridoxal-5'-phosphate dependent enzyme found throughout brain tissue, including both serotonergic and catecholaminergic neurons, where it converts 5-HTP to 5-HT and 3,4-dihydroxyphenylala-nine (DOPA) to dopamine, respectively (68). The affinity of this enzyme for 5-HTP is at least twice as high as it is for tryptophan, and thus 5-HTP is preferentially decarboxylated in serotonergic neurons (69). The decarboxylase is present in brain extracts in far greater amounts than tryptophan hydroxylase (70) and only trace amounts of 5-HTP are normally found in brain tissue (71) suggesting that 5-HTP can be decarboxylated almost as rapidly as it is formed. This also suggests that hydroxylation rather than decarboxylation is rate-limiting in serotonin formation. It thus seems reasonable to emphasize the hydroxylating reaction when studying the rate and regulation of serotonin biosynthesis in the brain.

A number of both direct and indirect in vitro assays of tryptophan hydroxylase have been developed. The in vitro methods are useful, for example for kinetic studies, but suffer from several disadvantages, such as the fact that isolated tryptophan hydroxylase is an extremely unstable enzyme which is dependent on numerous factors (61,67,72-75).

In vivo methods to estimate brain synthesis of serotonin include blocking of 5-hydroxyindole metabolism by inhibition of either the decarboxylating or the monoamine oxidating enzymes. Radioactive tracer techniques are alternative methods where high doses of interacting drugs are avoided (76), but these methods are expensive because of the high specific radioactivities required. Furthermore, the reliability of the radiolabelling techniques depends on the estimated loss of the labelled substance, or control of its metabolism (77, 78). Radiolabelling methods are, together with the imprecise method of determining urinary or cerebrospinal fluid levels of 5-HIAA (79), considered as steady-state methods compared to the non-steady-state methods of enzyme inhibition (78).

The non-steady-state methods all have the theoretical disadvantage of disturbing a finely-tuned system by blocking a part of it. Estimation of serotonin synthesis by MAO inhibition (80) may increase brain tryptophan concentrations and perturb the process they are being used to study. Accordingly, serotonin accumulation over time is found to be much higher the first 10 minutes following drug administration than at later times with this method. This may be due to a fall in tryptophan concentrations at longer time intervals, but could also result from the increased concentrations of serotonin that have been found to reduce serotonin synthesis in vivo (cf. ref.'s 71,76,81).

The decarboxylase inhibitors used for in vivo studies of serotonin synthesis are usually 3-hydroxybenzylhydrazine (NSD 1015), N-(3-hydroxybenzyl)-N'-methylhydrazine (NSD 1034) and N' (DL-seryl)-N''-(2,3,4-trihydrobenzyl)-hydrazine (Ro 4-4602) and are all derivatives of benzylhydrazine (82-84). The accumulation of 5-HTP over time after decarboxylase inhibition is a measure of in vivo tryptophan hydroxylating activity, the rate-limiting reaction in the biosynthesis of serotonin. The sensitivity of this method is largely determined by the method used to detect and quantify the 5-HTP formed. The capacity of chromatographic techniques today corresponds well to these demands for 5-HTP analysis in brain tissue (see e.g. ref. 85). The method with NSD 1015 produces only a very small increase, if any, in the level of brain tryptophan (81,86). NSD 1015 may also inhibit other enzymes, such as transaminases, that are involved in amino acid metabolism (87), but this possible effect is probably of minor concern when only 30 minutes of 5-HTP accumulation is used as a measure of serotonin formation. It can be objected that enough accumulated 5-HTP concentrations might result in negative feed-back on its own synthesis. However, it has been demonstrated that the accumulation of brain 5-HTP is linear for 45 minutes after administration of the decarboxylase inhibitor under normal conditions (84), and in PCS rats (28). In addition, 5-HTP is not supplied from sources outside the CNS or lost in significant amounts by diffusion out of the CNS during this time (88; Bengtsson, unpublished observations).

Additional methods for in vivo enzyme inhibition exist, including inhibition of tryptophan hydroxylase with e.g. p-chlorophenylalanine (PCPA) or alpha-propyldopacetamide, with measurement of the decline in 5-HT concentrations over time (89) or inhibition of 5-HIAA efflux from the CNS with e.g. probenecid with measurement of 5-HIAA accumulation over time (80). These methods give a rough estimate of the CNS serotonin synthesis rate. However, in relation to the rate-limiting process in serotonin biosynthesis, the use of decarboxylase inhibition seems most relevant today (78,90,91). Because it was imperative not to rely on just one in vivo method when interpreting experimentally-induced changes in serotonin synthesis rate (78) we also applied MAO inhibition in the present series of experiments (92). Furthermore, major changes in regional CNS serotonin turnover may be missed when the serotonin formation is measured only in the whole brain. In this respect, the observation of altered regional activity of tryptophan hydroxylation in the CNS of the rat (93-95) is very important. Therefore, we investigated regional CNS serotonin metabolism although this cannot exclude alterations within specific neuronal pathways or nuclei. The actual CNS dissection used in our studies is schematically displayed in Fig 1.

Regulation of serotonin synthesis in the brain

The processes that modulate serotonin turnover in vivo remain obscure (cf. ref. 78). The following discussion will be confined to possible regulatory steps in the CNS metabolism of serotonin, although regulatory effects on serotonin synthesis may also be exerted in functions such as transmitter storage, release and re-uptake, as well as pre- and post-synaptic receptor activation with short or long distance feed-back loops.

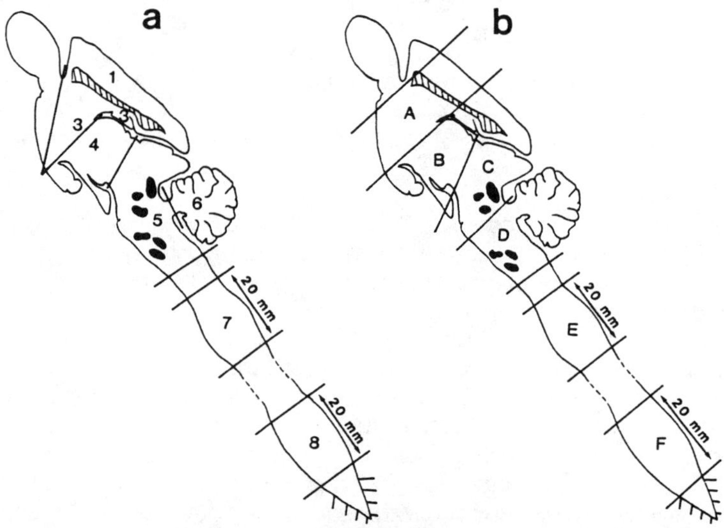

Fig. 1. Schematic illustration of the median sagital plane of the rat brain. Serotonergic nuclei are shown in the midbrain part. Lines indicate site of division. Dissection of the brain and spinal cord for HPLC determination of 5-hydroxyindole concentrations in different regions (a): (1) cortex, (2) striatum, (3) septum-hippocampus, (4) diencephalon, (5) mesencephalon-pons, (6) cerebellum, (7) proximal medulla spinalis, and (8) distal medulla spinalis. The striatal region (2) is not demonstrated in the figure. For fluorescence histochemistry (b): (A) frontal part of brain, (B) midbrain (rostral), (C) midbrain (caudal), (D) pons, (E) proximal medulla spinalis, and (F) distal medulla spinalis.

(Fig. 1 is reprinted from Bengtsson et al. (49) by the permission of the Elsevier Science Publishers B.V., Amsterdam.)

The calculated serotonin synthesis rate in intact neurons in vivo (or in synaptosomal or slice preparations in vitro) is less than one-tenth of the tryptophan hydroxylating activity that can be measured in vitro in supernatant preparations of the enzyme made from the same brain tissue and assayed under saturating conditions of substrate and cofactor (96). This indicates clearly that the activity of tryptophan hydroxylation within serotonergic neurons is severely restricted under normal conditions. Since the hydroxylation of tryptophan is the rate-limiting step in serotonin synthesis, this would be an obvious place to exert regulatory control (see Table 1).

CNS tryptophan concentrations are known to be subsaturating for tryptophan hydroxylating activity in vivo under normal conditions (97,98). Alterations in brain tryptophan content may therefore influence serotonin turnover. Various ways of administering exogenous tryptophan have been shown to increase CNS

Table 1. Ways of regulating CNS serotonin turnover in conjunction with hydroxylation of tryptophan

A) Regulation via CNS tryptophan concentration
B) Regulation of tryptophan hydroxylase
   1. Alterations in molecular oxygen supply
   2. Alterations in supply of reduced pterin (cofactor)
   3. Alterations in kinetic properties of tryptophan hydroxylase itself
C) Combinations of A and/or B

---

serotonin turnover (95,99,100). Thus, brain tryptophan concentration is probably important in the regulation of serotonin synthesis. However, the brain 5-HT and 5-HIAA concentrations correlate only "fairly well" with brain tryptophan (97).

Brain tryptophan levels may be high while brain serotonin and/or 5-HIAA concentrations are below normal or vice versa. As an example the brain serotonin synthesis rate may be enhanced by an increased tryptophan hydroxylating activity when tryptophan levels are severely depleted in experimentally-induced diabetes (101) or following chlorimipramine treatment (77), although the brain 5-HT and 5-HIAA concentrations may be unaltered compared with controls. A regulatory mechanism of serotonin synthesis is indicated that may buffer against severe substrate depletion. Conversely, when brain tryptophan levels are high in e.g. ammonia-infused PCS rats, the brain serotonin synthesis rate is normal or low (34). Thus, the transmitter availability may depend on other factors besides the brain tryptophan level. The brain serotonin synthesis rate in vivo still falls below in vitro tryptophan hydroxylating activity when tryptophan in excess is available (102), suggesting other factors of importance for the regulation of serotonin synthesis in vivo than brain tryptophan alone. In our view, it seems very unlikely that central serotonergic neurotransmission is entirely, or even mainly, under the control of gross changes in brain tryptophan content, which itself may be related only to the actual dietary composition. In a biological sense, regulation of tryptophan hydroxylation would seem more rational either by directly affecting tryptophan hydroxylase, or even more exquisite by primarily affecting the pterin cofactor. Further studies addressing the problem of brain serotonin synthesis regulation in vivo are therefore warranted.

Role of serotonin synthesis in the brain and functions of serotonin

Since little or none of the serotonin found in brain originates from peripheral tissues, we can assume that brain 5-HT synthesis is equivalent to brain 5-HT turnover. Thus, the brain 5-HT synthesis rate is a measure of the dynamic state of the neurons, and also possibly the functional state, as e.g. stimulation of the raphe nuclei increases the forebrain 5-HT turnover measured by NSD 1015 decarboxylase inhibiting technique (103-105). Major objections against the use of brain serotonin turnover to explain functional correlates in serotonergic neurotransmission can, however, also be pointed out. It was relatively early suspected that the brain synthesis of serotonin could result in both a small functionally active (i.e. participating in neurotransmission) pool of serotonin and a large functionally inactive pool of serotonin, both metabolized in the same way (cf. ref.'s 106,107). Supporting this theory was the discrepancy between turnover and behavioral alterations in the experimental situation. In addition, no method for determination of brain serotonin turnover can currently provide a "true" index of intraneuronal serotonin metabolism, although we have recently tried to elucidate this issue in the PCS rat with fluorescence histochemistry (49).

A fascinating observation suggests the coexistence of serotonin, substance P and thyrotropin-releasing hormone (TRH) in the same nerve (108). A related functional theory is based on the serotonergic nerve-terminal containing small vesicles with only 5-HT and larger vesicles with all three substances. These may perhaps be separately activated, and when a larger vesicle releases its content, the 5-HT and TRH will act on the postsynaptic receptor while the 5-HT and substance P may act on a presynaptic receptor (109). The substance P could thereby modulate 5-HT release, and possibly, also 5-HT turnover in the neuron.

BRAIN SEROTONIN IN RAT MODELS OF HEPATOCEREBRAL DYSFUNCTION

Previous studies on hepatocerebral dysfunction and brain serotonin have generally suggested that brain tryptophan and 5-HIAA concentrations are markedly increased, while the neurotransmitter 5-HT concentration is much less elevated. Brain 5-HT concentrations therefore do not appear to correlate with hepatocerebral dysfunction. Such a correlation may be observed if tryptophan/5-HT/5-HIAA (usually 5-HIAA/5-HT) ratios are calculated; this has previously been used as an indicator of brain serotonin turnover. However, measurement of absolute levels of tryptophan and tryptophan metabolites characterize a static event from which the dynamics of the system can not be inferred. Measurement of whole brain concentrations of tryptophan and its metabolites probably also obscures important regional variations.

In 1978 it was demonstrated in vitro in brain slices of PCS rats 6 weeks after operation that, although the 5-HT content was increased 30% compared with controls, the 5-HT formation from tryptophan was decreased by 35% whereas the formation of 5-HIAA was increased by 60%. This observation suggested that PCS causes a change in the activity of serotonin neurons (110). Estimation of serotonin formation in vivo has been conducted in PCS rats 4 weeks after operation, utilizing inhibition of 5-HIAA elimination from the brain by administration of probenecid (111). The authors observed that regional brain elevation of serotonin was present only to a minor degree in the brainstem, whereas the 5-HIAA accumulation in PCS rats was increased in all brain regions compared with controls. The suggestion was made that, in spite of the lack of change in 5-HT content, the brain 5-HT turnover was modified with possible implications for PSE. The results of our recent series of experiments further support the existence of a regional CNS increase in serotonin synthesis and turnover in rats with PCS ranging from one day to 6 months after operation, as well as in rats with CCl_4-induced cirrhosis (28,31,48,49,86, 92,112). This suggestion is validated by the use of the in vivo method with decarboxylase inhibition, thus allowing for estimation of the activity in the rate-limiting reaction of serotonin turnover. The reproducibility of the results, together with observations providing the same conclusion using MAO-inhibition (92), and increased brain histochemical serotonin fluorescence intensity following PCS (49), further contribute to the suggestion of an increased brain serotonin turnover associated with experimental chronic liver insufficiency.

Increments in regional brain serotonin turnover, but not in spinal cord preparations, were also unambiguously demonstrated to accompany 5 hours of the anhepatic state (42,52). Indeed, although the 5-HT levels in the brain were similarily (slightly) increased (as in PCS rats) compared with shamcontrols, 5-HTP accumulation, and thus serotonin turnover, was significantly increased in hepatectomized rats, not only compared to sham-operated controls but also compared to PCS rats (42,52). In contrast, regional brain serotonin turnover was lower in ammonia-infused PCS rats and in total liver ischemic rats

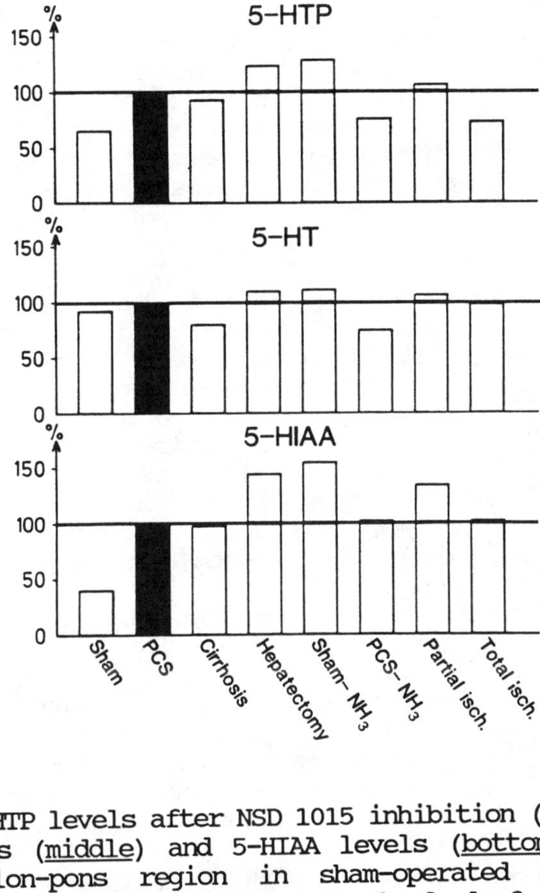

Fig. 2. 5-HTP levels after NSD 1015 inhibition (top), and 5-HT levels (middle) and 5-HIAA levels (bottom) in the mesencephalon-pons region in sham-operated rats and different rat models of hepatocerebral dysfunction in percent of the values measured in PCS rats (=100%).

than in "control" PCS rats. However, while the 5-HT level also was lower in ammonia-infused PCS rats (34) it was unchanged in total liver ischemia compared with PCS (38). The observations related above are summarized in Fig. 2. It is also apparent from Fig. 2 that, although brain 5-HIAA levels of ammonia-infused PCS rats and total liver ischemic rats were as elevated as those of PCS rats when compared with sham-operated controls, brain serotonin synthesis rates differed. Furthermore, although rats with partial liver ischemia had similar increases in brain serotonin turnover and 5-HT levels as PCS rats, the partially liver ischemic rats had higher 5-HIAA levels than PCS (38). The most extremely elevated brain serotonin turnover and 5-HIAA levels were found in sham-operated rats infused with ammonia, although these rats had similar 5-HT concentrations as compared to PCS rats (34). In addition, these ammonia-infused sham rats were neurologically normal (see below).

Similar increases in brain tryptophan and serotonin concentrations have been reported in liver insufficiency and sepsis (10,113). It has been suggested that an increased brain serotonin turnover may contribute to the cerebral dysfunction seen in both conditions (113,114). We therefore evaluated brain serotonin in a standardized sepsis model adopted from Hansson et al. (115), and were able to demonstrate that septic rats exhibit increased regional brain 5-HT and 5-HIAA concentrations (92). However, when evaluated by both decarboxylase and MAO-inhibition techniques in vivo, the septic rats had unaltered brain serotonin turnover compared with sham-operated controls (92).

The alterations in CNS serotonin turnover described above could be observed in all regions investigated. However, the changes were most pronounced in the brainstem (mesencephalon-pons) and in the cortex. This is of importance, since the mesencephalon-pons includes most serotonergic nuclei, and the cortical region reflects a major serotonergic projection area. We attempted to determine whether regional brain tryptophan hydroxylating activity is affected in the PCS rat (86). Brain tryptophan hydroxylating activity was saturated with a parenteral load of tryptophan. When sham-operated rats were tryptophan-loaded, serotonin turnover rose above levels measured for PCS rats. However, when PCS rats were tryptophan-loaded, serotonin synthesis rates in whole brain and in the mesencephalon-pons and cortex were even higher, suggesting that the rate-limiting enzyme in serotonin turnover is probably supranormally inducible in chronic PSE (86).

Many of the previous reports in the literature suggest a close relation between cerebral status in hepatic dysfunction and the alterations in brain serotonin. However, our recent results raise significant questions about the functional relevance of the increased brain serotonin turnover that accompanies hepatocerebral dysfunction. The indications for this lack of functional correlation were manifold, and additive results corroborated this assumption. For example, we were able to demonstrate no statistical correlation between increased brain serotonin synthesis rates and selective behavioral deficits observed in PCS rats (48). Anhepatic rats had even more increased brain serotonin turnover than PCS rats, whereas ammonia-infused and liver ischemic rats exhibited high, normal or low serotonin turnover; nevertheless, all these models manifested similar severe degrees of neurological dysfunction (34,38,42,51-53). In fact, if any difference in neurologic findings was obvious, it was that the anhepatic rats were less "comatose" than the others. In addition, ammonia-infused sham-operated rats with the overall highest brain serotonin turnover observed was not neurologically affected. Oral administration of branched-chain amino acids (BCAAs) to PCS rats normalized their increased brain serotonin metabolism, but without obvious effect on their impaired behavior (112). This feature was also observed in totally hepatectomized rats, where the brain serotonin turnover was normalized after BCAA-infusion without affecting the degree of coma (52). Cirrhotic rats exhibited a similarily increased regional brain serotonin turnover as seen in PCS rats; however, the cirrhotic rats showed no behavioral impairment comparable with that observed in PCS rats (31). Three week and 6 month PCS rats exhibited a similar increase in regional brain serotonin turnover (49). However, the 6 month PCS rats had progressive brain histopathological changes improvements in open field behavior despite when compared with 3 week PCS rats (49). This latter study also showed that the existing brain serotonin abnormality in PCS rats takes place primarily within the serotonergic neurons rather than in other neurons or other cellular compartments of the brain. In tryptophan-loaded rats (both PCS and sham-operated) high brain serotonin turnover was registered, although these animals were not different from non-loaded

PCS and sham-operated rats, when tested for behavioral abnormalities (86). Septic rats showed clear signs of encephalopathy, but these rats did not differ from controls with respect to regional brain serotonin turnover when investigated with two separate enzyme inhibition techniques in vivo.

According to present knowledge, it seems most likely that altered CNS intraneuronal serotonin turnover to the degree observed in PCS and anhepatic rats could have an impact on serotonin receptors and possible regulators of serotonin release such as substance P. Alterations in brain serotonin receptor binding properties and substance P concentration would be especially plausible if increased brain serotonin turnover were closely related to functional disturbances. However, we failed to observe any changes in brain 5-HT_1 or 5-HT_2 binding properties in PCS or totally hepatectomized rats (53,116). Thus, both the number of binding sites and the affinity for ^{3}H-5-HT and ^{3}H-ketanserin (a 5-HT_2 receptor antagonist), respectively, were unaltered in the brain of these animals. In addition, substance P concentrations in PCS rats were unchanged in all brain regions investigated (117).

Taken together, the findings summarized above suggest only a limited correlation between the increased brain serotonin turnover and the neurologic dysfunction observed in some types of hepatic insufficiency. It is quite possible that increased brain serotonin synthesis rates appear mainly because high brain tryptophan concentrations accompany most states of hepatocerebral dysfunction. The increase in serotonin formation may reflect a metabolic "event" in the serotonergic neurons rather than a close functional relation to central serotonergic neurotransmission, since no correlation with encephalopathy has been demonstrable in many different forms of liver dysfunction. The presence of a functionally inactive parallel metabolism of tryptophan to 5-HTP, 5-HT and 5-HIAA, besides that of a similarly metabolized but functionally active pool of serotonin, is possible (106, 107). This theory implies that behavioral abnormalities related to the central serotonergic system only occur when the local CNS concentrations of the neurotransmitter serotonin itself are altered. Accordingly, we believe there is a possibility of an increased brain tryptophan level and thereby serotonin synthesis simultaneous with liver failure, and that encephalopathy coincides with, rather than direct results from, this change in brain indoleamine metabolism.

Previous reports have demonstrated increased brain tryptophan and tryptophan metabolite concentrations in concert with different types of hepatocerebral dysfunction. However, our current studies suggest that an increased brain serotonin turnover in these conditions is plausible only when appropriate methodological considerations are met. Despite multiple different models of hepatocerebral dysfunction and varied neurochemical and behavioral assessments, we were able to demonstrate no real correlation between alterations in brain serotonin metabolism and neurologic status. Discrepancies between our findings and previous reports might relate to differences both in animal models of hepatocerebral dysfunction and in the neurochemical approach. The multifactorial nature of hepatocerebral dysfunction may include abnormalities in brain serotonin metabolism as an important part, although complex interactions may conceal a straightforward correlation. Many different conditions can result in hepatocerebral dysfunction, and multiple factors clearly play a role in its pathogenesis. This must be taken in account in experimental modelling and may have clinical implications in the future.

REFERENCES

1) Conn, H.O. & Lieberthal, M.M. The hepatic coma syndromes and lactulose. The Williams & Wilkins Co, Baltimore.
2) Tyce, G.M., Flock, E.V., Owen, C.A., Stobie, G.H.C. & David, C. 5-Hydroxyindole metabolism in the rat brain after hepatectomy. Biochem Pharmacol, 1967, 16, 979-992.
3) Baldessarini, R.J. & Fischer, J.E. Serotonin metabolism in rat brain after surgical diversion of the portal venous circulation. Nature New Biol, 1973, 245, 25-27.

4) Curzon, G., Kantamaneni, B.D., Winch, J., Rojas-Bueno, A., Murray-Lyon, I.M. & Williams, R. Plasma and brain tryptophan changes in experimental acute hepatic failure. J Neurochem, 1973, 21, 137-145.
5) Curzon, G., Kantamaneni, B.D., Fernando, J.C., Woods, M.S. & Cavanagh, J.B. Effects of chronic portocaval anastomosis on brain tryptophan, tyrosine and 5-hydroxy-tryptamine. J Neurochem, 1975, 24, 1065-1070.
6) Knott, P.J. & Curzon, G. Effects of increased rat brain tryptophan on 5-hydroxytryptamine and 5-hydroxyindoleacetic acid in the hypothalamus and other brain regions. J Neurochem, 1974, 22, 1065-1071.
7) Reichle, R.M. & Reichle, F.A. Effect of portacaval shunt and acute hepatic ischemia on brain and liver serotonin and catecholamines. Surg Forum, 1975, 26, 413-415.
8) Cummings, M.G., Soeters, P.B., James, J.H., Keane, J.M. & Fischer, J.E. Regional brain indoleamine metabolism following chronic portacaval anastomosis in the rat. J Neurochem, 1976, 27, 501-509.
9) Cummings, M.G., James, J.H., Soeters, P.B., Keane, J.M., Foster, J. & Fischer, J.E. Regional brain study of indoleamine metabolism in the rat in acute hepatic failure. J Neurochem, 1976, 27, 741-746.
10) Jellinger, K. & Riederer, P. Brain monoamines in metabolic (endotoxic) coma. A preliminary biochemical study in human postmortem material. J Neural Transm, 1977, 41, 275-286.
11) Ono, J., Huston, D.G., Dombro, R.S., Levi, J.U., Livingstone, A. & Zeppa, R. Tryptophan and hepatic coma. Gastroenterology, 1978, 74, 196-200.
12) Bloxam, D.L. & Curzon, G. A study of proposed determinants of brain tryptophan concentrations in rats after portocaval anastomosis or sham operation. J Neurochem, 1978, 31, 1255-1263.
13) Tricklebank, M.D., Smart, J.L., Bioxam, D.L. & Curxon, G. Effects of chronic experimental liver dysfunction and L-tryptophan on behaviour in the rat. Pharmacol Biochem Behav, 1978, 9, 181-189.

14) Tricklebank, M.D., Bloxam, D.L., Kantamaneni, B.D. & Curzon, G. Brain 5-hydroxytryptamine metabolism after portacaval anastomosis: Relationship with ambulation. Pharmacol Biochem Behav, 1981, 14, 259-262.

15) Kamata, S., Okada, A., Watanabe, T., Kawashima, Y. & Wada, H. Effects of dietary amino acids on brain amino acids and transmitter amines in rats with a portacaval shunt. J Neurochem, 1980, 35, 1190-1199.

16) Herlin, P.M., James, J.H., Gimmon, Z. & Fischer, J.E. Combined or individual administration of branched-chain amino acids following total hepatectomy in the rat: effects on amino acids and indoleamines in brain. J Parenter Enter Nutr, 1982, 6, 383-387.

17) Denis, J., Delorme, M.L., Boschat, M., Nordlinger, B. & Opolon, P. Respective roles of ammonia, amino acids, and medium-sized molecules in the pathogenesis of experimentally induced acute hepatic encephalopathy. J Neurochem, 1983, 40, 10-19.

18) Martin, J.R., Dedek, J. & Driscoll, P. Portacaval anastomosis in rats: effects on behavior and brain serotonin metabolism. Pharmacol Biochem Behav, 1983, 18, 269-272.

19) Jonung, T., Ramzy, A., Herlin, P., James, J.H., Edwards, L. & Fischer, J.E. Indole amines and amino acids in various brain regions after infusion of branched chain amino acids into hepatectomized rats. Eur Surg Res, 1985, 17, 83-90.

20) Mans, A.M. & Hawkins, R.A. Brain monoamines after portacaval anastomosis. Metabolic Brain Disease, 1986, 1, 45-52.

21) Sherlock, S., Summerskill, W.H.J., White, L.P. & Phear, E.A. Portal-systemic encephalopathy. Neurological complications of liver disease. Lancet, 1954, 11, 453-457.

22) Hoyumpa, J.r. .A.M. & Schenker, S. Perspectives in hepatic encephalopathy. J Lab Clin Med, 1982, 100, 477-487.

23) Ware, A.J., Agostino, A.N.D. & Combes, B. Cerebral edema: a major complication of massive hepatic necrosis. Gastroenterology, 1971, 61, 877-884.

24) Wilkinson, S.P., Hurst, D., Portmann, B. & Williams, R. Pathogenesis of renal failure in cirrhosis and fulminant hepatic failure. Postgrad Med J, 1975, 51, 503-505.

25) Ostrow, J.D. Medical management of fulminant hepatitis. In Gastrointestinal Emergencies (Eds. Clearfield HR and Dinoso VP), pp. 97-106, Grune & Stratton, New York, 1976.

26) Livingstone, A.S., Potvin, M., Goresky, C.A., Finlayson, M.H. & Hinchey, E.J. Changes in the blood brain barrier in the hepatic coma following hepatectomy. Gastroenterology, 1977, 73, 697-704.

27) Lee, S.H. & Fisher, B. Portacaval shunt in the rat. Surgery, 1961, 4, 668-672.

28) Bengtsson, F., Gage, F.H., Jeppsson, B., Nobin, A. & Rosengren, E. Brain monoamine metabolism and behavior in portacaval-shunted rats. Exp Neurol, 1985, 90, 21-35.

29) Proctor, E. & Chatamra, K. High yield micronodular cirrhosis in the rat. Gastroenterology, 1982, 83, 1183-1190.

30) Proctor, E. & Chatamra, K. Standardized micronodular cirrhosis in the rat. Eur Surg Res, 1984, 16, 182-186.

31) Bengtsson, F., Bugge, M., Vagianos, C., Jeppsson, B. & Nobin, A. Brain serotonin metabolism and behavior in rats with carbon tetrachloride-induced liver cirrhosis. Res. Exp. Med., 1987, 187, 429-438.

32) Gimmon, Z., James, J.H., Meyenfeldt, M.V. & Fischer, J.E. Opposing effects of prolonged ammonia and branched chain amino acid infusions on the accumulation of aromatic amino acids by brain. In Metabolism and clinical implications of branched chain amino and ketoacids (Halser, Williamson), pp. 487-492, Elsevier North Holland, 1981.

33) Rigotti, P., Jonung, T., James, J.H., Edwards, L.L., Peters, J.C. & Fischer, J.E. Effects of infusing branched chain amino acids and ammonium salts in rats after portacaval anastomosis. Arch Surg, 1985, 120, 1290-1295.

34) Bugge, M., Bengtsson, F., Nobin, A., Jeppsson, B., Hultberg, B., Jonung, T. & Herlin, P. Serotonin metabolism in the rat brain following ammonia administration. In: Advances in Ammonia Metabolism and Hepatic Encephalopathy. Soeters PB, Wilson JMP, Meijer AJ, Holm E (eds.). Exerpta Medica, International Congress Series 761, Elsevier Science Publishers B.V., Amsterdam, pp. 454-461, 1988.

35) Dodsworth, J.M., James, J.H., Cummings, M.C. & Fishcer, J.E. Depletion of brain norepinephrine in acute hepatic coma. Surgery, 1974, 75, 811-820.

36) Mans, A.M., Saunders, S.J., Kirsch, R.E. & Biebuyck, J.F. Correlation of plasma and brain amino acid and putative neurotransmitter alterations during acute hepatic coma in the rat. J. Neurochem., 1979, 32, 285-292.

37) Mazzioti, A., Bernardi, M., Antonini, L., Dioguardi, F.C., Bellusci, R., Papa, V., Tacconi, C., Gasbarrini, G., Cavallari, A. & Possati, L. Plasma amino acid patterns in experimental acute hepatic failure: Comparison between hepatectomy and liver devascularisation in pigs. Surgery, 1981, 90, 527-534.

38) Bugge, M., Bengtsson, F., Nobin, A., Jeppsson, B. & Herlin, P. The effect of liver ischaemia on brain monoamine synthesis in the rat. Res Exp Med, 1987, 187, 119-130.

39) Franco, M.M.D. & Bismuth, H. Necrose ischemique hepatique chez le rat. Influence de la necros sur l'evolution de l'insuffisance hepatique aique. Bordeaux Medical, 1977, 10, 1823-1826.

40) Bloch, P., Delorme, M., Rapin, J., Granger, A., Boschat, M. & Opolon, P. Reversible modifications of neurotransmitters of the brain in experimental acute hepatic coma. Surg Gynecol Obst, 1978, 146, 551-558.

41) Holmin, T., Alinder, G. & Herlin, P. A microsurgical method for total hepatectomy in the rat. Eur Surg Res, 1982, 14, 420-427.

42) Bugge, M., Bengtsson, F., Nobin, A., Jeppsson, B. & Herlin, P. Metabolism of monoamines in the brain after total hepatectomy in the rat. Exp Neurol, 1986, 94, 11-20.

43) Holmin, T. & Siesjö, B.K. The effect of porta-caval anastomosis upon the energy state and upon acid-base parameters of the rat brain. J Neurochem, 1974, 22, 403-412.
44) Schröder, R., Muller, O. & Bircher, J. The portacaval and splenocaval shunt in the normal rat. A morphometric and functional reevaluation. J Hepatology, 1985, 1, 107-123.
45) McLean, E.K., McLean, A.E.M. & Sutton, P.M. Instant cirrhosis. Br J Exp Pathol, 1969, 50, 502-506.
46) Vorobioff, J., Bredfeldt, J.E. & Groszmann, R.J. Increased blood flow throught the portal system in cirrhotic rats. Gastroenterology, 1984, 87, 1120-1126.
47) Fernandez-Munoz, D., Caramelo, C., Santos, J.C., Blanchart, A., Hernando, L. & Lopez-Novoa, J.M. Systemic and splanchnic hemodynamic disturbances in conscious ratsb with experimental liver cirrhosis without ascites. Am J Physiol, 1985, 249, G316-G320.
48) Bengtsson, F., Nobin, A., Falck, B., Gage, F.H. & Jeppsson, B. Portacaval shunt in the rat: selective alterations in behavior and brain serotonin. Pharmacol Biochem Behav, 1986, 24, 1611-1616.
49) Bengtsson, F., Bugge, M., Brun, A., Falck, B., Henriksson, K.-G. & Nobin, A. The impact of time after portacaval shunt in the rat on behavior, brain serotonin, and brain and muscle histology. J. Neurol. Sci., 1988, 83, 109-122.
50) Bures, J., Buresova, O. & Huston, J.P. Innate and motivated behavior. In Techniques and basic experiments for the study of brain and behavior, pp. 37-45, Elsevier North Amsterdam Holland, 1976.
51) Bugge, M., Bengtsson, F., Nobin, A., Holmin, T., Jeppsson, B., Hultberg, B., Falck, B. & Herlin, P. Amino acids and indoleamines in the brain after infusion of branched-chain amino acids to rats with liver ischemia. J. Parent. Ent. Nutr., 1986, 10, 474-478.
52) Bugge, M., Bengtsson, F., Nobin, A., Jeppsson, B. & Herlin, P. The turnover of brain monoamines after total hepatectomy in rats infused with branched chain amino acids. World J Surg, 1987, 11, 810-817.

53) Bugge, M., Bengtsson, F., Nobin, A., Hall, H., Wedel, I., Jeppsson, B. & Herlin, P. Brain serotonin concentrations and receptor binding following BCAA infusion in hepatectomized rats. J. Parent. Ent. Nutr., 1988, in press.
54) Masson, P. & Berger, L. Sur un novean mode de secretion intérné: La neurocrinié. C.R. Acad. Sci. (Paris), 1923, 123, 1748-1751.
55) Osborne, N.N. & Neuhoff, V. Identified serotonin neurones. In International Review of Cytology (Eds. Bourne GH, Danielli JF), pp. 259-290, Academic Press, New York, 1980.
56) Brodie, B.B., Bogdanski, D.F. & Bonomi, L. Formation, storage and metabolism of serotonin (5-hydroxytryptamine) and catecholamines in lower vertebrates. In Comparative Neurochemistry (Ed. Richter, D), pp. 367-378, Pergamon Press, Oxford, 1964.
57) Perkins, M.N. & Stone, T.W. Quinolinic acid: regional variations in neuronal sensitivity. Brain Res, 1983, 259, 172-176.
58) Smith, S.A., Marston, F.A.O., Dickson, A.J. & Pogson, C.I. Control of enzyme activities in rat liver by tryptophan and its metabolites. Biochem Pharmacol, 1979, 28, 1645-1651.
59) Melamed, E., Hefti, F. & Wurtman, R.J. L-3,4-dihydroxyphenylalanine and L-5-hydroxytryptophan decarboxylase activities in rat striatum: effect of selective destruction of dopaminergic or serotonergic input. J Neurochem, 1980, 34, 1753-1756.
60) Vamamoto, M., Chan-Palay, V. & Palay, S.L. Autoradiographic experiments to examine uptake, anterograde and retrograde transport of tritiated serotonin in the mammalian brain. Anat Embryol, 1980, 159, 137-149.
61) Friedman, P.A., Kappelman, A.H. & Kaufman, S. Partial purification and characterization of tryptophan hydroxylase from rabbit hind brain. J Biol Chem, 1972, 247, 4165-4173.
62) Fukushima, T. & Nixon, J.C. Analysis of reduced forms of biopterin in biological tissues and fluids. Anal Biochem, 1980, 102, 176-188.

63) Nixon, J.C., Lee, C.L., Abou-Donia, M., Fukushima, T., Nichol, C.A., Diliberto, J.r. .E. & Viveros, O.H. Induction of GPT-cyclohydrolase by the stimuli that increase tetrahydrobiopterin (BH4) in rat adrenal medulla and cortex. Soc Neurosci Abstr, 1980, 10, 643.

64) Nixon, J.C., Lee, C.L., Milstein, S., Kaufman, S. & Barthcolome, K. Neopterin and biopterin levels in patients with atypical forms of phenylketonuria. J Neurochem, 1980, 35, 898-904.

65) Kraut, H., Pabst, W., Rembold, H. & Wildemann, L. Uber das Verhalten des Biopterins in Saugetierorganismus. I. Bilanz und Wachstumersuche an Ratten. Z Physiol Chem, 1963, 332, 101-108.

66) Kettler, R., Bartholini, G. & Pletscher, A. In vivo enhancement of tyrosine hydroxylation in rat striatum by tetrahydrobiopterin. Nature (Lond), 1974, 249, 476-478.

67) Kaufman, S. Properties of the pterin-dependent aromatic amino acid hyroxylases. In Aromatic Amino Acids in the Brain, Ciba Foundation Symposium 22 (Eds. Wolstenhome GEW and Fitzsimons DW), pp. 85-108, Elsevier, Amsterdam, 1974.

68) Sourkes, T.L. Enzymology of aromatic amino acid decarboxylase. In Structure and Function of Monoamine Enzymes (Eds. Usdin E, Weiner N, Youdim MBH), pp. 477-495, Marcel Dekker Inc., New York, 1977.

69) Ichiyama, A., Nakamura, S., Nishizuka, Y. & Hayaishi, O. Enzyme studies on the biosynthesis of serotonin in mammalian brain. J Biol Chem, 1970, 245, 1699-1709.

70) Ichiyama, A., Nakamura, S., Nishizuka, Y. & Hayaishi, O. Tryptophan-5-hydroxylase in mammalian brain. Adv Pharmacol, 1968, 6A, 5-17.

71) Tappaz, M.L. & Pujol, J.F.M. Estimation of the rate of tryptophan hydroxylation in vivo: A sensitive microassay in discrete brain nuclei. J Neurochem, 1980, 34, 933-940.

72) Kuhn, D.M., Ruskin, B. & Lovenberg, W. Tryptophan hydroxylase. The role of oxygen, iron and sulfhydryl groups as determinants of stability and catalytic activity. J Biol Chem, 1980, 255, 4137-4143.

73) Kuhn, D.M., Vogel, R.L. & Lovenberg, W. Calcium-dependent activation of tryptophan hydroxylase by ATP and magnesium. Biochem Biophys Res Comm, 1978, 82, 759-766.
74) Hamon, M., Bourgoin, S., Hery, F. & Simmonet, G. Activation of tryptophan hydroxylase by adenosine triphosphate: magnesium and calcium. Mol Pharmacol, 1978, 14, 99-110.
75) Hamon, M., Bourgoin, S., Hery, F. & Simmonet, G. Phospholipid-induced activation of tryptophan hydroxylase from the rat brain stem. Biochem Pharmacol, 1978, 27, 915-922.
76) Morot-Gaudry, Y., Hamon, M., Bourgoin, S., Ley, J.P. & Glowinski, J. Estimation of the rats of 5-HT synthesis in the mouse brain by various methods. Naunyn Schmiedelberg's Arch Pharmacol, 1974, 282, 223-238.
77) Neckers, L.m., Biggio, G., Moje, E. & Meek, J.L. Modulation of brain tryptophan hydroxylase activity by brain tryptophan content. J Pharmacol Exp Ther, 1977, 201, 110-116.
78) Neckers, L.M. Serotonin turnover and regulation. In Biology of Serotonergic Transmission (Ed. Osborne NN), pp. 139-158, John Wiley & Sons Ltd., New York, 1982.
79) Costa, E. Appraisal of current methods to estimate the turnover rate of serotonin and catecholamines in human brain. Adv Biochem Psychopharmacol, 1972, 4, 171-183.
80) Neff, N.H., Lin, R.C., Negai, S.H. & Costa, E. Turnover rate measurements of brain serotonin in unanesthetized rats. Adv Biochem Psychopharmacol, 1969, 1, 91-109.
81) Carlsson, A. & Lindqvist, M. In vivo measurements of tryptophan and tyrosine hydroxylase activities in mouse brain. J Neural Transm, 1973, 34, 79-91.
82) Pletscher, A., Gey, K.F. & Burkard, W.P. Inhibitors of monoamine oxidase and decarboxylase of aromatic amino acids. In 5-Hydroxytryptamine and Related Indolealkylamines. Handbook of Experimental Pharmacology. Vol. XIX (EDS.Eichler O and Farah A), pp. 652-668, Springer Verlag, New York, 1966.

83) Carlsson, A. & Lindqvist, M. Accumulation of 5-hydroxytryptophan in mouse brain after decarboxylase inhibition. J Pharm Pharmacol, 1970, 22, 726-727.

84) Carlsson, A., Davis, J.N., Kehr, W., Lindqvist, M. & Atack, C.V. Simultaneous measurement of tyrosine and tryptophan hydroxylase activities in brain in vivo using an inhibitor of the aromatic amino acid decarboxylase. Naunyn-Schmiedeberg's Arch Pharmacol, 1972, 275, 153-168.

85) Hallman, H. Neurochemical studies on central monoamine neurons. Consequence of selective toxic lesions. PhD Thesis, Karolinska Institutet, Sweden, 1984, 14-21.

86) Bengtsson, F., Bugge, M., Johansen, K.H., Herlin, P. & Nobin, A. Brain tryptophan hydroxylating activity in experimental chronic portal-systemic encephalopathy. Submitted for publication, 1988.

87) Wurtman, R.J. Discussion on tryptophan and tyrosine hydroxylation in vivo. In Aromatic Amino Acids in the Brain, Ciba Foundation Symposium 22 (Eds. Wolstenholme GEW, Fitzsimons DW), pp. 128, Elsevier, Amsterdam, 1974.

88) Carlsson, A. The in vivo estimation of rates of tryptophan and tyrosine hydroxylation: Effects of alteration in enzyme environment and neuronal activity. In Aromatic Amino Acids in the Brain, Ciba Foundation Symposium 22 (Eds. Wolstenholme GEW, Fitzsimmons), pp. 117-125, Elsevier, Amsterdam, 1974.

89) Neff, N.H., Barrett, R.E. & Costa, E. Kinetic and fluorescent histochemical analysis of the serotonin compartments in rat pineal gland. Eur J Pharmacol, 1969, 5, 348-356.

90) Meek, J.L. & Neckers, L.M. Studies of serotonin turnover in discrete nuclei using HPLC. In Structure and Function of Monoamine Enzymes (Eds. Usdin E, Weiner N, Youdim M), pp. 799-809, Marcel Dekker Inc., New York, 1977.

91) van Wijk, M., Sebens, J.B. & Korf, J. Probenecid-induced increase of 5-hydroxytryptamine synthesis in rat brain, as measured by formation of 5-hydroxy-tryptophan. Pharmacology, 1979, 60, 229-235.

92) Bengtsson, F., Bugge, M., Hansson, L., Fyge, K., Jeppsson, B. & Nobin, A. Serotonin metabolism in the central nervous system following sepsis or portacaval shunt in the rat. J Surg Res, 1987, 43, 420-429.

93) Baumgarten, H.G., Victor, S.J. & Lovenberg, W. Effect of intraventricular injection of 5,7-dihydroxytryptamine on regional tryptophan hydroxylase of rat brain. J Neurochem, 1973, 21, 251-253.

94) Lovenberg, W. & Victor, S.J. Tryptophan hydroxylase of the central nervous system: effect of intraventricular 5,6- and 5,7-dihyroxytryptamine. In Advances in Biochemical Psychopharmacology (Eds. Costa E, Gessa GL, Sandler M), pp. 93-101, Raven Press Books Ltd., New York, 1974.

95) Meek, J.L. & Lofstrandh, S. Tryptophan hydroxylase in discrete brain nuclei: Comparison of activity in vitro and in vivo. Eur J Pharmacol, 1976, 37, 377-380.

96) Boadle-Biber, M.C. Biosynthesis of serotonin. In Biology of Serotonergic Transmission (Ed. Osborne, NN), pp. 63-94, John Wiley & Sons Ltd, New York, 1982.

97) Costa, E. & Meek, J.L. Regulation of biosynthesis of catecholamines and serotonin in the CNS. Annu Rev Pharamcol, 1974, 14, 491-511.

98) Young, S.N. & Sourkes, T.L. Tryptophan in the central nervous system: Regulation and significance. Adv Neurochem, 1977, 2, 133-191.

99) Fernstrom, J.D. & Wurtman, R.J. Brain serotonin content: Physiological dependence on plasma tryptophan levels. Science, 1971, 173, 149-150.

100) Fernstrom, J.D. & Wurtman, R.J. Brain serotonin content: Physiological regulation by plasma neutral amino acids. Science, 1972, 178, 414-416.

101) Trulson, M.E. & MacKenzie, R.G. Increased tryptophan hydroxylase activity may compensate for decreased brain tryptophan levels in streptozotocin diabetic rats. J Pharmacol Exp Ther, 1980, 212, 269-173. 102) Hamon, M., Bourgoin, S. & Youdim, M.B.H. Tryptophan hydroxylation in the central nervous system and other tissues. In Aromatic Amino Acid Hydroxylases and Mental Disease (Ed. Youdim MBH), pp. 233-297, John Wiley & Sons Ltd., New York, 1979.

103) Shields, P.J. & Eccleston, D. Effects of electrical stimulation of rat midbrain on 5-hydroxytryptamine synthesis as determined by a sensitive radioisotope method. J Neurochem, 1972, 19, 265-272.

104) Herr, B.E., Gallager, D.W. & Roth, R.H. Tryptophan hydroxylase: Activation in vivo following stimulation of central serotonergic neurons. Biochem Pharmacol, 1975, 24, 2019-2023.

105) Bourgoin, S., Oliveras, J.L., Bruxelle, J., Hemon, M. & Besson, J.M. Electrical stimulation of the nucleus raphe magnus in the rat. Effects on 5-HT metabolism in the spinal cord. Brain Res, 1980, 194, 377-389.

106) Grahame-Smith, D.G. Studies in vivo on the relationship between brain tryptophan, brain 5-HT synthesis and hyperactivity in rats treated with a monoamine oxidase inhibitor and L-tryptophan. J Neurochem, 1971, 18, 1053. 107) Grahame-Smith, D.G. Does the total turnover of brain 5-HT reflect the functional activity of 5-HT in brain? In Serotonin and Behavior (Eds. Barchas J, Usdin E), pp. 5-7, Academic Press, New York, London, 1973.

108) Johansson, O., Hökfelt, T., Pernow, B., Jeffcoate, S.L., White, N., Steinbusch, H.W.M., Verhofstad, A.A.J., Emson, P.C. & Spindel, E. Immunohistochemical support for three putative transmitters in one neuron: Coexistence of 5-hydroxytryptamine, substance P and thyrotropin releasing hormone-like immunoreactivity in medullary neurons projecting to the spinal cord. Neuroscience, 1981, 6, 1857-1881.

109) Hansen, S., Svensson, L., Hökfelt, T. & Everitt, B.J. 5-hydroxytryptamine-thyrotropin releasing homone interactions in the spinal cord: Effects on parameters of sexual behaviour in the male rat. Neurosci Lett, 1983, 42, 299-304.

110) Simert, G., Nobin, A., Rosengren, E. & Vang, J. Neurotransmittor changes in the rat brain after portacaval anastomosis. Eur Surg Res, 1978, 10, 73-85.

111) Moroni, F., Carla, V., Lombardi, G., Pellegrini, D., Carassale, G.L. & Cortesini, C. Excitatory amino acids, tryptophan metabolites, and endogenous neurotoxins in porta-caval shunted rats. In Advances in Hepatic Encephalopathy and Urea Cycle Diseases (Eds. Kleinberger G, Ferenci P, Riederer P, Thaler H), pp. 385-393, Karger, Basel, 1984.

112) Bengtsson, F., Nobin, A., Falck, B., Gage, F.H. & Jeppsson, B. Effect of oral branched chain amino acids on behavior and brain serotonin metabolism in portacaval shunted rats. World J Surg, 1987, 12, 246-253.

113) Freund, H.R., Muggia-Sullam, M., LaFrance, R., Holroyde, J. & Fischer, J.E. Regional brain amino acid and neurotransmitter derangements during abdominal sepsis and septic encephalopathy in the rat. Arch Surg, 1986, 121, 209-216.

114) Hasselgren, P.-O. & Fischer, J.E. Septic encephalopathy. Etiology and management. Intensive Care Med, 1986, 12, 13-16.

115) Hansson, L., Alwmark, A., Christensen, F., Jeppsson, B., Holst, E. & Bengmark, S. Standardized intraabdominal abscess formation with generalized sepsis: pathophysiology in the rat. Eur Surg Res, 1985, 17, 155-159.

116) Bengtsson, F., Bugge, M., Hall, H. & Nobin, A. Brain 5-HT_1 and 5-HT_2 binding properties following portacaval shunt in the rat. Submitted for publication, 1988.

117) Bengtsson, F., Ekman, R. & Nobin, A. Substance P-like immunoreactivity in the rat central nervous system following portacaval shunt. Surg Res Comm, 1987, 2, 65-69.

BIOGENIC AMINES IN HEPATIC ENCEPHALOPATHY: EVIDENCE FOR INCREASED SEROTONIN TURNOVER IN HUMAN BRAIN

Marcelle Bergeron, Tomás A. Reader, Gilles Pomier Layrargues and Roger F. Butterworth,
André-Viallet Clin. Res. Centre, Hôpital St-Luc, and Centre de recherche en sciences neurol., (U. of Montréal), Montréal, Québec, Canada H2X 3J4.

SUMMARY

Disturbances in the metabolism of monoamine neurotransmitters may be implicated in the development of hepatic encephalopathy (HE) in human liver disease. In order to evaluate this possibility, amino acid precursors of monoamines, monoamines and some of their metabolites were measured in homogenates of prefrontal cortex (PFCo) dissected from brain tissue obtained at autopsy from seven cirrhotic patients who died in hepatic coma and an equal number of control subjects, free from neurological, psychiatric and hepatic diseases, and matched for age and time interval from death to freezing of autopsied brain samples. Amino acids were measured using high performance liquid chromatography (HPLC) with fluorescence detection and monoamines were measured by HPLC with electrochemical detection. In brain tissue of cirrhotic patients, phenylalanine (PHE) and tyrosine (TYR) levels were found to be increased by 141% and 71% ($p<0.01$) respectively; concentrations of dopamine and its metabolites, 3-methoxytyramine (3-MT) and homovanillic acid (HVA) were unchanged. Levels of 5-hydroxytryptophan (5-HTP) and serotonin (5-HT) were likewise unchanged but levels of the 5-HT metabolite, 5-hydroxyindoleacetic acid (5-HIAA) were found to be increased by 72% ($p<0.05$) in PFCo of cirrhotic patients vs the controls. These results suggest that alterations of the 5-HT system may play an important role in the pathogenesis of certain neurological symptoms associated with HE in humans.

INTRODUCTION

Alterations of amino acid profiles have been reported for the plasma of humans with liver failure (Record et al., 1976; Rosen et al., 1977; Cascino et al., 1982) as well as in animal models of hepatic disease (Blitzer et al., 1978; Mans et al., 1984; Jeppsson et al., 1985). In particular, plasma aromatic amino acids (AAA) are increased as much as 2-4 fold in chronic liver disease in humans (Fischer et al., 1974, 1976b). Moreover, increased levels of brain aromatic amino acids (tyrosine, phenylalanine, tryptophan) have previously been reported following portacaval shunting (Curzon et al., 1975; Mans et al., 1984; Jeppsson et al., 1985). Alterations of AAA's may be of importance as they are direct precursors of the biogenic amines (DA and 5-HT) (Figures 1 and 2). Thus, altered monoamine neurotransmitter metabolism may be involved as a causal factor in hepatic encephalopathy (HE) (Fischer and Baldessarini, 1971, 1976; Munro et al., 1975). In several studies, increased brain 5-HT levels and turnover were found in experimental hepatic failure (Bengtsson et al., 1985; Bugge et al., 1987, 1988). However, results of studies on DA metabolism are inconsistant, with either increases (Yurdaydin et al., 1989), decreases (Kamata et al., 1980; Faraj et al., 1981) or no change, being reported (Curzon et al., 1975; Cuilleret et al., 1980; Herlin et al., 1983; Bengtsson et al., 1985; Mans and Hawkins, 1986; Bugge et al., 1987). Whether these changes are of importance in the development of HE in humans is still unclear. This is due in part, to the paucity of data on amino acid and monoamine neurotransmitter levels in brain tissue of patients with HE. Thus, we undertook a study to investigate the possible involvement of altered DA and/or 5-HT metabolism in chronic liver failure. We measured the concentrations of amino acid precursors (PHE; TYR), the monoamines DA and 5-HT as well as some of their metabolites in prefrontal cortex of autopsied brain tissue from cirrhotic patients who had died in hepatic coma and an equal number of age-matched control subjects. This region of the brain previously exhibited significant neuropathological changes in HE (Butterworth et al., 1987).

MATERIALS AND METHODS

Brain samples

Human brain tissue was obtained at autopsy from seven cirrhotic patients who died in hepatic coma and from seven control subjects free from neurological or psychiatric disorders. Subjects' age, sex and clinical diagnostic features as well as postmortem delay times are summarized in Table 1. Mean ages ranged from 49-67 years (mean, 56.4 ± 2.4 years) for controls and from 48-70 years (mean, 57.0 ± 2.8 years) for cirrhotic patients. The mean time interval from death to freezing of dissected autopsied material was 16.8 ± 3.3 h (range, 7.8-28.5 h) for controls and 14.3 ± 3.5 h (range, 3.5-23.5 h) for cirrhotics. The differences observed in ages and postmortem delay times between the two groups were not significant by Student's t test.

Brain dissections were performed according to a standardized dissection procedure based upon the atlas of De Armond et al., 1976. Prefrontal cortex was defined as the cerebral region comprised in the incurvation of the most rostral part of the frontal lobe (parts of Brodmann areas 10-11). A 1.0 cm transverse incision at this level gave rise to our samples which consisted largely of grey matter. Dissected material was stored as 10-50 mg samples at -70°C until the time of assay.

Amino acid measurements

Amino acid levels in brain tissue were measured fluorimetrically after HPLC separation of their o-phthalaldehyde (OPA) derivatives as previously described (Héroux and Butterworth, 1988) with some modifications. Tissue samples were homogenized in cold $HClO_4$ (0.6 M) containing norvaline as an internal standard, and were neutralized with $KHCO_3$

CATECHOLAMINE SYNTHESIS AND METABOLISM

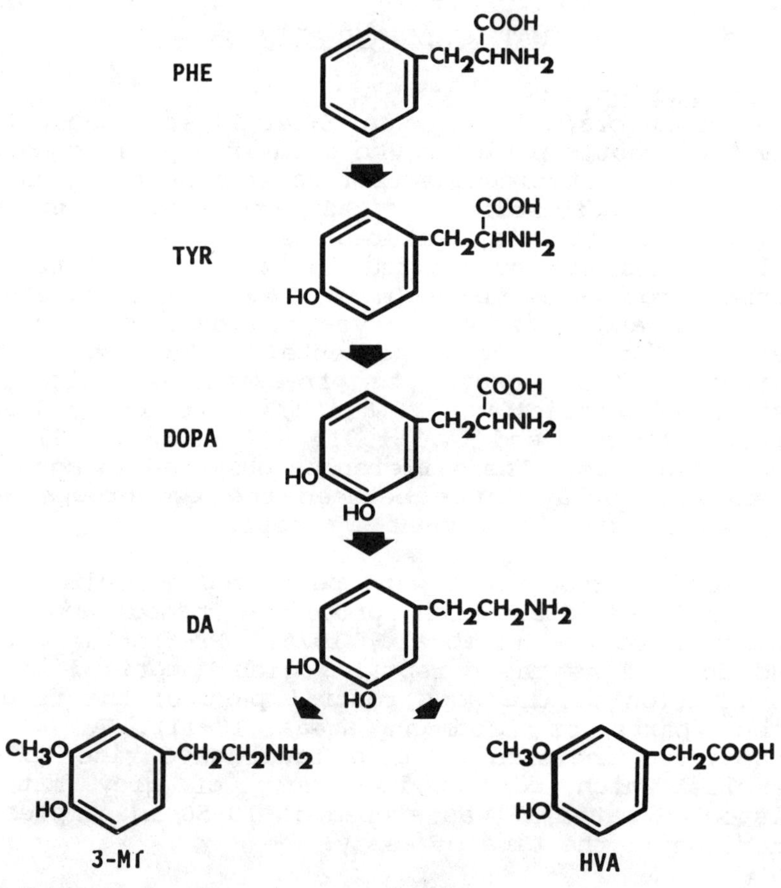

Figure 1. Catecholamine synthesis and metabolism. Abbreviations: Phenylalanine (PHE); Tyrosine (TYR); 3,4-dihydroxyphenylalanine (DOPA); Dopamine (DA); 3-methoxytyramine (3-MT); Homovanillic acid (HVA).

SEROTONIN SYNTHESIS AND METABOLISM

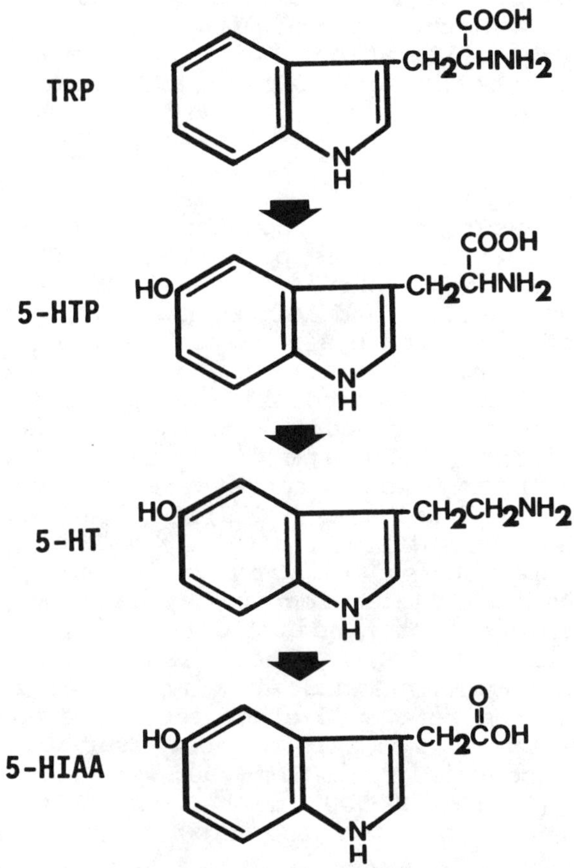

Figure 2. Serotonin synthesis and metabolism. Abbreviations: Tryptophan (TRP); 5-hydroxy-tryptophan (5-HTP); serotonin or 5-hydroxy-tryptamine (5-HT); 5-hyroxindoleacetic acid (5-HIAA).

(2.5 M). Aliquots (50 μl) of the brain amino acid extract were reacted at room temperature with OPA reagent in a total volume of 140 μl. 50 μl of the final solution were injected on the column for amino acid determination. The chromatograms were run with a linear gradient between methanol and a sodium acetate buffer (50 mM, pH 5.25 containing 2% THF) as follows: 25% to 50% methanol over 25 minutes and maintained at 50% methanol for an extra 15 minutes. The column was then washed with water for 10 minutes before re-equilibration (with 25% methanol: 75% sodium acetate buffer) for an additional 3 minutes before reuse.

Monoamine measurements

The measurement of dopamine, serotonin and some of their metabolites was performed by HPLC with ion-pairing and electrochemical detection as previously described (Reader and Grondin, 1987). Briefly, samples of dissected brain tissue were homogenized in cold monochloroacetic acid 0.1 N, pH 3.3 containing 2% (v/v) of an ascorbate oxidase solution (1.0 mg/ml in monochloroacetic acid 0.1 N, pH 3.3). After centrifugation (12,500 rpm for 45 min at 4°C), 200 μl of the resulting supernatants were injected into the column (5 μm particles, ODS). The pellets were dissolved overnight in 1.0 N NaOH for protein determinations (Lowry et al., 1951). Electrochemical detection was performed using an amperometric detector adjusted to a sensitivity of 5 nA full scale. The glassy-carbon working electrode was maintained at a potential of +750 mV versus the the reference electrode. The mobile phase was made up of monochloroacetic acid 0.1 N at pH 3.30-3.35 containing: EDTA (750 mg/l), octyl sodium sulfate (OSS: 300-500 mg/l) and 10% (v/v) of methanol. By varying the flow through the column (0.7-1.0 ml/min; 1,200-1,450 psi), the quantity of OSS and the temperature of the column (32-38°C), separation of the monoamines and their metabolites could be obtained within 50-60 minutes.

<u>Analysis of Data</u>

Statistical differences between groups were determined by unpaired Student's t test. A p level less than 5% was taken to indicate significance. Final results were expressed as micromole per gram of tissue (μmole/g wet weight) for amino acid concentrations and as nanogram per milligram of protein (ng/mg prot.) for monoamines and their metabolites.

TABLE 1: Summary of patient data

Control patients	Diagnosis	Sex	Age (years)	Time from death to freezing of tissue at -70°C (hr)
1	Cardiac disease	M	58	20
2	Carcinoma	M	61	8.25
3	Arteriosclerosis	M	49	28.5
4	Carcinoma	M	52	7.75
5	Arteriosclerosis	M	51	21.5
6	Leukemia	M	57	8
7	Carcinoma	M	67	23.25
		Mean:	56	16.8
Cirrhotic patients	**Diagnosis**	**Sex**	**Age (years)**	**Time from death to freezing of tissue at -70°C (hr)**
1	Hepatic cirrhosis	M	70	21
2	Hepatic cirrhosis	M	56	21
3	Hepatic cirrhosis	M	51	5
4	Hepatic cirrhosis	M	59	23.5
5	Hepatic cirrhosis	M	48	3.5
6	Hepatic cirrhosis	M	53	19
7	Hepatic cirrhosis	M	62	7
		Mean:	57	14.3

RESULTS

Concentrations of aromatic amino acids (PHE, TYR) in prefrontal cortex of autopsied brain tissue from cirrhotic patients and age-matched controls are shown in Table 2. Amino acid contentrations from control subjects agree well with the values previously reported for autopsied brain tissue from normal patients (Perry et al., 1971 a, b). In cirrhotic patients, amino acids levels were found to be significantly increased by 141% and 71% respectively ($p<0.01$) versus control values.

TABLE 2: Phenylalanine and tyrosine content of prefrontal cortex obtained at autopsy from control and cirrhotic patients

	n	Phenylalanine	Tyrosine
Controls	7	0.08 ± 0.02	0.21 ± 0.02
Cirrhotics	7	0.19 ± 0.03*	0.36 ± 0.02*

Values represent means ± SEM expressed in μ mole per gram wet weight. Significant differences from control values were determined by unpaired Student's t test: (*) $p<0.01$.

Concentrations of dopamine (DA), serotonin (5-HT) and their metabolites, 3-methoxytyramine (3-MT), homovanillic acid (HVA) and 5-hydroxyindole-acetic acid (5-HIAA) found in prefrontal cortex are shown in Table 3.

TABLE 3: Dopamine, serotonin and their metabolites in prefrontal cortex obtained at autopsy from control and cirrhotic patients

	n	Controls	Cirrhotics
DA	7	0.064 ± 0.039	0.062 ± 0.055
3-MT	7	0.012 ± 0.009	0.013 ± 0.005
HVA	7	13.519 ± 6.740	12.731 ± 6.024
5-HTP	7	0.497 ± 0.297	0.621 ± 0.476
5-HT	7	0.012 ± 0.007	0.027 ± 0.009
5-HIAA	7	1.250 ± 0.184	2.153 ± 0.365*

Values represent means ± SEM expressed in nanogram per milligram protein. Significant difference from control values were determined by unpaired Student's t test: (*) $p<0.05$.

Biogenic amine concentrations in brain tissue from control subjects are in good agreement with those previously reported for human prefrontal cortex (Korpi et al., 1986; Hardy et al., 1987; Wester, 1987).

No significant variations in the levels of DA or its metabolites 3-MT and HVA were observed in prefrontal cortex from cirrhotic patients compared to the controls. In addition, concentrations of 5-HTP (precursor of 5-HT) and 5-HT were similar in both groups of patients. However, the concentrations of the main metabolite of serotonin, 5-HIAA, were found to be significantly increased by 72% versus controls.

DISCUSSION

It has been repeatedly demonstrated that the aromatic amino acids (AAA) phenylalanine, tyrosine and tryptophan, are increased in plasma and brain tissue of humans with liver failure (Record et al., 1976; Cascino et al., 1982) and in animal models of both acute (Herlin et al., 1983; Jonung et al., 1985) and chronic (Mans et al., 1984; Jeppsson et al., 1985) liver dysfunction. Considering the fact that these amino acids are the direct precursors of the monoamine neurotransmitters (Figures 1 and 2), it has been postulated that an increased level of AAA in the brain may lead to a disturbance of catecholaminergic and serotoninergic neurotransmission (Fischer and Baldessarini, 1971). As catecholamines and 5-HT may play a role in the regulatory mechanisms of arousal and sleep (Jouvet, 1972), alterations of the metabolism of these neurotransmitters could account for the irregularities of sleep patterns, diurnal rythms and behavior observed clinically, in cirrhotic patients (Kurtz et al., 1972; Rehnstrom et al., 1977; Elsass et al., 1978) and experimentally, in animal models of chronic liver failure (Beaubernard et al., 1977; Bengtsson et al., 1985, 1987).

The results of the present study clearly demonstrate that hepatic encephalopathy resulting from chronic liver cirrhosis in man is accompanied by increased levels of AAA (PHE and TYR) in prefrontal cortex. Similar findings were previously reported in animal models following portacaval anastomosis (Battistin et al., 1980; Mans et al., 1984; Jeppsson et al., 1985). Moreover, the serotonin metabolite (5-HIAA) was increased in the prefrontal cortex of cirrhotic patients vs controls, suggesting that an increased brain 5-HT turnover may occur in patients with chronic liver failure and hepatic encephalopathy (HE). In contrast, the unchanged levels of DA, 3-MT and HVA found in these same patients suggest that the dopaminergic system in prefrontal cortex is largely unaffected by the increased levels of AAA associated with this disorder. Thus, these results do not support the hypothesis that HE is associated with a decrease in the levels of brain catecholamines

(Fischer and Baldessarini, 1976; Smith et al., 1978).

An overview of the literature (Table 4) reveals a conscensus regarding increased 5-HIAA levels in brain in chronic and acute liver failure. However, evidence for changes in 5-HT concentrations is less clear-cut. In chronic liver failure or following portacaval shunting, 5-HT levels have been found to be either increased (Curzon et al., 1975; Bengtsson et al., 1985, 1987; Mans and Hawkins, 1986; Bugge et al., 1988), decreased (Kamata et al., 1980) or unchanged (Cummings et al., 1976a; Bloxam et al., 1978; Moroni et al., 1984; Mans et al.,1987; present study). On the other hand, in experimental acute liver failure, results show increased 5-HT in the rat cerebral cortex after hepatectomy (Jonung et al., 1985; Bugge et al., 1987) or after liver ischemia (Cummings et al., 1976b), and in cerebral cortex of rats treated with thioacetamide (Yurdaydin et al., 1989). However, in cerebral cortex of pigs after liver ischemia, 5-HT content of cerebral cortex was found to be unchanged (Curzon et al., 1973; Delorme et al., 1981). In animal models of both acute and chronic liver failure using carbon tetrachloride treatment, increased 5-HT levels were observed (Knott and Curzon, 1975; Bengtsson et al., 1987.

Similarly, in chronic liver failure or following portacaval shunting, data on DA content is conflicting. Either no change (Curzon et al., 1975; Cuilleret et al., 1980; Bengtsson et al., 1985; Mans and Hawkins, 1986; present study) or decreases of DA (Kamata et al., 1980; Faraj et al., 1981) have been reported. In addition, DA levels in the rat after total hepatectomy have been reported to be unchanged in cerebral cortex (Herlin et al., 1983; Bugge et al., 1987) while in rats treated with thioacetamide, DA content was found to be increased (Yurdaydin et al., 1989). No changes in HVA levels were found in cerebral cortex of portacaval shunted rats (Mans et al., 1987) or in humans with chronic liver disease (present study).

Hence, there is evidence to suggest that a disturbance in the serotoninergic system, characterized by an increase in levels of 5-HIAA, is

TABLE 4: BRAIN MONOAMINES IN HEPATIC ENCEPHALOPATHY: A REVIEW OF THE LITERATURE

Chronic liver failure

Model	DA	HVA	5-HTP	5-HT	5-HIAA	Reference
RAT (PCS)	nc			↑	↑	Curzon et al., 1975
RAT (PCS)				nc	↑	Cummings et al., 1976a
RAT (PCS)				nc	↑	Bloxam et al., 1978
RAT (PCS)	↓			↓	↑	Kamata et al., 1980
RAT (PCS)				nc	↑	Moroni et al., 1984
RAT (PCS)	nc			↑		Bengtsson et al, 1985
RAT (PCS)	nc			↑	↑	Mans & Hawkins, 1986
RAT (PCS)		nc		nc	↑	Mans et al., 1987
RAT (PCS)			↑	↑	↑	Bugge et al., 1988
DOG (PCS)	↓					Faraj et al., 1981
RAT (CCl_4)			↑	↑	↑	Bengtsson et al., 1987
Human (cirrhosis)	nc					Cuilleret et al., 1980
Human (cirrhosis)	nc	nc	nc	nc	↑	Present study

Acute Liver Failure

Model	DA	HVA	5-HTP	5-HT	5-HIAA	Reference
RAT (hepatectomy)	nc					Herlin et al., 1983
RAT (hepatectomy)				↑	↑	Jonung et al., 1985
RAT (hepatectomy)	nc		↑	↑	↑	Bugge et al., 1987
RAT (liver ischemia)				↑	↑	Cummings et al., 1976b
RAT (thioacetamide)	↑			↑	↑	Yurdaydin et al., 1989
RAT (CCl_4)				↑	↑	Knott & Curzon, 1975
PIG (liver ischemia)				nc	↑	Curzon et al., 1973
PIG (liver ischemia)				nc	nc	Delorme et al., 1981

↑ increase; ↓ decrease; nc: no change

associated with liver failure. The elevated levels of 5-HIAA, accompanied by normal 5-HT levels reported in the present study suggest an <u>increase in 5-HT turnover</u>. Nevertheless, the possibility cannot be ruled out that the elevated levels of 5-HIAA could also be due to reduced 5-HIAA transport out of the brain. In fact, 5-HIAA exits the brain by an active transport system common to many aromatic carboxylic acids (Norton et al., 1967; Yuwiler et al., 1982; Moroni et al., 1986). It has been suggested recently that this transport system might be inhibited by elevated levels of ammonia observed in the brain during HE (Bugge et al., 1988). Further studies are needed in order to elucidate the functional relevance of 5-HT abnormalities in HE.

REFERENCES

Battistin, L., and Zanchin, G. (1980). The role of amino acids in hepatic encephalopathy. <u>In</u> Neurochemistry and Clinical Neurology, Alan R. Liss, Inc., New York, pp. 315-326.

Beaubernard, C., Salomon, F., Grange, D., Thangapregassam, M.J., and Bismuth, J. (1977). Experimental hepatic encephalopathy. Changes of the level of wakefulness in the rat with portacaval shunt. Biomedecine, 27: 169-171.

Bengtsson, F., Gage, F.H., Jeppsson, B., Nobin, A., and Rosengren, E. (1985). Brain monoamine metabolism and behavior in portacaval shunted rats. Exp. Neurol. 90: 21-35.

Bengtsson, F., Bugge, M., Vagianos, C., Jeppsson, B., and Nobin, A. (1987). Brain serotonin metabolism and behavior in rats with carbon tetrachloride-induced liver cirrhosis. Res. Exp. Med. 187: 429-438.

Bloxam, D.L., and Curzon, G. (1978). A study of proposed determinants of brain tryptophan concentrations in rats after portocaval anastomosis or sham operation. J. Neurochem. 31: 1255-1263.

Blitzer, B.L., Waggoner, J.G., Jones, E.A., Gralnick, H.R., Towne, D., Butler, J., Weise, V., Kopin, I.J., Walters, I., Teychenne, P.F., Goodman, D.G., and Berk, P.D. (1978). A model of fulminant hepatic failure in the rabbit. Gastroenterol. 74: 664-671.

Bugge, M., Bengtsson, F., Nobin, A., Jeppsson, B., and Herlin, P. (1987). The turnover of brain monoamines after total hepatectomy in rats infused with branched chain amino acids. World J. Surg. 11: 810-817.

Bugge, M., Bengtsson, F., Nobin, A., Jeppsson, B., Jonung, T., and Herlin, P. (1988). Serotonin metabolism in the rat brain following ammonia administration. In Advances in ammonia metabolism and hepatic encephalopathy, Soeters, P.B., Wilson, J.H.P., Meijer, A.J., and Holm, E. (eds.), Elsevier Science Publishers B.V., pp. 454-461.

Butterworth, R.F., Giguère, J.F., Lavoie, J., and Pomier Layrargues, G. (1987). Ammonia: Key factor in the pathogenesis of hepatic encephalopathy. Neurochem. Pathol. 6: 1-12.

Cascino, A., Cangiano, C., Fiaccadori, F., Ghinelli, F., Merli, M., Pelosi, G., Riggio, O., Rossi Fanelli, F., Sacchini, D., Stortoni, M., and Capocaccia, L. (1982). Plasma and cerebrospinal fluid amino acid patterns in hepatic encephalopathy. Digest. Dis. Sc. 27: 828-832.

Cuilleret, G., Pomier Layrargues, G., Pons, F., Cadilhac, J., and Michel, H. (1980). Changes in brain catecholamine levels in human cirrhotic hepatic encephalopathy. Gut, 21: 565-569.

Cummings, M.G., Soeters, P.B., James, J.H., Keane, J.M., and Fischer, J.E. (1976a). Regional brain indoleamine metabolism following chronic portocaval anastomosis in the rat. J. Neurochem. 27: 501-509.

Cummings, M.G., James, J.H., Soeters, P.B., Keane, J.M., Foster, J., and Fischer, J.E. (1976b). Regional brain study of indoleamine metabolism in the rat in acute hepatic failure. J. Neurochem. 27: 741-746.

Curzon, G., Kantamaneni, B.D., Winch, J., Rojas-Bueno, A., Murray-Lyon, I.M., and Williams, R. (1973). Plasma and brain tryptophan changes in experimental acute hepatic failure. J. Neurochem. 21: 137-145.

Curzon, G., Kantamaneni, B.D., Fernando, J.C., Woods, M.S., and Cavanagh, J.B. (1975). Effects of chronic porto-caval anastomosis on brain tryptophan, tyrosine and 5-hydroxytryptamine. J. Neurochem. 24: 1065-1070.

De Armond, S.J., Fusco, M.M, and Dewey, M.M. (1976). Structure of human brain: A photographic atlas., 2nd ed., Oxford University Press, New York.

Delorme, M.L., Denis, J., Nordlinger, M., Baschat, M., and Opolon, P. (1981). Relationship between plasma and brain tryptophan in pigs during experimental hepatic coma before and after hemodialysis with selective membranes. J. Neurochem. 36: 1058-1066.

Elsass, P., Lund, Y., and Ranek, L. (1978). Encephalopathy in patients with cirrhosis of the liver. A neuropsychological study. Scand. J. Gastroenterol. 13: 241-247.

Faraj, B.A., Camp, V.M., Ansley, J.D., Scott, J., Ali, F.M., and Malveaux, E.J. (1981). Evidence for central hypertyraminemia in hepatic encephalopathy. J. Clin. Invest. 67: 395-402.

Fischer, J.E., and Baldessarini, R.J. (1971). False neurotransmitters and hepatic failure. Lancet, 2: 75-80.

Fischer, J.E., Yoshimura, N., Aguirre, A., James, J.H., Cummings, M.G., Abel, R.M., and Deindoerfer, F. (1974). Plasma amino acids in patients with hepatic encephalopathy. Effects of amino acid infusions. Am. J. Surg. 127: 40-47.

Fischer, J.E., and Baldessarini, R.J. (1976a). Pathogenesis and therapy of hepatic coma. In Popper, H., Schaffner, F. (eds.), Progress in liver diseases, New York: Grune and Stratton: 363-397.

Fischer. J.E., Rosen, H.M., Ebeid, A.M., James, J.H., Keane, J.M., Soeters, P.B. (1976b). The effect of normalization of plasma amino acids on hepatic encephalopathy in man. Surgery, 80: 77-91.

Hardy, J.A., Wester, P., Backstrom, I., Gottfries, J., Oreland, L., Stenstrom, A., and Winblad, B. (1987). The regional distribution of dopamine and serotonin uptake and transmitter concentrations in the human brain. Neurochem. Int. 10: 445-450.

Herlin, P.M., James, J.H., Nachbauer, C.A., and Fischer, J.E. (1983). Effect of total hepatectomy and administration of branched-chain amino acids on regional norepinephrine, dopamine and amino acids in rat brain. Ann. Surg. 198: 172-177.

Héroux, M., and Butterworth, R.F. (1988). Reversible alterations of cerebral -aminobutyric acid in pyrithiamine-treated rats: Implications for the pathogenesis of Wernicke's encephalopathy. J. Neurochem. 51: 1221-1226.

Jeppsson, B., James, J.H., Edwards, L.L., and Fischer, J.E. (1985). Relationship of brain glutamine and brain neutral amino acid concentrations after portocaval anastomosis in rats. Eur. J. Clin. Invest. 15: 179-187.

Jonung, T., Ramzy, A., Herlin, P., James, J.H., Edwards, L., and Fischer, J.E. (1985). Indole amines and amino acids in various brain regions after infusion of branched chain amino acids into hepatectomized rats. Eur. Surg. Res. 17: 83-90.

Kamata, S., Okada, A., Watanabe, T., Kawashima, Y., and Wada, H. (1980). Effects of dietary amino acids on brain amino acids and transmitter amines in rats with a portacaval shunt. J. Neurochem. 35: 1190-1199.

Knott, P.J., and Curzon, G. (1975). Tryptophan and tyrosine disposition and brain tryptophan metabolism in acute carbon tetrachloride poisoning. Biochem. Pharmacol. 24: 963-966.

Korpi, E.R., Kleinman, J.E., Goodman, S.I., Phillips, I., DeLisi, L.E., Linnoila, M., and Wyatt, R.J. (1986). Serotonin and 5-hydroxyindoleacetic acid in brains of suicide victims. Arch. Gen. Psychiatry, 43: 594-600.

Kurtz, D., Zenglein, J.P., Imler, M., Girardel, M., Grinspan, G., Peter, B., and Rohmer, F. (1972). Etude du sommeil nocturne au cours de l'encéphalopathie porto-cave. Electroenceph. Clin. Neurophysiol. 33: 167-178.

Lowry , O.H., Rosebrough, N.J., Farr, A.L., and Randall, R.J. (1951). Protein measurements with Folin phenol reagent. J. Biol. Chem. 193: 265-275.

Mans, A.M., Biebuyck, J.F., Davis, D.W., and Hawkins, R.A. (1984). Portacaval anastomosis: Brain and plasma metabolite abnormalities and effect of nutritional therapy. J. Neurochem. 43: 697-705.

Mans, A.M., and Hawkins, R.A. (1986). Brain monoamines after portacaval anastomosis. Metab. Brain Dis. 1: 45-52.

Mans, A.M., Consevage, M.W., De Joseph, M.R., and Hawkins, R.A. (1987). Regional brain monoamines and their metabolites after portacaval shunting. Metab. Brain Dis. 2: 183-193.

Moroni, F., Carlà, V. Lombardi, G., Pellegrini, D., Carassale, G.L., and Cortesini, C. (1984). Excitatory amino acids, tryptophan metabolites, and endogenous neurotoxins in porta-caval shunted rats. In Advances in hepatic encephalopathy and urea cycle diseases, Kleinberger, G., Ferenci, P., Reiderer, P., and Thaler, H. (eds.), Karger, Basel, pp. 385-393.

Moroni, F., Lombardi, G., Carlà, V., Pellegrini, D., Carassale, G.L., and Cortesini, C. (1986). Content of quinolinic acid and other tryptophan metabolites increases in brain regions of rats used as experimental models of hepatic encephalopathy. J. Neurochem. 46: 869-874.

Munro, H.N., Fernstrom, J.D., and Wurtman, R.J. (1975). Insulin, plasma aminoacid imbalance, and hepatic coma. Lancet, 1: 722-724.

Norton, H.N., Tozer, T.N., and Brodie, B.B. (1967). Application of steady-state kinetics to studies of the transfer of 5-hydroxyindoleacetic acid from brain to plasma. J. Pharmacol. Exp. Therap. 158: 214-218.

Perry, T.L., Berry, K., Hansen S., Diamond, S., and Mok, C. (1971a). Regional distribution of amino acids in human brain obtained at autopsy. J. Neurochem. 18: 513-519.

Perry, T.L., Hansen, S., Berry, K., Mok, C., and Lesk, D. (1971b). Free amino acids and related compounds in biopsies of human brain. J. Neurochem. 18: 521-528.

Reader, T.A., and Grondin, L. (1987). Distribution of catecholamines, serotonin, and their major metabolites in the rat cingulate, piriform-entorhinal, somatosensory, and visual cortex: a biochemical survey using high-performance liquid chromatography. Neurochem. Res. 12: 1087-1097.

Rehnstrom, S., Simert, G., Hansson, J.A., Johnson, G., and Vang, J. (1977). Chronic hepatic encephalopathy. A psychometrical study. Scand. J. Gastroenterol. 12: 305-311.

Record, C.O., Buxton, B., Chase, R.A., Curzon, G., Murray-Lyon, I.M., and Williams, R. (1976). Plasma and brain amino acids in fulminant hepatic failure and their relationship to hepatic encephalopathy. Eur. J. Clin. Invest. 6: 387-394.

Rosen, H.M., Yoshimura, N., Hodgman, J.M., and Fischer, J.E. (1977). Plasma amino acid patterns in hepatic encephalopathy of differing etiology. Gastroenterol. 72: 483-487.

Smith, A.R., Rossi-Fanelli, F., Ziparo, V., James, J.H., Perelle, B.A., and Fischer, J.E. (1978). Alterations in plasma and CSF amino acids, amines and metabolites in hepatic coma. Ann. Surg., 187: 343-350.

Wester, P., Gottfries, J., and Winblad, B. (1987). Simultaneous liquid chromatographic determination of seventeen of the major monoamine neurotransmitters, precursors and metabolites. II. Assessment of human brain and cerebrospinal fluid concentrations. J. Chromatog. 415: 275-288.

Yurdaydin, C., Zimmermann, Ch., Pifl, Ch., Singer, E., Hortnagl, H., and Ferenci, P. (1989). Increased serotoninergic and noradrenergic activity in an experimental model of hepatic encephalopathy in rats. In Hepatic encephalopathy: Pathogenesis and treatment, Butterworth, R.F., and Pomier Layrargues, G. (eds.), Humana Press, N.J., USA.

Yuwiler, A., Bennett, B.L., and Geller, E. (1982). Is there a probenecid sensitive transport system for monoamine catabolites at the level of the brain capillary plexus? Neurochem. Res. 7: 1277-1285.

INCREASED SEROTONINERGIC AND NORADRENERGIC ACTIVITY IN AN EXPERIMENTAL MODEL OF HEPATIC ENCEPHALOPATHY IN RATS

C.Yurdaydin, Ch.Zimmermann, Ch.Pifl, E.Singer, H.Hörtnagl, P.Ferenci
University of Vienna, Austria

Lazarettgasse 14, A-1090 Vienna, Austria

ABSTRACT

The pathogenesis of hepatic encephalopathy (HE) is poorly understood. One of the main hypothesis concerning HE is the "false neurotransmitter hypothesis". To test this hypothesis in a well characterized model of HE, brain monoamines were measured in brains of rats with thioacetamide (TAA) induced acute liver failure. Rats were killed by decapitation and monoamines were measured in control rats, rats with HE and TAA-treated rats without HE. Concentrations of 5-hydroxytryptamine (5-HT), 5-hydroxyindoleacetic acid (5-HIAA), dopamine (DA), dihydroxyphenylacetic acid (DOPAC), noradrenaline (NA) and 3-methoxy-4-hydroxyphenylglycol (MHPG) were measured by high performance liquid chromatography in the frontal cortex, the hippocampus and the striatum. In the frontal cortex and hippocampus the concentrations of 5-HT, 5-HIAA, DA and MHPG were significantly increased and those of NA decreased in HE vs. controls. In contrast, in the striatum DA and DOPAC concentrations were unchanged. These data indicate the presence of an augmented noradrenergic but unchanged dopaminergic activity in HE. Therefore, these findings do not support the "false neurotransmitter hypothesis". Furthermore, the data suggest a high serotoninergic activity in HE. Whether these changes play a role in the mediation of HE has to be explored by further neuropharmacological studies.

INTRODUCTION

Despite intensive research efforts the pathogenesis of hepatic encephalopathy (HE) is not understood. Neurochemical studies in experimental HE may be useful to detect changes in neurotransmission mediating the evolution of HE. Based on such data two hypotheses of the pathogenesis of HE were formulated. One hypothesis implicates alterations of central monoamines in the pathogenesis of HE. According to the "false neurotransmitter hypothesis" (11) in liver failure neurons are depleted of noradrenaline (NA) and dopamine (DA) and on the other hand "false" catecholamine neurotransmitters such as octopamine and ß-phenylethanolamine accumulate. The occurence of these changes was explained by the altered ratio of the serum concentrations of aromatic (AAA) and branched chain amino acids (BCAA) resulting in an increased transport of AAA's from plasma into the brain. Furthermore, increased cerebral concentrations of glutamine (a consequence of ammonia fixation) may also facilitate the influx of AAA's into the brain (14). The increased concentrations of phenylalanine and tryptophan in the brain were suggested to inhibit tyrosine hydroxylase, the key enzyme of catecholamine synthesis. Since tryptophan is the precursor of serotonin, in addition to the changes in catecholaminergic neurotransmission an augmented serotonin synthesis may also contribute to the development of the HE (11).

Another hypothesis involves an activation of the inhibitory gamma aminobutyric acid (GABA)ergic neurotransmitter system. Recently, this system has been studied intensively by various neurochemical methods and no consistent changes were observed (10,19). In a well described model of HE in rats with thioacetamide (TAA)induced liver failure we also could not detect any changes in GABA-ergic neurotransmission (26). Therefore, in the present study the investigations were extended to the amineergic neurotransmitter systems. These systems were studied in various experimental models of HE by many investigators with inconsistent results, previously (2-5,8,13,15,20,21, 24,25). Such neurochemical studies in well established models of HE are necessary for better understanding of the functional alteration of the brain occuring in liver failure.

MATERIAL AND METHODS

Animal model

The model of HE used in this study was extensively characterized by clinical, light and electron microscopic investigations as well as by blood biochemistry, previously (26). Briefly, 300 mg/kg body weight TAA was administered by gavage to overnight fasted male Sprague-Dawley rats on two consecutive days. Control rats received 2 ml normal saline by gavage. Supportive therapy was given to prevent hypoglycemia and electrolyte disturbances. Rats were sacrificed by decapitation as soon as they developed stage III HE. After decapitation brains were quickly removed and frozen on dry ice. Brains were then dissected according to the atlas of König and Klippel (16). Frontal cortex, striatum and hippocampus were taken from coronal slices.

Measurements

5-hydroxytryptamine (5-HT), dopamine (DA), noradrenaline (NA) and their metabolites 5-hydroxyindolacetic acid (5-HIAA), dihydroxyphenylacetic acid (DOPAC) and 3-methoxy-4-hydroxyphenylglycol (MHPG), respectively, were measured by high performance liquid chromatography with electrochemical detection (9,22,23) in supernates of homogenates of these three brain regions of control rats, rats with HE and TAA-treated rats not developing HE within 84 hrs after TAA administration.

RESULTS

Results are shown in figures 1 and 2. 5-HT increased ($p < 0.05$) in rats with HE as compared to controls in the cerebral cortex and the hippocampus but not in the striatum. In TAA-treated rats without HE 5-HT concentrations decreased in the hippocampus ($p < 0.05$) whereas in other regions they did not differ from controls. In all three regions investigated, concentrations of 5-HIAA were increased ($p < 0.01$) both in TAA-treated rats with or without HE. In encephalopathic and non-encephalopathic TAA-treated rats DA concentrations in the frontal cortex and the hippocampus were higher as in controls. In rats with HE this increase was more pronounced than in those without HE ($p < 0.05$, $p < 0.01$). The increased concentrations of DA in the frontal cortex and hippocampus were associated with

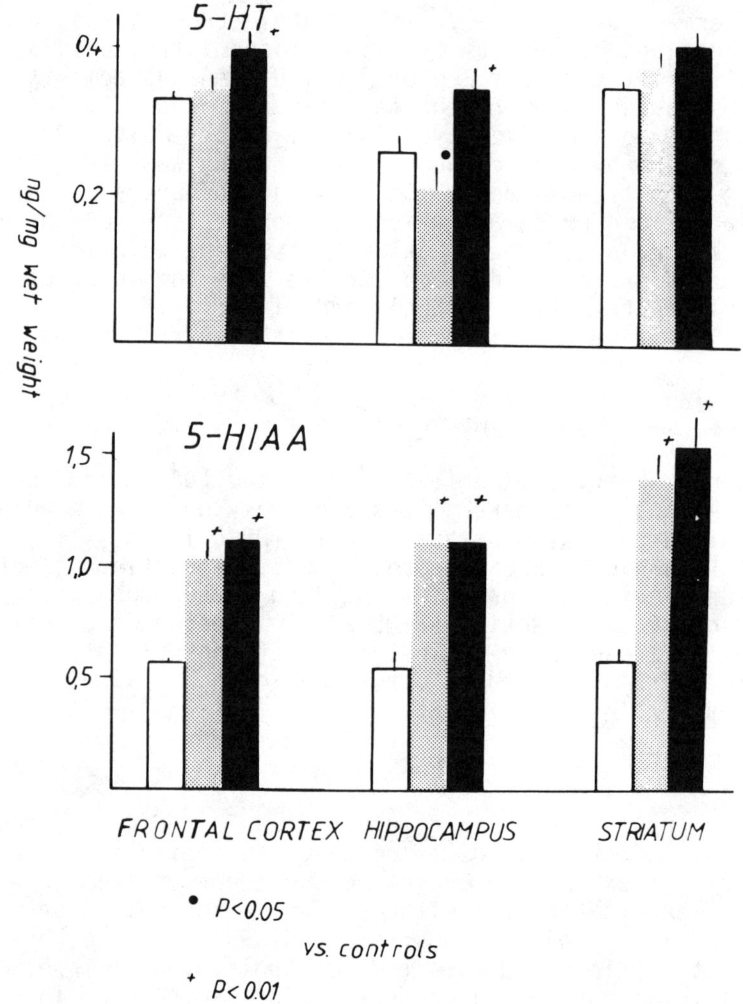

Fig. 1. Concentrations of 5-HT and 5-HIAA in various brain regions (open bars: controls; stippled bars: TAA rats without HE; closed bars: TAA rats with HE)

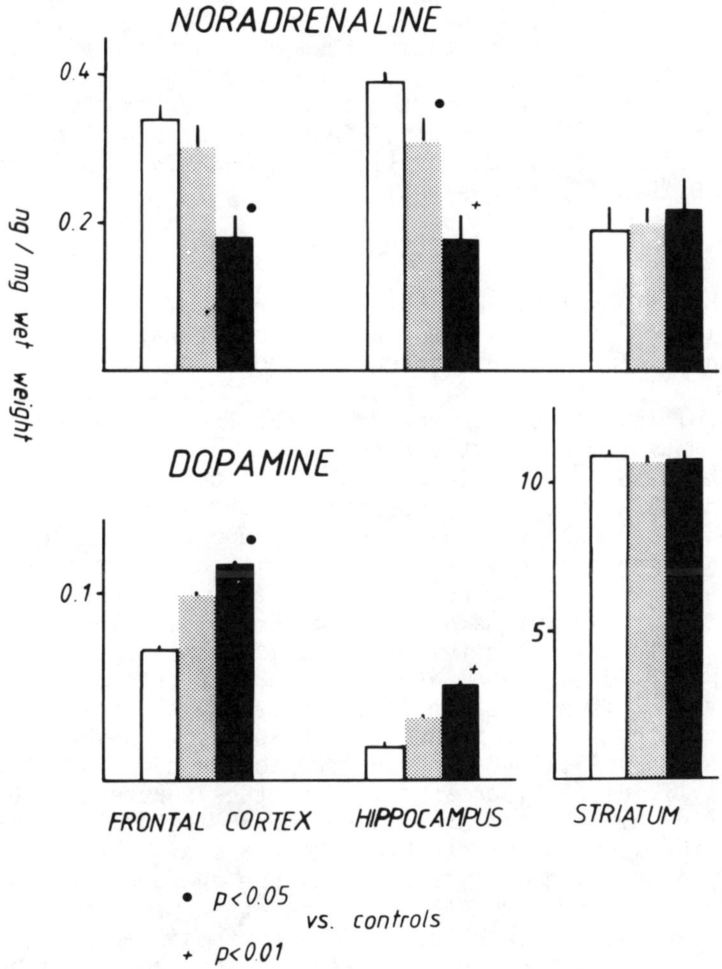

Fig. 2. Concentrations of NA and DA in various brain regions (open bars: controls; stippled bars: TAA rats without HE; closed bars: TAA rats with HE)

decreased of NA ($p < 0.01$) and increased of MHPG ($p < 0.01$) ($\bar{x} \pm$ SEM: frontal cortex: 0.22 ± 0.02; 0.32 ± 0.01; 0.38 ± 0.05, hippocampus: 0.32 ± 0.01; 0.40 ± 0.02; 0.66 ± 0.04 ng/mg wet weight in controls, TAA-treated rats without HE, and TAA-treated rats with HE, respectively). In rats without HE, NA was decreased only in the hippocampus ($p < 0.05$). In contrast, however, in the striatum, a region with rich dopaminergic innervation, DA levels were unchanged. In this region, the concentration of the DA metabolite DOPAC in controls, TAAtreated rats without and with HE did also not differ (1.77 ± 0.09; 1.57 ± 0.13 and 1.79 ± 0.25 ng/mg wet weight).

DISCUSSION

This study indicates that HE in rats with TAA-induced acute liver failure is associated with alterations of serotoninergic and noradrenergic neurotransmission.

5-HT levels were increased in the frontal cortex and hippocampus in TAA-treated rats with HE and 5-HIAA levels were increased in TAA-treated rats with or without HE in the three brain regions investigated. The magnitude of increase in 5-HIAA levels was more pronounced in rats with overt HE. These observed changes are in good agreement with the findings in other experimental models of HE (5,20,24) and in humans dying with HE (15,21). The most likely explanation for this finding is increased substrate availability due to augmented transport of tryptophan into the brain. This suggestion is supported by the dose dependent increase of 5-HT and 5-HIAA levels in several brain areas following exogenous tryptophan administration to rats (7,18). An increased 5-HT synthesis rate was observed in some experimental models of HE (2,3). The augmented transport of tryptophan into the brain may be mediated by hyperammonemia (1,12) or by the changed ratio of BCAA to AAA's in plasma (11). On the other hand, it is uncertain whether an increase in 5-HT synthesis actually produces an increased 5-HT release and thus an increased serotoninergic function. Recent data provide evidence that infusion of tryptophan in "naive" rats enhances 5-HT metabolism without affecting release (7,18). Alternatively, one may speculate that besides augmented 5-HT synthesis rate 5-HIAA increases as consequence of delayed transport of 5-HIAA out of the brain. Supporting this assumption, a slow disappearance of radioactivity out of the brain in hepatectomized rats after injection of labeled tryptophan was observed (24) and the probenecid (a drug which blocks the efflux of 5-HIAA out of the brain) induced increase of cerebrospinal fluid 5-HIAA was not different in cirrhotic patients with HE compared to those without HE (17). It is obvious that more studies on brain 5-HT metabolism in HE are needed before a final conclusion can be drawn about the functional importance of the central serotoninergic system in HE.

In addition, the present investigation demonstrates that central noradrenergic systems are affected in this animal model of HE. These observations are in good agreement with findings in other types of experimental HE (3,8, 13,25). One possible explanation for the observed decrease of NA levels associated with increased MHPG levels in frontal cortex and hippocampus is a strong increase in the firing rate of noradrenergic neurons. Increased noradrenergic impulse flow could also account for the concomitant increase of DA in these two brain areas. In brain areas with sparse dopaminergic innervation DA is mainly present as the immediate precursor of NA in noradrenergic terminals. Under conditions of increased noradrenergic function DA might accumulate due to inefficient ß-hydroxylation to NA in presynaptic granules. This assumption is supported by the finding that increase in noradrenergic impulse flow elicited by electrical stimulation of noradrenergic pathways results in a decrease of NA levels together with an increase of DA levels in the hippocampal formation of rats (22). No apparent explanation can be given why the firing rate of noradrenergic neurons is increased in HE. A potential mechanism of an increased NA release may be an enhanced GABA-ergic function, since GABA-ergic mechanisms appear to facilitate central noradrenergic transmission (6).

On the other hand, in HE the function of the dopaminergic system was not altered in the striatum, a brain area especially rich in dopaminergic innervation. Unchanged levels of DA or of its main metabolite DOPAC indicate a normal synthesis and release of this neurotransmitter. Similar observations were made in other models of HE (2,3,20). In contrast, in DA rich brain areas DA concentrations were decreased in some experimental models of HE (13) and were increased in human HE (4). The reason for these conflicting findings is unknown but may be explained by the different models of HE studied.

Taken together, the opposite changes of central noradrenergic and dopaminergic systems in this animal model of HE do not support the "false neurotransmitter hypothesis". Although, no measurements of octopamine or ß-phenylethanolamine were made an inhibition of catecholamine synthesis, the primary event predisposing for the accumulation of false neurotransmitters, was excluded. Whether these findings reflect primary or secondary compensatory mechanisms to counteract neural inhibition remains to be further explored. Furthermore, the role of the observed neurotransmitter changes for the development of HE should be investigated by neuropharmacological studies.

REFERENCES

1. Bachmann C, Colombo JP. Increased tryptophan uptake into the brain in hyperammonemia. Life Sci 1983; 33:2417-24

2. Bengtsson F, Gage FH, Jeppsson B, et al. Brain monoamine metabolism and behaviour in portacaval-shunted rats. Exp Neurol 1985; 90:21-35

3. Bugge M. Thesis Bulletin No 65, Department of Surgery, Lund University, Lund, 1987

4. Cuilleret C, Pomier-Layrargues G, Pons F, et al. Changes in brain catecholamine levels in human cirrhotic encephalopathy. Gut 1980; 21:565-9

5. Cummings MG, James JH, Soeters PB, et al. Regional brain study of indoleamine metabolism in the rat in acute hepatic failure. J Neurochem 1976; 27:741-6

6. Dennis T, Curet O, Nishikawa T, et al. Further evidence for, and nature of, the facilitatory GABA-ergic influence on central noradrenergic transmission. Naunyn Schmiedeberg's Arch Pharmacol 1985; 331: 225-34

7. DeSimoni MG, Sokola A, Fodritto F, et al. Functional meaning of tryptophan-induced increase of 5-HT metabolism as clarified by in vivo voltametry. Brain Res 1987; 411: 819-24

8. Dodsworth JM, James JH, Cummings MG, et al. Depletion of brain norepinephrine in acute hepatic coma. Surgery 1974; 75:811-20

9. Felice LJ, Felice JD, Kissinger PT. Determination of catecholamines in rat brain parts by reverse-phase ion-pair liquid chromatography. J Neurochem 1978; 31:1461-5

10. Ferenci P, Zieve L, Ebner J, et al. Postsynaptic gamma aminobutyric acid receptors in hepatic coma following portacaval shunt and hepatic artery ligation in the rat. Met Brain Dis 1987; 2:195-200

11. Fischer JE, Baldessarini RJ. False neurotransmitters and hepatic failure. Lancet 1971; 2:75-80

12. Grippon P, Lafitte MLP, Boschat M, et al. Evidence for the role of ammonia in the intracerebral transfer and metabolism of tryptophan. Hepatology 1986; 6:682-6

13. Herlin PM, James JH, Hachbauer CA, et al. Effect of total hepatectomy and administration of branched chain amino acids on regional norepinephrine, dopamine and amino acids in rat brain. Ann Surg 1983; 198:172-7

14. James JH, Jeppsson B, Ziparo V, et al. Hyperammonemia, plasma amino acid imbalance, and blood-brain amino acid transport: A unified theory of portal systemic encephalopathy. Lancet 1979; 2:772-5

15. Jellinger K, Riederer P, Kleinberger G, et al. Brain monoamines in human hepatic encephalopathy. Acta Neuropathol (Berl) 1978; 43:63-8

16. König JFR, Klippel RA. The rat brain. A stereotaxic atlas for the forebrain and lower parts of the brain stem. Krieger RE, ed. Huntington, New York, 1970

17. Lal S, Aronoff A, Garelis E, et al. Cerebrospinal fluid homovanilic acid, lactic acid and pH before and after probenecid in hepatic coma. Clin Neurol Neurosurg 1974; 77:142-54

18. Lookingland KJ, Shannon NJ, Chapin DS, et al. Exogenous tryptophan increases synthesis, storage, and intraneuronal metabolism of 5-hydroxytryptamine in the rat hypothalamus. J Neurochem 1986; 47:205-12

19. Maddison JE, Dodd PR, Johnston GAR, et al. Brain gamma aminobutyric acid receptor binding is normal in rats with thioacetamide-induced hepatic encephalopathy despite elevated plasma gamma aminobutyric acid-like activity. Gastroenterology 1987; 93:1062-8

20. Mans AM, Hawkins RA. Brain monoamines after portacaval anastomosis. Met Brain Dis 1986; 1:45-52

21. Riederer P, Jellinger K, Kleinberger G, et al. Oral and parenteral nutrition with L-valine: Mode of action. Nutr Met 1980; 24:209-17

22. Scatton B, Dennis T, Curret O. Increases in dopamine and DOPAC levels in noradrenergic terminals after electrical stimulation of the ascending noradrenergic pathways. Brain Res 1984; 298:193-6

23. Sperk G. Simultaneous determination of serotonin, 5-hydroxyindoleacetic acid, 3,4-dihydroxyphenylacetic acid and homovanilic acid by high performance liquid chromatography with electrochemical detection. J Neurochem 1982; 38:840-3

24. Tyce GM, Flock EV, Owen CA Jr, et al. 5-hydroxyindole metabolism in the brain after hepatectomy. Biochem Pharmacol 1967; 16:979-92

25. Tyce GM, Owen CA Jr. Administration of L-3,4-dihydroxyphenylalanine to rats after complete hepatectomy. Biochem Pharmacol 1979; 28:3271-8

26. Zimmermann Ch, Ferenci P, Pifl Ch, et al. Hepatic encephalopathy in thioacetamide induced acute liver failure in rats. Characterization of an improved model and study of amino acid-ergic neurotransmission. Hepatology 1989 (in press)

EXCITATORY AMINO ACIDS AND HEPATIC ENCEPHALOPATHY

Roger F. Butterworth[1], Joel Lavoie[1], Christine Peterson[2], Carl W. Cotman[2] and John C. Szerb[3] [1]Lab. of Neurochem., André-Viallet Clin. Res. Centre, Hôpital St-Luc, Montreal, [2]Dept. of Psychobiology, Univ. of California, Irvine,California and [3]Dept. of Physiology and Biophysics, Dalhousie University, Halifax, Nova Scotia.

SUMMARY

There is substantial evidence to suggest that glutamate and aspartate are excitatory neurotransmitters in the mammalian central nervous system. Glutamate content of brain is decreased in experimental models of acute and chronic hyperammonemia as well as in autopsied brain tissue from patients dying in hepatic coma resulting from fulminant hepatic failure or cirrhosis. Glutamate-related enzyme changes have also been observed in brain in both experimental and human hepatic encephalopathy. Exposure of hippocampal slices to pathophysiological concentrations of ammonia did not result in significant decreases of K^+-stimulated Ca^{2+}-dependent release of glutamate suggesting that the synthesis of neurotransmitter amino acid was not affected in hyperammonemia. However, evoked release of glutamate from hippocampal slices from 4-week portacaval shunted rats was <u>increased</u> 2-fold. Assessment of glutamate receptors by quantitative autoradiography revealed widespread reductions of the N-Methyl-D-Aspartate (NMDA) subclass of receptors. Taken together with the increased glutamate release of in this material, these results suggest that extracellular glutamate (or aspartate) may be increased in sustained hyperammonemia, probably as a result of loss of astrocytic integrity. Such changes could be of pathophysiological significance in hepatic encephalopathy.

INTRODUCTION

Glutamate (and aspartate) are dicarboxylic acidic amino acids found in relatively high abundance in the mammalian central nervous system. When applied to CNS neurons, they show potent excitatory effects and, over the last 20 years, convincing evidence has been accumulating to suggest that these amino acids are excitatory neurotransmitters. Immunocytochemical studies have revealed the existence of glutamate-like immunoreactivity in excitatory nerve terminals (Storm-Mathisen et al., 1983) and glutamate and aspartate are released from brain slice preparations in a calcium-dependent manner following electrical stimulation (Potashner 1978; Szerb, 1988).

Regulation at the glutamatergic synapse involves a functional interaction between astrocytes and neurons and the existence of the so-called "glutamate-glutamine cycle" shown schematically in Figure 1. Thus, glutamate released by presynaptic nerve terminals into the synaptic cleft interacts with postsynaptic glutamate receptors and is then effectively removed from the cleft by uptake (mainly into astrocytes) where it is converted into glutamine. The enzyme responsible for glutamine formation, namely glutamine synthetase (GS) is highly concentrated (and, perhaps, uniquely localized) in astrocytes (Norenberg, 1979). The glutamine formed in astrocytes is then either returned to the presynaptic neuron to serve as precursor for the neosynthesis of releasable glutamate or diffuses into the csf.

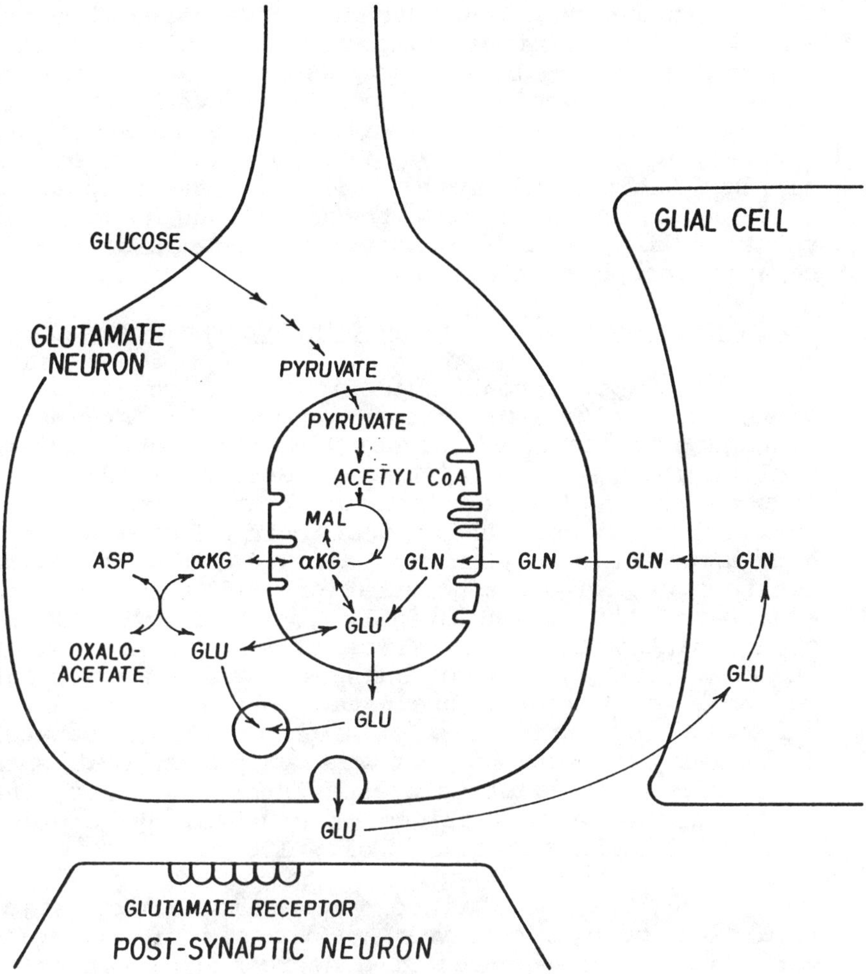

Figure 1: The glutamate-glutamine cycle: neuron-glial metabolic interrelations

During the last decade, several actions of ammonia on various aspects of glutamatergic regulation have been reported. These include actions of ammonia on glutamate synthesis from glutamine, on glutamate reuptake as well as on the generation of excitatory post-synaptic potentials per se (Butterworth et al., 1987). The purpose of this paper is to review these various reports in relation to ammonia's proposed pathogenetic role in hepatic encephalopathy.

Brain Glutamate and Aspartate in Hyperammonemia

Intraperitoneal injection of high doses of ammonium salts in rats results in decreased glutamate content of cerebral cortex and caudate nucleus (Moroni et al., 1983). Similar findings of reduced brain glutamate and aspartate were observed in cerebral cortex and spinal cord of the acutely hyperammonemic cat (Lin and Raabe, 1985). Following hepatectomy, brain concentrations of glutamate and aspartate have consistently been found to be significantly decreased (Tyce et al., 1971; Holmin et al., 1983). Such changes were observed in cerebral cortex and brainstem and were apparent prior to any reductions of high energy phosphates following hepatectomy. It was suggested that such alterations of glutamate/aspartate could be the cause of the manifestations of cerebral dysfunction in these animals (Holmin et al., 1983).

In the sustained hyperammonemic state resulting from portacaval anastomosis in the rat, various authors have reported decreased glutamate in whole brain (Cremer et al., 1975), cerebral hemispheres (Holmin and Siesjo, 1974; Hindfelt et al., 1977), cerebral cortex and brainstem (Butterworth and Giguère, 1986). As was the case in acute hyperammonemia, these reductions of brain glutamate were evident prior to any reductions of brain high energy phosphates (Holmin and Siesjo, 1974; Hindfelt et al., 1977).

Two independent investigations of autopsied brain tissue from patients dying in hepatic coma have provided evidence for reduced excitatory amino acids. In a study by Record et al., 1976, glutamate content of cerebral cortex from patients with fulminant hepatic failure was found to be decreased. In a more recent study, reductions of glutamate in prefrontal cortex, caudate nucleus and cerebellar vermis were observed in nine cirrhotic patients dying in hepatic coma (Lavoie et al., 1987a). Similar reductions of brain aspartate were also noted in this material.

Thus there is considerable evidence accumulated from studies of acute and chronic hyperammonemia as well as from biochemical investigations on postmortem tissue from patients to suggest that brain levels of excitatory amino acids are modified in hepatic encephalopathy.

However, it should be borne in mind that, in addition to its action as an excitatory neurotransmitter, glutamate is involved in intermediary metabolism (metabolic pool) as well as being the precursor of γ-aminobutyric acid (GABA), an important inhibitory CNS neurotransmitter (precursor pool of glutamate). Selective lesioning studies have shown that, in several brain regions, the neurotransmitter pool of glutamate is of the order of 20-30% of total tissue glutamate (Fonnum, 1985). Thus, studies of total tissue levels of excitatory amino acids are of limited value in the assessment of glutamatergic regulation in hepatic encephalopathy. For this reason, many recent studies have attempted to evaluate cell-specific (neuronal vs astrocytic) enzymes related to excitatory amino acids as well as their release and uptake properties in conditions of hyperammonemia. In addition, recent attempts have been made to measure excitator amino acid release and the integrity of glutamate receptors in hepatic encephalopathy.

Excitatory amino acid-related enzymes

A number of enzymes are involved in the synthesis and metabolism of glutamate and aspartate in brain. Some of thes enzymes are enriched in certain cell populations so that measurement of their specific activities reveals information concerning the integrity of certain neuronal and astrocytic populations. Glutamine Synthetase (GS, see Figure 1) is, as already discussed, an astrocytic enzyme responsible for glutamate's transformation into glutamine. Ammonia removal by brain depends to a very large extent on glutamine synthesis by GS (Cooper and Plum, 1987). GS is not induced in conditions of sustained hyperammonemia such as occurs following portacaval shunting (Cooper et al., 1985, Butterworth et al., 1988). On the contrary, GS activities were found to be significantly decreased in cerebral cortex of portacaval shunted rats leading to the suggestion that the capacity for ammonia removal by brain following portacaval anastomosis may be limited in certain brain regions (Butterworth et al., 1988 and Girard et al., this volume). Similar reductions of GS were reported in frontal cortex and caudate nucleus obtained at autopsy from cirrhotic patients dying in hepatic coma (Lavoie et al., 1987b). Hepatic encephalopathy resulting from chronic liver disease is associated with degenerative astrocytic changes (Alzheimer Type II astrocytosis) and it was suggested that the decreased GS activities observed were the result of astrocytic loss in this material. In support of this possibility, activities of a second glutamate-related enzyme, glutamate dehydrogenase (GDH), also enriched in astrocytes relative to neurons were also found to be reduced in autopsied brain tissue from cirrhotic patients with hepatic encephalopathy (Giguère et al., this volume). Furthermore, it has been suggested that GDH may be preferentially localized to astrocytes adjacent to glutamate nerve terminals (Aoki et al., 1987; Wenthold et al., 1987) so that selective loss of GDH-containing astrocytes could result in alterations of glutamatergic synaptic regulation.

Glutamic acid decarboxylase (GAD) is responsible for the synthesis of GABA from glutamate. GAD is localized in presynaptic GABA nerve terminals. Although some earlier studies suggested that GAD activities were altered in the brains of dogs with galactosamine-induced fulminant hepatic failure (Baraldi et al., 1984), more recent studies in other experimental models of hepatic encephalopathy did not confirm these findings (Butterworth and Giguère, 1986; Maddison et al., 1987; Roy et al., 1988). Furthermore, GAD activites in homogenates of several regions of brain obtained at autopsy from cirrhotic patients with hepatic encephalopathy were found to be within normal limits (Lavoie et al., 1987b). It therefore appears that presynaptic GABA nerve terminals are not modified, at least in hepatic encephalopathy associated with chronic liver disease or portal-systemic shunting. Thus, the reductions of brain glutamate consistently observed in experimental and human hepatic encephalopathy are probably not due (even in part) to loss of GABA neurons (ie: loss of the precursor pool of glutamate).

Excitatory amino acid release

A useful way to evaluate the size of the releasable pool of a neurotransmitter is to measure the amount of that substance released in a Ca^{2+}-dependent manner under depolarizing conditions. A popular preparation for such studies is the hippocampal slice. As part of a series of studies to assess the role of glutamatergic synaptic dysfunction in hepatic encephalopathy, we recently completed a series of studies of the effects of ammonia on K^+-stimulated, Ca^{2+}-dependent release of glutamate from rat hippocampal slices. Results are summarized in Figure 2.

As can be seen from the results summarized in Figure 2, exposure of hippocampal slices to pathophysiological concentrations of ammonia resulted in a small but significant increase in Ca^{2+}-dependent release of glutamate. It had previously been suggested by several investigators, including ourselves, that ammonia could affect glutamate synthesis in glutamate nerve terminals by inhibition of glutaminase (see Figure 1) (Hamberger et al., 1979; Theoret and Bossu, 1984; Butterworth et al., 1987). The present findings, however, do not support such a conclusion. The issue of whether or not ammonia exerts its neurotoxicity via

Figure 2

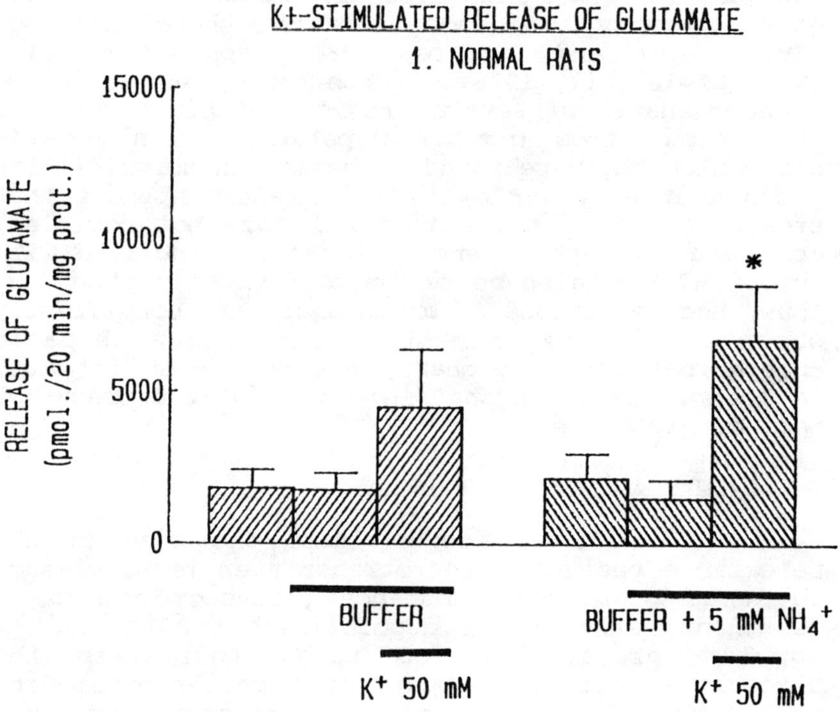

*p<0.05 compared to buffer-perfused values

inhibition of neurotransmitter glutamate synthesis is addressed in a separated chapter by Szerb in this volume. A previous study (Hamberger et al., 1979) reported decreased evoked release of glutamate following exposure of hippocampal slices to 3 mM NH_4^+. The reasons for these discrepancies are unclear at the present time.

In contrast to the lack of effects of ammonia exposure on evoked release of glutamate in hippocampal slices in vitro, results of preliminary studies reveal that slices from rats four weeks after portacaval anastomosis showed increased evoked-release of glutamate compared to sham-operated controls (compare values in Figures 2 and 3).

Figure 3

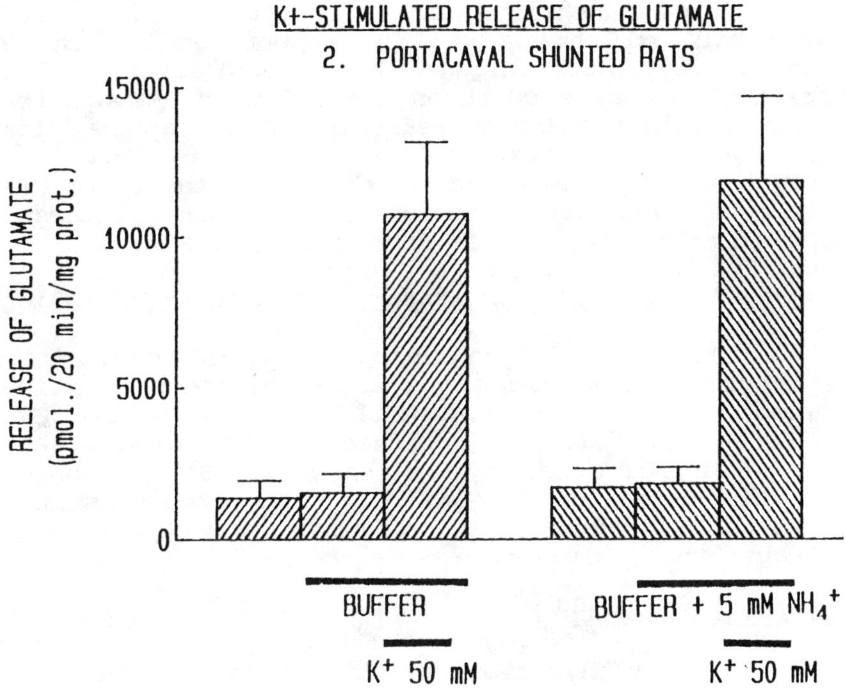

There are 2 possible explanations for these findings:

(i) Portacaval anastomosis results in increased synthesis of neurotransmitter glutamate.

(ii) Portacaval anastomosis results in decreased reuptake of released glutamate.

Whereas there is no evidence to date for possibility (i), there is substantial evidence that the effect may relate to reduced glutamate reuptake. As already mentioned, portacaval shunting results in astrocytic changes. In particular, glutamine synthetase and the capacity for ammonia removal is reduced (Butterworth et al., 1988). Since glutamatergic synaptic regulation relies to a significant degree on perineuronal astrocytes for reuptake of released glutamate (Figure 1), loss of astrocytic integrity could result in increased synaptic glutamate. In support of such a possibility, Norenberg et al. (1985) showed that primary astrocyte cultures treated with ammonia for 4 days exhibit a 35% decrease in uptake capacity for glutamate. Furthermore, an article by Mena and Cotman (1985) demonstrated that pathophysiological concentrations of ammonium ions block glutamate uptake into synaptosomes from rat forebrain.

Ammonia's effect on excitatory synaptic transmission

Ammonia intoxication impairs excitatory synaptic transmission in the rat hippocampal slice (Theoret et al., 1985) as well as in hippocampal CA_1 and CA_3 pyramidal cells (Theoret and Bossu, 1985) and in spinal cord *in vivo* (Raabe, 1988). Ammonium ion blocks the conduction of action potentials into presynaptic excitatory afferent nerve terminals (Raabe, this volume).

Glutamate Receptors

Electrophysiological studies suggest that the action of glutamate is mediated by multiple receptors. Using a quantitative autoradiographic approach, Monaghan et al. (1983) described the anatomical distribution in rat brain of the Kainate (KA), quisqualate (QA) and N-Methyl-D-Aspartate (NMDA) subclasses of glutamate receptors, so-named for their pharmacological sensitivity to displacement by kainic acid, quisqualic acid and N-Methyl-D-Aspartic acid respectively. Previous studies using *in vitro* ligand binding techniques revealed small but significant increases of total ^{3}H-glutamate binding to brain membrane preparations from rats following portacaval anastomosis (Butterworth et al., 1986) and from rabbits with fulminant hepatic failure following galactosamine administration (Ferenci et al., 1984). In view of these reports, and of the reported decrease of glutamate content and release from cerebral cortex or portacaval shunted rats (Butterworth and Giguère, 1986 and data in Figure 3), we studied by quantitative autoradiography, the densities of the KA, QA and NMDA subclasses of glutamate receptors in the brains of these animals. Densities of KA and QA sensitive glutamate receptors were similar in the brains of shunted and sham-operated rats. However, in the case of the NMDA subclass, densities of receptors were consistently found to be decreased by 20-40% in all brain structures investigated (Figure 4).

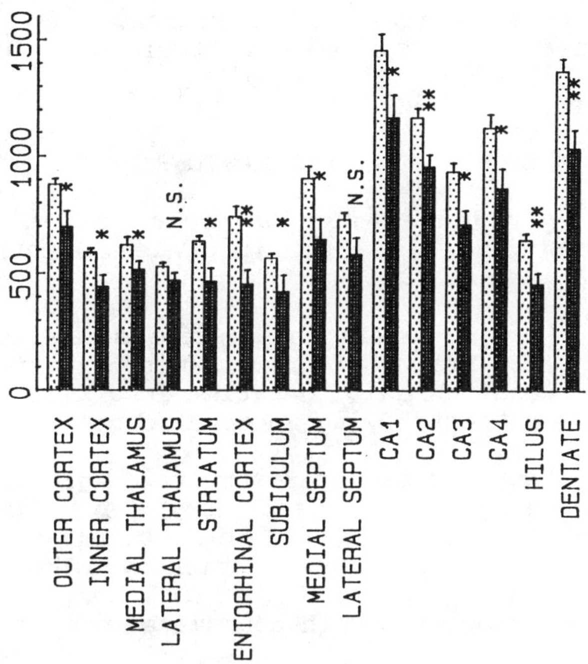

Figure 4: Densities of NMDA-displaceable ^{3}H-glutamate receptors in sham-operated ▭ and shunted ▦ rats.
*$p<0.05$, **$p<0.01$ compared to sham.

One possible explanation for the decrease of NMDA receptors in portacaval-shunted rat brain could relate to an up-regulation phenomenon resulting from increased extracellular glutamate or aspartate. Evoked release of glutamate from hippocampal slices from portacaval shunted rats is increased (Figure 3) and increases of extracellular aspartate have been reported in striatum of portacaval shunted rats using an in vivo cerebral dialysis technique (Tossman et al., 1987). It has been suggested that aspartate's synaptic action may be mediated by NMDA receptors (Cotman and Monaghan, 1987). Further studies are required in order to evaluate these possibilities.

CONCLUSIONS

Evidence for a role of glutamatergic dysfunction in hepatic encephalopathy may be summarized as follows:

(i) Brain levels of glutamate are decreased in the brains of animals with experimental acute and chronic hepatic insufficiency and in patients dying in hepatic coma.

(ii) Activities of astrocytic glutamate-related enzymes (GS and GDH) are decreased in experimental and clinical hepatic encephalopathy. Immunocytochemical studies show that GDH is enriched in astrocytes associated with glutamate nerve terminals.

(iii) Preliminary studies suggest that evoked release of glutamate from hippocampus of portacaval shunted rats is significantly increased.

(iv) Portacaval shunting leads to a generalized reduction in densities of the NMDA subclass of glutamate receptors in several brain structures.

REFERENCES

Aoki, C., Milner, T.A., Sheu, K.F.R., Blass, J.P. and Pickel, V.M. (1987). Regional distribution of astrocytes with intense immunoreactivity for glutamate dehydrogenase in rat brain: implications for neuron-glia interactions in glutamate transmission. J. Neurosci., 7, 2214-2231.

Baraldi, M., Zeneroli, M.L., Ventura, E. et al. (1984). Portal-systemic encephalopathy in dogs: changes in brain GABA receptors and neurochemical correlates. Adv. Hep. Enceph. and Urea Cycle Dis. (G. Kleinberger, P. Ferenci, P. Riederer and H. Thaler, eds), pp. 353-359, Karger, Basel.

Butterworth, R.F. and Giguère, J.F. (1986). Cerebral amino acids in portal-systemic encephalopathy: lack of evidence for altered -aminobutyric acid (GABA) function. Metab. Brain Dis., 1, 221-228.

Butterworth, R.F., Giguère, J.F., Samii, A. and Bergeron, M. (1986). Regional glutamate changes in experimental hepatic encephalopathy. Trans. Am. Soc. Neurochem., 17, 242.

Butterworth, R.F., Lavoie, J., Giguère, J.F., Pomier Layrargues, G. and Bergeron, M. (1987). Cerebral GABAergic and glutamatergic functions in hepatic encephalopathy. Neurochem. Pathol., 6, 131-144.

Butterworth, R.F., Girard, G. and Giguère, J.F. (1988). Regional differences in the capacity for ammonia removal by brain following portacaval anastomosis. J. Neurochem., 51, 486-490.

Cooper, A.J.L., Mora, S.N., Cruz, N.F. and Gelbard, A.S. (1985). Cerebral ammonia metabolism in hyperammonemic rats. J. Neurochem., 44, 1716-1723.

Cooper, A.J.L. and Plum, F. (1987). Biochemistry and physiology of brain ammonia. Physiol. Rev., 67, 440-519.

Cotman, C.W. and Monaghan, D.T. (1987). Organization of excitatory amino acid receptors: functional implications. In "Excitatory Amino Acid Transmission" (T.P. Hicks, D. Lodge, H. McLennan, eds), pp. 325-332, Alan R. Liss Inc., New York.

Cremer, J.E., Heath, D.F., Teal, H.M., Woods, M.S. and Cavanagh, J.B. (1975). Some dynamic aspects of brain metabolism in rats given a portocaval anastomosis. Neuropathol. and Appl. Neurobiol., 3, 293-311.

Ferenci, P., Pappas, S.C., Munson, P.J. and Jones, E.A. (1984). Changes in glutamate receptors on synaptic membranes associated with hepatic encephalopathy or hyperammonemia in the rabbit. Hepatol., 4, 25-27.

Fonnum, F. (1985). Determination of transmitter amino acid turnover. In "Neuromethods" (A.A. Boulton, G.B. Baker and J.D. Wood, eds), pp. 201-237, Humana Press, Clifton, N.J.

Hamberger, A., Hedquist, B. and Nystrom, B. (1979). Ammonium ion inhibition of evoked release of endogenous glutamate from hippocampal slices. J. Neurochem., 33, 1295-1302.

Hindfelt, B., Plum, F. and Duffy, T.E. (1977). Effect of acute ammonia intoxication on cerebral metabolism in rats with portacaval shunts. J. Clin. Invest., 59, 386-396.

Holmin, T., Agardh, C.D., Alinder, G., Herlin, P. and Hultberg, B. (1983). The influence of total hepatectomy on cerebral energy state, ammonia-related amino acids of the brain and plasma amino acids in the rat. Eur. J. Clin. Invest., 13, 215-220.

Holmin, T. and Siesjo, B.K. (1974). The effect of porta-caval anastomosis upon the energy state and upon acid-base parameters of the rat brain. J. Neurochem., 22, 403-412.

Lavoie, J., Giguère, J.F., Pomier Layrargues, G. and Butterworth, R.F. (1987a). Amino acid changes in autopsied brain tissue from cirrhotic patients with hepatic encephalopathy. J. Neurochem., 49, 692-697.

Lavoie, J., Giguere, J.F., Pomier Layrargues, G. and Butterworth, R.F. (1987b). Activities of neuronal and astrocytic marker enzymes in autopsied brain tissue from patients with hepatic encephalopathy. Metab. Brain Dis., 2, 283-290.

Lin, S. and Raabe, W. (1985). Ammonia intoxication: effects on cerebral cortex and spinal cord. J. Neurochem., 44, 1252-1258.

Maddison, J.E., Dodd, P.R., Morrison, M. and Johnson, G.A.R. (1987). Plasma GABA, GABA-like activity and the brain GABA-benzodiazepine receptor complex in rats with chronic encephalopathy. Hepatol., 7, 621-628.

Mena, E.E. and Cotman, C.W. (1985). Pathologic concentrations of ammonium ions block L-glutamate uptake. Exp. Neurol., 89, 259-263.

Monaghan, D.T., Holets, V.R., Toy, D.W. and Cotman, C.W. (1983). Anatomical distributions of four pharmacologically distinct ^{3}H-L-glutamate binding sites. Nature, 306, 176-179.

Moroni, F., Lombardi, G., Moneti, G. and Cortesini, C. (1983). The release and neosynthesis of glutamic acid are increased in experimental models of hepatic encephalopathy. J. Neurochem., 40, 850-854.

Norenberg, M.D. (1979). The distribution of glutamine synthetase in the rat central nervous system. J. Histochem. Cytochem., 27, 756-762.

Norenberg, M.D., Mozes, L.W., Papendick, R.E. and Norenberg, L.O.B. (1985). Effect of ammonia on glutamate, GABA and rubidium uptake by astrocytes. Ann. Neurol., 18, 149.

Potashner, S.J. (1978). Spontaneous and electrically-evoked release from guinea-pig cortex of endogenous amino acids labelled via metabolism of U-^{14}C-glucose. J. Neurochem. 31, 177-186.

Record, C.O., Buxton, B., Chase, R.A., Curzon, G., Murray-Lyon, I.M. and Williams, R. (1976). Plasma and brain amino acids in fulminant hepatic failure and their relationship to hepatic encephalopathy. Eur. J. Clin. Invest., 6, 387-394.

Roy, S., Pomier Layrargues, G., Butterworth, R.F. and Huet, P.M. (1988). Hepatic encephalopathy in cirrhotic and portacaval shunted dogs: lack of changes in brain GABA-uptake, brain GABA levels and brain GAD activity. Hepatol., 8, 845-849.

Storm-Mathisen, J. and Ottersen, O.P. (1983). Immunohistochemistry of glutamate and GABA. In "Glutamine, Glutamate and GABA in the Central Nervous System" (L. Hertz, E. Kvamne, E.G. McGeer and A. Schousboe, eds), pp. 185-201, Alan Liss, New York.

Szerb, J.C. (1988). Changes in the relative amounts of aspartate and glutamate released and retained in hippocampal slices during stimulation. J. Neurochem., 50, 219-224.

Theoret, Y. and Bossu, J.L. (1985). Effects of ammonium salts on synaptic transmission to hippocampal CA1 and CA3 pyramidal cells in vivo. Neurosci, 14, 807-821.

Theoret, Y., Davies, M.F., Esplin, B. and Capek, R. (1985). Effects of ammonium chloride on synaptic transmission in the rat hippocampal slice. Neurosci. 14, 798-806.

Tossman, U., Eriksson, S., Delin, A., Hagenfeldt, L., Law D. and Ungerstedt, U. (1983). Brain amino acids measured by intracerebral dialysis in portacaval shunted rats. J. Neurochem. 41 1046-1051.

Tyce, G.M., Flock, E.V. and Owen, C.A. (1971). Metabolism of glucose in brain after hepatectomy. Exp. Biol. Med., 4, 92-103.

Wenthold, R.J., Altschuler, R.A., Skaggs, K.K. and Reeks, K.A. 91987). Immunocytochemical characterization of glutamate dehydrogenase in the cerebellum of the rat. J. Neurochem., 48, 636-643.

ACTIVITIES OF GLUTAMATE-RELATED ENZYMES IN AUTOPSIED BRAIN TISSUE FROM CIRRHOTIC PATIENTS WITH HEPATIC ENCEPHALOPATHY

Jean-François Giguère, Anne-Marie Besnard, Joël Lavoie, Gilles Pomier Layrargues and Roger F. Butterworth, Lab. of neurochem., André-Viallet Clin. Res. Ctre, Hôpital St-Luc (U. of Montreal), Montréal, Québec, Canada H2X 3J4.

SUMMARY

Activities of glutamic acid decarboxylase (GAD), glutamine synthetase (GS), glutamate dehydrogenase (GDH) and α-ketoglutarate dehydrogenase (αKGDH) were measured in autopsied brain tissue from nine cirrhotic patients dying in hepatic coma and an equal number of appropriately-matched control patients who were free from hepatic, neurological or psychiatric diseases in all cases. Activities of GAD were unchanged in cirrhotic brain tissue suggesting no loss of integrity of presynaptic GABA nerve terminals in this disease. Similarly, activities of the mitochondrial enzyme αKGDH were within normal limits in brain tissue from cirrhotic patients. On the other hand GS and GDH activities were selectively decreased by 20-30% in caudate nuclei of these patients. Previous immunocytochemical studies suggest that both GS and GDH are enriched in astrocytes. Thus, loss of these enzymes is consistent with the marked Alzheimer type II astrocytosis characteristic of hepatic encephalopathy resulting from chronic liver disease and shown to be particularly severe in caudate nucleus. Selective loss of glial GDH activity in these brains could result in abnormalities of glutamatergic function in this disease.

INTRODUCTION

Studies of the activities of specific neuronal and astrocytic enzymes in autopsied brain tissue have proven useful for the understanding of pathophysiological mechanisms involved in numerous neurological and psychiatric disorders including Hungtington's Disease (Bird and Iversen, 1974) and Alzheimer's Disease (Perry et al., 1977). There is evidence derived mainly from studies with animal models, to suggest that hepatic encephalopathy may result from abnormalities of brain glutamate. Glutamate concentrations in brain are reduced following portacaval shunting in rats (Hindfelt et al., 1977; Giguère and Butterworth, 1984) as well as in autopsied brain tissue from cirrhotic patients with hepatic encephalopathy (Lavoie et al., 1987a). As part of a series of studies to evaluate the role of glutamate in the pathogenesis of hepatic encephalopathy, we have measured the activities of four enzymes involved in glutamate metabolism by brain. The enzymes studied were the following:

(i) α-ketoglutarate dehydrogenase (αKGDH), a mitochondrial, tricarboxylic acid cycle enzyme responsible for the transformation of α-ketoglutarate into succinyl CoA.

(ii) glutamate dehydrogenase (GDH), an enzyme enriched in astrocytes involved in the direct synthesis of glutamate from α-ketoglutarate.

(iii) glutamate decarboxylase (GAD), the enzyme (localized in GABA nerve terminals) responsible for GABA synthesis from glutamate.

(iv) glutamine synthetase (GS), an enzyme highly enriched (and possibly uniquely localized) in astrocytes, responsible for ammonia removal by brain and concomitant glutamine synthesis.

The metabolic inter-relationships between these enzymes are shown schematically in Figure 1.

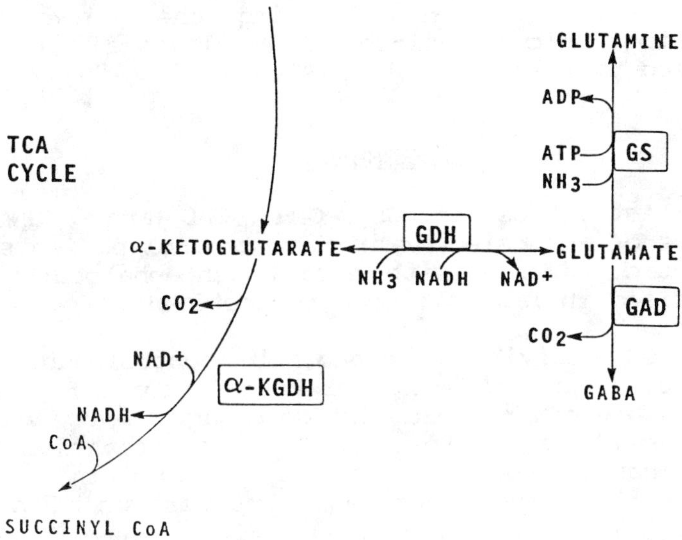

Figure 1. Enzymes involved in cerebral glutamate metabolism.

METHODS

Nine brains from cirrhotic patients dying in hepatic coma and nine control brains were obtained at autopsy. Subject ages, agonal status and postmortem delay times were rigorously matched and have been previously described (Lavoie et al., 1987a and accompanying paper by Bergeron et al., this volume). Dissected samples of prefrontal cortex, frontal cortex and caudate nuclei were obtained and were stored at -70°C until time of assay.

αKGDH activities were measured in tissue homogenates as previously described by Butterworth et al. (1986). GDH activities were measured spectrophotometrically by measuring the changes in optical density of NADH to NAD+ at 340 nm as described by Filla et al. (1986). GAD, was assayed in brain homogenates by a modified radiometric method whereby $^{14}CO_2$ from [1-^{14}C]-glutamate was monitored as described by Lavoie et al., 1987b. GS activities were measured using the ion-exchange chromatographic technique of Patel et al. 1983 as modified by Lavoie et al. 1987b.

RESULTS

Activities of GS, GDH, GAD and αKGDH in regions of autopsied brain from control patients and cirrhotic patients with hepatic encephalopathy are summarized in Tables 1 to 4 respectively.

GS activities ranged, in control material, from 0.77 μmole.h^{-1}.mg $protein^{-1}$ to 1.57 μmole h^{-1}.mg $protein^{-1}$ were in good agreement with a previous study (Carter, 1982). GS activities were significantly reduced (by 25%, $p<0.01$) in caudate nucleus of cirrhotic patients but were unchanged in cortical structures (Table 1).

GDH activities were found to be in good general agreement with those of a previous study (Filla et al., 1986). Enzyme activities in caudate nucleus were, as for GS, significantly reduced in autopsied brain tissue from cirrhotic patients (Table 2).

In contrast to GS and GDH, activities of GAD and αKGDH were similar in autopsied brain tissue from controls and cirrhotic patients (Tables 3 and 4).

Table 1

Glutamine Synthetase Activities in Autopsied Brain Tissue

Brain region	Controls (n)	Cirrhotics (n)
Prefrontal cortex	1.57±0.14+ (7)	1.49±0.11 (7)
Frontal cortex	1.08±0.11 (9)	1.02±0.12 (8)
Caudate nucleus	0.77±0.06 (9)	0.58±0.04 (9)*

+ (μmole h^{-1}.mg $protein^{-1}$)

*$p<0.01$ compared to control group

Table 2

Glutamate Dehydrogenase Activities in Autopsied Brain Tissue

Brain region	Controls (n)	Cirrhotics (n)
Prefrontal cortex	246.3±17.2 (7)+	233.3±14.0 (7)
Caudate nucleus	261.7±13.2 (8)	214.3±16.5 (9)*

+ (nmole.min^{-1}.mg·$protein^{-1}$)

*$p<0.02$ compared to control group

Table 3

Glutamic Acid Decarboxylase Activities in Autopsied Brain Tissue

Brain region	Controls (n)	Cirrhotics (n)
Prefrontal cortex	22.0±3.3^{+} (6)	19.4±5.8 (7)
Frontal cortex	18.4±3.4 (6)	15.3±3.6 (9)
Caudate nucleus	19.5±4.6 (6)	21.7±5.3 (9)

$^{+}$ nmol.h^{-1}.mg protein^{-1} NS

Table 4

α-Ketoglutarate Dehydrogenase Activities in Autopsied Brain Tissue

Brain region	Controls (n)	Cirrhotics (n)
Prefrontal cortex	2.32±0.41^{+} (7)	2.54±0.26 (7)
Frontal cortex	2.58±0.41 (9)	2.51±0.39 (9)
Caudate nucleus	2.56±0.33 (9)	2.47±0.38 (9)

$^{+}$ nmol.min^{-1}.mg protein^{-1} NS

DISCUSSION

Activities of the GABA-nerve terminal marker enzyme, GAD, were not found to be significantly altered in any of the three brain regions of cirrhotic patients with hepatic encephalopathy suggesting that presynaptic GABA neuronal integrity is maintained in this condition. GABA concentrations in the same material were previously shown to be within normal limits (Lavoie et al., 1987a) as were affinities and densities of high affinity GABA binding sites (Butterworth et al., 1988b). Taken together, these findings do not support the hypothesis that hepatic encephalopathy results from abnormalities of pre or postsynaptic cerebral GABA function as previously suggested (Schafer and Jones, 1982).

αKGDH activities were found to be within the normal reported range in brain tissue from cirrhotic patients. αKGDH is a thiamine-dependent mitochondrial enzyme and abnormalities in its activity in brain have been implicated in the pathogenesis of the Wernicke-Korsakoff Syndrome (Butterworth, 1986, 1989). All of the cirrhotic patients in the present study were alcoholics. It is well established that chronic alcoholism results in thiamine deficiency in man and thiamine deficiency results, in experimental animals, in reduced brain levels of αKGDH (Butterworth et al., 1986). The finding of normal activities of αKGDH in the brains of the cirrhotic patients in the present study, therefore, suggests that these patients were, surprisingly, not thiamine deficient.

The enzymes GDH and GS are responsible for ammonia removal by brain, with GS being, quantitatively, the more important of the two in this regard (Cooper and Plum, 1987). GS is highly enriched in astrocytes; so much so that it is often referred to as an "astrocytic marker enzyme". GS activities were significantly reduced by 25% in caudate nuclei of cirrhotic patients (Table 1). Neuropathological studies of the material used in the present study reveal severe astrocytic degenerative changes in caudate nuclei (Alzheimer type II astrocytosis) (Butterworth et al., 1987a). Small but significant reductions of GS have also been reported in the brains of rats following portacaval anastomosis (Butterworth et al., 1988a). Inhibition of GS results in depletion of the neurotransmitter pool of glutamate (Rothstein and Tabakoff, 1984) and it has been suggested that loss of astrocytic integrity may result in impaired glutamatergic function in hepatic encephalopathy (Butterworth et al., 1987b and chapter by Butterworth et al., this volume).

GDH activities are similarly decreased in caudate nuclei of cirrhotic patients dying in hepatic coma (Table 2). These findings are in agreement with a previous study in which GDH activities in brains of rats with subacute CCl_4-induced liver injury were found to be significantly decreased (Kisfaludy, 1975). The finding of decreased GDH in the brain in clinical and experimental hepatic encephalopathy may itself, also result in abnormalities of glutamatergic function. In conditions of sustained hyperammonemia, such as that resulting from chronic liver disease, GDH assumes a relatively greater importance for removal of ammonia from brain (Cooper and Plum, 1987). Furthermore, recent immunocytochemical studies have shown that GDH is selectively enriched in glial cells that are associated with glutamate nerve terminals (Aoki et al., 1987; Wenthold et al., 1987). These findings suggest that GDH (together with GS) is responsible for the metabolism of glutamate in astrocytes after its reuptake from the synaptic cleft (Aoki et al., 1987). Loss of astrocytic integrity and the accompanying loss of GS and GDH may thus be responsible for the glutamate changes observed in clinical and experimental hepatic encephalopathy (see chapter by Butterworth et al., this volume).

REFERENCES

Aoki C., Milner T.A., Sheu K.F.R., Blass J.P. and Pickel V.M. (1987). Regional distribution of astrocytes with intense immunoreactivity for glutamate dehydrogenase in rat brain: implications for neuron-glia interactions in glutamate transmission. J. Neurosci. 7, 2214-2231.

Bird E.D. and Iversen L.L. (1974). Huntington's chorea: postmortem measurement of glutamic acid decarboxylase, choline acetyltransferase and dopamine in basal ganglia. Brain, 97, 457-472.

Butterworth R.F. (1986). Cerebral thiamine-dependent enzyme changes in experimental Wernicke's Encephalopathy. Metab. Brain Dis. vol. 1, no. 3, 167-175.

Butterworth R.F., Giguère J.F. and Besnard A.M. (1986). Actitivities of thiamino-dependent enzyme changes in two experimental models of thiamine-deficiency encephalopathy. 2. α-ketoglutarate dehydrogenase. Neurochem. Res. vol. 11, no. 4, 567-577.

Butterworth R.F., Giguère J.F., Michaud J., Lavoie J.and Pomier Layrargues, G. (1987a). Ammonia: key factor in the pathogenesis of hepatic encephalopathy. Neurochem. Pathol. 6, 1-12.

Butterworth R.F., Lavoie J., Giguère J.F., Pomier Layrargues G. and Bergeron M. (1987b). Cerebral GABAergic and glutamatergic function in hepatic encephalopathy. Neurochem. Pathol. 6, 131-144.

Butterworth R.F., Girard G. and Giguère J.F. (1988a). Regional differences in the capacity for ammonia removal following portacaval anastomosis in the rat. J. Neurochem. 51, 486-490.

Butterworth R.F., Lavoie J., Giguère, J.F. and Pomier Layrargues G. (1988b). Affinities and densities of high-affinity [^{3}H]-muscimol (GABA-A) binding sites and of central benzodiazepine receptors are unchanged in autopsied brain tissue from cirrhotic patients with hepatic encephalopathy. Hepatology, 8, 1084-1088.

Butterworth R.F. (1989). Effects of thiamine deficiency on brain metabolism: implications for the pathogenesis of the Wernicke-Korsakoff Syndrome. Alcohol & Alcoholism, (in press).

Carter C.J. (1982). Loss of glutamine synthetase activity in the brain in Hungtington's Disease. The Lancet, 1, 782-783.

Cooper A.J.L. and Plum F. (1987). Biochemistry and physiology of brain ammonia. Physiol. Rev. 67, 440-519.

Filla A., De Michele G., Brescia Morra V., Palma V., Di Lauro A., Di Geronimo G. and Campanella G. (1986). Glutamate dehydrogenase: human brain in regional distribution and properties. J. Neurochem, 46, 422-424.

Giguère J.F. and Butterworth R.F. (1984). Amino acid changes in relation to function in experimental portal-systemic encephalopathy. Neurochem. Res. vol. 9, no. 9, 1309-1321.

Hindfelt B., Plum F. and Duffy T.E. (1977). Effect of acute ammonia intoxication on cerebral metabolism in rats with portocaval shunts. J. Clin. Invest. 59, 386-396.

Kisfaludy S. (1975). Effects of ammonia intoxication and liver injury on rat brain glutamate dehydrogenase and glutaminase activity. Acta Medica Acad. Sci. Hung. 32 279-287.

Lavoie J., Giguère J.F., Pomier Layrargues G. and Butterworth R.F. (1987a). Amino acid changes in autopsied brain tissue from cirrhotic patients with hepatic encephalopathy. J. Neurochem. 49, 692-697.

Lavoie J., Giguère J.F., Pomier Layrargues G. and Butterworth, R.F. (1987b). Activities of neuronal and astrocytic marker enzymes in autopsied brain tissue from patients with hepatic encephalopathy. Metabolic Brain Disease, 2, 283-290.

Patel A.J., Hunt A. and Tahourdin C.S.M. (1983). Regional development of glutamine synthetase activity in the rat brain and its association with the differentiation of astrocytes. Dev. Brain Res. 8, 31-37.

Perry E.K., Gibson P.H., Blessed Cr., Perry R.H. and Tomlinson B.E. (1977). Neurotransmitter enzyme abnormalities in senile dementia. J. Neurol. Sci. 34, 247-265.

Rothstein J.D. and Tabakoff B. (1984). Alteration of striatal glutamate release after glutamine synthetase inhibition. J. Neurochem 43, 1438-1446.

Schafer D.F. and Jones E.A. (1982). Hepatic encephalopathy and the γ-aminobutyric acid neurotransmitter system. The Lancet 1, 18-20.

Wenthold R.J., Altschuler R.A., Skaggs K.K. and Reeks K.A. (1987). Immunocytochemical characterization of glutamate dehydrogenase in the cerebellum of the rat. J. Neurochem. 48, 636-643.

DOES AMMONIA EXERT ITS NEUROTOXICITY THROUGH AN INHIBITION OF TRANSMITTER GLUTAMATE SYNTHESIS?

John C. Szerb

Department of Physiology and Biophysics

Dalhousie University, Halifax, N.S.
B3H 4H7, Canada.

INTRODUCTION

The central role of hyperammonemia in hepatic encephalopathy is now well recognized. In recent years, a number of excellent reviews have appeared on the metabolic effects of ammonia (Benjamin,1982; Kvamme,1983a; Cooper and Plum,1987). The purpose of this review is to examine one possible link between the metabolic and neurological effects of ammonia, namely the inhibition of glutaminase by ammonia. Glutaminase, mostly located in nerve terminals (Bradford and Ward, 1976), converts glutamine to glutamate and ammonia. There is extensive evidence to show that glutamine is synthesized in astroglia (Martinez-Hernandez et al., 1977) and then taken up by terminals where it is the major source of transmitter glutamate (Hamberger et al., 1979a; Szerb and O'Regan, 1985). Therefore, ammonia, by inhibiting glutaminase, can decrease the synthesis and release of the excitatory transmitter glutamate. The resulting impaired excitatory transmission could explain symptoms of CNS depression seen in hepatic encephalopathy. This mechanism of ammonia neurotoxicity, first proposed by Bradford and Ward (1975), has since been reaffirmed by a number of investigators (Benjamin 1981; Hamberger et al.,1981; Théoret et al.,1985).

To accept the validity of this hypothesis, it is necessary to establish the existence of such an inhibition, not only at the level of the isolated enzyme, but also under physiological conditions in nerve terminals, where glutaminase is compartmentalized and its activity regulated. Furthermore, since glutamate has both metabolic and transmitter functions (Fonnum,1984), the inhibition by ammonia of the replenishment of only the transmitter pool represents the necessary evidence.

DIFFERENCES IN THE SENSITIVITY OF DIFFERENT GLUTAMINASES TO AMMONIA INHIBITION

As reviewed by Kvamme (1983b), the predominant form of glutaminase in the CNS is the phosphate-activated glutaminase or PAG. The sensitivity to inhibitors of purified and membrane-bound PAG differs: while both forms are inhibited by one of the product of PAG, glutamate, the other product, ammonia, inhibits only the membrane-bound form, such as occurs in synaptosomes (Bradford and Ward,1976; Kvamme and Lenda,1982) and in brain homogenates (Benjamin,1981). In contrast to neuronal PAG, that in astrocytes in not inhibited by ammonia (Kvamme et al.,1982). However, most of the PAG in the central nervous system is localized in synaptosomal mitochondria (Bradford and Ward, 1976).

KINETICS OF INHIBITION OF GLUTAMINASE BY ITS PRODUCTS

In synaptosomes, glutaminase is found in two mitochondrial compartments: on the outer surface of the inner mitochondrial membrane and inside the inner mitochondrial membrane (Kvamme and Olsen, 1981). As a result of this compartmentation of glutaminase within mitochondria, the enzyme may be inhibited due to the accumulation of its products within this compartment. Since ammonia diffuses freely, glutamate is the likely product that accumulates and inhibits glutaminase activity in synaptosomes. This inhibition is seen after a delay, indicating that not the substrate of the enzyme, but its products inhibit the enzyme (Kvamme and Olsen, 1980). This product inhibition is stronger in the presence of 5 than 0.5 mM glutamine because more glutamate is formed from a high than a low concentration of glutamine (Kvamme and Olsen, 1980). The two studies reporting on the kinetics of inhibition by glutamate and ammonia of glutaminase activity (Benjamin, 1981; Kvamme and Lenda, 1982), gave essentially the same results, namely that ammonia depresses glutaminase activity in synaptosomes by a maximum of about 30-40% in the presence of 2 mM glutamine. On the other hand, the maximal depression by glutamate, in the presence of 2 mM glutamine was only 50 % according to Kvamme and Lenda (1982), but 80 % according to Benjamin (1981). In addition, Benjamin (1981) reported a competitive interaction between glutamine and ammonia, which may suggest that, when glutaminase is partly inhibited by accumulated glutamate, ammonia is less effective in producing inhibition. A further observation by Kvamme and Lenda (1982) was that the K_i of ammonia in the presence of 2 mM glutamine is approximately 0.2 mM, about the normal concentration of ammonia in the rat brain (Cooper and Plum, 1987).

Although synaptosomes are clearly a relevant model for studying the effect of ammonia on transmitter glutamate synthesis and release, important pieces of information are missing to decide whether ammonia indeed inhibits the synthesis of this fraction of glutamate. It is probable that in vivo and in more intact preparations, such as brain slices, the activity of glutaminase is regulated by its product, glutamate. This is shown by the increased synthesis of glutamate from glutamine, whenever depolarization in the presence of calcium causes an increased release of glutamate (Szerb, 1984). The increased synthesis can be suppressed by adding glutamate to the medium (Hamberger et al., 1979b). To make observations on synaptosomes relevant to the in vivo situation, it would be therefore necessary to show that ammonia inhibits glutaminase, not when it is maximally active, but also when its activity is partly or fully regulated by glutamate. Such a physiological condition can be demonstrated by showing an increase in synaptosomal glutaminase activity due to depolarization in the presence of calcium and an inhibition of this increase by ammonia.

EFFECT OF AMMONIA ON GLUTAMATE RELEASE FROM BRAIN SLICES

Brain slices have been a suitable model for both neurochemical and electrophysiological phenomena seen in vivo, but with the advantage that, unlike in vivo, the environment of the tissue can be controlled. Using slices of hippocampus, Hamberger et al. (1979c,1982) investigated the effects of 1-5 mM ammonium ions on the spontaneous and high potassium-induced release of glutamate. Surprisingly, they observed an increase in the spontaneous release of glutamate in the presence of ammonia, but a decrease in the high potassium-induced release. No details were provided on the calcium-dependence of the ammonia- and high K^+-induced release and therefore one cannot ascribe the observed changes in release to the transmitter pool of glutamate. To the contrary, their observation that ammonia severely depressed the incorporation of [^{14}C]acetate, makes it likely that ammonia affected primarily the viability of astroglia in the slices because mostly glia take up and utilize acetate (Muir et al., 1986). Paradoxically, brain slices from portacaval-shunted rats, showed a resistance to the effects of ammonia (Hamberger et al., 1980), although shunted rats develop an increased sensitivity to an ammonia load (Giguère and Butterworth, 1984).

In contrast to the above observations by Hamberger et al. on the effects of ammonia on release induced by a high concentration of potassium, the release of glutamate from slices of hippocampus induced by electrical stimulation was not affected at all by 5 mM ammonia (Butterworth et al., 1989, this volume). Nor had ammonia any effect on the electrically evoked release in slices prepared from porta-caval shunted rats. Since this release was almost totally suppressed by tetrodotoxin or in the presence of a reduced concentration of calcium, it must have come from nerve terminals that were depolarized by conducted action potentials. Therefore, these recent observations strongly suggest that ammonia does not interfere with the synthesis and release of transmitter glutamate.

EFFECT OF AMMONIA ON GLUTAMATE RELEASE IN VIVO

In agreement with these recent observations in brain slices on the absence of an effect of ammonium ions on synaptically released glutamate, the release of glutamate from the cortical surface in vivo is not decreased but rather increased by an acute injection of ammonium salts or in chronic portacaval shunted rats (Moroni al.,1983).

Neither was a decrease in the release of glutamate collected by microdialysis from the parietel cortex of portacaval shunted rats observed, and sham operated and shunted rats showed a similar increase in the amount of glutamate collected upon local depolarization with elevated potassium (Tossman et al., 1987). However, the efflux of a number of amino acids, other than glutamate, was greater in shunted than sham operated rats.

CONCLUSIONS

Based on data summarized here, it seems unlikely that ammonia interferes with synaptic transmission by inhibiting glutamate formation. These observations however, do not, by any means, rule out the possibility that ammonia interferes with synaptic transmission through other mechanisms. For instance, there is good evidence to indicate that ammonia, by inhibiting an outward chloride pump, eliminates the hyperpolarizing component of the IPSP, thus making inhibition less effective (Raabe, 1981; 1987). There is also evidence for a block in excitatory synaptic transmission in the rat hippocampus (Théoret et al.,1985; Théoret and Bossu, 1985). In addition, in the hippocampus, ammonia has a non-specific depressant effect on neuronal excitability and causes the release of potassium into extracellular spaces (Alger and Nicoll, 1983). These somewhat more subtle neurophysiological consequences of the metabolic effects of ammonium ions deserve now our attention.

REFERENCES

Alger B.E. and Nicoll R.A. (1983) Ammonia does not selectively block IPSPs in rat hippocampal pyramidal cells. J. Neurophysiol. **49**, 1381-1391.

Benjamin A.M. (1981) Control of glutaminase activity in rat brain cortex in vitro: influence of glutamate, phosphate, ammonium, calcium and hydrogen ions. Brain Res. **208**, 363-377.

Benjamin A.M. (1982) Ammonia. In: Handbook of Neurochemistry, 2nd Ed. (Lajtha A., ed.) Vol.1, pp.117-137, Plenum Press, New York.

Bradford H.F. and Ward H.K. (1975) Glutamine as a metabolic substrate for isolated nerve-endings:

inhibition by ammonium ions. Biochem.Soc. Trans. **3**,1223-1226.

Bradford H.F. and Ward H.K. (1976) On glutaminase activity in mammalian synaptosomes. Brain Res. **110**, 115-125.

Butterworth R.F., Lavoie J., Peterson C., Cotman C.W. and Szerb J.C. Excitatory amino acids in hepatic encephalopathy. In: Hepatic Encephalopathy: Pathophysiology and Treatment. (Butterworth R.F. and Pomier Layrargues G. eds.) Humana Press.

Cooper A.J.L. and Plum F. (1987) Biochemistry and physiology of brain ammonia. Physiol.Rev. **67**,440-519.

Fonnum F. (1984) Glutamate: a neurotransmitter in the mammalian brain. J.Neurochem. **31**, 1173-1181.

Giguère J.-F. and Butterworth R.F. (1984) Amino acid changes in regions of CNS in relation to function in experimental portal-systemic encephalopathy. Neurochem. Res. **9**, 1309- 1321.

Hamberger A., Chiang G.H., Nylen, E.S., Scheff S.W. and Cotman C.W. (1979a) Glutamate as a CNS transmitter I. Evaluation of glucose and glutamine as precursors for the synthesis of preferentially released glutamate. Brain Res. **168**, 531-541.

Hamberger A., Chiang G.H., Sandoval E. and Cotman C.W. (1979b) Glutamate as a CNS transmitter. II. Regulation of synthesis of the releasable pool. Brain Res. **168**, 531-541.

Hamberger A., Hedquist B., and Nyström B. (1979c) Ammonium ion inhibition of evoked release of endogenous glutamate from hippocampal slices. J.Neurochem. **33**, 1295-1302.

Hamberger A., Hedquist B., Lundborg H. and Nyström B. (1980) Hippocampal glutamate release after porta cava anastomosis: reduced sensitivity to ammonia inhibition. J. Neurosci. Res. **5**, 313-322.

Hamberger A., Jacobsson I., Molin S.-O., Nyström B. and Sandberg M. (1981) Regulation of glutamate biosynthesis and release by pathophysiological levels of ammonium ions. In: Glutamate as a Neurotransmitter (DiChiara G. and Gessa G.L., eds.) Adv. Biochem.Psychopharmac. **27**,115-126, Raven Press, N.Y.

Hamberger A., Lindroth P. and Nyström B. (1982) Regulation of glutamate biosynthesis and release in vitro by low levels of ammonium ions. Brain Res. **237**, 339-350.

Kvamme E. (1983a) Ammonia metabolism in the CNS. Progr. Neurobiol. **20**,109-132.

Kvamme E. (1983b) Glutaminase (PAG). In: Glutamine, Glutamate, and GABA (Hertz L., Kvamme E., McGeer E.G. and Schousboe A., eds.) pp. 51-67. A.R.Liss, New York.

Kvamme E. and Lenda K. (1982) Regulation of glutaminase by exogenous glutamate, ammonia and 2-oxoglutarate in synaptosomal enriched preparation from rat brain. Neurochem. Res. **7**, 667-678.

Kvamme E. and Olsen B.E. (1980) Substrate mediated regulation of phosphate-activated glutaminase in nervous tissue. Brain Res. **181**, 228-233.

Kvamme E., and Olsen B.E. (1981) Evidence for the compartmentation of synaptosomal phosphate-activated glutaminase. J. Neurochem. 36, 1916-1923.

Kvamme E., Svenneby G., Hertz L., and Schousboe A. (1982) Properties of phosphate activated glutaminase in astrocytes cultured from mouse brain. Neurochem. Res. **7**, 761-770.

Martinez-Hernandez A., Bell K.P. and Norenberg M.D. (1977) Glutamine synthetase: glial localization in the brain. Science **195**, 1356-1358.

Moroni F., Lombardi G., Moneti G. and Cortesini C. (1983) The release and neosynthesis of glutamic acid are increased in experimental models of hepatic encephalopathy. J.Neurochem. **40**, 850-854.

Muir D., Berl S. and Clarke D.D. (1986) Acetate and fluoroacetate as possible markers for glial metabolism in vivo. Brain Res. **380**, 336-340.

Raabe W.A. (1981) Ammonia and disinhibition in cat motor cortex by ammonium acetate, monofluoroacetate and insulin-induced hypoglycemia. Brain Res. **210**, 311-322.

Raabe W.A. (1987) Synaptic transmission in ammonia intoxication. Neurochem. Pathol. **6**,145-166.

Szerb J.C. (1984) Storage and release of endogenous and labelled GABA formed from [^{3}H]glutamine and [^{14}C]glucose in hippocampal slices: effect of depolarization. Brain Res. **293**, 293-303.

Szerb J.C. and O'Regan P.A. (1985) Effect of glutamine on glutamate release from hippocampal slices induced by high K^+ or by electrical stimulation: Interaction with different Ca^{2+} concentrations. J.Neurochem. **44**,

1724-1731.

Théoret Y. and Bossu J.-L. (1985) Effect of ammonium salts on synaptic transmission to hippocampal CA1 and CA3 pyramidal cells in vivo. Neuroscience **14**, 807-821.

Théoret Y., Davies M.F., Esplin B. and Capek R. (1985) Effects of ammonium chloride on synaptic transmission in the rat hippocampal slice. Neuroscience **14**, 798-806.

Tossman U., Delin A., Eriksson L.S. and Ungerstedt U. (1987) Brain cortical amino acids measured by intracerebral dialysis in portacaval shunted rats. Neurochem. Res. **12**, 265-269.

BETA-ENDORPHIN AND OPIATE RECEPTOR CHANGES IN ACUTE AND CHRONIC MODELS OF HEPATIC ENCEPHALOPATHY

M.L. Zeneroli, M. Baraldi*, E. Ventura,
P. Zanoli*, C. Vezzelli, A.M. Russo

Clinica Medica III e Cattedra di Farmacoligia e Farmacognosia*, Universita' di Modena, Italy

SUMMARY

In order to gain information on the mechanisms which induce a supersensitivity to morphine administration in patients with acute and chronic liver failure, the levels of the endogenous opioid beta-endorphin was studied in plasma and in two brain areas of rats and dogs at two different levels of encephalopathy. In the dog model of portal-systemic encephalopathy the binding characteristics of opiate receptors in hypothalamus and cortex were studied. Herein we report that during the development of encephalopathy there are changes in the levels of beta-endorphins and in the opiate receptor densities which taken overall might explain the altered responce to morphine administration in man.

INTRODUCTION

Patients with acute and chronic liver diseases are supersensitive to the administration of sedative-analgesic drugs. The injection of these drugs has been indicated as the second most common precipitating factor of episodes of encephalopathy occuring in liver cirrhosis (1). In this context it has been pointed out by Laidlaw et al. (2) that the administration of subthreshold doses of morphine can be used as a provocative test in the assessment of the early stages of hepatic encephalopathy. Although the molecular mechanisms of the morphine-precipitated episodes of hepatic coma were unknown, the electroencephalographic changes described in patients with liver disease treated with morphine have been attributed to an increased sensitivity of the brain (2).

Few studies have been performed in order to demonstrate a contribution of neuroendocrine alterations in the development of hepatic encephalopathy. Panerai et al (3) have described a decreased presence of beta-endorphins in the hypothalamus of rats with liver disease due to chronic carbon-tetrachloride administration and a reduced level of plasma beta-endorphin in patients with alcoholic liver cirrhosis (4). Changes in the presence of neurotransmitters or neuropeptides at the neuronal level usually affect the intrinsic characteristics of receptors. Binding studies performed on whole brain homogenates of rats in severe stage of galactosamine-induced hepatic encephalopathy showed a reduced presence of opiate receptors labelled by ^{3}H-Naloxone (5) while a similar study performed in a rabbit model of galactosamine-induced hepatic encephalopathy failed to find any change (6). The finding that the severe stage of encephalopathy due to fulminant hepatic failure in rats was associated with a loss of opiate receptors tested in a whole brain homogenate prompted us to study the levels of beta-endorphin in plasma and in two different brain areas of rats with fulminant hepatic failure during mild and severe stage of encephalopathy in order to find a correlation, if any, with the receptor changes. Moreover we decided to extend a previous study (7) performed on both beta-endorphin levels and opiate receptor characteristics in a mild stage of encephalopathy obtained in dogs treated with dimethylnitrosamine and subsequently submitted to porta-caval shunt.

MATERIAL AND METHODS

Hepatic encephalopathy was induced in Sprague-Dawley rats by intraperitoneal injection of galactosamine-HCl as previously described (8,9). Rats were sacrificed in mild and in severe stage of encephalopathy which were assessed by behavioral observation and visual evoked potential recordings.

The chronic model was obtained in dogs treating the animals with low doses of dimethylnitrosamine (DMNA) in order to acheive a light liver damage before to perform a portacaval shunt end-to-side as previously described (7,10). Dogs were sacrificed by application of a percussion gun to the cervical vertebrae in mild (N.6) and severe (N.3) stage of encephalopathy assessed by behavioral observation. In this model the use of visual evoked potential recordings was prevented both for the intrinsic difficulty to leave in place for long time permanently implanted electrodes and for the noises in the recordings due to interposition of thick muscles on the head whith electrodes positioned on the skin. In parallel 3 dogs treated only with DMNA and 3 dogs sham-operated were used as appropriate controls.

After decapitation the brains were rapidly removed and dissected according to the method of Glowinski and Iversen (11). Specimens of hypothalamus and frontal cortex were immediately placed in tubes containing 2 ml of boiling acetic acid for 15 min in order to inactivate the beta-endorphin cleaving. The tissues were than homogenized and centrifuged at 12,000 x g for 10 min. Beta-endorphins were assayed by radioimmunoassay in the supernatants as previously described (7).

The brain areas were then immediately frozen and kept at -20° C until used for binding studies which were performed as previously described (7).

Blood samples were collected in all animals for the assay of beta-endorphins in plasma.

RESULTS

As shown in Table 1, hepatic encephalopathy due to fulminant hepatic failure in rats is associated with no changes of circulating beta-endorphins in the mild stage and a significant increase in the severe stage. The concentrations of beta-endorphins tested in parallel in the hypothalamus and in the cortex during the mild and the severe stage of hepatic encephalopathy (Table 2) shows a different pattern in the two areas. In fact during the mild stage of encephalopathy the levels of beta-endorphins declined significantly in both areas. In the severe stage there was a recovery in the hypothalamus, which contains the cell bodies where this peptide is synthesised, while in the cortex beta-endorphins were found still decreased.

Beta-endorphin concentrations in serum of dogs in a mild stage of portal-systemic encephalopathy were found unchanged in comparison with controls but they were significantly increased in the severe stage (Table 3). The assay of beta-endorphin immunoreactivity performed in the

Table 1
BETA-ENDORPHIN CONCENTRATIONS IN PLASMA OF CONTROL RATS AND OF RATS WITH MILD AND SEVERE STAGES OF GALACTOSAMINE-INDUCED HEPATIC ENCEPHALOPATHY (HE)

Conditions	Beta-endorphin (fmol/ml)
Controls	11.1 ± 1.4
Mild HE	12.9 ± 2.1
Severe HE	17.6 ± 1.3*

Values are reported as mean ± SD of 8 determinations done in duplicate. Student's t-Test: *p < 0.05 vs controls.

hypothalamus and in the cortex during the mild and the severe stage of portal-systemic encephalopathy showed, as reported in Table 4, a significant decrease in both areas in the mild stage. On the other hand in the severe stage the levels of beta-endorphin recovered to normal values in the hypothalamus while remaining at significant lower levels than those of controls in the cortex.
The binding characteristics of opiate receptors studied in

Table 2
BETA-ENDORPHIN CONCENTRATIONS IN HYPOTHALAMUS AND CORTEX OF CONTROL RATS AND OF RATS WITH MILD AND SEVERE STAGES OF GALACTOSAMINE-INDUCED HEPATIC ENCEPHALOPATHY (HE)

Conditions	Beta-endorphin (pmol/mg prot.)	
	Hypothalamus	Cortex
Controls	17.0 ± 2.8	8.3 ± 1.3
Mild HE	9.9 ± 1.5**	5.2 ± 0.5*
Severe HE	16.2 ± 2.2	4.8 ± 0.6*

Values are reported as mean ± SD of 6 deteminations done in duplicate. Student't-Test: *p < 0.05, **p < 0.001 vs controls.

Table 3
BETA-ENDORPHIN CONCENTRATIONS IN PLASMA OF SHAM OPERATED OR DIMETHYLNITROSAMINE (DMNA) TREATED DOGS AND OF DOGS TREATED WITH DMNA PLUS PORTA-CAVAL SHUNT IN MILD AND SEVERE STAGES OF PORTAL-SYSTEMIC ENCEPHALOPATHY (PSE)

Conditions	Beta-endorphin (fmol/ml)
Sham operated	13.0 ± 2.4
DMNA treated	12.8 ± 1.6
Mild PSE	11.1 ± 3.0
Severe PSE	20.7 ± 2.3*

Values are reported as mean ± SD. Student't-Test: *p < 0.01 vs sham operated.

dogs in two brain areas which have a high concentration of opiate binding sites, the hypothalamus and the frontal cortex, are reported in Figure 1. Scatchard plot analysis performed by averaging the values of 3-6 saturation curves of ^{3}H-Naloxone binding to crude synaptic membrane preparations shows an increased presence of opiate receptors in

Table 4
BETA-ENDORPHIN CONCENTRATIONS IN HYPOTHALAMIC AND CORTICAL TISSUE OF SHAM OPERATED OR DIMETHYLNITROSAMINE (DMNA) TREATED DOGS AND OF DOGS TREATED WITH DMNA PLUS PORTA-CAVAL SHUNT IN MILD AND SEVERE STAGES OF PORTAL-SYSTEMIC ENCEPHALOPATHY (PSE)

Conditions	Beta-endorphin (pmol/mg prot.)	
	Hypothalamus	Cortex
Sham operated	24.3 ± 1.8	14.2 ± 2.6
DMNA treated	24.1 ± 1.9	13.7 ± 1.4
Mild PSE	14.8 ± 2.3*	8.7 ± 0.6*
Severe PSE	20.3 ± 2.0	6.3 ± 0.7*

Values are reported as mean ± SD. Student's t Test: *p < 0.001 vs sham operated.

both tested areas of dogs with mild encephalopathy, without any change in the affinity constants. An increased number of opiate recognition sites was detected in the cortex of dogs with mild and severe stages of encephalopathy. In contrast, in the hypothalamus the alterations of opiate receptors found in the mild and severe stages of encephalopathy underwent two different patterns showing a significant increase in the mild stage of encephalopathy and a decrease in the severe stage when compared with the values of control dogs.

DISCUSSION

Our results indicate that beta-endorphin concentrations are unchanged in plasma of both rats and dogs during the mild stage of encephalopathy. There is, however, an increase of beta-endorphins in both the acute and in the chronic models when the severe stage of encephalopathy develops. This finding seems to suggest that the turnover of beta-endorphin undergoes significant changes mainly in the severe stages of encephalopathy. It has been documented that dopamine exerts an inhibitory action on beta-endorphin secretion and that dopamine antagonists increase plasma beta-endorphins (12). Since decreased presence of catecholamines and reduced density of dopamine receptors was described in these models of encephalopathy only in the severe stage (13,14), it is likely to surmise that the increase in circulating beta-endorphin in this stage of encephalopathy may result from a reduced dopaminergic inhibitory tone on pituitary beta-endorphin releasing cells. The increased levels of beta-endorphin in plasma could be related also to several endocrine changes which have been described in experimental (15) and human liver cirrhosis (16), such as an increase in estrogens which have been described as stimulating the release of beta-endorphin (17).

The decreased concentrations of beta-endorphin in both hypothalamus and cortex measured during the mild stage of encephalopathy in rats and in dogs could result from either decreased synthesis or increased release of this peptide and it seems very hard to discriminate between them based on our present knowledge on the turnover of the endogenous peptides in the brain. However it seems difficult to attribute this decrease of beta-endorphin in brain areas to an increased release from the pituitary

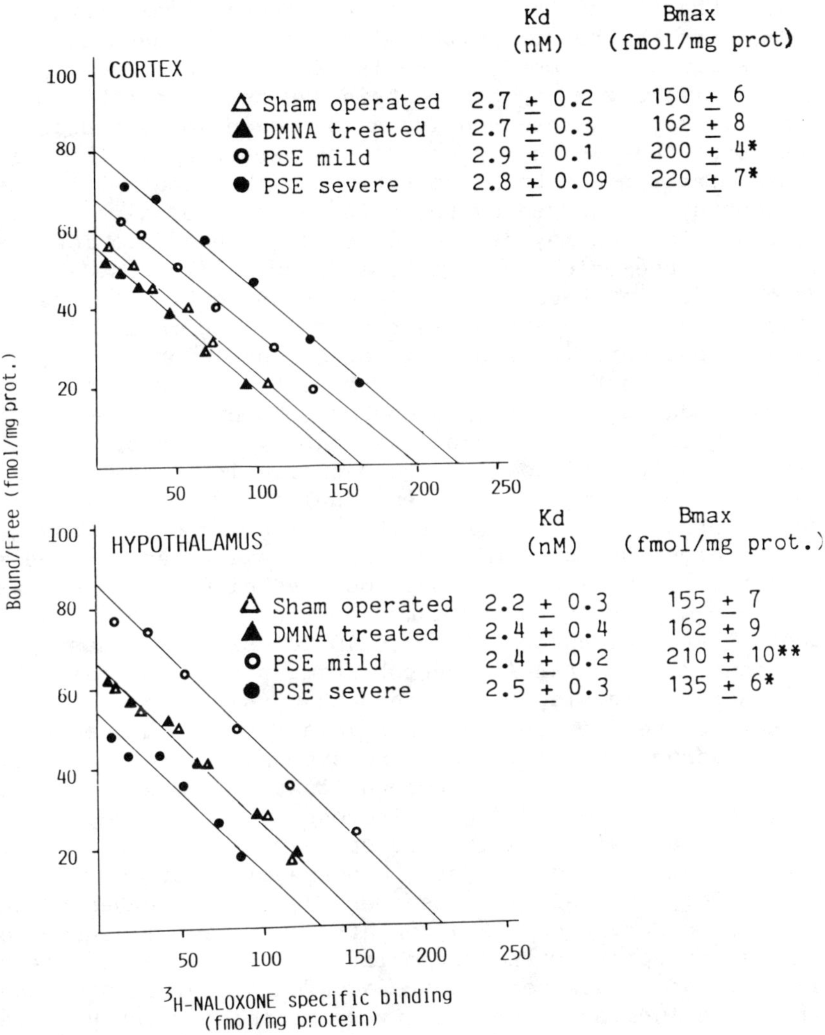

Figure 1. Scatchard plot analyses of ^{3}H-Naloxone high affinity binding to synaptic membrane preparations from cortical and hypothalamic tissues of dogs. Each point represents the mean value of 3-6 saturation curves performed in triplicate. Kd and Bmax values are reported as mean ± SD. Student's t-Test: *p < 0.01, **p < 0.001 vs sham operated and dymethylnitrosamine (DMNA) treated dogs.

since the plasma levels of beta-endorphin, as stated above, are unchanged in the mild stage of encephalopathy. The suggested reduced synthesis cannot be ascribed to a general derangement of protein synthesis in the brain since, amongst other considerations de novo synthesis of the proteins which express opiate receptors in the same brain areas seems to be operative at this stage of encephalopathy as judged by their increase in binding sites. Several factors may be considered as responsible for the reduced presence of beta-endorphin levels herein described. In this regard it must be mentioned that a hyperprolactinemic state has been described in liver cirrhosis patients in mild stage of encephalopathy (18,19). From these data one could surmise that also in these animal models of encephalopathy an hyperprolactinemic state would result in a lowering of endogenous opiate levels (18). Although the decrease in beta-endorphin during the mild stage of encephalopathy might be the result of several neurotransmitters and endocrine changes which occur during fulminant hepatic failure and in decompensated liver cirrhosis, one could attribute the decrease in beta-endorphin to the increased functional activity of the GABA-A receptor unit (5,9,10,20) since it has been documented that GABA-A receptor stimulation decreases the hypothalamic levels of beta-endorphin (3).

Wathever the mechanism responsible for the decrease in beta-endorphin may be, we found that at least in the mild stage of portal-systemic encephalopathy in dogs this derangement is associated with increased opiate receptors in the tested brain areas. This up-regulation of opiate receptors can be regarded as a supersensitivity phenomenon by analogy with similar phenomena described after drug-induced denervation (21) or after chronic blockade of opiate receptors with antagonists (22). This increased density of opiate receptors has a functional significance given the increased sensitivity to morphine which has been decribed (22). Moreover after chronic treatment with opiate antagonists (23) a reduced level of beta-endorphin associated with an increase of opiate receptors in both hypothalamus and frontal cortex have been observed (24). Based on these considerations we can surmise that the up-regulation of opiate receptors derives from a disuse supersensitivity due to a reduction of endogenous neuropeptides or to a decrease of endogenous opiate receptor inhibitors such as zinc (25), which decreases in the brain during encephalopathy (26).

From the functional point of view the reported data on the status of the opiatergic system in the mild stage of encephalopathy seems to be of peculiar importance in order to explain the ability of morphine to precipitate episodes of coma.

It is however noteworthy that in the severe stage of encephalopathy the alterations of the opiatergic system become difficult to understand since in the hypothalamus but not in the cortex there is an apparent recovery in the level of beta-endorphin in the rat model and an increase in the dog model associated with a decreased density of opiate receptors. This last finding seems to point out that it is imperative to assay the levels of endogenous neuropeptides and the functional status of their receptors in different brain areas. The observation that in the severe stage there is a different pattern of beta-endorphin in the two tested areas suggests that caution is needed in interpreting data from studies of beta-endorphin levels and opiate receptors in the brain as a whole.

In conclusion we have demonstrated that, using acute and chronic animal models, hepatic encephalopathy is associated with an altered presence of beta-endorphin and opiate receptors which might contribute to the increased sensitivity to morphine administration and favor the reduced reaction to pain stimuli which characterizes hepatic encephalopathy. The described changes in the functional status of the opiatergic system could be regarded as a secondary event due to changes of neurotransmitter systems such as the GABA-benzodiazepine receptor complex which take place early in the development of encephalopathy and of the dopaminergic system which undergoes to a reduced functional activity in the severe stage. The increased density of opiate receptors in the mild stage of portal-systemic encephalopathy must be considered as an additive component favoring the neurological abnormalities which characterize this syndrome and it could reinforce the generalized depression of the central nervous system attributed to an imbalance between inhibitory and excitatory systems in the brain (13).

REFERENCES

1. Conn H.A. and Lieberthal M.M., The hepatic coma syndrome and lactulose, Williams and Wilkinson, Baltimore, 1979: 1–45.

2. Laidlaw., Read A.E. and Sherlock S., Morphine tolerance in hepatic cirrhosis, Gastroenterology, 1961; 40: 389-396.
3. Panerai A.E., Salerno F., Baldissera F. et al., Brain beta-endorphin concentrations in experimental chronic liver disease, Brain Res., 1982; 247-: 188-190.
4. Panerai A.E., Martini A., De Rosa A., Salerno F., Di Giulio A.M. and Mantegazza P., Plasma beta-endorphin and met-enkephalin in physiological and pathological conditions, In: E. Costa and M. Trabucchi (Eds), Regulatory peptides: from molecular biology to function. Raven Press, New York, 1982: 139-149.
5. Baraldi M. and Zeneroli M.L., Experimental hepatic encephalopathy - A model to study a distinct high affinity GABA receptors. In: I. Hanin (Ed), Dynamics of neurotransmitter function, Raven Press, New York, 1984: 349-350.
6. Ferenci P. Pappas S.C., Munson P.J. et al, Changes in the status of neurotransmitter receptors in a rabbit model of hepatic encephalopathy, Hepatology, 1984; 4: 186-191.
7. Zeneroli M.L., Ventura E., Pinelli G. et al., Opiate receptors and beta-endorphin levels in brain areas of dogs with portal-systemic encephalopathy, J. Hepatol., 1985; 1:619-627.
8. Zeneroli M.L., Penne A., Parrinello G. et al., Comparative evaluation of visual evoked potentials in experimental hepatic encephalopathy and in pharmacological induced coma-like states in rat. Life Sci., 1981; 28: 1507-1515.
9. Baraldi M. and Zeneroli M.L., Experimental hepatic encephalopathy - Changes in gamma-aminobutyric acid receptors, Science, 1982; 216: 427-431.
10. Baraldi M., Zeneroli M.L., Ventura E., et al., Portal-systemic encephalopathy in dogs - Changes in brain GABA receptors and neurochemical correlates, In: G. Kleinberger, P. Ferenci, P. Riederer and H. Thaler (Eds), Advances in hepatic encephalopathy and urea cycle diseae, Karger, Basel, 1984: 353-359.
11. Glowinski J. and Iversen L.L., Regional studies of catecholamines in rat brain - The disposition of ^{3}H-norepinephrine, ^{3}H-dopamine and ^{3}H-dopa in various regions of the brain, J. Neurochem., 1966; 13: 655-659.
12. Murburg M., Paly D., Wilkinson C.W. et al., Haloperidol increases plama beta-endorphin-like im-

munoreactivity and cortisol in normal human males, Life Sci., 1986; 39: 373-379.
13. Baraldi M., Zeneroli M.L, Ricci P. et al, Down-regulation of striatal dopamine receptors in experimental hepatic encephalopathy, Life Sci., 1983; 32: 1417-1425.
14. Zeneroli M.L., Ventura E., Pinelli G. et al., Striatal dopamine receptors in a dog model of portal-systemic encephalopathy, J. Hepatol., 1985; 1: 156 (A 311).
15. Milligan S.R. and Sarna G.S., Effect of porto-caval anastomosis and chronic underfeeding on the hypothalamic-pituitary gonadal axis in the rat, J. Endocr. 1981; 88: 39-47.
16. Sherlock S., Diseases of the liver and biliary system, 5th edition, Blackwell Scientific Publ, Oxford, 1975.
17. Wardlaw S.L. and Frantz A.G., Brain beta-endorphin during pregnancy, parturition and postpartum period, Endocrinology, 1983; 113: 1664-1688.
18. Panerai A.E., Sawynok J., La Bella F.S. et al., Prolonged hyperprolactinemia influences beta-endorphin and met-enkefalin in the brain, Endocrinology, 1980; 106: 1804-1808.
19. Langer M., Masala A., Rovasio P. et al., Basal and dynamic pituitary secretion in hepatic encephalopathy. In: G. Kleinberger, P. Ferenci, P. Riederer and H. Thaler (Eds), Advances in hepatic encephalopathy and urea cycle diseae, Karger, Basel, 1984: 611-618.
20. Baraldi M., Zeneroli M.L., Ventura E. et al., Supersensitivity of benzodiazepine receptors in hepatic encephalopathy due to fulminant hepatic failure in the rat: reversal by a benzodiazepine antagonist, Clin. Sci., 1984; 67: 167-175.
21. Gardner E. L., Zukin C.R.S. and Makman M., Modulation of opiate receptors binding in striatum and amygdala by selective mesencephalic lesions, Brain Res., 1980; 194: 232-239.
22. Tang A.H. and Collins R.J., Enhanced analgesic effects of morphine after chronic administration of naloxone in the rat, Europ. J. Pharmacol. 1978; 47: 473-474.
23. Ragavan V.V., Wardlaw S.L., Kreek M.J. et al., Effect of chronic naltrexone and methadone administration on brain immunoreactive beta-endorphin in the rat, Neuroendocrinology, 1983; 37: 266-268.
24. Tempel A., Gardner E.L. and Zukin R.S., Visualization of opiate receptor up-regulation by light microscopy autoradiography, Proc. Nat. Acad. Sci. (USA), 1984;

81: 3893-3897.
25. Baraldi M., P. Zanoli, Benelli A. et al, Neurobehavioral, neuroendocrine and neurochemical effects of zinc supplementation in rats,In: R. Schwartz and Y. Ben-Ari (Eds), Excitatory amino acids and epilepsy, Plenum Press, New York, 1986: 571-585.
26. Baraldi M., Caselgrandi E., Borella P et al., Decrease of brain zinc in experimental hepatic encephalopathy, Brain Res. 1983; 258: 170-172.

C. Treatment of Hepatic Encephalopathy

Update of Existing Therapeutic Approaches

TREATMENT OF HEPATIC ENCEPHALOPATHY. NEW PERSPECTIVES ON OLD IDEAS.

Gilles Pomier Layrargues, Jean-François Giguère, Joël Lavoie and Roger F. Butterworth, André-Viallet Clin. Res. Centre, Hôpital Saint-Luc and Université de Montréal, Montreal, Quebec, Canada H2X 1P1

ABSTRACT

Hepatic encephalopathy (HE) was characterized in humans 35 years ago and hyperammonemia was proposed from the start as the main factor involved in the pathogenesis of this complication. Thereafter, different therapies were tried in HE, all aimed at reducing blood ammonia. Surprisingly some of them were considered to be efficient in spite of the fact that controlled studies on their efficacy were lacking. Lactulose and neomycin are the best examples of these treatments. On the other hand, other treatments (such as arginine and ornithine) targetted at stimulation of the urea cycle were at first advocated, then neglected in spite of the fact that no scientific data were provided to deny their efficacy. The 1970's and 1980's saw the emergence of new pathogenetic theories such as the "false neuro-transmitter" and the "GABA" hypotheses. At present, even the keenest supporters of these theories agree that ammonia remains a major determinant in the pathogenesis of HE. Therefore, it may be profitable to reassess the older treatments in the light of new methods developped to evaluate their efficacy. The first step is to determine the natural history of hepatic encephalopathy with multivariate analysis using hepatic function, the extent of portosystemic shunting and the nature of precipitating factors as variables. Such studies allow a stratification according to prognostic variables for future therapeutic trials. The second step is to improve the methods used to evaluate the grade of HE. Clinical assessment of HE is not standardized and is probably biased by poor inter-observer agreement. Measurement of blood ammonia should be performed under strictly defined conditions. New techniques such as evoked potentials should be adequately validated. Moreover, the design of future controlled randomized clinical trials should be markedly improved to allow valid conclusions. The studied population should be carefully defined; the sample size should be adequate since HE resolves itself spontaneously in 40 - 50% of the cases. Finally, the conclusions should rely on adequate statistical analysis which accounts for the influence of confounding variables in the evolution of this still mysterious hepato-neurological entity.

INTRODUCTION

Hepatic encephalopathy (HE) was characterized in humans more than 35 years ago (Adams and Foley, 1953). Its general mechanisms were then hypothesized as following: toxic gut-derivated substances bypass the liver due to hepatocellular dysfunction, portacaval anastomosis or both, then gain access to the brain where they induce disturbances of cerebral function. This concept is still considered to be valid.

Recent advances in the treatment of liver diseases would change the relative incidence of the different clinical presentations of HE. Acute HE associated with fulminant hepatitis either resolves itself spontaneously or worsens to hepatic coma which is best treated by liver transplant (Bismuth et al, 1987). On the other hand, the so-called chronic HE mainly associated with surgical portacaval shunt will probably disappear in the future, because sclerotherapy has replaced this operation in the treatment of bleeding varices. Therefore acute HE in patients with chronic liver diseases is the most frequent clinical presentation and will be the subject of this discussion.

Progress in the treatment of HE in cirrhotic patients has been limited during recent years. This is not surprising as the pathogenesis is still unknown despite a succession of different hypotheses.

PATHOGENETIC HYPOTHESES

Ammonia (MacDermott et al, 1953), synergism between ammonia, mercaptans and fatty acids (Zieve et al, 1974) false neuro-transmitters (Fisher and Baldesarini, 1971), GABA (Schafer and Jones, 1982) and more recently endogenous benzodiazepines (Mullen et al, 1988) have been successively suggested as the clue for neurological impairment in HE. None of these hypotheses have been fully convincing according to experimental evidence or clinical demonstration. Amazingly, the ammonia hypothesis, proposed many years ago, is gaining new popularity (Butterworth et al, 1987). The pros and cons of this theory are summarized on table 1.

TABLE 1
Pathophysiological Role for Ammonia in Hepatic Encephalopathy

Evidence for

- Blood and csf ammonia levels are frequently elevated in cirrhotic patients with encephalopathy.
- Hepatic encephalopathy results in neuropathological damage of a similar nature (Alzheimer type II astrocytosis) to that encountered in shunted rats after an ammonia load and in patients with hyperammonemia resulting from inherited defects of urea cycle enzymes.
- Ammonia removal in brain relies on glutamine formation. Cerebrospinal fluid and brain glutamine concentrations are found to be elevated in patients with hepatic encephalopathy and in rats following portocaval anastomosis.

Evidence against

- Blood ammonia is not elevated in all patients with encephalopathy.
- No exact correlation exists between blood ammonia levels and the grade of encephalopathy.

It is worth noting that all the so-called conventional treatments of HE aim at decreasing blood ammonia levels. Most of them are considered to be standard whereas a careful analysis of the existing litterature does not allow definite conclusions on their efficacy due to inadequate study design criteria. Methodology in clinical research is a relatively new area; it is now recognized that the natural evolution of the disease should be well known, and the criteria used to assess the treatment efficacy must be clearly defined and reproducible.

NATURAL HISTORY OF HEPATIC ENCEPHALOPATHY

The spontaneous evolution of HE has been poorly described in the past due to the uncontrolled use of various treatments. In most of the studies, lactulose or neomycin are considered to be the standard treatment for the control group.

In recent years, 3 studies reported the outcome of a group of cirrhotics with HE without any specific treatment in most of them (Michel et al, 1980; Wahren et al 1983; Michel et al 1985). These studies were double blind trials with a placebo group.

The inclusion criteria and the rate of improvement and survival are summarized in table 2. The studied population was made up of patients with severe liver dysfunction as evidence by a high mortality rate (20-39%). Nevertheless, spontaneous improvement of HE was not unusual (33 - 48%). This finding clearly demonstrates that the evaluation of a new treatment should include a control group. Moreover, the spontaneous clearing of neurological symptoms is probably higher in cirrhotics with a moderate impairment in liver function; as a matter of fact, the determinants of the severity of an HE episode rely on the degree of liver failure, the extent of porto-systemic shunting, the nature of precipitating factors, age, the etiology of cirrhosis, and associated diseases. All these parameters must be rigorously matched in a prospective placebo-controlled study to allow valid conclusions to be drawn.

TABLE 2
Evolution of Acute Hepatic Encephalopathy in Cirrhotic Patients Treated with Placebo

AUTHOR	MICHEL(1980)	WAHREN(1983)	MICHEL(1985)
Number of patients	38	25	34
Inclusion criteria	All stages	Stages II,III,IV	All stages
Exclusion criteria	Spontaneous clearing of HE or death within 48 hours; cardiac or respiratory failure; septic shock; anemia; uncontrolled GI bleeding		
Outcome (5 days)			
HE improvement %	37	48	33
Mortality (%)	39	20	21

EVALUATION OF HEPATIC ENCEPHALOPATHY

To evaluate the effect of a new treatment on the neurological function in cirrhotic patients with HE, reliable parameters are needed to assess the degree of HE. Clinical staging into four stages has been adopted by most of the people (Zieve, 1982). However, this classification is partly subjective and the inter-observer agreement has never been measured. Recently, Sanford and Saul (1988) proposed a new semi-quantitative classification which seems to be more precise and less subjective, but again has never been tested with regards to its reproducibility. There is clearly a place for additional studies on this critical issue.

The trail making test has been advocated as a quantitative index of impairment of the mental function (Conn, 1977), but it is non-specific and needs the collaboration of the patient; obviously it cannot be used in the deepest stage of HE. Blood ammonia levels have been used as a part of a "PSE index" (Atterbury et al, 1978). However, there are still many causes of inaccuracy in blood ammonia determinations due to the site of sampling (arterial or venous, blood or plasma), the sampling conditions (influence of digestive state and exercise), and the technique of ammonia assay. Therefore, up to now, the usefulness of this parameter remains unsure.

EEG abnormalities are non specific but might be of some help in excluding localized neurological diseases. Moreover, marked discrepancies do exist between clinical and EEG's staging, which denies any practical value to these parameters in the evaluation of the evolution of HE. Evoked potentials have been proposed recently as a good index of the deepness of HE; visual evoked potentials have severe limitations due to the possible influence of unrelated factors such as integrity of visual tract and cooperation of the patient(Sandford and Saul, 1988). Somatosensory and auditory evoked potentials are still under evaluation (Chu et al, 1988).

Therefore, controversies in the litterature with regards to different therapeutic modalities are not surprising, as there is no general agreement on how to assess the evolution of HE reliably and reproducibly.

EXISTING TREATMENT OF HEPATIC ENCEPHALOPATHY

The analysis of the litterature will focus on drugs used to decrease blood ammonia levels. Different therapeutic approaches based on other pathogenic hypothesis are disccused elsewhere in this book.

Mechanisms of action of the blood ammonia lowering drugs are summarized in table 3. The analysis of the litterature reveals quite surprising findings. Drugs which act on the gut are widely used in clinical practice despite a poorly demonstrated efficacy. On the other hand, the use of drugs which act on the liver have been neglected whereas their inefficacy has not been established unequivocally.

TABLE 3
Drugs Used in the Treatment of Hyperammonemia (Mechanisms of Action)

"Liver drugs"	
. Stimulation of urea cycle	Arginine, ornithine
. Stimulation of glutamine synthesis	L glutamate
. Diversion of nitrogen to alternative nitrogen products	Benzoate Phenylacetate
"Gut drugs"	
. Decrease ammonia synthesis by gut bacteria	Neomycin Metronidazole
. Increase ammonia elimination in the stool	Lactulose Lactitol

"Liver drugs". The most widely studied drugs aimed at decreasing the blood ammonia levels do so by stimulating the urea cycle: 2 precursors of urea have been tested, arginine and ornithine. Both drugs succeeded in lowering hyperammonemia in animals and cirrhotic patients (Zieve et al, 1986; Tobe et al 1961; Najarian et al, 1956). Uncontrolled observations suggested that ornithine infusions might improve HE in cirrhotic patients with acute HE but the efficacy of this drug has not been clearly established in humans. A recent study in rats made comatose by an ammonia load and awakened by ornithine infusions provide the basis for further clinical studies (Zieve et al, 1986). A controlled study in 1958 (Reynolds et al, 1958) failed to detect any therapeutic value of arginine infusion in cirrhotic patients with HE. However, as shown by a very high incidence of liver failure as a precipitant (87%) and a poor survival rate (28%), the majority of the patients were in end-stage liver failure and had probably lost the metabolic efficacy of the urea cycle. These results cannot be extrapolated to cirrhotics with HE and with a moderately impaired liver function.

Another therapeutic approach is to divert nitrogen from urea synthesis to alternative waste nitrogen products. Administration of sodium benzoate will divert nitrogen from ammonia to glycine and allow excretion as hippurate. Sodium phenylacetate combines with glutamine to form phenylacetylglutamine leading to excretion of two molecules of ammonia in the urine (Balistreri and Shubert, 1987). These treatments have been tested in congenital hyperammonemia due to defects in urea cycle enzymes, but their effects on HE in cirrhotic patients have only been assessed in a few cases of chronic HE (Mendenhall et al, 1986). This is an open field for further clinical studies.

"Gut drugs". These drugs are the most widely used in the treatment of HE and are considered to be treatments of choice in the litterature. The general principles are to decrease ammonia synthesis by gut bacteria or to increase ammonia elimination in the stool. Antibiotics were first tried and amongst them neomycin remains the most popular. Neomycin is able to decrease blood ammonia levels in cirrhotic patients (Pirotte et al, 1974) where its clinical efficacy is assumed (Dawson et al, 1957) but is not demonstrated by adequate controlled studies. More recently, metronidazole has been advocated as a useful therapeutic agent (Morgan et al, 1983) but the studies were performed only in patients with chronic HE.

Following poorly-designed clinical studies suggesting the efficacy of neomycin, lactulose was proposed in the 1960's as a new therapeutic agent; unfortunately it was compared to neomycin as many authors felt it unethical to use placebo given the "proven efficacy" of neomycin. In fact the assumption that lactulose is an efficient treatment of HE is based on one placebo-controlled study published in 1970 (Simmons et al, 1970). The group treated with lactulose was made up of 10 cirrhotic patients but clinical evolution could be evaluated in only 4 patients. In the following years, lactulose was found to be as efficient as neomycin used as a reference treatment (Atterbury et al, 1978; Orlandi et al, 1981). Thereafter, almost all the comparative studies used lactulose-treated patients as a control group when evaluating new agents such as lactitol (Morgan et al, 1987; Heredia et al, 1987).

In summary, lactulose (or lactitol) and neomycin (or

metronidazole) are perhaps effective in the treatment of acute HE in cirrhotic patients; however the possibilites do exist that reversal of the neurological symptoms are mainly related to a decreased amount of ingested protein, cleansing of the bowel or correction of the precipitating factors.

SUGGESTION FOR THE FUTURE

Before reassessing older treatments, or evaluating the efficacy of new therapeutic approaches, certain guidelines to set up an adequate clinical trial should be borne in mind. These criteria have been reviewed recently (Der Simionian et al, 1982) and are summarized in table 4.

TABLE 4

Criteria to Assess the Validity of Clinical trials. (from Der Simionian et al, 1982)

- Eligibility
- Admission before allocation
- Random allocation
- Method of randomization
- Patient's blindness to outcome
- Blind assessment of outcome
- Treatment complication
- Loss to follow-up
- Statistical analysis
- Statistical methods
- Power

Failure to meet these requirements led to unconclusive results in the past litterature with regards to the treatment of acute HE. Such studies should be double blind, as methods of assessment are partly subjective; statistical methods and analysis should be performed adequately and the power should be estimated precisely. For example, if the research hypothesis states that HE improvement would occur in 40% of the control group and in 70% of the treated patients, for a 0.05 alpha risk, one should include 62 or 48 patients per group for a beta risk of 0.1 and 0.2. We should also answer the following question: is it ethical to include a placebo group in a clinical trial on acute HE in 1988? If the answer is no, what is the best treatment currently available?

Finally even if this prerequisite seems to be obvious, clinical trials in the future should rely on a good rationale based on clear-cut experimental evidences for a new pathogenetic hypothesis.

REFERENCES

Adams R.D. and Foley J.M. (1953). The neurological disorder associated with liver disease, in Metabolic and Toxic Diseases of the Nervous System (Merritt H. and Hare C.C. eds), vol. 32, pp. 198-237, Williams and Wilkins, Baltimore, M.D.

Atterbury C.E., Maddrey W.C., and Conn H.O. (1978). Neomycin, sorbitol and lactulose in the treatment of acute portal-systemic encephalopathy. A controlled double blind clinical trial. Dig. Dis. Sci. 23, 398-406.

Balistreri W.F. and Schuber T.W.K. (1987). Liver disease in infancy and childhood in "Disease of the Liver" 6th Ed. (Schiff L. and Schiff E.R., eds), pp 1337-1426, Lippincott, Philadelphia, P.A.

Bismuth H., Samuel D., Gugenheim J., Castaing D., Bernuau J., Rueff B and Benhamou J.P. (1987). Emergency liver transplantation for fulminant hepatitis. Ann. Intern. Med. 107,337-341.

Butterworth R.F., Giguère J.F., Michaud J., Lavoie J., and Pomier-Layrargues G. (1987). Ammonia: key factor in the pathogenesis of hepatic encephalopathy. Neurochem. Pathol. 6,1-12.

REFERENCES (cont'd)

Chu N.S., and Yang S.S. (1988). Portal-systemic encephalopathy: alterations in somatosensory and brainstem auditory evoked potentials. J. Neurol. Sci. 84,41-50.

Conn H.O. (1977). Trail making and number-connection tests in the assessment of mental state in portal systemic encephalopathy. Am. J. Dig. Dis. 22,541-550.

Dawson A.M., MacLaren J.M. and Sherlock S. (1957). Neomycin in the treatment of hepatic coma. Lancet 2, 1263-1268.

DerSimonian R, Charette B.A., McPeek B., and Mosteller F. (1982). Reporting on methods in clinical trials. New. Engl. J. Med. 306,1332-1337.

Fisher J.E., and Baldessarini R.J. (1971). False neurotransmitters and hepatic failure. Lancet 2,75-80.

Heredia D., Caballeria J., Arroyo, V., Ravelli G., and Rodès J. (1987). Lactitol versus lactulose in the treatment of acute portal systemic encephalopathy. A controlled trial. J. Hepatol. 4:293-298.

McDermott W.V., and Adams R.D. (1954). Episodic stupor associated with an Eck fistula in the human with particular reference to the metabolism of ammonia. J. Clin. Invest. 33, 1-9.

Mendenhall C.L., Rouster S., Marshall L., and Weesner R. (1986). A new therapy for portal systemic encephalopathy. Am. J. Gastroenterol. 81,540-543.

Michel H., Bories P., Aubin J.P., Pomier-Layrargues G., Bauret P., and Bellet-Herman H. (1985). Treatment of acute hepatic encephalopathy in cirrhotics with a branched-chain amino acids enriched versus a conventional amino acids mixture. A controlled study of 70 patients. Liver 5,282-289.

REFERENCES (cont'd)

Michel H., Solere M., Granier P., Cauvet J.P., Bali J.P. and Bellet-Herman H. (1980). Treatment of cirrhotic hepatic encephalopathy with L Dopa. A controlled trial. Gastroenterology 79,207-211.

Morgan M.Y., and Hawley K.E. (1987). Lactitol vs lactulose in the treatment of acute hepatic encephalopathy in cirrhotic patients: a double-blind randomized trial. Hepatology 7,1278-1284.

Morgan M.H., Read A.E., and Speller D.C. (1983). Treatment of hepatic encephalopathy with metronidazole. Gut 23,1-7.

Mullen K.D., Martin J.V., Mendelson W.B., Bassett M.L. and Jones E.A. (1988). Could an endogenous benzodiazepine ligand contribute to hepatic encephalopathy. Lancet 1,457-459.

Najarian J.S. and Harper H.A. (1956). A clinical study of the effect of arginine on blood ammonia. Am. J. Med. 21,832-842.

Orlandi F., Freddara U., Candelaresi M.T., Morettini A., Corazza G.R., Di Simone A., Dubrilla G., and Cavallini G. (1981). Comparison between neomycin and lactulose in 173 patients with hepatic encephalopathy. A randomized clinical study. Dig. Dis. Sci. 26,498-506.

Pirotte J., Guffen M., and Devos J. (1974). Comparative study of basal arterial ammonemia and of orally-induced hyperammonemia in chronic portal systemic encephalopathy treated with neomycin, lactulose and an association of neomycin and lactulose. Digestion 10,435-444.

Reynolds T.B., Redeker A.G., and Davis P. (1958). A controlled study of the effects of L arginine on hepatic encephalopathy. Am. J. Med. 25,359-367.

Sandford N.L., and Saul R.E. (1988). Assessment of hepatic encephalopathy with visual evoked potential compared with conventional methods. Hepatology 8,1094-1098.

REFERENCES (cont'd)

Schafer D.F., and Jones E.A. (1982). Hepatic encephalopathy and the gamma-aminobutyric acid neurotransmitter system. Lancet 1,18-19.

Simmons F., Goldstein H., and Boyle J.D. (1970). A controlled clinical trial of lactulose in hepatic encephalopathy. Gastroenterology 59,827-832.

Tobe B.A. (1961). Observations on the use of L arginine and L glutamate in the treatment of hepatic encephalopathy. Can. Med. Assoc. J. 85,591-603.

Zieve L., Doizaki W.M., and Zieve F.J. (1974). Synergism between mercaptans and ammonia or fatty acids in the production of coma: a possible role for mercaptans in the pathogenesis of hepatic coma. J. Lab. Clin. 83,16-28.

Zieve L. (1982). Hepatic encephalopathy in Diseases of the Liver, 5th Ed (Schiff L., and Shiff E.R., eds) p. 433-459, Lippincott, Philadelphia, P.A.

Zieve L., Lyftogt C., and Raphael D. (1986). Ammonia toxicity: comparative protective effects of various arginine and ornithine derivatives, aspartate , benzoate and carbanyl glutamate. Metab. Brain Dis. 1,25-35.

TREATMENT OF HEPATIC ENCEPHALOPATHY

WITH LACTULOSE AND ANTIBIOTICS

Fredrick L. Weber, Jr., M.D.

Case Western Reserve University School of Medicine
Veterans Administration Medical Center
Cleveland, Ohio 44106

Lactulose and antibiotics, particularly neomycin, presently constitute standard therapy for both acute and chronic hepatic encephalopathy. This section will review information relating to their mechanisms of action and clinical data demonstrating their efficacy. An issue of considerable practical importance is whether these agents might have an additive effect if used together; the limited available evidence concerning this point will also be reviewed.

ANTIBIOTICS

Mechanism of Action

The initial rationale for the use of antibiotics in the treatment of hepatic encephalopathy was to inhibit bacterial enzymatic reactions that produced ammonia from urea and other nitrogenous compounds in the colonic lumen (Fisher and Faloon, 1957). One estimate of the amount of ammonia formed from various sources in the colon indicated that 42% arose from ureolysis, 9% from the metabolism of arterial glutamine and 49% from metabolism of other nitrogen containing compounds in colonic lumen (Weber and Veach, 1979). Neomycin, which is poorly absorbed from the gastrointestinal tract and is primarily effective against aerobic bacteria, has been the standard antibiotic used in treatment of hepatic encephalopathy. McBeth et al. (1965) demonstrated that neomycin was effective in reducing fecal urease activity when enzyme activity in stool was expressed as either concentration or as total amount excreted per day. Neomycin caused a 70% reduction in urea degradation rate in cirrhotic patients (Weber et al. 1982), but tetracycline and sulfa had no demonstrable effect on urea breakdown (Walser and Bodenlos, 1959).

Antibiotics also inhibited proteolytic activity of the intestinal microflora and their subsequent production of ammonia as McBeth et al. (1965) demonstrated with the administration of neomycin. Anaerobic bacteria have the most proteolytic activity (Vince and Burridge, 1980), and antibiotics primarily effective against anaerobic bacteria such as metronidazole and vancomycin have also been recently found to be effective in the treatment of hepatic encephalopathy (Morgan et al., 1982; Tarao et al., 1985).

The effectiveness of antibiotics in treating hepatic encephalopathy might not be dependent on a reduction in the blood ammonia concentration. Although most data has indicated that neomycin reduced the blood ammonia concentration, Morgan et al. (1982) found that treatment with either neomycin or metronidazole improved patients' encephalopathy but had little effect on blood ammonia. Phear et al. (1956) found that administration of the amino acid methionine precipitated hepatic encephalopathy without any demonstrable change in blood ammonia, whereas pretreatment with chlortetracycline reversed this effect of methionine. A number of potentially toxic bacterial metabolites might be inhibited by antibiotic treatment.

Clinical Trials

Early observations indicated that neomycin was particularly effective in reducing the blood ammonia concentration and in treating hepatic encephalopathy (Dawson et al., 1957; Fisher and Faloon, 1957; Najarian et al., 1959). These initial studies did not meet rigorous modern standards of being double-blinded and controlled, but the initial evidence was compelling and subsequent clinical efficacy so apparent that antibiotics, neomycin in particular, assumed a major role in the treatment of hepatic encephalopathy.

More recently, antibiotics have been evaluated in the treatment of both acute and chronic hepatic encephalopathy in double-blinded, controlled studies where protein intake was sufficiently controlled to permit comparisons. Conn et al. (1977) evaluated neomycin in patients with chronic hepatic encephalopathy compared to a no treatment control period. They found that neomycin significantly improved encephalopathy and the blood ammonia concentration when compared to control. Morgan et al. (1982) found that both neomycin and metronidazole were equally effective when compared to a no treatment control period in 18 patients with chronic encephalopathy. In acute hepatic encephalopathy, treatment trials using control periods are not available, but neomycin has been compared to other regimens and found to be at least equally effective. Atterbury et al. (1978) compared neomycin 6 g/day plus sorbitol with lactulose plus placebo tablets and found that both regimens were equally effective. Uribe et al. (1981) compared neomycin (3 g/day) plus starch enemas to placebo tablets and lactulose enemas, and found these treatments were also equally effective. Another partially-blinded, controlled study found an equal response to treatment with neomycin and lactulose in a large group (N=173) of patients with both acute and chronic hepatic encephalopathy (Orlandi et al., 1981).

Neomycin therapy is occasionally associated with significant toxicity, especially during long term use. Diarrhea may be caused by changes in the bacterial flora due to Staphylococcal overgrowth or Clostridium difficile colitis (Tisdale et al., 1960; Bolton et al., 1979). Malabsorption has been caused by neomycin associated with both precipitation of intraluminal bile salts and direct mucosal damage (Jacobson et al., 1960). Neomycin has caused deafness without associated renal failure (Berk and Chalmers, 1970). Neomycin is also nephrotoxic and has been associated with renal failure although a direct cause and effect relationship has never been precisely proven in cirrhotic patients because of the frequency with which they develop uremia due to functional renal failure and acute tubular necrosis. After oral administration, neomycin is partly absorbed from the gastrointestinal tract. In one study absorption averaged 0.85% of the administered dose resulting in a urinary concentration >10 mcg/ml which was bacteriocidal for many urinary pathogens (Breen et al., 1976).

LACTULOSE AND RELATED CARBOHYDRATES

Mechanism of Action

Ingelfinger (1964-1965) first proposed the possible utility of lactulose in treating hepatic encephalopathy. He suggested that it might promote the growth of lactobacilli which contain little urease activity and decrease the numbers of putrefactive bacteria within the intestinal lumen. Bircher et al. (1966) demonstrated that lactulose was indeed an effective therapeutic agent in treating hepatic encephalopathy; however, its mechanism of action remains controversial.

Postulated Mechanisms for the Effect of Lactulose

Qualitatively alter bacterial species in colon

Cathartic effect

Luminal acidification - Trapping and excretion of NH_4^+

Stimulate bacterial proliferation and incorporation of nitrogen

Inhibit bacterial catabolism and production of ammonia (effect of reduced pH and/or presence of metabolizable carbohydrate)

Reduced production of ammonia caused by inhibition of glutamine transport into the enterocyte

Lactulose is a synthetic disaccharide consisting of galactose and fructose which is neither broken down nor absorbed in the small intestine since the mammalian intestine contains no endogenous lactulases. When lactulose reaches the colon it is metabolized by a number of bacterial species. Hoffman et al. (1964) found that most aerobic gram negative rods, with the exception of Escherichia coli, did not metabolize lactulose to any appreciable extent; however, lactulose was extensively metabolized by a number of anaerobic species. With most bacteria, lactic acid was the major identified metabolite along with smaller amounts of formic and acetic acids. Perman et al. (1981) observed that after an initial rise, there was a subsequent fall in breath hydrogen production with chronic lactulose administration, and suggested that the acidic intraluminal environment inhibited subsequent utilization of lactulose by colonic bacteria. Further studies indicated that colonic bacteria continued to metabolize lactulose when it was given chronically although the flora altered their metabolic pathways and digested lactulose with greater efficiency (Florent et al., 1985).

The initial suggestion that lactulose may qualitatively alter the bacterial flora has been tested in patients but has not been confirmed. An increase in lactobacilli has been observed in some patients receiving lactulose, but these alterations have shown an inconsistent correlation with clinical response (Bircher et al., 1971). There have been no consistent quantitative changes in ammonia producing bacterial species such as bacteroides or enterococci (Zeegen et al., 1970; Conn and Floch, 1970; Vince et al., 1974).

In clinical use, lactulose acts as a cathartic. Occasional reports suggest that other cathartics are as effective as lactulose in treating hepatic encephalopathy, but the majority of studies indicate that lactulose is superior to alternative cathartics (Weber, 1981). We have found that although lactulose and magnesium sulfate caused a comparable increase in stool nitrogen, only lactulose caused a fall in the urea production rate indicating that less ammonia nitrogen was entering the portal system (Weber and Fresard, 1981).

We have used urea production and degradation rates to assess the effects of lactulose in reducing the production of urea precursor nitrogen, presumably ammonia nitrogen within the intestinal lumen (Weber, 1979; Weber and Fresard, 1981). In stable cirrhotic patients, lactulose caused a fall in stool pH to a mean value of 5.5 and a 25% reduction in the urea production rate. This drop in the urea production rate was mainly accounted for by a large increase in fecal nitrogen excretion and a corresponding reduction in urinary urea excretion. An additional component of the reduction in the urea production rate was a modest fall in the urea degradation rate which was primarily accounted for by a fall in the serum urea concentration. Only an occasional patient showed a significant drop in intestinal urea clearance (urea degradation rate/plasma urea) which would be more likely to directly reflect a change in bacterial ureolysis than the urea degradation rate.

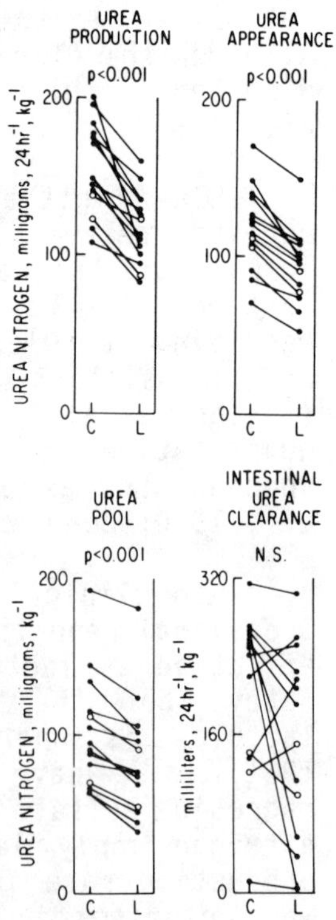

FIGURE LEGEND: Effects of lactulose on urea production, urea appearance (urinary urea), urea pool, and intestinal urea clearance in cirrhotic patients. p-Values indicate significant changes by paired t-test. NS= no significant change, C= control period, L = lactulose period.

Another hypothesis has been that the acidic intracolonic environment caused by lactulose metabolism would trap ammonia in its ionized form and permit its evacuation in the stool. The transport of ammonia across the intestinal mucosa is at least partially dependent on non-ionic diffusion, and some (Down et al., 1972; Castell and Moore, 1971) but not all studies (Wolpert et al., 1970) found that an acidic intracolonic environment trapped ammonia or least retarded its transport from the intestinal lumen. The viability of this hypothesis was diminished by the finding that fecal dialysates did not show any increase in fecal ammonia following lactulose administration (Zeegan et al., 1970; Agostini et al., 1972). Since lactulose caused approximately 3.5 g less nitrogen to enter the portal system of cirrhotic patients, an increase in fecal ammonia should have been readily apparent if it accounted for this large reduction in portal ammonia (Weber, 1979). Other studies have suggested that lactulose actually lowered the ammonia concentration in the colonic lumen (Lebek and Luginbuhl, 1988; Bircher et al., 1971).

Providing a fermentable carbohydrate source to the colonic bacterial flora has caused profound alterations in nitrogen metabolism. As Mason (1974) emphasized, studies in ruminates indicated that metabolizable carbohydrates increased nitrogen incorporation into intestinal bacteria. In human studies the ingestion of metabolizable dietary fiber increased fecal nitrogen excretion which was primarily contained in fecal bacteria (Stephen and Cummings, 1980). With lactulose administration the major portion of the increased fecal nitrogen excreted was also found in the bacterial fraction of stool (Weber et al., 1987). Lactulose or other carbohydrates might also inhibit ammonia generation by fecal bacteria because of a reduction in fecal pH and/or the presence of a metabolizable carbohydrate (Vince et al., 1973; Vince et al., 1978; Vince and Burridge, 1980). These dual effects, an increase in ammonia nitrogen incorporation into fecal bacteria and a reduction in ammonia production by fecal bacteria, were likely to be the mechanisms whereby lactulose reduced the amount of ammonia entering the portal system.

Effect of Lactulose on Fecal Composition (n=6)

	Control	Lactulose
Total stool		
Weight	161 ± 43	417 ± 57[a]
Solids	22 ± 2	52 ± 6
Nitrogen	1.1 ± 0.2	2.8 ± 0.4[a]
pH	7.1 ± 0.3	5.6 ± 0.1[a]
Fiber fraction		
Weight	1.3 ± 0.3	4.0 ± 1.0[a]
Nitrogen	0.04 ± 0.02	0.04 ± 0.01
Bacterial fraction		
Weight	11.1 ± 1.3	12.3 ± 3.2[a]
Nitrogen	0.52 ± 0.14	1.38 ± 0.21[a]
Soluble fraction		
Weight	9.4 ± 1.7	24.3 ± 3.1[a]
Nitrogen	0.58 ± 0.08	1.36 ± 0.23[a]

All results except pH expressed as mean ± SEM g/day.

Nitrogen intake was 12.5 ± 0.4 g/day. [a]p <0.025 by paired T-test.

Other types of metabolizable carbohydrates have been shown to act similarly to lactulose even though they did not acidify the stool. Vegetable protein diets were more effective than animal protein diets in the treatment of patients with hepatic encephalopathy (Greenberger et al., 1977). Our evidence indicated that the fiber component of vegetable protein diets was the active agent since it promoted increased fecal excretion of nitrogen that was primarily contained in the bacterial fraction of stool. These fiber diets also reduced the urea production rate which was consistent with a reduced entry of ammonia nitrogen into the portal system. No acidification of stool was noted on fiber diets indicating that their mechanism of action did not depend on an acidic intraluminal environment (Weber et al., 1986).

Clinical Trials

A large number of clinical trials have demonstrated the efficacy of lactulose in the treatment of hepatic encephalopathy, and 30 of these studies have been recently reviewed (Orlandi et al., 1988). Few of the trials have actually been double-blind, randomized, and controlled.

One large multicenter double-blind, controlled trial in 33 cirrhotic patients with chronic hepatic encephalopathy found that treatment with either lactulose or neomycin was significantly better than placebo. Using a PSE index to assess clinical response, lactulose was somewhat more effective than neomycin (Conn et al., 1977). In studies of acute hepatic encephalopathy, lactulose has usually been compared with neomycin therapy (which was the former standard treatment). One multicentered, double-blind, controlled trial found that neomycin and lactulose were equally effective in the treatment of acute portal systemic encephalopathy which represents the consensus from other studies (Atterbury et al., 1978). The largest controlled study compared lactulose with neomycin in 173 patients with both acute and chronic encephalopathy. This partially-blinded study also found that both lactulose and neomycin were equally effective (Orlandi et al., 1981). As noted by Orlandi et al. (1988), until 1981 neomycin was the standard to which other therapies including lactulose were compared. Since that time, lactulose has assumed the status of standard therapy to which other agents have been compared.

Part of the appeal of lactulose is that it has been relatively free of serious side effects. Severe hypernatremia has been reported, usually in the setting of profuse diarrhea and hemoconcentration resulting from continuous administration of inappropriately large doses of lactulose (Conn, 1988). Up to 15% of patients receiving lactulose have also reported significant nausea, flatulence, cramping abdominal pain, and diarrhea. The severe cramping and flatulence frequently observed early in the course of lactulose therapy has generally resolved after a few days, consistent with more efficient metabolism of lactulose by the enteric flora (Florent et al., 1985). Also, the timing of lactulose's cathartic effect may be unpredictable which can interfere with compliance.

Second Generation Lactuloses

Lactulose is not the only disaccharide that has been effective in the treatment of hepatic encephalopathy. Other disaccharides that are not efficiently absorbed by the small intestine have similar treatment effects. Welch et al. (1974) demonstrated that lactose was effective in the treatment of encephalopathy in a lactase deficient individual. Subsequent controlled trials by Uribe and coworkers similarly demonstrated efficacy of oral lactulose in the treatment of both acute and chronic encephalopathy in lactase deficient individuals (Uribe, 1980; Uribe et al., 1987) as well as the general usefulness of lactose enemas (Uribe et al., 1981; 1987). The treatment effects of lactose alone may explain the prior observation that the ingestion of dairy protein improved hepatic encephalopathy when compared to animal protein (Fenton et al., 1966).

Lactitol is an alternative disaccharide which has been more recently studied. This sugar is derived by reducing the glucose moiety of lactose. The resulting product is considerably more pure than commercial lactulose which contains significant amounts of lactose, galactose and other sugars. Lebek and Luginbuhl (1988) found some differences in the bacterial species that metabolized lactulose vs. lactitol, but the mechanism of action of the two disaccharides was similar. Clinical studies have generally found that lactitol was at least as effective as lactulose in the treatment of both acute and chronic encephalopathy. Morgan and Hawley (1987) found that lactitol caused a significantly more rapid improvement within the first 72 hours than lactulose in patients with acute encephalopathy. Many patients reportedly preferred lactitol to lactulose because it was less sweet tasting. There has been a trend toward less frequent nausea, unpredictable diarrhea and intestinal cramping with lactitol than with commercial lactulose, although there has been no statistically significant reduction in the incidence of these side effects (Morgan et al., 1987; Morgan and Hawley, 1987; Uribe et al., 1987). In some patients experiencing side effects from either lactitol or lactulose, either drug may be a useful alternative for the other. Whether lactitol becomes the drug of choice will be in part dependent on its cost. At present, it is not available in the United States.

Crystalline or highly purified lactulose has also been recently evaluated as an alternative therapy. Purified lactulose can avoid the side effects that would be caused by other disaccharides and monosaccharides that contaminate commercial lactulose. Rossi-Fanelli et al. (1988) demonstrated in a cross-over study that crystalline lactulose caused significantly fewer side effects than commercial lactulose. Otherwise, either lactulose preparation appeared to be therapeutically equivalent. In patients who are unable to tolerate lactulose,the use of crystalline lactulose would have appeal, although presently the availability of this compound is limited and its cost is uncertain.

COMBINATION THERAPY WITH LACTULOSE AND ANTIBIOTICS

A question of considerable practical importance is whether combining lactulose with an antibiotic such as neomycin or metronidazole can give a better therapeutic effect than single agent therapy alone. There is reason to believe that a significant percentage of patients will respond better to a combination regimen but the data on this point is mixed.

Theoretical Basis for Combination Therapy

In considering a rationale for combination therapy, it must be emphasized that the beneficial effects of lactulose depend on its metabolism by the colonic flora (Conn et al., 1977). Can lactulose be metabolized by a subset of intestinal bacteria which is resistant to neomycin and some other antibiotic of choice? Antibiotics that inhibit the growth of lactulose metabolizing bacteria might be effective in their own right, but at the expense of abolishing the effects of lactulose. Alternatively the effectiveness of neomycin could be reduced by the metabolism of lactulose since an acidic environment reduces the activity of the aminoglycoside antibiotics (Sabath and Tuftegard, 1974). In fact, because of the variability of the colonic flora between individuals, probably no two patients will have the same response to combination therapy.

A number of bacterial species capable of metabolizing lactulose are likely to be resistant to neomycin. Although the facultative anaerobes are almost uniformly sensitive to neomycin (Lorian, 1986), anaerobic organisms out number aerobes by a factor of 100 to 10,000 (Simon and Gorbach, 1984). Anaerobes are uniformly resistent to neomycin (Ristuccia, 1984) because transport of the aminoglycosides into bacteria is an energy dependent process which requires molecular oxygen (Hancock, 1981). Anaerobes would thus be likely to survive and metabolize lactulose in the presence of neomycin. Metronidazole, on the other hand, might be expected to inhibit a larger proportion of the lactulose metabolizing bacteria because of its effectiveness against anaerobes.

Metabolic Data Concerning Additive Effects of Lactulose and Antibiotics

Data has been conflicting as to whether there were additive metabolic effects of lactulose and neomycin when administered to patients with cirrhosis. Monitoring stool pH has been one way of assessing the interaction of these drugs. If the stool remained acidic after the introduction of an antibiotic, lactulose metabolizing bacteria probably survived and one might have expected an additional effect of the antibiotic. On the other hand, if the stool became neutral, lactulose metabolizing bacteria had probably been eliminated by the antibiotic. A third possible outcome was that the stool pH would rise, although not to neutrality, indicating that the capacity of the colonic flora to metabolize lactulose was only partially impaired. Conn and Lieberthal (1979) studied 22 patients who either received lactulose or neomycin followed by combination treatment. Approximately one-third of the group demonstrated each of the three postulated outcomes on combination therapy: a) stool pH remained acidic, b) pH rose to neutrality or c) pH rose but remained within the acidic range. Our laboratory reported 6 patients whose stool pH was acidic during lactulose therapy and showed no change after the addition of 4 g/day of neomycin (Weber et al., 1982). Uribe (1988) found that stools remained acidic in 40 patients initially treated with either lactulose or lactitol followed by 3 g/day of neomycin.

Contradictory findings were reported by Patil et al. (1987). Using a radiotelemetry device to monitor colonic pH, they found that in 3 normal subjects, neomycin (2 to 4 g/day) abolished the fall in the right colonic pH caused by lactulose. They also found that in higher concentrations, neomycin inhibited lactulose metabolism in a fecal incubation system. One singular aspect of these studies was that lactulose alone did not significantly reduce the fecal pH of these patients indicating that they received a subtherapeutic dose of lactulose.

We have used the assessment of urea production and degradation rates in cirrhotic patients to assess the metabolic effects of combination therapy (Weber et al., 1982). Six stable cirrhotic patients were studied during treatment periods lasting 8 to 9 days. Lactulose was given in doses to produce 2 to 4 bowel movements per day and then neomycin in a dose of 4 g/day was added to lactulose. The addition of neomycin caused a 70% decrease in the urea degradation rate which entirely accounted for a 14% reduction in the urea production rate. The intestinal clearance rate also fell by 68%. Therefore, neomycin clearly inhibited urea breakdown, but there was no evidence that neomycin reversed the effects of lactulose since stool nitrogen excretion remained high and urinary urea excretion remained low. As previously mentioned, stool pH was acidic during the administration of lactulose and showed no rise when neomycin was added.

Effects on Urea Metabolism after Neomycin was Added to Lactulose

	Lactulose	Lactulose+ neomycin	Change
Urea production (mg N)[a]	119 ± 10	102 ± 8	-17 ± 4[b]
Urea appearance (mg N)[a]	92 ± 8	94 ± 8	+2 ± 1
Urea degradation (mg N)[a]	27 ± 4	9 ± 2	-18 ± 4[b]
Plasma urea (mg N/dl)	14 ± 2	14 ± 3	--
Urea pool (mg N/kg)	76 ± 11	74 ± 13	-2 ± 3
Renal urea Clearance (ml)[a]	665 ± 65	695 ± 74	+30 ± 28
Intestinal urea Clearance (ml)[a]	182 ± 32	58 ± 15	-123 ± 24[c]

[a] *Results are expressed 24 hr* $^{-1} \cdot kg^{-1}$. [b] *p<0.01.*

[c] *p<0.005. N = nitrogen.*

We have further analyzed the output of fecal solids in their distribution in fiber bacterial insoluble fraction of stool (Weber et al., 1988). Nitrogen in the bacterial fraction was significantly reduced after the introduction of neomycin from 1.4 ± 0.2 to 1.0 ± 0.1 g/day. This relatively small reduction in bacterial nitrogen indicated that a significant proportion of the colonic flora probably remained viable after the administration of neomycin.

A study by Pirotte et al. (1974) also indicated that lactulose and neomycin may act additively to reduce ammonia production. In their study, 9 cirrhotic subjects were evaluated during a control period and during periods when they received lactulose alone (45 g/day), neomycin alone (1 g/day), and a combination of lactulose and neomycin in the same dosages. Both lactulose and neomycin decreased the fasting ammonia concentration when compared to control, but ammonia was further reduced during combination therapy when compared to either lactulose or neomycin alone. Combination therapy also caused lower peak ammonia concentrations after an ammonia load than seen with single agent therapy with either lactulose or neomycin. One criticism of this study is that both agents were given in doses which would generally be considered substandard.

Available data does therefore suggest that the combination of lactulose with an antibiotic such as neomycin can cause additive metabolic effects on nitrogen metabolism that may be beneficial to some patients with hepatic encephalopathy.

Therapeutic Effects of Combination Therapy

There is little clinical information on the effects of combination therapy in the treatment of hepatic encephalopathy itself. Imler et al. (1971) found that neither lactulose (45 g/day) nor neomycin (2 g/day) individually normalized the mental state of two patients with hepatic encephalopathy, but a combination of lactulose and neomycin did cause further improvement in mental status, blood ammonia and EEGs. In a preliminary report, Uribe (1988) described 40 patients who initially received either lactose, lactulose or lactitol followed by 3 g/day of neomycin. Neomycin did not cause any improvement in the PSE index, but whether or not there were subgroups of patients that improved was not indicated.

We have studied eight patients with stable cirrhosis and mild (no greater than grade I) encephalopathy who had an inadequate response to lactulose alone when given in doses to produce 2 to 4 acidic bowel movements per day. Patients were then systematically evaluated during a control period, a lactulose period, and a period when neomycin (4 g/day) was added to lactulose; each period lasted at least 7 days. Patients were closely monitored on a metabolic ward and received a constant protein intake of either 1 g or 0.6 g/kg body weight depending on individual tolerance. Liver tests, blood electrolytes and renal function were stable throughout the study, and patients were free of infection, gastrointestinal bleeding, renal failure and electrolyte abnormalities. Blood ammonia (Seligson and Hirahara, 1957), Number Connection Tests (Conn, 1977) and electroencephalograms (Parsons-Smith et al., 1967) were all performed and evaluated by blinded observers on two separate occasions at the end of each period.

Lactulose alone had no significant effect on any parameter except stool pH when compared to the control period. After neomycin was added to lactulose, two of the eight patients showed no further improvement in encephalopathy or in blood ammonia concentrations, and both of these patients developed less acidic stools. The other six patients whose stools remained acidic had significant improvement in blood ammonia levels and Number Connection Tests when compared to the control period. In four of these six patients, EEGs also showed substantial improvement on combination therapy when compared to the lactulose or control period.

Additive Effect of Lactulose and Neomycin in Chronic Hepatic Encephalopathy (n=6)

	Control	Lactulose	Lactulose + Neomycin
Blood ammonia (mg nitrogen per dl)	175±13	161±19	129±11*
Number Connection Tests (seconds)	56±3	51±4	45±4*
Stool pH	7.0±0.2	5.3±0.3*	5.3±0.3*

**Significantly different from control by analysis of variance and multiple comparison procedures.*

A rationale does exist for the use of combination therapy, and when there has been an incomplete response to lactulose alone, this study supports the thought that combination therapy may be more effective. In this regard, there is the most experience with neomycin among the antibiotics and there is a theoretical rationale to suggest that this relatively narrow spectrum antibiotic is more likely to permit lactulose-metabolizing bacteria to survive than metronidazole. If patients are treated with combination therapy stool pH should be monitored, and if one observes a significant rise after the antibiotic is added, combination therapy should probably be discontinued. In this situation, one would most likely be left with neomycin as the sole effective therapeutic agent. Depending on the assessment of clinical response, one may stop the antibiotic and restart lactulose or continue neomycin alone since individual patients may have a better response to one or the other agent.

CONCLUSION

The conventional therapy of acute and chronic hepatic encephalopathy which has consisted of antibiotics and more recently of nonabsorbable disaccharides, lactulose in particular, has centered around a manipulation of the colonic microflora. Both antibiotics and lactulose do reduce the blood ammonia concentration, and this correlates reasonably well with clinical improvement in these patients. Antibiotics and nonabsorbable carbohydrates also have the potential to alter many other aspects of bacterial nitrogen metabolism. Whether the beneficial effects of lactulose and antibiotics are mediated by a reduction in other nitrogenous compounds that normally enter portal blood remains to be determined.

REFERENCES

Agostini L, Down PF, Murison J, and Wrong OM. (1972) Faecal ammonia and pH during lactulose administration in man: comparison with other cathartics. *Gut* **13**, 859-866.

Atterbury CE, Maddrey WC, and Conn HO. (1978) Neomycin-sorbitol and lactulose in the treatment of acute portal-systemic encephalopathy. *Am J Dig Dis* **23**, 398-406.

Berk DP and Chalmers T. (1970) Deafness complicating antibiotic therapy of hepatic encephalopathy. *Ann Int Med* **73**, 393-396.

Bircher J, Müller J, Guggenheim P, Haemmerli UP. (1966) Treatment of chronic portal-systemic encephalopathy with lactulose. *Lancet* **1**, 890-892.

Bircher J, Haemmerli UP, Trabert E, Largiader F, and Mocetti T. (1971) The mechanism of action of lactulose in portal-systemic encephalopathy. Nonionic diffusion of ammonia in the canine colon. *Rev Éurop Etudes Clin et Biol* **16**, 352-357.

Bircher J, Haemmerli UP, and Scollo-Lavizzavi G, and Hoffman K. (1971) Treatment of chronic portal-systemic encephalopathy with lactulose. Report of six patients and review of the literature. *Am J Med* **51**, 148-159.

Bolton RP. (1979) Clostridium difficile-associated colitis after neomycin treated with metronidazole. *Br Med J* **2**, 1479-1480.

Breen KJ, Bryant RE, Levinson JD, and Shenker S. (1972) Neomycin absorption in man. Studies of oral and enema administration and effect of intestinal ulceration. *Ann Int Med* **76**, 211-218.

Castell DO and Moore EM. (1971) Ammonia absorption from the human colon. The role of nonionic diffusion. *Gastroenterology* **60**, 33-42.

Conn HO and Floch MH. (1970) Effects of lactulose and lactobacillus acidophilus on the fecal flora. *Am J Clin Nutr* **23**, 1588-1594, 1970.

Conn HO. (1977) Trailmaking and number connection tests in the assessment of mental state in portal-systemic encephalopathy. *Am J Dig Dis* **22**, 541-550.

Conn HO, Leevy CM, Vlahcevic AR, Rodgers JB, Maddrey WC, Seeff L, and Levy LL. (1977) Comparison of lactulose and neomycin in the treatment of chronic portal systemic encephalopathy. A double-blind controlled trial. *Gastroenterology* **72**, 573-583.

Conn HO and Lieberthal MM. (1979) Lactulose and neomycin, in *The Hepatic Coma Syndromes and Lactulose*, pp. 340-345. Williams and Wilkins Co., Baltimore.

Conn HO. (1988) Adverse reactions and side effects of lactulose and related agents, in *Hepatic Encephalopathy Management with Lactulose and Related Carbohydrates* (Conn H.O. and Bircher J., eds), pp. 199-206. Medi-Ed Press, East Lansing.

Dawson AM, McLaren J, and Sherlock S. (1957) Neomycin in the treatment of hepatic coma. *Lancet* **2**, 1262-1268.

Down PF, Agostini L, Murison J, and Wrong OM. (1972) The interrelations of faecal ammonia, pH and bicarbonate: evidence of colonic absorption of ammonia by nonionic diffusion. *Clin Sci* **43**, 101-114.

Fenton JC, Knight EJ, and Humpherson PL. (1966) Milk and cheese diet in portal-systemic encephalopathy. *Lancet* **1**, 164-166.

Fisher CJ and Faloon WW. (1957) Blood ammonia levels in hepatic cirrhosis: their control by the oral administration of neomycin. *New Engl J Med* **256**, 1030-1035.

Florent C, Flourie B, Leblond A, Rautuneau M, Bernier JJ, and Rambaud JC. (1985) Influence of chronic lactulose injection on the colonic metabolism of lactulose in man. *J Clin Invest* **75**, 608-13.

Greenberger NJ, Carley J, Schenker S, Bettinger I, Stamnes C, and Beyer P. (1977) Effect of vegetable and animal protein diets in chronic hepatic encephalopathy. *Am J Dig Dis* **22**, 845-855.

Hancock RE. (1981) Aminoglycoside uptake and mode of action with special reference to streptomycin and gentamycin. I. Antagonists and mutants. *J Antimicrob Chemother* **8**, 249-276.

Hoffmann K, Mossell DAA, Korus W, and vandeKamer JH. (1964). Investigation on the mode of action of lactulose (β-galactosido-fructose) in human intestine. *Klin Wschr* **42**, 126-130.

Imler M, Kurtz D, and Bockel R. (1971) Etude comparative du traitment de l'encephalopathie porto-cave par le lactulose, les bacilles lactiques et les antibiotiques. *Thérapeutique* **47**, 237-248.

Ingelfinger F. (1964-65) Editorial comments, in *Year Book of Medicine* (Beeson PB et al., eds.), pp. 591-592. Year Book Medical Publishers, Chicago.

Jacobson ED, Chodos RB, and Faloon WW. (1960) An experimental malabsorption syndrome induced by neomycin. *Am J Med 28*, 524-533.

Lebek G and Luginbuhl M. (1988) Effect of lactitol and lactulose on human intestinal flora in portal-systemic encephalopathy, in *Hepatic Encephalopathy Management with Lactulose and Related Carbohydrates* (Conn HO and Bircher J, ed.), pp. 271-281. Medi-Ed Press, East Lansing.

Lorian V. (1986) *Antibiotics in Laboratory Medicine*, Second ed. Williams and Wilkins, Baltimore.

Macbeth WAAG, Kass E, and McDermott WV Jr. (1965) Treatment of hepatic encephalopathy by alteration of intestinal flora with lactobacillus acidophilus. *Lancet* 1, 399-403.

Mason VC. (1974) An explanation of the effect of lactulose in the treatment of hepatic encephalopathy (letter). *Gastroenterology* 66, 1271.

Morgan MH, Read AE, Speller DC. (1982). Treatment of hepatic encephalopathy with metronidazole. *Gut* 23, 1-7.

Morgan MY and Hawley KE. (1987) Lactitol vs. lactulose in the treatment of acute hepatic encephalopathy in cirrhotic patients: a double-blind, randomized trial. *Hepatology* 7, 1278-1284.

Morgan MY, Hawley KE, and Stambuk D. (1987) Lactitol versus lactulose in the treatment of chronic hepatic encephalopathy. A double-blind, randomized, cross-over study. *J Hepatol* 4, 236-244.

Najarian JS, Jew J, Dakin RL, Harper HA, Quinnell CM, and McCorkle HJ. (1959) Control of ammonia production in the colon with neomycin enemas. *AMA Arch Surg* 78, 844-850.

Orlandi F, Freddara U, Candelaresi MT, Morettini A, Corazza GR, DiSimone A, Dobrilla G, and Cavallini G. (1981) Comparison between neomycin and lactulose in 173 patients with hepatic encephalopathy. A randomized clinical study. *Dig Dis Sci* 26, 498-506.

Orlandi F, Brunelli E, Benedetti A, and Macarri G. (1988). Clinical trials of lactulose therapy in hepatic encephalopathy, in *Hepatic Encephalopathy Management with Lactulose and Related Carbohydrates* (Conn HO and Bircher J, eds.), pp. 191-198. Medi-Ed Press, East Lansing.

Parsons-Smith BG, Summerskill WHJ, Dawson AM, and Sherlock S. (1967) The electroencephalograph in liver disease. *Lancet* **2**, 867-871.

Patil DH, Westaby D, Mahida YR, Palmer KR, Rees R, Clark ML, Dawson AM, and Silk DB. (1987) Comparative modes of action of lactitol and lactulose in the treatment of hepatic encephalopathy. *Gut* **28**, 255-259.

Perman JA, Molders S, and Olson AC. (1981) Role of pH in production of hydrogen from carbohydrates by colonic bacterial flora. Studies in vivo and in vitro. *J Clin Invest* **67**, 643-650.

Phear EA, Ruebner B, Sherlock S, and Summerskill WHJ. (1956) Methionine toxicity in liver disease and its prevention with chlortetracycline. *Clin Sci* **15**, 93-117.

Pirotte J, Guffens JM, and Devos J. (1974) Comparative study of basal arterial ammonemia and of orally-induced hyperammonemia in chronic portal systemic encephalopathy, treated with neomycin, lactulose, and association of neomycin and lactulose. *Digestion* **10**, 435-444.

Ristuccia AM. (1984) Aminoglycosides, in *Antimicrobial Therapy* (Ristuccia AM and Cunha BA, eds.), pp. 305-328. Raven Press, New York.

Rossi-Fanelli F, Cangiano C, Ceci F, Familiari M, Muscaritoli M, Riggio O, and Cascino A. (1988) Long-term treatment with crystalline lactulose, in *Hepatic Encephalopathy Management with Lactulose and Related Carbohydrates* (Conn HO and Bircher J, eds.), pp. 267-270. Medi-Ed Press, East Lansing.

Sabath LD and Tuftegard I. (1974) Rapid microassays for clindamycin and gentamycin when present together and the effect of pH and of each other on the antibacterial activity of the other antimicrobial agents. *Chemother* **6**, 54-59.

Seligson P and Hirahara K. (1957) Measurement of ammonia in whole blood, erythrocytes, and plasma. J Lab Clin Med **49**, 962-974.

Simon GL and Gorbach SL. (1984) Intestinal flora in health and disease. *Gastroenterology* **86**, 174-193.

Stephen AM and Cummings JH. (1980) Mechanism of action of dietary fiber in the human colon. *Nature* **284**, 283-284.

Tarao K, Ikeda T, Hayashi K, and Sakurai A. (1985) Clinical significance of anaerobic gram negative bacteria in hepatic encephalopathy and therapeutic effects of vancomycin hydrochloride on hepatic encephalopathy. *Jap J Gastroenterol* **82**, 1875-1883.

Tisdale WA, Fenster LF, and Klatskin G. (1960) Acute Staphylococcal enterocolitis complicating oral neomycin therapy in cirrhosis. *New Engl J Med* **263**, 1014-1016.

Uribe M. (1980) Treatment of chronic portal-systemic encephalopathy with lactose in lactase-deficient patients. *Dig Dis Sci* **25**, 924-928.

Uribe M, Moreno BJ, and Lewis H. (1981) Lactose enemas plus placebo tablets vs. neomycin tablets plus starch enemas in acute portal systemic encephalopathy. *Gastroenterology* **81**, 101-106.

Uribe M, Campollo O, Vargas F, Ravelli GP, Mundo F, Zapata L, Gil S, and Garcia-Ramos G. (1987) Acidifying enemas (lactitol and lactose) vs. nonacidifying enemas (tap water) to treat acute portal systemic encephalopathy: a double blind randomized clinical trial. *Hepatology* **7**, 639-643.

Uribe M, Toledo H, Perez F, Vargas F, Gil S, Garcia-Ramos G, Ravelli GP, Guevara L. (1987) Lactitol, a second generation disaccharide for treatment of chronic portal-systemic encephalopathy. A double-blind, crossover, randomized clinical trial. *Dig Dis Sci* **32**, 1345-1353.

Uribe M. (1988) Management of portal-systemic encephalopathy with lactose, in *Hepatic Encephalopathy Management with Lactulose and Related Carbohydrates* (Conn HO and Bircher J, eds.), pp. 283-300. Medi-Ed Press, East Lansing.

Vince A, Dawson AM, Park M, and O'Grady F. (1973) Ammonia production by intestinal bacteria. *Gut* **14**, 171-177.

Vince A, Zeegen R, Drinkwater JE, O'Grady F, and Dawson AM. (1974) The effect of lactulose on the faecal flora of patients with hepatic encephalopathy. *J Med Microbiol* **7**, 163-168, 1974.

Vince A, Killingley M, and Wrong OM. (1978) Effect of lactulose on ammonia production in a fecal incubation system. *Gastroenterology* **74**, 544-549, 1978.

Vince AJ and Burridge SM. (1980) Ammonia production by intestinal bacteria: the effects of lactose, lactulose, and glucose. *J Med Microbiol* **13**, 177-191.

Walser M and Bodenlos LJ. (1959) Urea metabolism in man. *J Clin Invest* **38**, 1617-1626.

Weber FL Jr. (1979) The effect of lactulose on urea metabolism and nitrogen excretion in cirrhotic patients. *Gastroenterology* **77**, 518-523.

Weber FL Jr and Veach GL. The importance of the small intestine in gut ammonium production in the fasting dog. (1979) *Gastroenterology* **77**, 235-240.

Weber FL Jr. (1981) The use of lactulose in the treatment of portal-systemic encephalopathy, in *Nutritional Pharmacology* (Spiller GA ed.), pp. 217-253. Alan R. Liss Inc., New York.

Weber FL Jr and Fresard KM. (1981) Comparative effects of lactulose and magnesium sulfate on urea metabolism and nitrogen balance in cirrhotic subjects. *Gastroenterology* **80**, 994-998.

Weber FL Jr, Fresard KM, and Lally BR. (1982) Effects of lactulose and neomycin on urea metabolism in cirrhotic subjects. *Gastroenterology* **82**, 213-217.

Weber FL Jr, Minco D, Fresard KM, and Banwell JG. (1985) Effects of vegetable diets on nitrogen metabolism in cirrhotic subjects. *Gastroenterology* **89**, 538-44.

Weber FL Jr, Banwell JG, Fresard KM, and Cummings JH. (1987) Nitrogen in fecal bacterial, fiber, and soluble fractions of patients with cirrhosis: Effects of lactulose and lactulose plus neomycin. *J Lab Clin Med* **110**, 259-263.

Welch JD, Cassidy D, Prigatano GP, and Gunn CG. (1974) Chronic hepatic encephalopathy treated with oral lactose in a patient with lactose malabsorption. *New Engl J Med* **291**, 240-241.

Wolpert E, Phillips SF, and Summerskill WHJ. (1970) Ammonia production in the human colon. Effects of cleansing, neomycin and acetohydroxamic acid. *New Engl J Med* **283**, 159-164.

Zeegen R, Drinkwater J, Fenton J, Vince A, and Dawson AM. (1970) Some observations on the effects of treatment with lactulose on patients with chronic hepatic encephalopathy. *Quart J Med* **39**, 245-263.

ACKNOWLEDGMENT: These studies were supported in part by the Veterans Administration Medical Research Service. Patricia Howell provided excellent secretarial assistance.

NUTRITIONAL THERAPY OF PORTAL-SYSTEMIC ENCEPHALOPATHY: THE BRANCHED CHAIN AMINO ACID STORY

Harold O. Conn, M.D.
V.A. Medical Center West Haven
Yale University School of Medicine
New Haven, CT 06510 U.S.A.

Ljusk Siw Eriksson, M.D.
Huddinge University Hospital
Karolinska Institute
Huddinge, Sweden

Fools rush in where angels fear to tread.
Alexander Pope

A frequently asked question of hepatologists is, "What is the real value of branched chain amino acids (BCAAs) in portal-systemic encephalopathy (PSE)?". Despite the publication of a number of randomized clinical trials (RCTs), the answer remains uncertain. BCAAs were introduced for the treatment of PSE on the basis of several observations. First, many investigators had reported that plasma concentrations of BCAA (leucine, valine, isoleucine) are greatly reduced in patients with PSE, ranging from 40 to 80% of normal levels (1-4). Second, plasma levels of the aromatic amino acids (AAA), tyrosine, phenylalanine and tryptophan, have been shown to be elevated (25% to 200% above normal concentrations) (1-3,5). This imbalance results in an abnormally low plasma BCAA/AAA ratio (1.0 to 1.5), which is normally >3.0. [The ratio is calculated by totalling the molar concentrations of leucine, valine and isoleucine and dividing that sum by that sum of the molar concentrations of phenylalanine, tyrosine plus tryptophan (2)]. Third, BCAAs, which are metabolized mainly in muscle, are benign, noncomagenic substances that can be administered to cirrhotic patients without toxicity (2). Fourth, AAAs, which are metabolized primarily in the liver are said to result in the synthesis of false neurotransmitters such as octopamine and phenylethanolamine, and the reduction of norepinephrine and dopamine levels in the brain (6). A reasonable hypothesis involving glutamine synthesis was proposed to explain these diverse findings (6). It was postulated that the presence of increased intake of nitrogenous substances in cirrhotic patients results in an increase in ammonia levels in blood and brain and, consequently, in elevated glutamine concentrations in the blood, cerebrospinal fluid and brain. Glutamine synthesis "protects" central nervous tissues from the toxic effects of high levels of ammonia. The increased intracerebral levels were postulated to induce an efflux of glutamine from the brain to the plasma in exchange for plasma amino acids. Since BCAAs and AAAs share the same carrier system for transport across the blood brain barrier (8), the plasma imbalance leads to increased availability of and the preferential uptake of AAAs and, presumably to the enhancement of the synthesis of false neurotransmitter substances, such as octopamine, in the brain (9). Injury to the blood brain barrier which occurs in patients with acute hepatic necrosis (10) and in experimental animals (11,12) and which was postulated to exist in cirrhotic patients, could contribute to the influx of AAAs. Fifth, in healthy subjects, the administration of BCAAs increases plasma levels of BCAA's and induces a decrease in phenylalanine and tyrosine in the blood (14-16). Similar findings have been reported in cirrhotic patients (3,13,17). Indeed, infusions of BCAAs in cirrhotic patients reduce the increased uptake of phenylalanine and tyrosine by the brain and may enhance the uptake of BCAAs (3). Finally, administration of BCAA-enriched, AAA-depleted mixtures of amino acids have been postulated to restore normal plasma levels of the amino acids and normal central nervous system neurotransmission (13).

Randomized, controlled therapeutic trials to assess the efficacy of BCAA therapy and/or supplementation have been undertaken by a number of investigators. Our analysis of the RCTs was undertaken to try to answer as objectively as possible the question, "Is BCAA therapy effective in PSE?". The authors, both of whom had conducted such RCTs (18-20), have assessed each RCT independently.

Seven RCTs in which BCAAs had been administered to patients with acute PSE and four in which they had been given to patients with chronic PSE that had been identified and published in peer-reviewed journals are the basis of this analysis. In each study the experimental treatment had been compared with control therapy which varied from study to study. Six of the 11 were double-blind studies (3 in acute and 3 in chronic PSE trials). All seven acute studies were parallel comparisons as was one of the investigations of chronic PSE. The other three were crossover comparisons of the two therapies. The duration of the investigations in acute PSE ranged from 3 to 14 days and in chronic PSE from 11 to 56 days. Appropriate randomization had been performed in all 11 investigations. More than 80 variables were assessed for each study. Critical requirements of ideal RCTs (21) that had not been observed in these investigations are commented upon.

BCAA therapy in acute hepatic encephalopathy

Rossi-Fanelli et al. from Rome reported their results in 40 patients with grade 3 or 4 PSE (17) (Table 1). The patients, one-third of whom had alcoholic cirrhosis, had PSE of various causes. One-third had been precipitated by bacterial infection, one-fifth by hemorrhage from varices and the remaining 40% from miscellaneous causes. Seventeen patients in the control group received 400 gm glucose intravenously plus lactulose and 17 in the treatment group a commercial BCAA solution (BS 692) (Table 2) for four days. If either therapy failed, both were given together.

Table 1

USE OF BCAA'S IN ACUTE HE

First Author	No.	Episodes	Therapy	BCAA
Rossi-Fanelli	40	17	GLUC(G)-LACT(L)	0
		17	G-BS 692	57
Wahren	50	25	G-L	0
		25	G-L-BCAA	40
Cerra	75	35	G-NEO(N)	0
		40	G-F080	28
Michel	70	34	G-L-AA	15
		36	G-L-BCAA	28
Strauss	29	16	G-N	--
		16	G-F080	--
Fiaccadori	48	16	G-L	0
		16	G-BS 666	22
		16	G-BS 666-L	22
Vilstrup	65	33	G-L	0
		32	G-L-Pf8	29

G=glucose ; L=lactulose ; N=Neomycin ; BS692, F080, Pf8 are commercially available BCAA-enriched solutions of amino acids. Underlined first authors indicate double-blind trials

Mental state was restored more frequently with BCAA (70%) than with lactulose (47%), a difference that is not statistically significant (NS) (Table 3). Mortality was similar in the two groups (18% vs 24%; NS). The BCAA:AAA ratio which was 1.2-1.3 in the two groups before therapy rose to normal values with BCAA, but remained low with lactulose therapy. Blood ammonia levels fell in both groups, more with lactulose than with BCAA.

The authors concluded that BCAA therapy was as effective as lactulose. The design of the study, however, does not permit such a conclusion. Many of the patients had PSE precipitated by infection and received antibiotic therapy. Since antibiotic therapy may prevent the effects of lactulose, this group was at a disadvantage. Similarly, many of the patients in both groups had PSE precipitated by gastrointestinal hemorrhage, and the control of PSE in such cases is determined more by the efficacy of the treatment of the bleeding than by the treatment of PSE, per se. Finally, the randomization had not provided similar groups of patients; the control group included two to three times as many patients with alcoholic cirrhosis as did the BCAA group. Although BCAA therapy seemed effective, it is not clear that the BCAA treatment was responsible for the clinical improvement.

Wahren et al. studied 50 cirrhotic patients from France and Sweden almost all of whom had grade 3 or 4 PSE (19) (Table 1). The control group received 100 gm glucose and lipid intravenously and the BCAA group the same substances plus 40 gm pure BCAAs intravenously, i.e., no other amino acids were given. Treatment was administered for a minimum of five days. The two groups, three-fourths of whom had alcoholic cirrhosis, were similar demographically except that those in the BCAA group were somewhat older ($p<0.05$). More of the control group had PSE precipitated by bacterial infection.

One-half the episodes in both groups improved (Table 3). The mortality rate was higher in the BCAA group (40%) than in the control group (20%; NS). The causes of death were similar. The

Table 2

COMPARISON OF COMPOSITION OF
BCAA SOLUTIONS (gm per liter)

Solution	BS 962	KV	F080	MOD AA	BS-666	Pf8
Isoleucine	9	2	9	9	9	12
Leucine	11	14	11	11	11	14
Valine	8.4	4	8.4	8.4	8.4	10
Tyrosine	0	0	0	0	0	0
Phenylalanine	0	0	1	1	1	0.4
Methionine	0	0	1	1	1	0.8
Tryptophan	0	0	0.7	0.8	0.8	0.2
Total nitrogen	4.5	3.2	12.5	12.5	12.5	14

BCAA:AAA ratio was low in both groups (1.0-10.2) and remained so in the control group, but rose to 2.0 in the BCAA group ($p<0.01$). Blood ammonia levels did not change. After therapy the EEGs were similar in the two groups.

In this short term study the only nitrogenous substance taken by the BCAA group was 40 g of BCAAs; the control group received none at all. Despite the presumed negative nitrogen balance, the control group showed a lower mortality than the BCAA-treated group. This study shows that pure BCAAs do not reverse the PSE, although they can restore the plasma BCAA:AAA ratio toward normal. Thus, this investigation does not support the BCAA hypothesis.

Cerra et al. in a multicentered, American RCT studied 75 patients with severe decompensated cirrhosis and acute PSE (grade 2 to 4) (Table 1) who received either neomycin plus intravenous glucose (control group) or Hepatamine (F080) (Table 2) plus glucose (BCAA group) (22). The study was conducted as a simple comparison, but if the patients did not respond during the first three days of treatment they were switched to the other therapy. No oral nutriments were given as long as the PSE persisted, after which diet was given ad lib. Treatment was continued for up to 14 days.

Encephalopathy and EEGs improved in both groups, and was said to do so faster and to a greater degree in the BCAA group. Nitrogen balance remained grossly negative in the control group, but became positive in the BCAA group. The BCAA:AAA ratio was abnormally low in both groups and remained so in the control group, but became normal in the BCAA group. The number of deaths was greater in the BCAA group during the first 3 days than in the control group (9 vs 6), but lower from day 4 to 14 (2 vs 3 or 3 vs 7) (Table 4). After the study terminated on day 14, there were fewer deaths in the BCAA than the control group (2 vs 9) during the remainder of the hospital admission (Table 4). The total number of deaths were fewer in the BCAA group ($p<0.05$). The authors concluded that BCAA therapy is more effective than neomycin.

A study that starts as a parallel comparison but becomes a cross-over study is difficult to evaluate. It was further complicated by the dropping out of 16 patients with severe medical complications. The authors' mortality analysis, which is based on intention to treat, includes a number of deaths that occurred after the study was concluded, but before the patient was discharged from the hospital. Reanalysis of the data based on the treatment received shows no difference in survival with the two therapies. We conclude that no real benefits of BCAA therapy have been proved in this investigation. Analysis of the improvement in PSE is even more complex, since early dropouts and different nitrogen intakes make it almost impossible for the reader to calculate.

Michel et al. in a multicentered trial in France studied 70 patients most of whom had alcoholic cirrhosis (23) (Table 1). The control group received a conventional commercial mixture of amino acids intravenously and the treatment group of BCAA-enriched, AAA-depleted solution (MOD AA) (Table 2). Both were given in progressively increasing amounts. Both received glucose and lipid supplements intravenously. Treatment was continued from 5 days. The groups were similar in demographic characteristics and in the causes of PSE.

BCAA:AAA ratios increased in the BCAA group but were unchanged in the control group. However, encephalopathy and EEG improved in about one-third of both groups (Table 3). An equal number of deaths (20%) occurred in each group.

The authors concluded that BCAA-supplemented amino acids were ineffective in PSE.

Strauss et al. in Sao Paulo in an unblinded study compared neomycin therapy with Hepatic Aid (F080) (Table 2) in 32 episodes of PSE in 29 cirrhotic patients (24) (Table 1). Sixteen episodes were treated in each group. The majority of the patients had alcoholic cirrhosis that tended to be more severe in the BCAA group. The encephalopathy was of similar degree in the two groups. The control group received 10 gm dietary protein per day plus neomycin; the BCAA group received the same diet plus the BCAA solution intravenously (Table 2).

After five days of therapy, PSE had been eradicated in 88% of both groups, but improvement appeared to be more rapid in the BCAA group (Table 3). Unfortunately, the etiology of the encephalopathy differed in the two groups and may have accounted for the differences in the speed of response. Deaths occurred in two patients in each group. Thus, the two groups appeared to respond equally to both therapies.

Table 3

EFFICACY OF THERAPY ACUTE HE

First Author	Days Rx	Response HE Control	Response HE BCAA	Mortality Control	Mortality BCAA
Rossi-Fanelli	4	8/17	12/17	24%	18%
<u>Wahren</u>	5	12/25	14/25	20%	40%
<u>Cerra</u>	14	--	--	28%	37%
Michel	5	10/34	12/36	21%	19%
Strauss	5	14/16	14/16	12%	12%
Fiaccadori	7	10/16	15/16	0%	0%
<u>Vilstrup</u>	16	17/33	17/32	34%	30%

Underlined first authors indicate double-blind trials.

Fiaccadori et al. in Parma evaluated 48 patients with cirrhosis of various types and acute HE (25) (Table 1). The patients were divided into three groups of 16 patients each who received lactulose and glucose, a BCAA-enriched, AAA-depleted amino acid solution (BS-666) (Table 2) and glucose or lactulose, BS 666 and glucose. Treatment was administered for seven days.

Treatment resulted in disappearance of coma in 62%, 94% and 100%, respectively ($p<0.01$) (Table 3). The PSE Index and EEG parallelled mental state but the differences between the three therapies are not statistically different. Blood ammonia levels decreased in the two BCAA groups, but not with lactulose alone. The BCAA:AAA ratios improved in

Table 4 CALCULATIONS AND RECALCULATIONS OF MORTALITY DATA IN RCT OF CERRA ET AL.

Treatment	No. of patients	No. of deaths during study Day 0-3	Day 4-14	Total	No. of deaths after study	Total no. of deaths Dropouts included	Dropouts excluded
Calculated by Intent to treat (Cerra et al:							
Neomycin	35	6	3	13	9	22*	16*
Crossed over to BCAA (n=7)		--	4				
				NS	p<0.05	p<0.05	p<0.01
BCAA	40	9	2	12	2	14*	9*
Crossed over to Neomycin (n=4)		--	1				
Calculated by treatment received (Cerra et al, modified by Eriksson and Conn):							
Neomycin	35	6	3	10	9	19	13
Crossed over BCAA (n=4)		--	1				
				NS	p<0.05	NS	NS
BCAA	40	9	2	15	2	17	8
Crossed over from Neomycin (n=7)		—	4				

*Figures used by Cerra et al. in calculating mortality

both BCAA groups. No deaths occurred during the period of treatment, but six in the lactulose group and one in the BCAA group died afterward. The authors' conclusion that BCAA therapy of acute HE is superior to lactulose can be challenged on several grounds.

The lactulose group received no source of nitrogen and, presumably, remained in negative nitrogen balance throughout the study period, a deficiency that could affect outcome adversely in that group. Furthermore, the clinical assessment of degrees of encephalopathy is a subjective determination, that is open to error in unblinded studies. Finally, the lack of significant improvement in either EEG or PSE does not support the investigators' opinion.

Vilstrup et al. in Copenhagen presented a preliminary report of a double-blind study in 77 patients that compared glucose plus lactulose with a BCAA-enriched mixture of amino acids (Pf 8) in glucose (Table 2) plus lactulose (26) (Table 1). The encephalopathy response rates and the mortality rates were similar (Table 3). The authors conclude that BCAAs are not beneficial.

<u>BCAA therapy in chronic PSE</u>

Schafer et al. in a crossover trial compared the effects of glucose, milk protein diet or a mixture of proteins enriched with BCAA in 8 patients with mild chronic PSE (27) (Table 5). Treatment with each therapy continued for 4 to 8 weeks during which 45 gm protein was given per day. The control protein contained 20% BCAA and the BCAA-enriched protein 34%. About half the patients received lactulose throughout.

The overall degree of PSE was not recorded, but the Number Connection Test improved during BCAA therapy. EEGs, blood ammonia levels and plasma BCAA:AAA ratios did not change in any group. The milk protein diet induced abdominal discomfort and diarrhea, which were attributed to lactase deficiency, so that patients consumed smaller amounts of protein in the casein group.

This investigation showed no benefits of BCAA supplementation.

Eriksson et al. performed a double-blind, crossover investigation that compared supplements of pure BCAA with supplements of carbohydrate in cirrhotic patients with mild PSE (18) (Table 5). All patients received a basal protein diet taken ad lib (40-100 gm per day) for the 2 to 14 day periods of study.

Table 5

USE OF BCAA'S IN CHRONIC HE

First Author	Type	Duration (days)	No.	Alcoholic
Schafer	Crossover	28-56	8	2 (25%)
Eriksson	Crossover	14	7	2 (29%)
McGhee	Crossover	12	4	—
Horst	Parallel	28	12 (PROT) 14 (BCAA)	5 (42%) 5 (36%)

Underlined first authors indicate double-blind trials.

Although the plasma BCAA:AAA ratio was restored to normal by the BCAA supplements, neither mental state, a battery of psychometric tests nor EEGs showed significant improvement.

This investigation, which studied the effects of pure BCAAs in restoring the normal BCAA:AAA relationship, showed no beneficial effects although the BCAA:AAA ratios were restored to normal.

McGhee et al. compared a casein diet that contained 19% BCAA with Hepatic Aid which contained 28% BCAA in 4 protein-intolerant patients with very mild PSE (28) (Table 5). Each period in this crossover study lasted 10 days.

No significant improvement occurred with either treatment as determined by psychometric tests and EEGs. Both diets induced positive nitrogen balance. Plasma BCAA levels did not change, but AAA concentrations decreased. This study, which contained too few patients, suggests that a casein diet and Hepatic Aid supplements are about equally effective in not improving PSE.

Horst et al. performed a double-blind, parallel comparison in 26 protein-intolerant patients with chronic, stable, minimal PSE (20) (Table 5). Both groups received 20 gm of dietary protein throughout. The control group was given progressively increasing amounts of mixed dietary protein until, by the fourth week they were taking 80 gm per day plus a carbohydrate solution. The BCAA group received Hepatic Aid in a carbohydrate solution that was identical in calories, nitrogen content, volume and osmolar concentration to the control group (Table 2). The study was continued for four weeks or was terminated whenever a patient's mental state deteriorated to grade 2 PSE or worse.

Seven of 14 patients in the protein group developed HE compared to 1 of 12 in the BCAA group ($p<0.01$). The mean mental state grade increased significantly in the protein group but remained constant or improved slightly in the BCAA group. One patient died in the protein group; none died in the BCAA group. Nitrogen balance became strongly positive in both groups. Although the BCAA:AAA ratios increased with BCAA therapy and not with the protein diet, no significant changes in ammonia levels or EEGs were noted.

This investigation suggests that a BCAA-enriched, AAA-depleted solution of amino acids is less comagenic than dietary protein and restores negative nitrogen balance to normal equally well.

DISCUSSION

These 11 investigations are diverse in experimental design and execution and subjective in interpretation. The number of patients are small, probably too small to detect any but the most gross changes, and too small to exclude false-negative results. Variations in the reasons for exclusion of potential patients in the individual investigations make the findings difficult to apply. Differences in the types of cirrhosis studied, in the ages of the patients, in the severity of the encephalopathy and in the causes of the coma have created a playground for the innocent and a quicksand swamp for the unwary. The absence of nitrogenous nutriments in some groups, the presence of inadequate amounts in others and a surplus in still others renders comparison of these investigations difficult. Despite a surprising similarity of the intravenous commercial solutions (Table 2), the types, amounts and methods of administration of BCAAs differ sufficiently to prevent unifying conclusions. It should be pointed out that the administration of <u>pure BCAAs</u> used in two investigations (18,19) addresses a different question from that posed in the other studies in which mixtures of other amino acids were administered simultaneously.

Methodologic differences in assessing the presence or degree of encephalopathy are so variable and subjective in these studies as to challenge whatever conclusions are reached. It is difficult to determine whether subgroups are responsible for the findings observed in the total groups. It is impossible to integrate the conclusions from this diversity of studies. A metaanalysist would be driven mad!

How do these studies influence the interpretation of the BCAA hypothesis?

Several findings require important conceptual changes in the hypothesis. First, it has been shown that the reductions in the BCAA/AAA ratio parallel the severity of cirrhosis and not the presence or degree of PSE (29-32). Furthermore, the typical amino acid pattern becomes progressively more abnormal as the volume of portal-systemic shunting increases in the absence of PSE (33). Thus, the hepatic encephalopathy amino acid pattern correlates better with the presence and severity of cirrhosis rather than with the presence or absence of PSE, *per se*. The low BCAA, high AAA plasma aminogram can be more accurately termed the "cirrhosis amino acid pattern".

Second, BCAA/AAA ratios do not necessarily become more normal when hepatic encephalopathy is successfully treated. In patients with PSE(29) and dogs with postportacaval shunt encephalopathy (34) the BCAA/AAA molar ratios have remained unchanged despite recovery from PSE.

Third, the postulated decrease in the brain uptake of BCAA when the BCAA/AAA ratio is reduced has not been confirmed. In cirrhotic patients with depressed plasma BCAA levels, the uptake of each of the three BCAAs by the brain was normal in the basal state (3). After intravenous infusions of the BCAAs the uptake was greater in cirrhotic patients than in normal subjects (3).

Fourth, the postulated decreased catecholamine levels in the brain have not been found in cirrhotic patients with PSE(7).

Fifth, failure of the administration of pure BCAAs to improve PSE despite normalization of BCAA levels and improvement in BCAA/AAA ratios (18,19,23) indicates that the reduction in plasma BCAAs *per se* is not responsible for PSE.

Does the failure of therapy with these BCAA-enriched supplements invalidate the BCAA false neurotransmitter hypothesis? This ingenious hypothesis is based on a series of clinical observations that are consistent with the theory. Although these contradictory therapeutic results do not support this hypothesis, they do not completely invalidate it. It appears likely that the abnormal BCAA/AAA ratio in PSE is not causally related to the induction of PSE, although it could conceivably play a contributory role. The finding in one study that BCAA enriched, AAA depleted dietary supplements are better tolerated than dietary protein (20) tends to support the hypothesis. However, virtually all attempts to confirm this hypothesis by treating patients with BCAAs or BCAA-enriched supplements have failed. We conclude, therefore, that there is no mandate for the use of branched chain amino acids in the therapy of human PSE.

REFERENCES

1 Walshe, J.M. (1953). Disturbances of amino acid metabolism following liver injury. A study by means of paper chromatography. Q.J. Med., 22, 483-507.

2 Fischer, J.E., Yoshimura, N., Aguirre, A. et al. (1974). Plasma amino acids in patients with hepatic encephalopathy. Effect of amino acid infusions. Am. J. Surg., 127, 40-47.

3 Sato, Y., Eriksson, S., Hagenfeldt, L. et al. (1981). Influence of branched-chain amino acid infusion on arterial concentrations and brain exchange of amino acids in patients with hepatic cirrhosis. Clin. Physiol., 1, 151-165.

4 Iob, V., Coon, W.W., Sloan, M. (1967). Free amino acids in liver, plasma and muscle of patients with cirrhosis of the liver. J. Surg. Res., 7, 41-43.

5 Iber, F.L., Rosen, H., Levensen, S.M. et al. (1957). The plasma amino acids in patients with liver failure. J. Lab. Clin. Med., 50, 417-425.

6 James, J.H., Ziparo, V., Jeppsson, B. et al. (1979). Hyperammonaemia, plasma amino acid imbalance and blood-brain amino acid transport: a unified theory of portal-systemic encephalopathy. Lancet, 2, 772-775.

7 Cuilleret, G., Pomier-Layrargues, G., Pons F. et al. (1980). Changes in brain catecholamine levels in human cirrhotic hepatic encephalopathy. Gut., 21, 565-569.

8 Oldendorf, W.H., Szabo, J. (1976). Amino acid assignment to one of three blood-brain barrier amino acid carriers. Am. J. Physiol., 230, 94-98.

9 Fernstrom, J.D., Wurtman, R.J. (1972). Brain serotonin content: physiological regulation by plasma neutral amino acids. Science, 178, 414-416.

10 Ware, A.J., D'Agostino, A.N., Combes, B. (1971). Cerebral edema: a major complication of massive hepatic necrosis. Gastroenterology, 61, 877-884.

11 James, J.H., Escourrou, J., Fischer, J.E. (1978). Blood brain neutral amino acid transport activity is increased after portacaval anastomosis. Science 200, 1395-1397.

12 Livingstone, A.S., Potvin, M., Goresky, C.A. (1977). Changes in the blood-brain barrier in hepatic coma after hepatectomy in the rat. Gastroenterology, 73, 697-704.

13 Fischer, J.E., Rosen, H.M., Ebied, A.M. et al. (1976). The effect of normalization of plasma amino acids on hepatic encephalopathy in man. Surgery, 80, 77-91.

14 Sherwin, R.S. (1978). Effect of starvation on the turnover and metabolic response to leucine. J. Clin. Invest., 6, 1471-1481.

15 Hagenfeldt, L., Eriksson, S., Wahren, J. (1980). Influence of leucine on arterial concentrations and regional exchange of amino acids in healthy subjects. Clin. Sci., 59, 173-181.

16 Eriksson, S., Hagenfeldt, L., Wahren, J. (1981). A comparison of the effects of intravenous infusion of individual branched-chain amino acids on blood amino acid levels in man. Clin. Sci., 60, 95-100.

17 Rossi-Fanelli, F., Angelic, M., Cangiano, C. et al. (1981). Effect of glucose and/or branched-chain amino acid infusion in plasma amino acid imbalance in chronic liver failure. J. Ent. Paren. Nutr., 5, 414-419.

18 Eriksson, L.S., Persson, A., Wahren, J. (1982). Branched chain amino acids in the treatment of chronic hepatic encephalopathy. Gut, 23, 801-806.

19 Wahren, J., Denis, J., Sesurmont, P. et al. (1984). Is intravenous administration of branched chain amino acids effective in the treatment of hepatic encephalopathy. A multicenter study. Hepatology, 4, 475-480.

20 Horst, D., Grace, N.D., Conn, H.O. et al. (1984). Comparison of dietary protein with an oral branched chain-enriched amino acid supplement with chronic portal-systemic encephalopathy: a double-blind controlled trial. Hepatology, 4, 279-287.

21 Chalmers, T.C., Smith, H. Jr., Blackburn, B. et al. (1981). A method for assessing the quality of a randomized control trial. Contr. Clin. Trials, 2, 31-49.

22 Cerra, F.B., Cheung, N.K., Fischer, J.E. et al. (1985). Disease-specific amino acid infusion (F080) in hepatic encephalopathy: a prospective, randomized, double-blind controlled trial. JPEN, 9, 288-295.

23 Michel, H., Bories, P., Aubin, J.P. et al (1985). Treatment of acute hepatic encephalopathy in cirrhotics with a branched-chain amino acid enriched versus a conventional amino acid mixture. Liver, 5, 282-289.

24 Strauss, E., dos Santos, W.R., daSilva, E.C. et al. (1986). Treatment of hepatic encephalopathy: a randomized clinical trial comparing a branched chain enriched amino acid solution to oral neomycin. Nutr. Supp. Svcs, 6, 18-21.

25 Fiaccadori, F., Ghinelli, F., Pedretti, G. et al. (1985). Branched-chain enriched amino acid solutions in the treatment of hepatic encephalopathy: a controlled trial. Ital. J. Gastroenterol., 17, 5-10.

26 Vilstrup, H., Gluud, C., Hardt, F. et al. Branched chain enriched amino acid versus glucose treatment of hepatic encephalopathy: a double-blind study of 65 patients with cirrhosis. Hepatology (in press).

27 Schafer, V.K., Winthe, M.B., Ukida, M. et al. (1981). Influence of an orally administered protein mixture enriched in branched chain amino acids on the chronic hepatic encephalopathy (CHE) of patients with liver cirrhosis. Z. Gastroentérologie, 19, 356-362.

28 McGhee, A., Henderson, J.M., Millikan, W.J. et al. (1983). Comparison of the effects of Hepatic-Aid and a casein modular diet on encephalopathy, plasma amino acids, and nitrogen balance in cirrhotic patients. Ann. Surg., 197, 288-293.

29 Cascino, A., Cangiano, C., Calcaterra, V. et al. (1978). Plasma amino acids imbalance in patients with liver disease. Am. J. Dig. Dis., 23, 591-598.

30 Hagenfeldt, L., Eriksson, L.S., Wahren, J. (1983). Amino acids in liver disease. Proc. Nutr. Soc., 42, 497-502.

31 Morgan, M.Y., Milson, J.P., Sherlock, S. (1978). Plasma ratio of valine, leucine and isoleucine to phenylalanine and tyrosine in liver disease. Gut, 19, 1068-1073.

32 McCullough, A.J., Czaja, J., Jones, J.D., Go, V.L.W. (1981). The nature and prognostic significance of serial amino acid determinations in severe chronic active liver disease. Gastroenterology, 645-652.

33 Sherwin, R., Joshi, P.H., Hendler, R. et al. (1974). Hyperglucagonemia in Laennec's cirrhosis: the role of portal-systemic shunting. N. Engl. J. Med., 290, 239-242.

34 Schafer, K., Ukida, M., Steffen, C. et al. (1985). Effect of ammonia on plasma and cerebrospinal fluid amino acids in dogs with and without portacaval anastomoses. Res. Exp. Med., 185, 33-44.

NUTRITION, DIET AND HEPATIC ENCEPHALOPATHY

Misael Uribe, M.D., DSc.
Liver Unit. National Institute of Nutrition
Mexico 14000, D.F., MEXICO

INTRODUCTION

The connection between hepatic encephalopathy (HE), diet and nitrogen metabolism is undisputed but its exact nature remains contentious (1). Excess of protein intake has been associated to deterioration or appearance of HE and in the opposite protein restriction is an established measure for the treatment of patients with HE.

The management of HE may be considered an important part of the therapy for chronic portal systemic HE. This syndrome is a common feature in advanced liver disease (ALD). We will refer in this chapter to the dietary treatment of chronic HE. The treatment of acute and fulminant liver failure is described elsewhere in this book.

ORAL TOLERANCE TO NUTRIENTS IN ADVANCED LIVER DISEASE

In most of ALD patients malnourishment is a frequent finding and although these patients require adequate nutritional support to sustain hepatic function, there is a sharply limited ability to tolerate most nutrients (2).

Tolerance to Carbohydrates

Glucose intolerance may frequently occur in ALD and hypoglycemia may be cause of death in acute liver failure. The incidence of diabetes has been reportedly increased in patients with cirrhosis,on the other hand the association between diabetes and cirrhosis ranges from 4.4% to 12.9% in autopsy statistics and from 7.5% to 34.6% in clinical studies (3,4). Related to liver conditions, glucose regulatory hormones may vary. Hence, hyperinsulinemia has been observed in cirrhosis. This abnormality is probably due to decreased hepatic degradation (5). With the worsening of the mental state, insulin levels fail to further increase in cirrhotics, in sharp contrast with that observed during acute HE of fulminant hepatic failure where up to a ten fold increase of insulin levels have been reported (6). The relative fall in insulin in the presence of possible increased liver cell damage and portal systemic shunts, can be only explained on the basis of reduced secretion, just as it happens in other catabolic states like starvation.

Elevated glucagon blood levels have been described in patients with ALD only when portal systemic shunting is present (7). Some have shown hyperglucagonemia as a feature of hepatocellular damage. This abnormality seems to be associated to higher production since the kidney, not the liver, is the major site for glucagon degradation. It is believed that hyperglucagonemia closely correlated with encephalopathy grades since it increases progresively with the deterioration of the mental state (7).

Tolerance to Lipids

A possible role for short chain fatty acids in the induction of HE was first suggested by Muto and Takahashi (8).They caused reversible coma in animals after acute infusion of fatty acids, further, using gas chromatography they were able to demonstrate that plasma levels of fatty acids were double in HE than in cirrhotic non encephalopathic patients.

More recently Takai et al (9) administered a fat emulsion and total parenteral nutrition (TPN) in 68 cirrhotic patients who underwent surgical treatment for esophageal varices, 28 patients received TPN plus the fat emulsion and 40 cases received TPN without fat. From this study the authors concluded that administration of fat solution did not worsen the results of liver function tests and cumulative nitrogen balance was comparable. The authors suggested that tolerance to fat emulsion in ALD was adequate and these solution can be used if required, during post-operative periods in patients with liver disease. Surprisingly no complications like cholestasis or fibrosis were reported secondary to TPN.

On the clinical side, no specific recommendation have been suggested as regard to limit triglycerides or cholesterol (unless underlying hyperlypidemia). From the therapeutic side it is interesting the in vitro finding that in assays incubated with blood, the production of C_4 - C_6 fatty acids was completely inhibited by lactulose, whereas the production of acetate was increased. Lactulose thus "detoxified" the profile of short chain fatty acids produced in the presence of blood and proteins. Invasive techniques and in vivo experiments are however required to confirm this interesting study (10).

Tolerance to Proteins

Of the three nutrients, the most relevant is the intolerance to dietary protein which often precipitates or uncovers HE. This is particularly troubling because protein plays the leading role in hepatic regeneration.

The connection between protein intake and HE precipitation was first observed in dogs in 1893, after constructing an Eck's fistula, Hahn administered a meat load and observed that the animals presented incoordination and changes in behavior. Since then the syndrome was called "meat intoxication"(11). Discontinuation of meat was followed by the recovery of normal behavior. Humans with ALD and naturally or surgically-induced portal systemic shunts resemble the "meat intoxication syndrome". When an excessive protein load (meat mainly) is administered patients may develop clinical signs of encephalopathy, thus reproducing animal experiments performed almost a century ago.

According with these previous experiments, recommendations were made to restrict meat intake and to cleanse the intestine to prevent HE. Patients with advanced liver disease and protein-calorie malnutrition are in negative nitrogen balance. Deprivation of protein leads to further depletion of liver protein stores, breakdown of cellular (muscle) protein and RNA, and conversion of polysomes to free ribosomes. These events occur mainly in the rough endoplasmic reticulum where albumin is synthesized (12). At our Institutions a recent survey demonstrated that up to 50% of the cirrhotic patients had protein-calorie malnutrition (13).

Cirrhosis is by definition a catabolic disease. Its clinical features include muscle wasting and several biochemical abnormalities such as hypoproteinemia, mainly hypoalbuminemia, and an abnormal plasma aminoacid pattern. Several studies have demonstrated that positive nitrogen balance can be achieved in cirrhotic patients at dietary protein intake of 0.8-1.0 g per kg per day. A recent work and hypothesis suggested that nitrogen balance and ultimately the nutritional state should improve and nitrogen requirements should be lowered in cirrhotic patients; if this early reaction of starvation could be prevented, this could be done by supplying protein and calories in frequent small meals during the day and the evening hours (14). However patients with ALD and chronic HE are faced with a nutritional dilemma (Figure 1); they need dietary protein to achieve a positive nitrogen balance but protein intake often precipitates or accentuates encephalopathy.

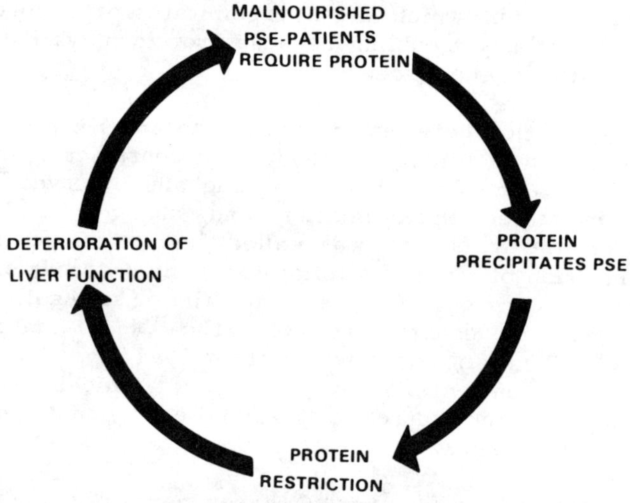

Figure 1

The Vicious Circle of Protein Restriction in Patients with Advanced Liver Disease

Several concepts (some proven and some unproven) regarding protein tolerance to specific types of protein in humans are generally accepted and used to make dietary recommendations for cirrhotic patients. The goal of dietary management is to reverse the catabolic state, to induce a positive nitrogen balance, to provide an adequate amount of Kcal. and to supplement with vitamins and minerals when indicated. (Table 1)

The intrinsic content of toxins like ammonia in different foods has been stressed (15) although it is not clear whether diet or intestinal generation of ammonia is more important (16).

Factors Influencing the Tolerance to Proteins

The amount, type and composition (AA, fiber, NH_4 etc.) of protein are factors related to protein tolerance. Amount is obviously an important factor. In fact the mental state of a cirrhotic patient can often be titrated by the amount of meat administered and tolerated without precipitation of HE. In approximately 10-15% of acute HE cases the precipitating factor is excessive protein intake. Further, gastrointestinal bleeding (which is equivalent to a high protein load) causes up to 20% of cases of acute HE (17).

Table 1
Dietary Recommendations in Advanced Liver Disease

Nutrients	Recommendations	Cautions
Carbohydrates	**Provide enough-achieve 50-60% of total Kcal**	**Do not exceed in diabetic-cirrhotics**
Lipids	**Provide enough to achieve 25-35% of total Kcal.**	**Special limitation: hyperlipidemia (PBC, biliary obstruction**
Proteins	**Provide 0.5 - 1.0 g/kg according with toleran ce. Prefer vegetable and dairy over meat protein**	**Caution with vegetable diets, may decrease mineral absorption**
Vitamins	**B complex in alcoholics and fat soluble vitamins in biliary obstruction (PBC, Biliary damage, etc)**	**Do not exceed vitamin A, it may cause liver damage**
Minerals	**Zn, oligoelements (after see Zn levels)**	**May be reduced by vegetable diets and by alcoholism**
Fiber	**Supplement with 10 - 20 g/day**	**May cause bloating satiety, in diabetics may cause hypoglycemia**

Several authors believe that besides quantity, quality is also an important factor to improve protein tolerance (18). Others claim that tolerance is related to other factors. For instance, initially it has been assumed that the advantageous therapeutic effect of vegetable protein as compared to animal protein could be explained by protein composition (mainly aminoacids and ammonia content or their low capacity to generate it in the human intestine by bacterial degradation). However there were no studies able to demonstrate a relationship between dietary aminoacid content and plasma aminoacids or ammonia levels. The main content and effects of various proteins are shown in Table 2.

Table 2
Putative Content and Effects of Various Proteins

CONTENT					
	NH_4	Mercaptans	AAA	Fat	Fiber
Blood protein	++++	+++	+++	+	–
Meat protein	+++	+++	+++	+++	–
Egg protein	+++	++	+++	+++	–
Dairy protein	++	++	++	++	–
Vegetable protein	++	++	++	+	++

EFFECTS			
	Laxative	Colonic Acidification	N, Stool Increased
Blood protein	+++	NO	NO
Meat protein	–	NO	NO
Egg protein	–	NO	NO
Dairy protein	++ (*)	YES *	YES
Vegetable protein	++ (**)	YES **	YES

* In lactose intolerant individuals
** Depends of fiber contents
AAA = Aromatic aminoacids

In our (19,20) own studies (Figures 2,3) we observed similar NH_4 levels and clinical HE improvements regardless the use of 40 g meat protein and 40-80 grams of vegetable protein. Vegetable diets were rich in branched chain aminoacids but also rich in methionine (Table 3). During vegetable diets the standard treatment was discontinued (neomycin plus milk of magnesia).

DIETARY FIBER, HE AND DIABETES

At our institution, which acts as a National referral center for metabolic and nutritional diseases, the association of diabetes and cirrhosis is very high, up to 18%. The frequent concurrence of PSE and diabetes prompted us to evaluate in a controlled trial the effects of vegetable protein diets, which are rich in fiber, in a group of patients with both diseases (20). The study was conducted in a controlled, crossover fashion in eight cirrhotic-diabetic patients with chronic PSE and uncontrolled diabetes. After a basal period of stabilization, the patients received either meat protein (0.8 g per kg per day) plus neomycin and milk of magnesia or the same amount of vegetable protein supplemented with psyllium fiber to reach a daily intake of 35 g fiber per day. The characteristics of both diets are shown (Table 4).

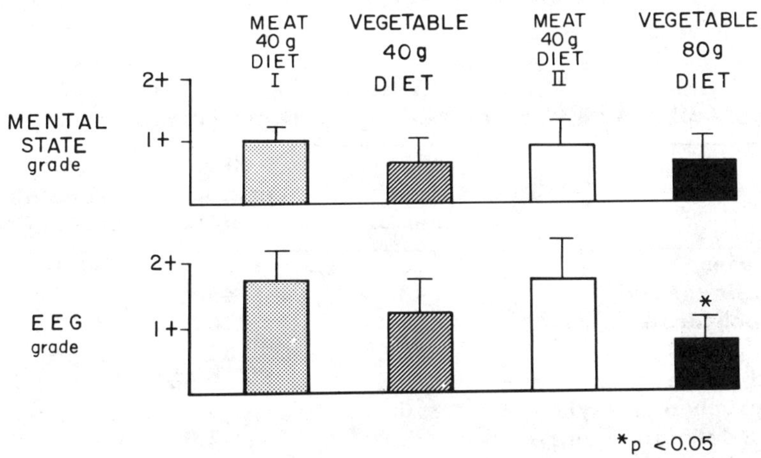

Figure 2
Mental State and EEG Grade in a Group of Ten HE Patients During Meat and Vegetable Protein Diets,after Uribe et al Dig Dis Sci 1982; 27: 119-126

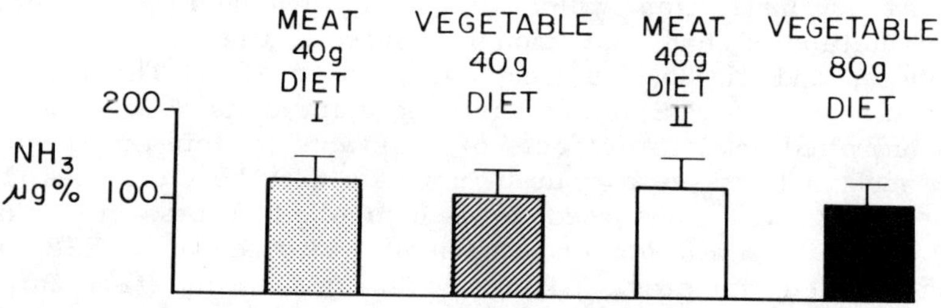

Figure 3
Arterial Blood Ammonia in a Group of Ten HE Patients During Meat and Vegetable Protein Diets,after Uribe et al Dig Dis Sci 1982; 27: 119-126

Table 3
Amino Acid Content (per day) of Diets Administered

	40 g Meat Protein Diet	40 g Vegetable Protein Diet	80 g Vegetable Protein Diet
Calories	1600	1555	1662
Protein (g/day)	43	45	88
Carbohydrate (g/day)	230	270	223
Fat (g/day)	56	33	46
Methionine (g/day)	0.95	0.52	1.1
Tryptophan (g/day)	0.51	0.33	1.0
Phenylalanine (g/day)	2.0	2.0	3.9
Tyrosine (g/day)	1.6	1.1	2.2
Valine (g/day)	2.5	2.3	4.2
Leucine (g/day)	3.4	3.8	6.7
Isoleucine (g/day)	2.2	2.2	4.4
Molar ratio BCAA/AAA	3.60	3.60	3.48

Dig Dis Sci 1982; 27: 119-126.

Table 4
Amino Acid Content (per day) of Diets Administered

	Meat Protein Diet (40 g)	Vegetable Protein Diet (40 g)
Calories	1600	1558
Protein (g/day)	43 (11.0)	45 (11.5)
Carbohydrate (g/day)	230 (57.5)	235 (60.5)
Fat (g/day)	56 (31.5)	49 (28.0)
Methionine (g/day)	0.95	0.52
Tryptophan (g/day)	0.51	0.33
Phenylalanine (g/day)	2.0	2.0
Tyrosine (g/day)	1.6	1.1
Valine (g/day)	2.5	2.3
Leucine (g/day)	3.4	3.8
Isoleucine (g/day)	2.2	2.2
Molar ratio BCAA/AAA	3.60	3.60
Fiber (g/day)	8.32	15.7
Hemicellulose (Metamucil)	0.0	20.0
Total Fiber/day (g)	8.32	35.7

Gastroenterology 1985; 88: 901–907

The results of the study demonstrated that both diets were equally effective in controlling PSE (Table 5). No patient deteriorated during the vegetable protein diet, eventhough neomycin and cathartics were discontinued. The beneficial effect of fiber supplementation in PSE was apparent.

Table 5

	Basal	MPST	VPDF
Mental state (grade)	0.2 ± 0.4	0.2 ± 0.4	0.2 ± 0.4
Asterixis (grade)	0.12 ± 0.5	0.2 ± 0.2	0
EEGs (grade)	0.25 ± 0.2	0.4 ± 0.4	0.4 ± 0.4
NH_4 (ug%)	151 ± 12	129 ± 35	151 ± 33
NCT (s)	67 ± 7	56 ± 15	66 ± 23

MPST = Meat Protein Plus Standard Therapy (neomycin + milk of magnesia.

VPDF = Vegetable Protein Diet Plus Fiber.
Gastroenterology 1985; 88: 901 – 907

As regard to metabolic control of diabetes, fasting glucose levels were 204± 86 mg per dl after the meat protein diet and 127± 8 mg per dl after the vegetable protein diet ($p < 0.02$), eventhough patients received similar amounts of hypoglycemic agents. The individual results are shown graphically in Figures 4-5.

This study was the first to suggest that oral supplementation with fiber may modify the tolerance to a protein load. The same phenomenon happensin regard to lipids and carbohydrates and these concepts were applied to improve blood levels of cholesterol and glucose.

DOES FIBER DETERMINE TOLERANCE TO NITROGEN IN HE?

The assumption that fiber may exert a beneficial effect in the treatment of PSE has recently been tested. The study was performed in 20 patients with cirrhosis and chronic PSE. After a stabilization period of one week, two isocaloric (35 Kcal per kg per day), isonitrogenous (0.8 g per kg per day) diets were administered for two weeks. The diets were similar in aminoacid composition, but were different in fiber content. Ten patients were randomized to receive meat and vegetable (amaranth) protein containing a mean of 28 g of fiber per day. Ten other patients received a similar diet that contained only 12 g of fiber per day. The results (Table 6) demonstrated that fiber was associated with better protein tolerance.In our study, nitrogen balance was similar although fecal elimination of nitrogen was significantly increased by the vegetable protein diet. Both the number and weight of the stools increased with the administration of vegetable protein enriched with fiber. Although in both groups the BCA/AAA molar ratio was found to be low, improvement in mental state and PSE after fiber supplementation was not associated with changes in the plasma aminoacid pattern.

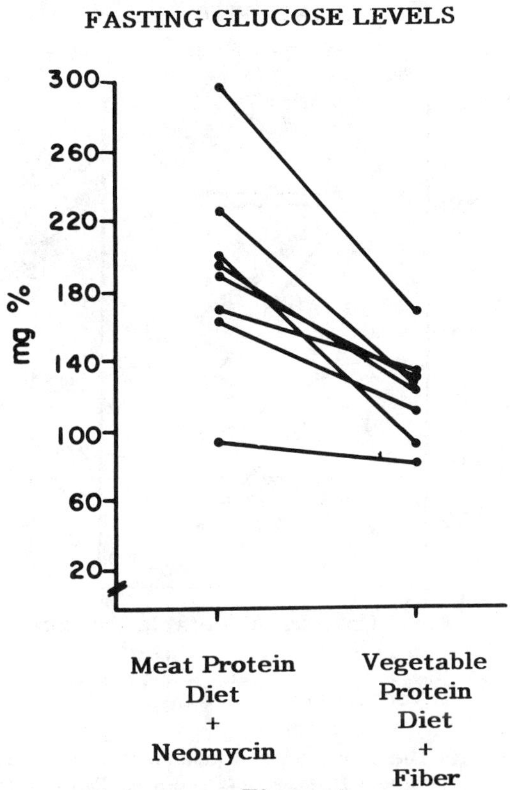

Figure 4

Fasting Glucose Levels in a Group of 8 Cirrhotic Diabetic Patients receiving in Cross-Over Fashion a 40 g Meat Protein Diet plus Neomycin or 40 g vegetable Protein Diet plus Fiber. Fasting Glucose Improved in All After the Vegetable Diet.
After Uribe et al Gastroenetrology 1985; 88: 901-907.

NITROGEN METABOLISM DURING FIBER SUPPLEMENTATION

Weber et al (22) compared the effects of a vegetable protein diet with a meat protein diet. They studied nitrogen metabolism in a group of six nonencephalopathic cirrhotic patients. The vegetable protein diets produced a fall in urea production rates associated with a fall in the urinary urea output. They also produced an increase in fecal nitrogen output. These investigators confirmed the findings that these changes were not associated

TREATMENT

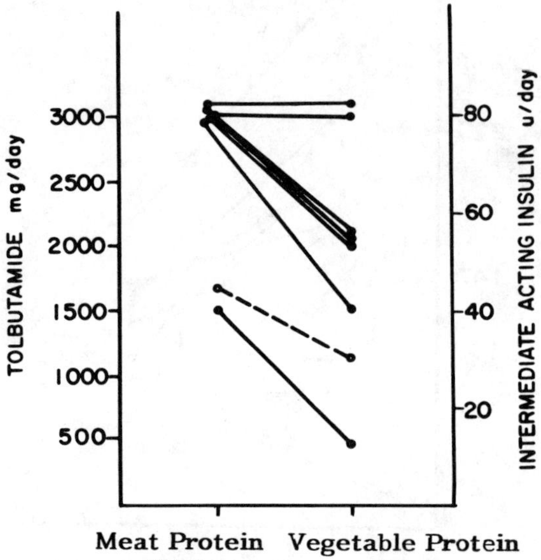

Figure 5

The Figure Depicts the Dose of Tolbutamide Received by the Eight cirrhotic-Diabetes Patients (shown in Figure 4). The Improvement of Fasting Glucose Levels was not Associated to Larger Tolbutamide Dosage.
After Uribe et al Gastroenterology 1985; 88: 901-907.

with any significant modification in the plasma aminoacid pattern. Weber et al concluded that the effect of vegetable protein diets on nitrogen metabolism can be atributed to the increase in dietary fiber intake.

On the other hand enhanced bacterial uptake of ammonia in the presence of carbohydrate energy (like soluble fiber) may explain why blood ammonia falls without a concomitant increase in fecal ammonia excretion but with an increase in fecal (bacterial) nitrogen and why blood ammonia levels and fecal ammonia excretion remain low during lactulose therapy despite the rise in colonic pH during passage through the large bowel (23). This explains not only the clinical effects of lactulose

Table 6
Amaranth (High Fiber) versus Meat Protein (Low Fiber) in Chronic PSE

	High Fiber Diet		Low Fiber Diet	
Kcal/kg	35	± 5	34	± 5
Protein g/kg	0.81	± 0.12	0.80	± 0.17
Fat g/kg	1.3	± 0.20	1.45	± 0.19
CH g/kg	5.13	± 0.38*	4.0	± 0.52
Fiber g/kg	0.46	± 0.05**	0.19	± 0.09
Total/day, mean	(28 g)		(12 g)	
N mg/kg	136	±17	132	±11

*** P < 0.001; ** P<0.01**
Same content of AA, except tyrosine (Lower in HFD)

but also those of lactose (24) and lactitol (25), disaccharides that in vivo are able to lower NH_4 levels in patients with HE. (Figures 6 and 7).

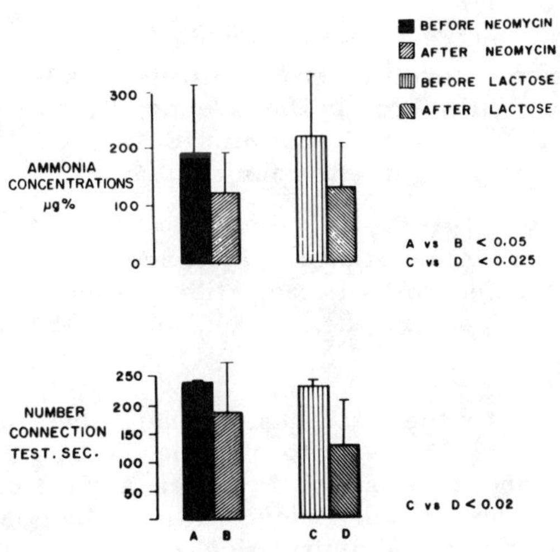

Figure 6

Ammonia Blood Levels and Number Connection Test Performance in a Group of 10 Patients Receiving Lactose Enemas or Starch Enemas plus Neomycin. After Uribe et al Gastroenterology 1981; 81: 101-106

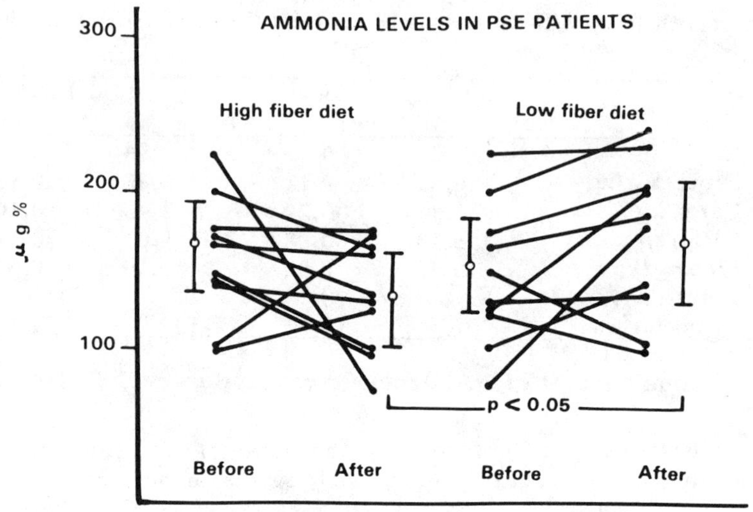

Figure 7
Blood Ammonia Levels after Equinitrogenous Isocaloric High Fiber or Low Fiber Protein Diets Administered to Groups of 10 Cirrhotic Patients with Chronic HE.
After Uribe et al Gastroenterology (in Press).

There is now substantial evidence that some dietary polisaccharides, notably dietary fiber, escape absorption in the small bowel and are then broken down in the large intestine of man.

It has recently been suggested that fiber administration like disaccharides, is able to provide substrate for colonic fermentation and that breath hydrogen expiration and plasma acetate levels are equally elevated in volunteers after both lactulose and pectin administration (26). However, the peak time of both hydrogen expired acetate, propionate, butyrate, and plasma acetate levels occurred during the first two hours with disaccharides and after 8 to 20 hours with pectin. Significant increments in hydrogen and acetate levels lasted 6 and 24 hours after the disaccharide and the fiber, respectivelly. Finally, Jacobs and coworkers in a study performed in rats demonstrated that colonic fermentation and colonic acidifcation occurred after the aministration of psyllium fiber. The same study suggested that in the large intestine the fermentable fiber (pectin and guar) were also stimulants to cell (colonic mucosa) proliferation (27). The similar effects of disaccharides on clinical use for the treatment of HE and the supplementation with dietary fiber are shown in Table 7.

Table 7
Amaranth versus Meat Protein in Chronic PSE

	PSE Parameters			
	High Fiber		Low Fiber	
	Before	After	Before	After
NCT, sec	**69 ± 5**	**55 ± 7***	**73 ± 11**	**75 ± 7**
EEG, grade	**0.7 ± 0.3**	**0.2 ± 0.3**	**0.8 ± 0.4**	**1.5 ± 0.6***
NH_4 ug%	**165 ± 31**	**134 ± 31**	**154 ± 31**	**170 ± 40**
Molar ratio BCAA/AAA*	**1.8 ± 0.3**	**1.8 ± 0.3**	**1.5 ± 0.3**	**1.5 ± 0.3**

*** $p < 0.05$**

WORLDWIDE TREND TO REDUCE FIBER INTAKE

From the philosophical point of view, it seems unfortunate that over the last two centuries humans have radically changed their alimentary habits and diets (Figure 8). There are few doubts that inhabitants of many "developed" countries in which dietary fiber intake is very low, due in part to modern food processing, suffer from a variety of disorders, including obesity, hyperlipidemia, hypertension, diverticular disease, cardiac disease, constipation, and perhaps, even neoplastic diseases like colonic cancer. Paradoxically, in poor undeveloped countries in which high fiber diets are consumed, these diseases are less common.

The development of new food products to meet the needs of the 20th century consumers in developed countries seems to forget the historical lesson as regard to nutrients intake. Cholesterol, meat and fat are among the foods more commonly suggested that taken in excess may be dangerous for health even without the presence of liver disease while avoiding food high in cholesterol and getting enough fiber in the diet make nutritionally sound advise.

Inhibition of Fiber Absorption by Disaccharidase Inhibition

A novel approach to treat HE by inducing carbohydrate maldigestion is the oral administration of disaccharide inhibitors. In preliminary studies acarbose and a N.substituted derivative of valiolamine (AO-128) were able to induce disaccharide malabsorption. The dietary unabsorbed oligosaccharides and their breakdown products by intestinal microflora are accumulated in the intestine which induce acidic diarrhea and flatulence. Acarbose inhibits amylase,

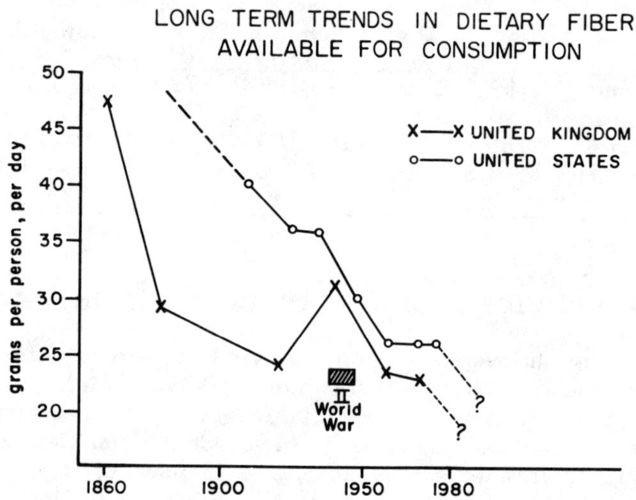

Figure 8

Trends in Dietary Fiber Available for Consumption in Two "Developed" Countries

maltase and sucrase, whereas AO-128 inhibits only maltase and sucrase (29). In preliminary (unpublished studies) carried out in volunteers the administration of AO-128 given during 7 days was able to increase fecal weight from 100 g to 300 g per day, reduce pH of stools from 6.5 to 5.5 grade and cause a sharp increase of stools lactic and acetic acids. We had the chance to observe these changes in our own ongoing trial with AO-128 for the treatment of chronic HE.

BIBLIOGRAPHY

1. Uribe M.: Dietary managements of portal systemic encephalopathy. In Hepatic Encephalopathy. H.O. Conn and J. Bircher. Eds. Ed. Med. Press East Lansing Michigan P. 1988.

2. Maier K.P.: Progress in the treatment of portal systemic encephalopathy (PSE). J Hepatol 1987;5:355-361.

3. Creutzfeld W., Frerich H., Schinger K. et al.: Liver disease and diabetes mellitus. In: Popper H., Schaffner F. Eds. Progress in Liver Disease. Vol. III, New York: Grune and Stratton, 1970,371-407.

4. Uribe M., Dibildox M., Malpica S., et al.: Beneficial effect of vegetable protein diet supplemented with psyllium plantago in patients with hepatic encephalopathy and diabetes mellitus. Gastroenterology 1985; 88:901-907.

5. Johnston D.G., Alberti K., Faver O.K.: Hyperinsulinism of hepatic cirrhosis: Diminished degradation or hypersecretion?. Lancet 1977;1: 10-12.

6. Chase R.A., Sullivan S., Bloom S.R. et al.: Insulin, glucagon and aminoacid inbalance in fulminant hepatic failure. Gut 1977;18:953.

7. Marchesini G., Forlani G., Zoli M. et al.: Dig Dis Sci 1979;24:594-600.

8. Muto Y., Takahashi Y.: Gas chromatographic determination of plasma short chain fatty acids in diseases of the liver. J. Jap. Soc Intern Med 1964;53: 828-839.

9. Takai T., Nagayama M., Okuno M. et al.: Postoperative total parenteral nutrition with fat emulsions for patients with esophageal varices. Journal Jap Surg Soc 1988;89:325-335.

10. Brobech P., Sandvad H., Holtung K.: Lactulose detoxifies in vitro short-chain fatty acid production in colonic contents induced by blood: Implications for hepatic coma. Gastroenterology 1988;94: 750-754.

11. Hahn M., Massen O., Nencki M. et al.: Die Eck'sche fistelzwishen der unturen hohlvene und pfortades und inhre foigen dur den organismus. Arch Exp Pathol Pharmackol 1893; 32: 161-210.

12. Achord J.L.: Malnutrition and the role of nutritional support in alcoholic liver disease. Am J Gastroenterol 1987; 82: 1-7

13. Yañez M.A., Palacios N., Uribe M.: Estado Nutricional de la muestra de población de pacientes cirróticos de la clínica de hígado del Instituto Nacional de la Nutrición. Rev gastroenterol (Mex) 1986; 51:307.

14. Swart G.R., Berg J.L., Wattimena J.L.D. et al.: Elevated protein requirements in cirrhosis of the liver investigated by whole body turnover studies. Clin Sci 1988; 75: 101-107.

15. Rudman D., Smith III R.B., Salam AA et al.: Ammonia content of food. Am J Clin Nutr 1973; 14: 171-177.

16. Vince A., Dawson N., Park N., et al.: Ammonia production by intestinal bacteria. Gut 1973; 14: 171-177.

17. Conn H.O.: Pathogenesis of portal systemic encephalopathy. In the Hepatic Coma Syndromes and lactulose. Chapter 2,Williams and Wilkins. Baltimore 1979 pp 46-121.

18. Greenberger B.J., Carley J., Schenker S. et al.: Effect of vegetable and animal protein diets in chronic hepatic encephalopathy. Am J Dig Dis 1977;22: 845-855.

19. Uribe M., Marquez MA., Garcia Ramos G., et al.: Treatment of chronic portal systemic encephalopathy with vegetable and animal protein diets. A controlled crossover study. Dig Dis Sci 1982; 27: 119-126.

20. Uribe M., Dibildox M., Malpica S., et al.: Beneficial effect of vegetable protein diet supplemented with psyllium plantago in patients with hepatic encephalopathy and diabetes mellitus. Gastroenterology 1985; 88: 901-907.

21. Uribe M., García Compean D., Vargas F., et al.: Fiber improves tolerance to protein in Hepatic Encephalopathy. Gastroenterology in press.

22. Weber F.L., Mincs D., Fressard K. et al.: Effects of vegetable diets on nitrogen metabolism in cirrhotic subjects. Gastroenterology 1985;89: 538-544.

23. Doumes A.C., Van der Burg G.J., Bouman A.A.: Lactitol, lactulose and blood ammonia. Lancet 1987 II; 3429-30 (Letter).

24. Uribe M., Marquez M.A., Garcia Ramos, G.: Treatment of chronic portal systemic encephalopathy with lactose in lactase deficient patients. Dig Dis Sci 1980; 25: 924-928.

25. Uribe M., Toledo H., Pérez F., et al.: Lactitol a second generation disaccharide for treatment of chronic portal systemic encephalopathy. A double blind, crossover, randomized clinical trial. Dig Dis Sci 1987; 32: 1345-1953.

26. Pomare E.W., Branch W.J., Cummings, J.M.: Carbohydrates fermentation in the human colon and its relation to acetate concentration in venous blood. J Clin Invest 1985; 74: 1448-1454.

27. Lupton J., Codei D., Jacobs L.: Long term effects of fermentable fibers on rat colonic pH and epithelial cell cycle. J Nutr 1988; 118:840-845.

28. Bursey R.G.: Diet modification and the development of new food products. Ann Intern Med 1983; 98: 872-874.

29. Ruppen M., Hagel J., Feuerbach W. et al.: Fate and effects of the α-glucosidase inhibitor acarbose in humans. Gastroenterology 1988; 95: 93-99.

PARTIAL PORTAL DECOMPRESSION: INITIAL CLINICAL EXPERIENCE WITH SMALL-STOMA PORTACAVAL SHUNT

Kaj Johansen, MD, PhD

From the Department of Surgery, Harborview Medical Center, University of Washington School of Medicine, Seattle, Washington, USA

ABSTRACT

To test the hypothesis that partial portal decompression to treat variceal hemorrhage will diminish subsequent encephalopathy, 50 consecutive patients were studied following small-stoma (10-12 mm) side-to-side portacaval shunt (PCS). During follow-up averaging 26 months, 6 patients (12%) died. Four patients (8%) re-bled, only 1 from varices. All patients (100%) had patent shunts at subsequent angiography and/or ultrasonography. Despite consistent (100%) post-shunt reversal or stagnation of portal flow on duplex scan, encephalopathy on clinical and psychometric grounds was observed in only 3 patients (6%).

This study suggests that small-stoma PCS can be performed with reliably low rates of rebleeding and encephalopathy. That encephalopathy was rare in the absence of hepatic portal perfusion incriminates other factors besides portal flow in the genesis of post-shunt hepatic failure.

INTRODUCTION

Standard portacaval shunt (PCS) effectively controls hemorrhage from esophageal varices, and reduces deaths from such bleeding, in cirrhotics with portal hypertension (1-4). However, long-term survivors of these procedures are at substantial risk for accelerated liver failure, increased deaths from which virtually neutralize the reduced mortality from hemorrhage conferred by successful shunt (1-4). In addition, patients surviving standard PCS demonstrate a distressingly high incidence - 20-100% - of portal-systemic encephalopathy (PSE). These complications have led to a general disenchantment with standard PCS for the long-term management of variceal hemorrhage, and have directly stimulated development of a number of radiologic (5), endoscopic (6), and non-shunt operative (7,8) techniques intended to control variceal bleeding.

The unacceptably high rates of liver failure and chronic PSE following standard PCS also have caused Warren (9) and others (10,11) to reassess the underlying pathophysiology of post-shunt mortality and morbidity, especially following total portal diversion. Their studies have led to operative modifications designed to effect variceal decompression while preserving hepatic portal perfusion. Although each of these operations - the distal splenorenal shunt (9), the left gastric-caval shunt (10), and the portacaval prosthetic H-graft (11) - is flawed by problems which render it impractical in a substantial proportion of variceal bleeders, all have been found to reduce the incidence and severity of post-shunt hepatic deterioration.

A characteristic but previously-underemphasized aspect of all these recent modifications of the standard Eck fistula concept is that each provides only <u>incomplete</u> or <u>partial</u> portal decompression, primarily via a limited-caliber, low-flow shunt. To test the hypothesis that <u>partial</u> decompression of the portal system might provide durable protection against recurrent variceal bleeding, yet mitigate post-shunt hepatic dysfunction, a prospective trial of small-stoma side-to-side PCS has been conducted since 1981 in a consecutive series of patients with variceal bleeding at Harborview Medical Center, a teaching hospital of the University of Washington, Seattle, USA. This report details the outcome of the first fifty patients in this trial.

Materials and Methods

Fifty consecutive patients who had survived side-to-side PCS for at least thirty days were followed up for periods of one to forty-six months (mean, 26 months)(Table I).

Patients undergoing other operations for bleeding varices were excluded from analysis, as were those requiring emergency portal decompression. Forty-two (84%) were alcoholics, six (12%) clearly had various forms of non-alcoholic liver cirrhosis, and 2 (4%) had cirrhosis of indeterminate origin. Child classification at the time of operation included Child A = 8, Child B = 38, and Child C = 4. No patient had a history of chronic PSE prior to operation.

TABLE I

PRE-OPERATIVE CHARACTERISTICS OF STUDY POPULATION (n=50)

Mean age	52.6 (range 29-75)
Sex	31 M/19 F
Cause of cirrhosis:	
alcoholic liver disease	42 (84%)
post-necrotic	6 (12%)
indeterminate	2 (4%)
Child classification:	
A	8
B	38
C	4

During each operation the portacaval anastomosis was carefully calibrated to 10-12 mm in the long axis, in a fashion designed to result in a shunt gradient (portal pressure minus vena cava pressure) approximating 10 mm Hg. Actual operative values varied from 8.0 to 15.3 mm Hg (average 10.4 mm Hg) after opening the shunt. In six cases the anastomosis was modified by suture or metal clip in order to produce a shunt gradient of at least 10 mm Hg.

Patients were followed up on an outpatient basis; episodes of rebleeding were documented, and bleeding sites identified endoscopically. Crude and actuarial survival statistics were computed. Patient evaluation included clinical, neurologic, and psychometric assessment [Reitan trail-making test (12)] for signs and symptoms of chronic PSE. Shunt patency was serially evaluated by angiographic techniques, by means of ultrasonic-Doppler duplex scanning (13), or at post-mortem examination. Angiographic studies included shunt manometry. Where appropriate, differences in results were analyzed for statistical significance using Chi-square analysis.

Results

During follow-up six patients (12%) died, three from hepatic failure (Table II). Four patients (8%) rebled from various sources: one patient (2%) was documented to be bleeding from small esophageal varices despite a patent PCS. Using transfemoral angiography (n=24), ultrasonic duplex-scanning (n=42), both modalities (n=23), or post-mortem examination (n=3), shunt patency was documented in all 50 patients (100%). Follow-up shunt manometry in 24 patients demonstrated maintenance of "relative" portal hypertension; shunt gradients averaged 9.0 mm Hg (range 0.0-15.5), an insignificant (p=0.10) change from values measured at the time of operation. All patients (100%) studied by Doppler or by angiographic means were found to have reversed or stagnant portal vein flow in the hepatic limb of that vessel: no patient had ultrasonographic evidence for post-shunt portal perfusion of the liver.

TABLE II

POST-OPERATIVE CHARACTERISTICS OF STUDY POPULATION
(n=50)

Mortality	6 (12%)
due to liver failure	3 (6%)
Rebleeding	4 (8%)
from varices	1 (2%)
Shunt patency	50 (100%)
Shunt manometry (average) (range 0-15.5)	9.0 mm Hg
Resumption of alcohol	16/42 (38%)
NEW ENCEPHALOPATHY	3 (6%)

A new onset of encephalopathy substantial enough to require intervention clearly occurred in three patients (6%) in the absence of other known causes of post-shunt neurologic deterioration (sepsis, narcotic or sedative misuse, gastrointestinal bleeding, etc.). Two patients required institutionalization, and the third, treated with lactulose and diminished dietary protein intake, was restricted to activities of daily living. One of these patients underwent angiographic shunt manometry, and was found to have a portacaval shunt gradient of 0.

All other patients (94%) returned to independent function not discernibly different from their pre-operative lifestyles. Among 42 alcoholic patients 16 (38%) are known to have consumed alcohol during the follow-up period.

Discussion

Hepatic cirrhosis, primarily secondary to alcoholism, is the ninth leading cause of death in the United States, and ranks fourth among causes of mortality among middle-aged Americans (14). Esophageal varices secondary to the portal hypertension of cirrhosis are the leading cause of death from gastrointestinal hemorrhage; the likelihood of bleeding again, and dying from such bleeding, is overwhelming following an episode of variceal bleeding (15). The economic costs of this lethal complication are staggering (16).

Portal-systemic shunting procedures as therapeutic maneuvers to treat variceal hemorrhage were resurrected by Whipple 40 years ago (17). Standard PCS has been demonstrated to be highly effective at preventing further variceal hemorrhage, and is technically relatively easily performed. Unfortunately, all long-term studies of standard PCS have demonstrated an unacceptably high incidence of post-shunt hepatic failure and encephalopathy, with rates ranging from 20 to 100% (1-4, 18). This has been felt to be due both to the deprivation of portal blood flow to the liver by total shunts (19), as well as the diversion into the central circulation of various toxic nitrogenous products absorbed from the gut - most notably ammonia. Randomized trials of standard PCS have demonstrated, at best, only trends toward improved survival when compared with medical management (1-4, 18).

The demonstration that standard PCS halts variceal bleeding only at the cost of accelerated hepatic deterioration and PSE led to the concept of "selective" portal-systemic shunting, most notably gastrosplenic venous decompression via the left renal vein as described by Warren (9). Experience has demonstrated that "selective" shunts have been effective at diminishing variceal bleeding, and have a reasonable early record of maintaining prograde portal flow. However, the operation has remained technically challenging, especially for those surgeons confronted only occasionally with the problem of variceal bleeding. Further, especially in alcoholics (20), a substantial number of patients develop reversal or stagnation of portal flow or actual portal vein thrombosis during late followup (21, 22). Whether selective shunts prolong survival in comparison with total shunts remains unclear (18).

Notably, however, all series of patients undergoing selective shunt demonstrate a markedly lower incidence of PSE, especially of the severe or incapacitating type (9, 10, 18, 20). While this favorable result had initially been ascribed to the distal splenorenal shunt's preservation of hepatic portal perfusion, this explanation is inconsistent with the finding that a substantial proportion of these patients lose hepatic portal perfusion over time (21, 22). Rather than preserving hepatic portal perfusion, an alternate explanation for these patients' protection against post-shunt PSE, one considered recently by Warren (20) and Rikkers (23), may relate to the maintenance of "relative" mesenteric venous hypertension by these patients' small-caliber, low-flow shunts. Why might persistently-elevated splanchnic venous pressures result in a reduced risk of PSE?

Orloff and colleagues demonstrated in dogs undergoing PCS of various diameters that protein loads were better tolerated in those animals with a diminutive anastomosis (24). Clinical (23, 28) and experimental (24-28) data have suggested that the absorption of ammonia (24-28) and D-xylose (23) from the gut is a function of mesenteric venous pressure, such that complete portal decompression markedly increases absorption of such materials. Further, Rikkers has demonstrated that enhanced absorption of galactose following shunt is associated with a significantly higher likelihood of developing PSE (23).

These data permit the suggestion that various neurotoxins which might cause PSE are absorbed from the gut as a function of mesenteric venous pressure. It might follow that a shunting technique which adequately decompresses esophageal varices (to prevent further bleeding), but maintains "relative" mesenteric venous hypertension (to diminish toxin absorption and thus the likelihood of developing PSE), might prove to be an "optimal" portal-systemic shunt. The current clinical trial has been conducted to test this hypothesis.

Proof that a limited-stoma PCS can indeed be effective in preventing recurrent variceal hemorrhage while diminishing chances of developing PSE would rest on several determinations. One would be that constructing a small-caliber shunt does not result in unacceptably high rates of shunt thrombosis and recurrent variceal hemorrhage. Results of the current series suggest that construction of a small-orifice PCS can indeed be associated with acceptably low rates of shunt thrombosis (0%) and variceal rebleeding (2%). A second variable would be the demonstration that "relative" mesenteric venous hypertension, established intra-operatively under general anesthesia in a sometimes-hypovolemic patient, can be maintained chronically in the post-operative period. Follow-up angiographic shunt manometry in 24 of the patients in the current series suggests that the shunt gradient measured at the time of operation <u>can</u> be maintained post-operatively at a value near that measured during initial construction.

A third important variable would relate to Rikkers' suggestion (23) that maintenance of mesenteric venous hypertension diminishes toxin absorption from the gut. If this contention is correct, a shunt which maintains elevated splanchnic venous pressures should result in a low incidence of post-operative PSE. The current clinical trial resulted in clearly worsened neurophyshiatric status in only three patients (6%). In view of the hypothesis under examination, it is interesting (although not persuasive) that the single encephalopathic patient undergoing shunt manometry had lost his "protective" shunt gradient; his portal pressure had equilibrated with that of his inferior vena cava.

Thus, this trial, while flawed by a too-brief follow-up period and the absence of a concurrent control group, nevertheless is the first PCS series with post-shunt PSE rates low enough to be competitive with those of various forms of selective shunt (10, 20) or non-shunt (7, 8) operations. This series corroborates the apparently excellent early results of Sarfeh, Rypins and coworkers, who have studied the use of a small-diameter prosthetic interposition PCS in variceal bleeders (11), and those of Nespoli <u>et</u> <u>al</u>, who have observed <u>no</u> neurologic deterioration up to one year post-operatively in a small series of patients undergoing a calibrated PCS which reduces pre-shunt portal pressure by 30% (29).

The current trial supports the general impression that most PCS patients who survive the immediate post-operative period tend to enjoy a reasonable prognosis; only six patients (12%) in this series died during follow-up. It is noteworthy that only three (6%) expired as a direct consequence of hepatic failure.

An unexpected result of this trial arose from the utilization of ultrasonic duplex scanning (13) to demonstrate post-operative PCS patency. This technique, which combines real-time B-mode ultrasound and pulsed Doppler velocity flowmetry, images the blood vessels of interest, as well as the shunt itself, and can demonstrate (by virtue of high flow velocities and turbulence in the region of the shunt) that the anastomosis is patent. Of novel physiologic significance is the ultrasonographic demonstration that, even with a small shunt orifice (averaging 11 mm in long axis), <u>all</u> patients (100%) had reversed (hepatofugal) flow in the hepatic limb of the portal vein following PCS.

The longstanding belief that continued post-shunt portal perfusion of the liver is necessary to preserve hepatic function (19) has led to innumerable innovative (although vain) attempts to provide simultaneous variceal decompression and hepatic portal perfusion - demonstrably a hemodynamic improbability. Either immediately [as in the mesocaval shunt (29)] or over time (as occurs in many patients undergoing distal splenorenal shunt (21, 22)] hepatic portal perfusion is lost as splanchnic venous blood under high pressure flows preferentially toward the low-pressure shunt.

Experimental work by Castaing et al (25) and Hess and coworkers (27), as well as the pivotal clinical work of Burchell and colleagues (31), has suggested that maintenance of <u>total</u> hepatic blood flow, mostly via the hepatic artery in cirrhotics (31), rather than perfusion of the liver by "hepatotrophic factors" carried in the portal vein from the upper abdominal viscera (19), is the major determinant of survival and avoidance of liver failure following portal decompression. That the patients in the current series demonstrated a chronic encephalopathy rate of 6% and a late hepatic failure mortality rate of 6% in the presence of invariably hepatofugal portal flow supports the contention that factors other than portal perfusion of the liver are operative in post-shunt hepatic and neurologic dysfunction. Further experimental and clinical corroboration of these results may validate partial portal decompression via a small-calibre PCS as an "optimal" choice of shunt in the operative management of bleeding esophageal varices.

REFERENCES

1. Jackson FC, Perrin EB, Felix WR, Smith AG. A clinical investigation of the portacaval shunt. V. Survival analysis of the therapeutic operation. Ann. Surg. 1971; 174:672-701.

2. Mutchnick MG, Lerner E, Conn HO. Portal-systemic encephalopathy and portacaval anastomosis: a prospective controlled investigation. Gastroenterology 1974; 66:1005-1010,

3. Reynolds TB, Donovan AJ, Mikkelsen WP, Redeker AG, Turrill FL, Weiner JM. Results of a 12-year randomized trial of portacaval shunt in patients with alchoholic liver disease and bleeding varices. Gastroenterology 1981; 80:1005-1011.

4. Rueff B, Prandi D, Degos F, Sicot J, Degos JD, Sicot C, Maillard JN, Fauvert R, Benhamou JP. A controlled study of therapeutic portacaval shunt in alcoholic cirrhosis. Lancet 1976; 1:655-657.

5. Widrich WC, Johnson WC, Robbins AH, Nabseth DC. Esophagogastric coronary vein occlusion. Arch. Surg. 1978; 113:331-338.

6. Clark AW, McDougall BRD, Westaby, Williams R. Prospective controlled trial of injection sclerotherapy in patients with cirrohosis and recent variceal hemorhage. Lancet 1980; 2:552-554.

7. Johnston GW. Treatment of bleeding varices by oesophageal transection with the SPTV gun. Ann. R. Coll. Surg. Eng. 1977; 59:404-408.

8. Sugiura M, Futagawa S. A new technique for treating esophageal varices. J. Thorac. Cardiovasc. Surg. 1973; 66:677-685.

9. Warren WD, Zeppa R, Fomon JJ. Selective transsplenic decompression of gastroesophageal varices by distal splenorenal shunt. Ann. Surg. 1967; 166:437-455.

10. Inokuchi K, Kobayashi, Ogawa Y, Saku M, Nagasue N, Iwaki A. Results of left gastric vena caval shunt for esophageal varices; analysis of 100 clinical cases. Surgery 1975; 78:628-635.

11. Sarfeh IJ, Rypins EB, Fardi M, Conroy RM, Mason GR, Lyons KP. Clinical implications of portal hemodynamics after small-diameter portacaval H-graft. Surgery 1984; 96:223-229.

12. Conn HO. Trailmaking and number connection tests in the assessment of mental state in portal systemic encephalopathy. Am. J. Dig. Dis. 1977; 22:541-550.

13. Ackroyd N, Gill R, Griffiths K, Kossoff G, Reeve T. Duplex scanning of the portal vein and portasystemic shunts. Surgery 1986; 99:591-597.

14. Vital Statistics of the United States, U.S. Printing Office, 1984.

15. Smith JC, Graham DY. Variceal hemorrhage: a critical evaluation of survival analysis. Gastroenterology 1982; 82:968-973.

16. Chung R, Lewis JW. Cost of treatment of bleeding esophageal varices. Arch. Surg. 1983; 118:483-485.

17. Whipple AO. The problem of portal hypertension in relation to the hepatosplenopathies. Ann. Surg. 1945; 122:449-475.

18. Busuttil RW, Brin B, Tompkins RK. Matched control study of distal splenorenal and portacaval shunts in the treatment of bleeding esophageal varices. Am. J. Surg. 1979; 138:62-70.

19. Starzl TE, Porter KA, Francavilla JA. A hundred years of the hepatotrophic controversy. Hepatotrophic factors. Ciba Symposium, Amsterdam, Elsevier (1978), 111-138.

20. Warren WD. Control of variceal bleeding; reassessment of rationale. Am. J. Surg. 1983; 145:8-16.

21. Nabseth DC. The distal splenorenal shunt: an enigma (editorial). Am. J. Surg. 1981; 141:579-580.

22. Maillard JN, Flamant YM, Hay JM, Chandler JG. Selectivity of the distal splenoreual shunt. Surgery 1979; 86:663-670.

23. Rikkers LF. Portal hemodynamics, intestinal absorption, and post-shunt encephalopathy. Surgery 1983; 94:126-133.

24. Orloff MJ, Wall MH, Hickman EB, Neesby T. Influence of stomal size of portacaval shunts on peripheral blood ammonia levels. Ann. Surg. 1963; 158:172-181.

25. Castaing D, Beaubernard C, Ariogul O, Gigou M, Franco D, Bismuth H. Liver atrophy and encephalopathy after portacaval shunt in the rat. Eur. J. Surg. Res 1982; 14:192-202.

26. Harrison ID, Hardy-Smith A, Shields R. Ammonia absorption from the canine colon after portacaval shunt. Br. J. Surg. 1977; 64:851-856.

27. Hess F, Jerusalem C, Willemen A. Altered portal pressure secondary to portacaval and portasystemic shunts in the rat: the effect on liver function and intestinal integrity. Eur. J. Surg. Res. 1982; 14:286-297.

28. Sein MM, Hardy-Smith A, Shields R. The effect of portacaval shunt on colon transfer of ammonia in cirrhosis of the liver. Gut 1974; 15:837-843.

29. Nespoli A, Marini A, Infante M, Berra E, Coviello E, Della Salla S, Spinnler H. Assessment of neuropsychologic derangement after portocaval shunt (PCS) versus azygos-portal disconnection (APD) in selected cirrhotics (abstract). J. Hepatol. 1987; 4, Suppl. 1:S-38.

30. Smith RB, Warren WD, Salam AA, Millikan WJ, Ansley JD, Galambos JT, Kutner M, Bain RP. Dacron interposition grafts for portal hypertension; an analysis of morbidity correlates. Ann. Surg. 1980; 192:9-18.

31. Burchell AR, Moreno AH, Panke WF, Nealon TF. Hepatic artery flow improvement after portacaval shunt: a single hemodynamic clinical correlate. Ann. Surg. 1976; 184:289-302

Potential Therapeutic Use of Benzodiazepine Antagonists

BENZODIAZEPINE-ANTAGONIST (RO 15-1788) FLUMAZENIL, ANEXATE[R], PHARMACOKINETICS AND THERAPEUTIC APPLICATIONS

R. Meier and K. Gyr

Department of Internal Medicine, Division of Gastroenterology, Kantonsspital Liestal, University of Basel, CH-4410 Liestal, Switzerland

1. INTRODUCTION

The excellent anticonvulsant, anxiolytic and hypnotic effect is the reason why benzodiazepines are among the most widely used drugs throughout the world. In the mid-seventies, Haefely and co-workers first described that benzodiazepines enhance the action of GABA (Gamma-amino-butyric-acid), the major inhibitory neurotransmitter in the brain (1,2). The pharmacological and therapeutic effect of benzodiazepines is mediated by specific high affinity binding sites in the brain which have been identified as benzodiazepine receptors (BZR). These receptors are an integral part of the GABA receptor complex (3-6).

2. GABA RECEPTOR COMPLEX AND BENZODIAZEPINE RECEPTOR

Two GABA receptors have been described in the brain ($GABA_A$ and $GABA_B$, s. Table I). The $GABA_A$ receptor regulates the opening of the chloride channels, the $GABA_B$ response is linked to potassium (K^+), calcium (Ca^{++}) and cAMP-mediated processes. $GABA_B$ receptors are much less frequent than $GABA_A$ receptors, they are not given further consideration in this paper. Only $GABA_A$ receptors are linked to BZR.

Neurons operating with GABA as neurotransmitter are widespread in the brain. The receptor complex can be visualized autoradiographically on brain sections in vitro using radiolabelled high affinity ligands that either bind specifically to the GABA-binding site (^{3}H muscimol) or the BZR

Table I:
Characteristics of $GABA_A$ and $GABA_B$ receptors

Parameter	$GABA_A$ receptor	$GABA_B$ receptor
Molecular entity	GABA-gated chloride channel	GABA response linked to K^+ and Ca^{++} or amplification of some cAMP-mediated responses
Function	Mediates the main synaptic response to GABA in the CNS	Inhibition of release of certain neurotransmitters and other less well defined responses
Specificity	GABA effect is mimicked by muscimol, blocked by bicuculline	GABA effect is mimicked by baclofen (not by muscimol), blocked by phaclofen (not by bicuculline)
Allosteric modulators	Benzodiazepine receptor ligands, barbiturates, some convulsants	Unknown

Möhler and Richards 1988 (94)

site (^3H flumazenil)(7,8). The in vivo distribution of the BZR binding sites is very similar after intravenous injections.

Up to one third of all synapses in the brain are thought to be GABA-ergic. The receptor density is high in the olfactory bulb, cerebral cortex, cerebellum, pallidum, hippocampus, substantia nigra and inferior colliculus. So far we do not know of an association between particular localization of the BZR and particular pharmacological effects. In all cases synaptically released GABA exerts its effect mainly by interacting with the $GABA_A$ receptor complex located on the subsynaptic membrane of the effector neuron. These receptors trigger the gating of the channel for chloride ions. Two molecules of GABA have to bind to the receptor for the channel to open. The regulation in the conductance of chloride ions at the cell membrane is the basis of GABA-ergic synaptic inhibition.

The BZR is a modulatory unit through which its ligands can alter the GABA-mediated channel gating process. The therapeutically used benzodiazepines enhance the effect of submaximal amounts of GABA in a synapsis, however, the maximal response to GABA cannot be intensified further by these agents. This may explain the high degree of their therapeutic safety.

Figure 1: GABA/Benzodiazepine receptor/chloride inophore complex

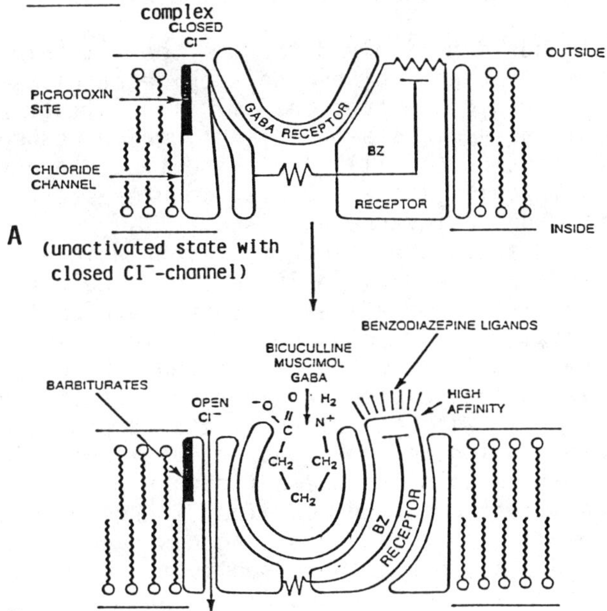

Reproduced from The Liver: E.A. Jones 1988 (95)

Figure 1 shows schematically the GABA/BZR chloride inophore complex in the surface membrane of a postsynaptic neuron. It contains receptors for GABA, benzodiazepines and barbiturate ligands. At the barbiturate site bind: barbiturates (through which the channel for chloride ions and thus the action of GABA can be blocked).

At the GABA site bind: GABA, other GABA-receptor agonists, e.g. muscimol, and GABA-receptor antagonists, e.g. bicucullin.

At the benzodiazepine site bind: BZR agonists (benzodiazepines), BZR antagonists (flumazenil) and BZR inverse agonists (e.g. beta-carbolines).

The activation of the receptor is associated with opening of the chloride channel. Through this chloride flux the cell membrane will be changed from a depolarized to a hyperpolarized state.

3. BENZODIAZEPINE RECEPTOR LIGANDS

The BZR ligands can widely differ in their intrinsic efficacies from the classical benzodiazepine agonists, which enhance GABA-ergic synaptic transmission (9), to the inverse agonists (e.g. beta-carbolines) which reduce GABA-ergic transmission (10,11). The third type are antagonists such as flumazenil. The antagonist counteracts the effect of agonists and of inverse agonists.

Figure 2 shows the spectrum of intrinsic activities of these ligands at the BZR. Positive signs indicate agonistic intrinsic activity, negative signs indicate an inverse intrinsic activity and 0 shows no intrinsic activity. As outlined above, agonists at the BZR include the classical tranquillizing benzodiazepine. They enhance the synaptic response to GABA with the clinical effect of reducing anxiety, muscle tension, convulsions and vigilance. In the presence of inverse agonists the efficiency of GABA-ergic synaptic transmission is reduced. They induce pharmacological effects opposite to those of BZR agonists such as anxiety attacks, convulsions and increased muscle tension. These compounds demonstrate that the BZR is unique in that it can shift GABA responses in both directions. Full agonists and full inverse agonists represent the extremes of an entire spectrum of BZR ligands. Between these two extremes, there are compounds with partial agonists, partial inverse ago-

Figure 2: **Spectrum of intrinsic activities of Benzodiazepine-receptor ligands**

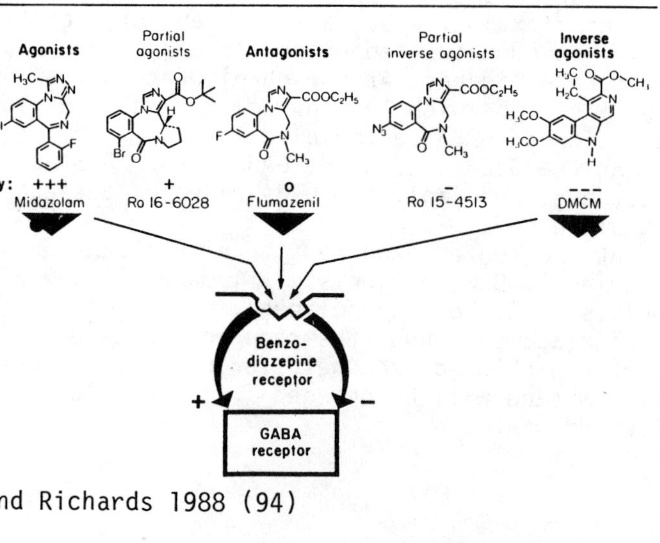

Möhler and Richards 1988 (94)

nists and antagonist properties.

4. BENZODIAZEPINE RECEPTOR ANTAGONIST

The first highly specific antagonist in therapeutic use is Ro 15-1788 (flumazenil, ANEXATE^{R}). Chemically it is an imidazobenzodiazepine. Flumazenil interacts exclusively with the BZR in the brain. The interaction of flumazenil is highly specific, it only binds to the central type of receptors. It has no relevant affinity to the peripheral benzodiazepine binding sites (12). Flumazenil is a competitive antagonist. Its action is fully surmountable and reversible. The ligand receptor complex associates and dissociates spontaneously. The lifespan of such a complex lies between a few milliseconds and perhaps a few seconds. The pharmacological effect of flumazenil depends upon the number of BZR that can be occupied by flumazenil according to the mass-action-law and depends on the affinities and concentrations of the ligands. The in vivo interaction of flumazenil with other BZR ligands can be visualized by the positron emission tomography technique (PET) in the living brain (13,14). Figure 3 shows the rapid disappearance of radioactive labelled flumazenil (^{11}(C) flumazenil) from brain receptors in human volunteers after the injection of another ligand (here nonradioactive flumazenil)(15).

Figure 3: Disappearance of ^{11}C flumazenil from the human brain after injection of cold flumazenil

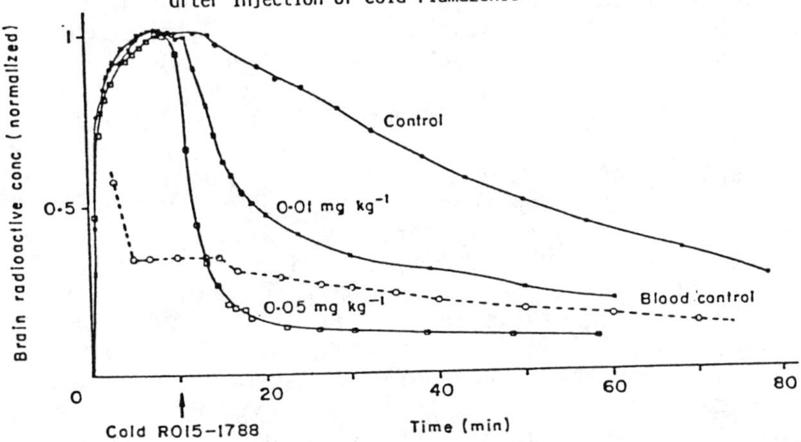

Möhler and Richards 1981 (15)

4.1. Antagonistic Effect of Flumazenil

Flumazenil can block all benzodiazepine effects as shown in animals (16,19) and in man (20-22). These effects were investigated on administration of flumazenil given before, simultaneously or after a benzodiazepine agonist. When the antagonist is given before or after the agonist, the agonistic effect is inhibited. An intravenous dose of 5 mg flumazenil is capable of inhibiting or attenuating the central effects of a subsequent dose of $<$6 mg Midazolam. At doses in the range of 6-10 mg Midazolam reversed the antagonistic effect of flumazenil (23,24). Flumazenil antagonizes of course also the effect of simultaneously administered benzodiazepine agonists. The degree and duration of its effect depend on the doses of antagonist and agonist. After oral intake, it appears that the maximum antagonistic effect is attained with a dose of about 100 mg. Higher oral doses produce neither a distinct intensification nor a prolongation of the antagonistic effect, and may in fact show a mild agonistic effect. After intravenous administration, flumazenil is effective at lower doses than after oral administration probably due to the high first-pass effect (25-37). When the antagonist is administered after the agonist, an intravenous dose of 0.25 to 0.5 mg flumazenil is sufficient. And the effect is extremely rapid in onset. The duration of antagonistic effect is dependent upon the dose of the antagonist, the type and the dose of the previously administered benzodiazepine agonist and the interval between benzodiazepine agonist and flumazenil administration. The duration of the antagonistic effect is limited to 3-4 hours due to the rapid hepatic elimination (38-49).

4.2. Intrinsic Effect of Flumazenil

The specifity of flumazenil is very high, non-specific effects occur only under special conditions. Flumazenil has a minimal positive intrinsic efficacy, as is the case, for instance, for the intrinsic sympathomimetic activity of beta-sympatholytica. Under experimental conditions, flumazenil has weak agonistic and virtually no inverse agonistic activity. In clinical investigations, after a dosage about 10 times higher than therapeutic antagonistic doses, a slight central stimulating activity (inverse agonistic) and in higher doses a central depressant (agonistic) effect was found. Higgitt et al. observed with an oral dose of 30 mg some inverse agonistic and with a high dose of 100 mg some agonistic effects by measuring EEG and psychophysiological changes (50).

Subjective agonistic effects were seen in two other studies: restlessness, tiredness and dizziness after 400 mg orally (51) and mild dizziness after 0.1 mg/kg intravenously (42).

Furthermore, agonistic effects consisting of a feeling of fear and a motor restlessness were found in some volunteers after 5 mg intravenous flumazenil (52).

Placebo-controlled EEG studies showed decreased amplitude of auditory evoked potentials and decreased alpha and theta power after 5 mg flumazenil intravenously (49,52,53). Epileptic patients previously treated with only partial success with high doses of benzodiazepines responded well to slow intravenous injections of 2.5 mg flumazenil. The deeply sedated, precomatous patients awaked promptly and spike and wave activity in the EEG as well as the convulsions disappeared. Scollo-Lavizzari proposed therefore flumazenil as an anticonvulsent drug (54).

A reversal of slow wave sleep (55) and an arousal effect in patients with hepatic encephalopathy were observed after flumazenil (56-60). These effects could be explained by a modulation of an increased GABA-ergic activity, maybe under the influence of an endogenous BZR agonist (90).

While the BZR ligands differ widely in their intrinsic activities, the potency is determined by a common parameter: the affinity to the BZR. The duration of action of benzodiazepine receptor ligands is not linked to receptor events, but is determined by the rate of metabolism and elimination.

4.3. Pharmacokinetics of Flumazenil

Pharmacokinetic investigations (61) have shown that flumazenil is a potent antagonist with a complete and direct metabolism resulting into inactive metabolites.

4.3.1. Absorption. 200 mg of oral flumazenil is rapidly absorbed (absorption half-life time 0.3 hours). The peak plasma concentration is observed after 20 to 90 minutes. Oral bioavailability is low (16%) because of a high hepatic clearance and a consequent first pass metabolism. Absorption of flumazenil from the GI-tract into the portal tract is nearly complete in man. The excretion profile is almost identical following an oral or single intravenous dose of ^{14}C-labelled flumazenil (90-95% in urine, 5-10% in faeces). Mostly flumazenil is intravenously administered because of the short elimination half-life time and the need of a rapid onset of action (62).

A plasma concentration of 10 to 20 μg/l flumazenil can effectively reverse benzodiazepine-induced central nervous system depression. This can be achieved with a bolus of 2.5 mg and maintained for 1 to 2 hours (63).

4.3.2. Distribution and elimination. The plasma protein binding is low (40%). The distribution is rapid as we see by the initial steep decline in the plasma concentration time profile after a single intravenous dose of 2.5mg flumazenil (Fig. 4). Flumazenil is extensively metabolized as only less than 0.2% of the dose is found unchanged in the urine. Renal clearance is negligible. The high total clearance is to be attributed to hepatic clearance. Hepatic elimination is rapid as shown by short elimination half-life time (0.9 hours ± 0.2 (mean ± SD)) and a high total plasma

Figure 4: Plasma concentration time profile of flumazenil after a single intravenous injection of 2,5 mg

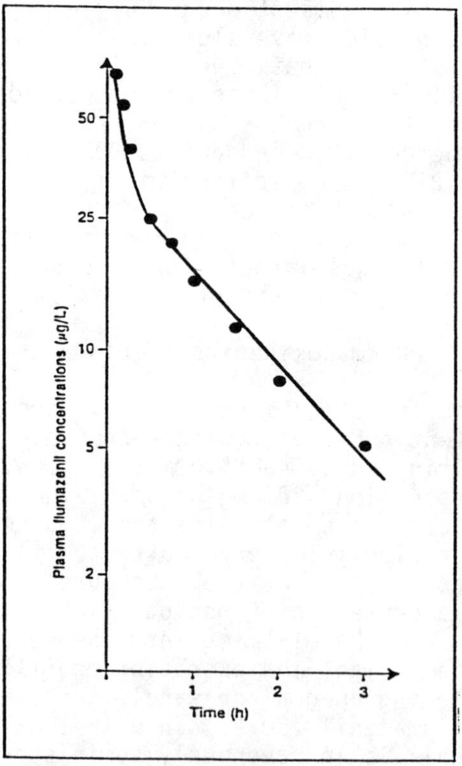

Klotz and Kanto 1988 (65)

and blood clearance. The hepatic blood clearance approaches nearly the rate of hepatic blood flow. In severe liver insufficiency, the plasma half-life time is elevated (G. Pomier, personal communication).

Table II gives the ranges of the pharmacokinetic parameters for flumazenil in humans. The high volume of distribution (0.6-1.6 l/kg) indicates that the distribution exceeds the plasma compartment and that there is considerable tissue uptake. The high blood/plasma ratio indicates that flumazenil is also taken up by the blood cells (63).

The basic pharmacokinetic parameters are more or less independant of the dose (range 2-100 mg). Increasing doses are accompanied by corresponding increases in the area under the curve (AUC). In the broad dose range applications a linear disposition kinetic was found. Furthermore, little inter- and intraindividual variability was observed (64). A clinical effect usually can be observed within one minute following flumazenil injections. Flumazenil is administered stepwise in low doses. The minimal effective dose is 0.1mg.

When prolonged antagonism is needed, multiple injections or a continuous infusion are necessary. A therapeutic steady state plasma concentration can be achieved by an infusion rate of 1.5 mg/h (65).

As shown before, the uptake and the regional brain distribution of flumazenil were investigated by PET. In healthy volunteers the drug was initially distributed according

Table II:
Range of expected pharmacokinetic parameters for flumazenil in humans

Parameter	Range
$t_{½}$ (hours)	0.7 - 1.3
CL_p (L/h)	31 - 78
Vd_{ss} (L/kg)	0.6 - 1.6
Free fraction (%)	54 - 64
Blood/plasma ratio	0.8 - 1.3

Abbreviations: $t_{½}$ = elimination half-life;
CL_p = plasma clearance;
Vd_{ss} = volume of distribution

Klotz and Kanto 1988 (65)

to cerebral blood flow and taken up by grey matter structures. The binding of ^{11}C-labelled flumazenil is reduced by approximatively 90% within 10 minutes after injection of a high dose of unlabelled drug. Cerebral activity decreased with a half-life time of about 30 minutes (66). As far as we know, flumazenil is not metabolized in the brain and the water-soluble metabolites of flumazenil in the blood do not pass the blood brain barrier (67).

4.3.3. Metabolism. So far, three inactive metabolites have been identified. Flumazenil and its metabolites are completely eliminated within 48-72 hours.

4.3.4. Drug interactions. The pharmacokinetics and dynamics of flumazenil are not affected by co-administration of benzodiazepines (68, 69) or alcohol (70).

5. SAFETY

Flumazenil is a very safe compound. Even high doses, e.g. 600 mg orally or 100 mg intravenously, were well tolerated locally and systemically (32). No negative influences on respiratory or cardiovascular parameters were found. Withdrawal symptoms in patients dependent on benzodiazepine derivatives have been observed only seldom after flumazenil.

6. THERAPEUTIC INDICATIONS

The present therapeutic indications of flumazenil are restricted to special clinical situations (Table III).

Flumazenil antagonizes effectively the central effects of benzodiazepines with regard to sedation and anterograde amnesia in patients undergoing short sedation, e.g. endoscopy or dental surgery (71-73). Postsurgical patients are

Table III: Therapeutic indications

- Antagonism of benzodiazepine sedation
- Antagonism of benzodiazepine effects in general anaesthesia
- In emergency medicine and intensive care
  - Benzodiazepine intoxication
  - Differential diagnosis of intoxication and coma
  - Neurological evaluation of sedated patients
  - Weaning from mechanical ventilation
- Paradoxical reaction to benzodiazepines
- Epilepsy ?
- Hepatic encephalopathy ?

awaking from benzodiazepine-induced anesthesia or sedation within 5 minutes instead of 30 to 120 minutes for spontaneous awaking (21, 74-76). The quick arousal decreases the risks of pulmonary obstruction during the postoperative period.

Well documented is also the use of flumazenil for benzodiazepine intoxications. Benzodiazepines are frequently taken drug in case of self-induced poisoning. The reversal of pure benzodiazepine-overdose takes less than 5 minutes (77,78). Also in self-induced miscellaneous drug poisoning a prompt response to flumazenil was found (76-82). A careful clinical observation of such patients is necessary to avoid renewed sedation when the effect of flumazenil is fading off.

It may be useful to apply the antagonist as a primary diagnostic tool for the differential diagnosis of intoxications and other causes of coma (77,86). A benzodiazepine intoxication can be confirmed or excluded immediately.

Furthermore, flumazenil can be used in patients who require long-term sedation. Sometimes it is necessary e.g. to interrupt sedation in order to evaluate the patient's neurological condition or to terminate sedation when weaning from mechanical ventilation is needed (87-89). In such conditions flumazenil was completely effective in reversing benzodiazepine-induced sedation.

There is also some evidence in favour of using flumazenil for the reversal of paradoxical reactions to benzodiazepines (89).

The possible indication in epilepsy was already mentioned. Flumazenil was also tried in hepatic encephalopathy. There is evidence for an increased GABA-ergic tone in hepatic encephalopathy due to a possible endogenous benzodiazepine ligand or another unknown GABA-ergic stimulating factor. An increased GABA-ergic neurotransmission in hepatic encephalopathy may be antagonized by several antagonists at the GABA/BZR complex. This was shown using partial inverse benzodiazepine agonists such as CGS 8216 and Ro 15-4513 in rats (90) and antagonists such as flumazenil in rats and rabbits (91) and in man (56-60). Also after administration of bicuculline, a GABA-receptor antagonist, and isopropylbicyclophosphate, a chloride ionofore blocker, an amelioration was found (91). The finding of an endogenous benzodiazepine ligand with agonistic function in hepatic encephalopathy by Mullen and co-workers (92) offers an explanation for the clinical improvement observed after BZR antagonists.

Up to now, case reports of 40 patients with 44 episodes

of hepatic encephalopathy (grade I to IV) treated with flumazenil in a dose of 0.2 to 5 mg have been presented. The reports describe remission in 3 episodes, improvement in 28 episodes, no changes in 8 episodes, deterioration in 1 episode and 4 episodes were not evaluable (56-60 and personal communications).

Further studies in animals and controlled clinical trials of this class of drugs in management of hepatic encephalopathy are required.

We are indebted to F. Hoffmann-La Roche for their assistance in preparing this manuscript and to Mrs. J. Gass for her secretarial help.

7. REFERENCES

1) Haefely W., Kulcsar A., Möhler H. et al.: Possible involvement of GABA in the central actions of benzodiazepines. Adv. Biochem. Psychopharmacol. 1975; 14 : 131-151
2) Haefely W., Polc P.: Physiology of GABA enhancement by benzodiazepines and barbiturates. In: Olsen RW, Venter JC, eds. Benzodiazepine-GABA Receptors and Chloride Channels: Structure and Function Properties. New York: Alan R Liss, 1986; 97-133
3) Sigel E., Barnard E.A.: A γ-aminobutyric acid/benzodiazepine receptor complex of bovine cerebral cortex: purification and partial characterization. J. Biol. Chem. 1983; 2658 : 6965-6971
4) Schoch P., Häring P., Takacs B. et al.: A GABA/benzodiazepine receptor from bovine brain: purification, reconstruction and immunological characterization. J. Recept. Res. 1984; 4 : 189-200
5) Häring P., Stähli C., Schoch P. et al.: Monoclonal antibodies reveal structural homogeneity of γ-aminobutyric acid/ benzodiazepine receptors in different brain areas. Proc.Natl. Acad. Sci USA 1985; 82 : 4837-4841
6) Schofield P.R., Darlison M.G., Fujita N. et al.: Sequence and functional expression of the $GABA_A$ receptor shows a ligand-gated receptor super-family. Nature 1987; 328 : 221-227
7) Richards J.G., Möhler H., Haefely W.: Mapping benzodiazepine receptors in the CNS by radiohistochemistry and immunohistochemistry. In: Panula P, Päivärinta H, Soinita S, eds. Neurology and Neurobiology Vol. 16, Neurohistochemistry, Modern Methods and Applications. New York: Alan R. Liss, 1986 : 629-677
8) Richards J.G., Glinz R., Schoch P. et al.: New trends in mapping benzodiazepine receptors. In: Biggio G, Costa E, eds. Proceedings of 5th Capo Boi Conference on Neurosciences, May 1987, Advances in Biochemical Psychopharmacology. New York: Raven Press 1988 (in press)
9) Haefely W.: Actions and interactions of benzodiazepine agonists and antagonists at GABA-ergic synapses. In: Bowery N.G., ed. Actions and Interactions of GABA and Benzodiazepines. New York: Raven Press 1984; 263-285
10) Braestrup C., Schmiechen R., Nielsen M. et al.: Interaction of convulsive ligands with benzodiazepine receptors. Science 1982; 216 : 1241-1243
11) Polc P., Bonetti E.P., Schaffner R. et al.: A three-state model of the benzodiazepine receptor explains the interactions between the benzodiazepine antagonist Ro 15-1788,

benzodiazepine tranquillizers, β-carbolines and phenobarbitone. Arch. Pharmacol. 1982; 321 : 260-264
12) Haefely W.: The preclinical pharmacology of Flumazenil. Eur. J. Anaesth. 1988; 2 : 25-36
13) Persson A., Ehrin E., Eriksson L. et al.: Imaging of (^{11}C)-labelled Ro 15-1788 binding to benzodiazepine receptors in the human brain by positron emission tomography. J. Psychiat. 1985; Res. 4 : 609-622
14) Samson Y., Hantraye P., Baron J.C. et al.: Kinetics and displacement of ^{11}C- Ro 15-1788, a benzodiazepine antagonist, studied in human brain in vivo by positron tomography. Eur. J. Pharmacol. 1985; 110 : 247
15) Möhler H., Richards J.G.: Agonist and antagonist benzodiazepine receptor interaction in vitro. Nature 1981; 294 : 763-764
16) Hunkeler W., Möhler H., Pieri L. et al.: Selective antagonist of benzodiazepines. Nature 1981; 290 : 514-516
17) Bonetti E.P., Pieri L., Cumin K. et al.: Benzodiazepine antagonist Ro 15-1788: neurological and behavioural effects. Psychopharmacology 1982; 78 : 8-18
18) Möhler H., Burkard W.P., Keller H.H. et al.: Benzodiazepine antagonist Ro 15-1788: binding characteristics and interaction with drug-induced changes in dopamine turnover and cerebellar cGMP levels. J. Neurochem. 1981; 37 : 714-722
19) Polc P., Laurent J.P., Scherschlicht R. et al.: Electrophysiological studies on the specific benzodiazepine antagonist Ro 15-1788. Nanyn Schmiedeberg's Arch. Pharmacol. 1981; 316 : 317-325
20) Radakovic D.: Expérience clinique acquise avec l'Anexate (Ro 15-1788) en anesthésiologie. MPH Editions - JPBT 1987; 24 : 40-42
21) Lauven P.M., Ebeling B.J. Stoeckel, H. et al.: Wirksamkeit des Benzodiazepin-Antagonisten Ro 15-1788 nach einer Anästhesie mit Flunitrazepam-Einleitung. Anästh. Intensivther. Notfallmed. 1986; 21 : 311-314
22) Geller E., Niv D., Matzkin C. et al.: The antagonism of midazolam sedation by Ro 15-1788 in 50 postoperative patients. Anesthesiology 1985; 63 : A369
23) Gath I.: Electrophysiological and clinical aspects of benzodiazepine antagonist Ro 15-1788 and Ro 15-3505. Final study report. Research Report B-113'659, January 23, 1984
24) Gath I., Weidenfeld J., Collins G.I. et al.: Electrophysiological aspects of benzodiazepine antagonists, Ro 15-1788 and Ro 15-3505. Br. J. clin. Pharmacol. 1984; 18 : 541-547

25) Downie W.W.: Ro 15-4230: First study in man of ad hoc combination. Dose-response curve using 20, 30 and 40 mg Ro 15-1788 against a fixed dose (4 mg) of Ro 11-3128. Research Report B-95'505, January 12, 1981
26) Downie W.W.: Ro 15-4230: Further study of the ad hoc combination using 8 mg Ro 11-3128 and 200 mg Ro 15-1788. Research Report B-95'506, February 2, 1981
27) Darragh A., Scully M., Lambe R. et al.: Investigation in man of the efficacy of a benzodiazepine antagonist, Ro 15-1788. Lancet 1981/II; 8-10
28) Darragh W., Lambe R., Brick I. et al.: Antagonism of the central effects of 3-methyl-clonazepam. Br. J. clin. Pharmacol 1982; 14 : 871-872
29) Darragh A.: A study of the efficacy of Ro 15-1788 as an antagonist of the sedative effects of Ro 11-3128 in healthy volunteers. Research Report B-77'175, September 1, 1981
30) O'Boyle C., Darragh A., Lambe R. et al.: Antagonism of the sedative effects of meclonazepam in man by Ro 15-1788. Irish J. Med. Sci. 1983; 152 : 210-211
31) Darragh A.: A study of the efficacy of Ro 15-1788 (600 mg) as an antagonist of the sedative effects of Ro 11-3128 (20 mg) in healthy volunteers. Research Report B-77'176, September 3, 1981
32) Darragh A., Lambe R., Brick I. et al.: Reversal of benzodiazepine-induced sedation by intravenous Ro 15-1788. Lancet 1981; No. 8254, 1042
33) Darragh A., Lambe R., O'Boyle C. et al.: Blockade of the central effects of diazepam by a novel benzodiazepine antagonist. Research Report B-95'511, August 18, 1981
34) Darragh A., Lambe R., O'Boyle C. et al.: Antagonism of diazepam-induced sedation in man. Research Report B-95'512, August 25, 1981
35) Darragh A., Lambe R., Kenny M. et al.: Ro 15-1788 antagonizes the central effects of diazepam in man without altering diazepam bioavailability. Br. J. clin. Pharmacol. 1982; 14 : 677-682
36) Darragh A., O'Boyle C., Lambe R. et al.: Antagonism of the central effects of diazepam in man by Ro 15-1788. A novel benzodiazepine antagonist. Irish J. Med. Sci. 1982; 151 : 90
37) O'Boyle C., Darragh A., Lambe R. et al.: Ro 15-1788 antagonises diazepam induced amnesia without altering bioavailability. Irish J. Med. Sci. 1982; 151 : 415
38) Ziegler W.H., Schalch E.: Antagonism of benzodiazepine-induced sedation in man. In: Sleep 1982, 6th Eur. Congr. Sleep Res., Zürich 1982, Karger, Basel 1983 : 427-429

39) Doenicke A., Suttmann H., Kugler J. et al.: Pilot study of a benzodiazepine antagonist. Br. J. Anaesthesia 1984; 54 : 1131-1133

40) Doenicke A., Suttmann H., Kapp W. et al.: Zur Wirkung des Benzodiazepin-Antagonisten Ro 15-1788. Anaesthesist 1984; 33 : 343-347

41) Doenicke A.: Modern trends in the investigation of new hypnotics in anesthesia. In: Hindmarch I. et al. eds. Sleep, Benzodiazepines and Performance. Psychopharmacology 1984 (Suppl. 1), Springer, Berlin 1984 : 119-132

42) Forster A., Rouiller M., Morel D. et al.: Double-blind randomized study evaluating a specific benzodiazepine antagonist. Anesthesiology 1983; 59 : A375

43) Lupolover R.: Evaluation of the efficacy and safety, tolerance, of Ro 15-1788 compared to Ro 15-3505 i.v. administered in normal healthy volunteers. Research Report B-113'654, June 10, 1983

44) Müller T., Ziegler W.H., Schalch E. et al.: Minimal effective doses of the benzodiazepine antagonist Ro 15-1788 under midazolam steady-state conditions. Research Report B-113'108, May 10, 1986

45) Ziegler W.H., Schalch E., Crevoisier C. et al.: Ro 15-1788: Investigation of the antagonistic effect of Ro 15-1788 in a midazolam steady-state model. Research Report B-113'107, May 10, 1986

46) Lauven P.M., Stoeckel H., Schwilden H. et al.: Application of a benzodiazepine antagonist Ro 15-1788 under steady-state conditions of midazolam. Anesthesiology 1982; 57 : A325-A326

47) Schwilden H., Stoeckel H., Lauven P.M. et al.: Action of a benzodiazepine antagonist during midazolam infusion in steady-state. Quantitative EEG studies. Anesthesiology 1982; 57 : A326

48) Lauven P.M., Schwilden H., Stoeckel H. et al.: The effects of a benzodiazepine antagonist Ro 15-1788 in the presence of stable concentrations of midazolam. Anesthesiology 1985; 63 : 61-64

49) Laurian S., Schöpf J.: Influence of Ro 15-1788 on the auditory evoked potentials. Preliminary results. Research Report B-42'389, December 6, 1982

50) Higgitt A., Lader M., Fonagy P.: The effects of the benzodiazepine antagonist Ro 15-1788 on psychophysiological performance and subjective measures in normal subjects. Psychopharmacology 1986; 89 : 395

51) Darragh A., Lambe R., O'Boyle C. et al.: Absence of central effects in man of the benzodiazepine antagonist Ro

15-1788. Psychopharmacology 1983; 80 : 192-195
52) Schöpf J., Laurian S., Le P.K. et al.: Intrinsic activity of the benzodiazepine antagonist Ro 15-1788 in man: an electrophysiological investigation. Pharmacopsychiatry 1984; 17 : 79-84
53) Laurian S., Schöpf J. Le P.K. et al.: The intrinsic pharmacological effect of Ro 15-1788. In: 7th World Congress of Psychiatry, July 11-16, 1983; 402, Abstr. No. P184, Society of Austrian Neurologists and Psychiatrists, Austria
54) Scollo-Lavizzari G.: The anticonvulsant effect of benzodiazepine antagonist, Ro 15-1788: An EEG study in 4 cases. Eur. Neurol. 1984; 23 : 1-6
55) Ziegler G., Ludwig L., Fritz G.: Effect of the specific benzodiazepine antagonist Ro 15-1788 on sleep. Pharmacopsychiatry 1986; 19 : 200-201
56) Scollo-Lavizzari G., Steinmann E.: Reversal of hepatic coma by benzodiazepine antagonist Ro 15-1788. Lancet 1985/I; 1324
57) Bansky G., Meier P.J., Ziegler W.H. et al.: Reversal of hepatic coma by benzodiazepine antagonist Ro 15-1788. Lancet 1985/I; 1324-1325
58) Bansky G., Meier P.J., Riederer R. et al.: Wirkung des Benzodiazepinantagonisten Flumazenil (Anexate[R]) bei hepatischer Enzephalopathie. Schweiz. med. Wschr. 1988; 23 : 8
59) Grimm G., Lenz K., Kleinberger G. et al.: Ro 15-1788 improves coma in 4 out of 5 patients with fulminant hepatic failure; verification by long latency auditory and somatosensory evoked potentials. J. Hepatol. 1987; 4 (Suppl. 1) : 21
60) Meier R., Gyr K.: Treatment of hepatic encephalopathy (HE) with the benzodiazepine antagonist flumazenil: a pilot study. Eur. J. Anesth. 1988, Suppl. 2 : 139-146
61) Roncari G., Zumbrunnen R., Ziegler W.H. et al.: Ro 15-1788: Mass balance study in man following single intravenous and oral administration of ^{14}C-Ro 15-1788. Research Report No. B-104'497, October 17, 1985
62) Roncari G., Ziegler, W.H., Guentert T.W.: Pharmacokinetics of the new benzodiazepine antagonist Ro 15-1788 in man following intravenous and oral administration. Br. J. clin. Pharmacol. 1986; 22 : 421-428
63) Klotz U., Ziegler G., Reimann I.W.: Pharmacokinetics of the selective benzodiazepine antagonist Ro 15-1788 in man. Eur. J. clin. Pharmacol. 1984; 27 : 115-117
64) Roncari G., Timm U., Zell M. et al.: Dose proportionality study of Ro 15-1788 in man following intravenous infusion of 2, 10, 40 and 100 mg of the parent drug. Research

Report No. B-113'226, February 13, 1986
65) Klotz U., Kanto J.: Pharmacokinetics and clinical use of flumazenil (Ro 15-1788). Clin. Pharmacokinetics 1988; 14 : 1-12
66) Samson Y., Hantraye P., Baron J.C. et al.: Kinetics and displacement of (^{11}C) Ro 15-1788, a benzodiazepine antagonist, studied in human brain in vivo by positron tomography. Eur. J. Pharmacol. 1985; 110 : 247-251
67) Shinotoh H., Yamasaki T., Inoue O. et al.: Visualization of specific binding sites of benzodiazepines in human brain. J. Nucl. Med. 1986; 27 : 1593-1599
68) Klotz U., Duka T., Dorow R. et al.: Flunitrazepam and Iormetazepam do not affect the pharmacokinetics of the benzodiazepine antagonist Ro 15-1788. Br. J. clin. Pharmacol. 1985a; 19 : 95-98
69) Klotz U., Ziegler G., Ludwig L. et al.: Pharmacodynamic interaction between midazolam and a specific benzodiazepine antagonist in humans. J. clin. Pharmacol. 1985b; 25 : 400-406
70) Klotz U., Ziegler G., Rosenkranz B. et al.: Does the benzodiazepine antagonist Ro 15-1788 antagonize the action of ethanol? Br. J. clin. Pharmacol. 1986; 22 : 513-520
71) Jensen S., Knudsen L., Kirkegaard L.: Flumazenil used in the antagonizing of diazepam and midazolam sedation in outpatients undergoing gastroscopy. Eur. J. Anaesth. 1988; 2 : 161-166
72) Rosenbaum N.L., Hooper P.A.: The use of flumazenil as an antagonist to midazolam in intravenous sedation for dental procedures. Eur. J. Anaesth. 1988; 2 : 183-190
73) Holloway A.M., Logan, D.A.: The use of flumazenil to reverse diazepam sedation after endoscopy. Eur. J. Anaesth. 1988; 2 : 191-198
74) Raeder J.C., Nilsen O.G., Hole A. et al.: The use of flumazenil after total i.v. anaesthesia with midazolam in out-patients. Eur. J. Anaesth. 1988; Suppl. 2 : 257-264
75) Wolff J.: Flumazenil for post-operative recovery after general anaesthesia. Eur. J. Anaesth. 1988; 2 : 239-249
76) Nilsson A., Persson M.P., Hartvig P.: Effects of flumazenil on postoperative recovery after total intravenous anaesthesia with midazolam and alfentanil. Eur. J. Anaesth. 1988; 2 : 251-256
77) Hofer P., Scollo-Lavizzari G.: Benzodiazepine antagonist Ro 15-1788 in self-poisoning: diagnostic and therapeutic use. Arch. Intern. Med. 1985; 145 :663-664
78) Geller E., Niv D., Silbiger A. et al.: Ro 15-1788 a benzodiazepine antagonist in the treatment of 34 intoxicated patients. Anesthesiology 1985; 63 (Suppl.), abstr. 157

79) Bismuth C., Baud F.J., Fournier P.E. et al.: Antagonistic Ro 15-1788 dans l'intoxication volontaire par benzodiazepines. Valeur diagnostique et therapeutique. Etude sur 40 cas. Ann. Méd. int. 1986; 137 : 363-364
80) Hruby K., Prischl F., Smetana R. et al.: Spezifische Therapie der akuten Benzodiazepinintoxikation mittels Flumazenil (Ro 15-1788). Wien Med. Wschr. 1987; 137 : 179-183
81) Jensen S., Knudsen L., Lonka L. et al.: Benzodiazepine intoxication treated with Anexate[R] (Ro 15-1788). Acta Anaesthesiol. Scand. 1987; 31, suppl. 86, abstr. 196
82) O'Sullivan G.F., Wade D.N.: Flumazenil in the management of acute drug overdose with benzodiazepine and other agents. Clin. Pharmacol. Ther. 1987; 42 : 254-259
83) Baehrend S., Höjer J.: Flumazenil in self-induced benzodiazepine poisoning. Eur. J. Anaesth. 1988; 2 : 287-293
84) Lheureux P., Askenasi R.: Double-blind study of Anexate[TM] in benzodiazepine intoxication. Eur. J. Anaesth. 1988; 2 : 300-304
85) Rouzioux J.M., Ducluzeau R., Baud F.J. et al.: The use of Anexate[TM] (Ro 15-1788) in acute self-poisoning with benzodiazepines alone or combined with other drugs: results of a therapeutical trial. Eur. J. Anaesth. 1988; 2 : 305-309
86) Mapelli A., Bellinzona G., Lorini F.L. et al.: Diagnostic and therapeutic use of flumazenil in emergency medicine. Eur. J. Anaesth. 1988; 2 : 295-299
87) Ritz R.: Clinical use of Ro 15-1788 in patients given high therapeutic doses of benzodiazepines in ICU. In Symposium on the Benzodiazepine Antagonist Ro 15-1788 (Anexate) at 4th World Congress on Intensiv and Critical Care Medicine. Jerusalem, June 23-28, 1985
88) Kleinberger G., Grimm G., Laggner A. et al.: Weaning patients from mechanical ventilation by benzodiazepine antagonist Ro 15-1788. Lancet 1985; 2 : 268-269
89) Ricou B., Forster A., Bruckner A. et al.: Clinical evaluation of a specific benzodiazepine antagonist (Ro 15-1788). Br. J. Anaesth. 1986; 58 : 1005-1011
90) Baraldi M., Zeneroli M.L., Ventura E. et al.: Supersensitivity of benzodiazepine receptors in hepatic encephalopathy due to fulminant hepatic failure in the rat: reversal by benzodiazepine antagonist. Clin. Sci. 1984; 67 : 167-175
91) Bassett M.L., Mullen K.D., Skolnick P. et al.: Amelioration of hepatic encephalopathy by pharmacologic antagonism of the $GABA_A$-benzodiazepine receptor complex in a rabbit model of fulminant hepatic failure. Gastroenterology 1987; 93 : 1069-1977

92) Mullen K.D., Mendelson W.B., Martin J.V. et al.: Could an endogenous benzodiazepine ligand contribute to hepatic encephalopathy? Lancet 1988; I : 457-459
93) Gammal S.H., Geller D., Skolnick P. et al.: Unequivocal amelioration of hepatic encephalopathy (HE) by benzodiazepine receptor antagonists (BZA) in a rat model of fulminant hepatic failure (FHF). Gastroenterology 1987; 92 : 1734
94) Möhler H., Richards J.G.: The benzodiazepine receptor: a pharmacological control element of brain function. Eur. J. Anaesth. 1988; 2 : 15-24
95) Arias I., Jakoby W.B., Poper H. et al.: The liver, biology and pathobiology. 2nd Edition, Raven Press, New York 1988

EFFECTS OF THE BENZODIAZEPINE ANTAGONIST FLUMAZENIL IN HEPATIC ENCEPHALOPATHY IN MAN

Georg Bansky and Peter J. Meier
Division of Clinical Pharmacology, Dept. Medicine, University Hospital, CH-8091 Zurich

ABSTRACT

We tested the proposed role of increased γ-aminobutyric acid (GABA) neurotransmission in the pathogenesis of hepatic encephalopathy by acute intravenous administration of the benzodiazepine antagonist flumazenil to 14 patients with hepatic encephalopathy due to cirrhotic liver disease. Flumazenil administration induced variable and transient, but distinct improvements of the mental status in 71 % of patients. Grade IV hepatic encephalopathy converted to grade II in 4 patients and to grade III in 2 patients. Although all patients with less advanced encephalopathy (3 grade III, 1 grade II) did also improve, their responses were clinically less impressive. The arousal effect occurred in less than 1 minute and lasted for 1 to 2 hours. Furthermore, it was accompanied by a significant increase of the mean EEG frequency from 4.2 to 5.2 cycles per second. Among 8 patients who could ultimately be discharged from the hospital 7 responded to flumazenil, whereas no patient dying within 48 hours after flumazenil testing showed any arousal effect. These results strongly favour a prominent pathogenetic role of increased GABAergic tone in hepatic encephalopathy in man and suggest that positive flumazenil response might be of prognostic value in predicting short-term survival in severe hepatic encephalopathy.

INTRODUCTION

Although the pathogenesis of hepatic encephalopathy is essentially unknown, convincing experimental evidence has recently been provided for activation of the GABA-ergic (gamma-aminobutyric acid) inhibitory neurotransmitter system playing a dominant role in the development of hepatic encephalopathy (1-4). The GABA receptors are localized at the postsynaptic neurons and consist of chemically gated chloride channels that have additional non-competitive binding sites for benzodiazepines and barbiturates (5,6). In suitable animal models, hepatic encephalopathy and benzodiazepine receptor agonist induced encephalopathy have been reported to be similar in terms of clinical and behavioural features as well as with respect to electrophysiological changes as assessed by visual evoked responses (7,8). Furthermore, the benzodiazepine antagonists CGS 8216 and flumazenil (Ro 15-1788) have been shown to induce clinical and electrophysiological improvements of encephalopathy in rabbits (8) and rats (9,10) with fulminant hepatic failure. Similar preliminary observations have recently also been reported by us as well as others in single cases of patients with acute liver failure (11,12), cirrhosis (13) and chronic intractable portal-systemic encephalopathy (14). In the present study we have extended our original observation (13) to 14 patients with grade II to IV encephalopathy due to decompensated cirrhotic liver disease.

PATIENTS AND METHODS

The personal characteristics and clinical features of the 14 patients included in this study are summarized in Table 1. Cirrhotic liver disease was histologically proven by biopsy or autopsy in 13 patients and evidenced clinically and by sonography in one patient (R.H.). In 11 patients the cirrhosis developed as a consequence of chronic ethanol abuse with one of them (R.S.) being also HB_s-antigen carrier. In two patients (E.L. and J.T.) chronic active hepatitis due to HBV-infection progressed to cirrhosis. In patient J.T. hepatocellular carcinoma also occurred. One woman (E.M.) had cryptogenic cirrhosis. All patients exhibited signs of portal hypertension such as ascites, esophageal varices and/or splenomegaly. Patient E.K. underwent a distal splenorenal shunt operation four years before the actual hospitalization. All patients were admitted to the hospital because of acute deterioration of their residual liver function with rapidly progressing encephalopathy. Hepatic coma was precipitated in 7 patients (i.e. E.S., J.D., E.M., O.B., B.Q., E.L., and M.K.) by acute gastrointestinal bleeding with one patient (E.L.) having also been treated with oxazepam (15 mg) and chlordiazepoxide (50 mg i.v.) four and two days before the flumazenil trial, respectively. Another patient (R.M.) was treated with chlordiazepoxide before hospitalization because of impending delirium tremens. This was the only patient with a measurable benzodiazepine level in serum (chlordiazepoxide 0.3 ug/ml). All patients exhibited abnormal liver parameters in serum including total bilirubin (97 ± 64 umol/l [mean ± SD]; normal < 22), AST (143 ± 77 U/l; normal 10 - 60), alkaline phosphatase (173 ± 94 U/l; normal 30 - 115), albumin (26 ± 9 g/l; normal 36 - 51), vitamin K resistent prothrombin time (39 ± 10 %; normal 100) and venous ammonia (168 ± 83 umol/l; normal 16 - 48). Serum creatinine levels were normal in 9 patients (53 - 115 umol/l), slightly increased up to 147 umol in 3 patients (F.B., E.S., J.D.) and moderately elevated in patients R.S. and M.K. to 540 and 304 umol/l, respectively. Routine toxicological screening was performed in blood samples for the presence of ethanol, benzodiazepines and barbiturates. The widely used benzodiazepine agonists midazolam, flunitrazepam, diazepam and the active metabolite desmethyldiazepam were specifically looked for by sensitive high pressure liquid chromatography. Neurologic evaluation included examination of the mental status, pupils, ocular movements, optic fundi, speech, strength of arm and leg muscles, muscle tone, tendon and plantar reflexes and of pain and vibration sensation in hands and feet. In low grade encephalopathy psychometric testing was performed by the Reitan number connection test. The hepatic encephalopathy was graded into <u>stage I</u> (mild confusion but preserved orientation to time and place; adequate reaction to simple verbal commands; ability to recall past events; virtual normal number connecting test), <u>stage II</u> (drowsiness, orientation to place but not to time; deficits in reacting to verbal commands and in past memory;

slowed or lost ability to perform the number connecting test), stage III (somnolence but arousable, marked confusion with disorientation to place and time, inadequate reactions to verbal commands, amnesia, no communication possible), and stage IV (coma). In addition, a conventional 16-lead scalp electroencephalogram (needle electrodes; 10 - 20 system; bipolar longitudinal and transverse rows; 12-channel EEG) was continuously recorded during the test period starting 10 minutes before the first flumazenil administration. The EEG was independently analyzed by two neurologists without knowledge of the clinical response.

All patients received full treatment with low protein diet, lactulose and some also with neomycin (E.S., E.M.), diuretics (E.K., E.S., J.D., R.M.), vitamins (E.K., R.M.), ranitidine (J.D., E.M., R.M., E.L.) and antibiotics as required (R.S., E.K., F.B., J.D., E.M., R.M.). Patient R.M. was on respirator because of respiratory failure (aspiration pneumonia) and patient R.S. was treated with hemofiltration because of anuria due to hepatorenal syndrome. All patients had an intravenous line through which 0.9 % saline or 5 % dextrose was infused. Seven patients were treated in the intensive care unit where their vital functions could continuously be monitored. After the acute clinical situation had been stabilized (on average 1.9 days [range 1 - 7] after hospital admission) the coma stage was evaluated (see above) and the EEG recording initiated. After 10 minutes a bolus of 0.2 mg flumazenil was injected. Subsequently the psychoneurological status of the patients was reevaluated. If no significant improvements were observed, a second 0.2 mg dose of flumazenil was administered after 3 min and the patients reexamined. In some patients further doses of flumazenil were administered in 3 min intervals (dose increments of 0.4 mg, 0.8 mg and 1 to 2 mg thereafter). The total cumulative flumazenil doses applied are summarized in Table 1. Besides repeated psychoneurological testing the EEG was recorded during the whole test period lasting between a minimum of 13 min (patient J.D.) and a maximum of 34 min (patient E.S.). After the test periods the patients were continued to be clinically evaluated in hourly intervals for up to 6 hours.

RESULTS

As summarized in Table 1 flumazenil injection exerted an arousal effect in 9 out of 14 patients with stage III or IV hepatic encephalopathy. In addition, patient R.S. was able to perform the number connection test in 190 seconds only after flumazenil although no further improvement of the stage II encephalopathy could otherwise be observed. Thus, 71 % of all patients demonstrated some clinical improvement of the mental status after administration of flumazenil. The most unequivocal arousal effects were observed in 4 patients with coma grade IV who converted to grade II encephalopathy. All patients reacted within 20 to 30 seconds after flumazenil administration. They opened their eyes, were oriented to place, adequately reacted to verbal commands and, with the exception of patient R.M. (respirator) were able to communicate with the physician, nurse or relatives. In one woman (J.D.) the same arousal effect was reproduced 6 hours later when the first dose effect had abated and the patient was again in deep coma. In two other patients the grade IV coma improved to grade III encephalopathy. Both these patients initially did not react to acoustic stimuli, but they opened their eyes and tried to spell words when being called by name after having received flumazenil. However, they remained markedly confused and disoriented to place and time, and no communication was possible despite the injection of cumulative flumazenil doses of 2.9 and 8.2 mg, respectively (Table 1). Three patients with grade III encephalopathy converted to grade II encephalopathy. In all flumazenil responders the clinical improvement was paralleled by an increase of the mean EEG frequency from 4.2 cps (cycles per second) before to 5.2 cps after flumazenil. There was no overlap of the 95 % confidence interval and the differences between the mean EEG frequencies were highly significant (normal distribution of data). None of the patients demonstrated any undesirable side effects of flumazenil administration. On average the minimal effective flumazenil dose was 0.4 mg and the arousal effects lasted between one and two hours. No further improvements were observed with higher flumazenil doses. Similarly, the 4 patients with grade IV coma, who did not react to low doses, also showed no clinical

Table 1: Effects of Flumazenil in Cirrhotic Patients with Hepatic Encephalopathy

PATIENT'S INITIALS	AGE/SEX	ETIOLOGY OF CIRRHOSIS (Signs of Portal Hypertension)*	DOSE OF FLUMAZENIL MED+	DOSE OF FLUMAZENIL TOTAL	GRADE OF ENCEPHALOPATHY FLUMAZENIL Before	GRADE OF ENCEPHALOPATHY FLUMAZENIL After
R.S.	38/M	Alcoholic (A) HB_s	0.4	0.4	II	II (I)
E.K.	44/M	Alcoholic (V)	0.4	4.5	III	II
F.B.	51/M	Alcoholic (SM)	0.8	1.6	III	II
E.S.	67/F	Alcoholic (A)	0.4	9.6	III	II
J.D.	58/F	Alcoholic (A,V)	0.2	0.2	IV	II
M.M.	51/F	Alcoholic (A,V)	0.4	1.6	IV	II
E.M.	49/F	Unknown (V)	0.4	0.4	IV	II
R.M.	43/F	Alcoholic (V)	0.4	0.4	IV	II
R.H.	37/F	Alcoholic (A)	0.2	2.9	IV	III
O.B.	75/M	Alcoholic (V)	0.4	8.2	IV	III
B.Q.	51/M	Alcoholic (V)	0.4	4.0	IV	IV
E.L.	67/M	Chronic Active Hepatitis (V) HB_s	0.4	4.0	IV	IV
M.K.	52/F	Alcoholic (V)	0.4	3.6	IV	IV
J.T.	66/M	Chronic Active Hepatitis (A,V) HB_s	0.4	1.6	IV	IV

* Portal hypertension was suspected from the presence of either splenomegaly (SM), ascites (A) and/or esophageal varices (V).

+ MED: Minimal effective dose.

improvement with increased flumazenil doses (Table 1).

In the small number of patients studied no factors predicting a favourable response to flumazenil could be identified. Hence, no significant differences between responsers and nonresponders were found with respect to duration of encephalopathy, sex distribution, etiology of cirrhosis, biochemical serum parameters including ammonia, reaction to painful stimuli and the presence of pyramidal signs. Interestingly however, 3 out of 4 nonresponding patients died within 2 days after flumazenil testing because of liver failure (B.Q., J.T.) and/or refractory hypotension and uremia (M.K.). In contrast, among the flumazenil responders 7 patients could ultimately be discharged from the hospital while only one patient died in hepatic coma (O.B.) and two others of additional causes such as hemorrhagic shock (R.S.) and systemic tuberculosis (F.B.). In the latter three patients death occurred significantly later than in nonresponding patients, namely 13, 7 and 8 days after the flumazenil trial, respectively. Thus, unspecific (e.g. hypoxic) and irreversible structural brain damage might have contributed to the lack of flumazenil efficacy in the nonresponding patients with early fatal outcome.

DISCUSSION

In the present study we extended our previous observation of an acute arousal effect of flumazenil in a 58 year old woman with cirrhotic liver disease and grade IV encephalopathy (patient J.D. in Table 1; ref. 13) to 14 cirrhotic patients with hepatic encephalopathy stage II to IV. The results demonstrate transient but distinct clinical and electrophysiologic improvements in 71 % of flumazenil treated patients. The flumazenil induced arousal effects were greatest in 7 patients with coma grades III to IV who converted to grade II encephalopathy (Table 1). Maximal neuropsychiatric improvements lasted between one to two hours and the patients' mental status deteriorated to the original coma grades within 6 hours after flumazenil testing. The immediate response and transient nature of the pharmacological effects are strong arguments in favour of a specific flumazenil effect and argue against the possibility that the arousal reactions resulted from spontanous fluctuations of the underlying disease process itself. Furthermore, with the exception of patient R.M. (Table 1) none of the flumazenil responders showed any evidence for previous benzodiazepine intake or the presence of benzodiazepines in serum. Hence, the observed flumazenil efficacy strongly supports the concept that increased GABA-ergic inhibitory neurotransmission plays a crucial role in the pathogenesis of hepatic encephalopathy (1-4).

The reasons for the lack of any flumazenil response in 4 patients with advanced grade IV encephalopathy are not known. Since three of these patients died within 48 hours after flumazenil administration, it is possible that the nonresponding patients suffered from cerebral edema and/or hypoxia already during the test period. That patients with cerebral edema are unlikely to respond to flumazenil is evidenced by animal studies (15). Thus, flumazenil appears only to be effective during the reversible metabolic phase of hepatic encephalopathy. Furthermore, a flumazenil induced arousal reaction might be positively correlated with short term survival, since 7 out of 10 flumazenil responders could ultimately be discharged from the hospital and only one patient (O.B.) died of liver failure 13 days after flumazenil administration. Whether flumazenil testing will indeed have any prognostic significance in patients with hepatic encephalopathy remains to be established in future prospective and placebo controlled studies with a larger number of patients.

Although this study does not elucidate the mechanism(s) involved in the mediation of increased GABA-ergic tone in hepatic encephalopathy, it is tempting to speculate that the flumazenil induced arousal effects resulted from displacement of an endogenous ligand with benzodiazepine agonist-like properties from benzodiazepine receptors (16). In fact, recent observations suggest that cerebrospinal fluid from rabbits and human beings with hepatic encephalopathy contain a heat-stable substance with an apparent molecular weight of less than 10000 daltons that competitively inhibits binding of benzodiazepines to their receptors (17,18). Displacement of an endogenous substance with benzodiazepine agonist like properties from $GABA_A$ receptor complexes would be consistent with the observed potency of low flumazenil doses (Table 1).

In conclusion, the present study demonstrated variable and transient, but distinct arousal effects of the benzodiazepine antagonist flumazenil in 10 out of 14 patients with hepatic encephalopathy grade II to IV due to chronic (cirrhotic) liver disease. These observations are consistent with previous findings in animals (8-10) and with preliminary reports in patients with acute liver failure (11,12) and chronic intractable portal-systemic encephalopathy (14). If the efficacy of benzodiazepine antagonists in certain forms and/or stages of hepatic encephalopathy can be validated in randomized controlled studies, new standards and possibilities will have been reached for both pathophysiologic research as well as the clinical management of patients with hepatic encephalopathy.

REFERENCES

1) Jones EA, Schafer DF, Ferenci P, et al. The neurobiology of hepatic encephalopathy. Hepatology (Baltimore) 1984; 4: 1235-42.
2) Zeneroli ML. Hepatic encephalopathy. Experimental studies in a rat model of fulminant hepatic failure. J. Hepatol. 1985; 1: 301-12.
3) Hoyumpa AM. The unfolding GABA story. Hepatology (Baltimore) 1986; 6: 1042-44.
4) Schafer DF. Hepatic coma: Studies on the target organ. Gastroenterology 1987; 93: 1131-34.
5) Stephenson FA. Understanding the $GABA_A$ receptor: a chemically gated ion channel. Biochem. J. 1988; 249: 21-32.
6) Study RE, Barker JL. Diazepam and (-)-pentobarbital: Fluctuation analysis reveals different mechanisms for potentiation of γ-aminobutyric acid responses in cultured central neurons. Proc. Natl. Acad. Sci. USA 1981; 78: 7180-84.
7) Schafer DF, Pappas SC, Brody LE, et al.Visual evoked potentials in a rabbit model of hepatic encephalopathy. I. Sequential changes and compar-

isons with drug-induced comas. Gastroenterology 1984; 86: 540-45.

8) Bassett ML, Mullen KD, Skolnick P, Jones EA. Amelioration of hepatic encephalopathy by pharmacologic antagonism of the $GABA_A$-benzodiazepine receptor complex in a rabbit model of fulminant hepatic failure. Gastroenterology 1987; 93: 1069-77.

9) Baraldi M, Zeneroli ML, Ventura E, et al. Supersensitivity of benzodiazepine receptors in hepatic encephalopathy due to fulminant hepatic failure in the rat: reversal by a benzodiazepine antagonist. Clin. Sci. 1984; 67: 167-75.

10) Gammal SH, Geller D, Skolnick P, et al. Unequivocal amelioration of hepatic encephalopathy (HE) by benzodiazepine receptor antagonists (BZA) in a rat model of fulminant hepatic failure (FHF). (Abstract) J. Hepatol. 1987; 4 (suppl. 1): S17

11) Scollo-Lavizzari G, Steinmann E. Reversal of hepatic coma by benzodiazepine antagonist (Ro 15-1788). (Letter) Lancet 1985; i:1324.

12) Grimm G, Lenz K, Kleinberger G., et al. Ro 15-1788 improves coma in 4 out of 5 patients with fulminant hepatic failure; verification by long latency auditory and somatosensory evoked potentials (Abstract) J. Hepatol. 1987; 4 (suppl. 1): S21.

13) Bansky G, Meier PJ, Ziegler WH, et al. Reversal of hepatic coma by benzodiazepine antagonist (Ro 15-1788). (Letter) Lancet 1985; i: 1324-1325.

14) Ferenci P, Grimm G, Gangl A. Successful longtime treatment of chronic hepatic encephalopathy with a benzodiazepine antagonist. (Abstract) Hepatology (Baltimore) 1987; 7: 1064.

15) Traber PG, Granger DR, Blei AT. Brain edema in rabbits with galactosamine-induced fulminant hepatitis. Gastroenterology 1986; 91: 1347-56.

16) Mullen KD, Martin JV, Mendelson WB, et al. Could an endogenous benzodiazepine ligand contribute to hepatic encephalopathy? Lancet 1988; i: 457-59.

17) Mullen KD, Martin JV, Bassett ML, et al. Hepatic encephalopathy: a syndrome modulated by an endogenous benzodiazepine ligand? (Abstract) Hepatology (Baltimore) 1986; 6: 1221.

18) Mullen KD, Szanter KM, Galloway PG, Kaminsky K. CSF of patients with hepatic encephalopathy (HE) contains significant benzodiazepine (BZ) binding activity. Correlation with post mortem cortical BZ binding studies. (Abstract) Hepatology (Baltimore) 1987; 7: 1103.

TREATMENT OF HEPATIC ENCEPHALOPATHY WITH THE BENZODIAZEPINE ANTAGONIST FLUMAZENIL

Peter FERENCI and Georg GRIMM

1st Dept. of Gastroenterology and 1st Dept. of Medicine
University of Vienna, Austria
Lazarettgasse 14, A-1090 Vienna

ABSTRACT

Recently it was suggested that hepatic encephalopathy (HE) is mediated by an increased GABA-ergic tone. The logical consequence of such a hypothesis is the use of antagonists of the $GABA_A$-benzodiazepine receptor for treatment of HE. The experience with these drugs in human HE is limited. In order to get an estimate of the efficacy of this type of drug in humans 20 consecutive episodes of HE in 17 patients with acute or chronic liver failure were treated with flumazenil. All patients entered into the study where encephalopathic for up to 120 hours and failed to respond to conventional therapy. After an observation period of 8 hours 15 mg of flumazenil were infused intravenously over 3 hours. Before and after treatment patients were examined neurologically and the Glasgow coma scale was calculated. In addition somatosensory evoked potentials were recorded. In 5 out of 11 episodes in 10 patients with fulminant hepatic failure and in 7 out of 9 episodes in 7 patients with cirrhosis an unequivocal amelioration of HE was observed. The response to treatment occured very rapidly. After stopping treatment in 8 out of these 12 episodes HE worsened again after 2 to 4 hours. The favourable clinical response was also documented by improvement of somatosensory evoked potentials. In 5 of the 8 episodes not reacting to flumazenil patients had signs of increased intracranial pressure. These findings indicate that flumazenil may be valuable in treatment of acute HE occuring in fulminant hepatic failure or in decompensated cirrhosis.

Furthermore a patient with portal systemic encephalopathy refractory to standard therapy following extensive liver resection and construction of a portocaval shunt was treated with flumazenil 25 mg twice daily. Before treatment she was encephalopathic with 12 attacks of coma within 2 years. On treatment with flumazenil all signs of HE abated in spite of unrestricted dietary protein intake. After one year treatment with flumazenil was stopped. Two days later she became comatous again and remained encephalopathic for the following 3 months. Since reinstitution of flumazenil she is well for 16 months of follow up without any signs of HE. She does not require a protein restricted diet. The favourable response of this patient to treatment with flumazenil suggests that flumazenil may be also valuable in long term management of portal systemic encephalopathy.

INTRODUCTION

Recent studies indicated that hepatic encephalopathy (HE) may be associated with an increased tone of the gamma-aminobutyric acid (GABA)-ergic inhibitory neurotransmitter system (1,2,3). Interaction of GABA with its postsynaptic receptor induces the generation of an inhibitory postsynaptic potential. Other proteins closely associated with the GABA receptor and the chloride ionophore are receptors for benzodiazepines and barbiturates (4). These pharmacological agents exert their sedative effects by interaction with their respective binding sites on the $GABA_A$-benzodiazepine receptor complex by augmenting the GABA mediated chloride fluxes into postsynaptic neurons (4). Therefore, the logical consequence of a hypothesis implicating an increased GABA-ergic tone in HE was the study of effects of GABA- and benzodiazepine antagonists in HE. Indeed, administration of GABA or of benzodiazepine antagonist (bicuculline, flumazenil) to animals with fulminant hepatic failure was followed by a transient clinical improvement in HE associated with a normalization of abnormal visual evoked responses (5,6). The clinical use of currently available pure GABA-antagonists is precluded because of their seizure inducing properties. In contrast, flumazenil was successfully used in few cases of human HE. The published anectodal reports (7-13) do not allow an assessment of the efficacy of this treatment. The types of patients treated, the doses of the drug given, the duration of treatment and the methods used to document improvement of HE varied widely within the various reports. In this report our experience with this drug in HE occuring in fulminant liver failure or in patients with cirrhosis using a standardized dose of flumazenil is summarized. Patients were monitored by clinical neurological examination and by recording of somatosensory evoked potentials (SEPs). In addition, the effects of longterm treatment with flumazenil in a patient with portosystemic encephalopathy (PSE) are described.

STUDY DESIGN

Patient selection

Only patients in whom intake of benzodiazepines or of other sedatives within the preceeding week was excluded with certainty, were considered for this study. In addition, in none of the 11 urine samples obtained prior to treatment were traces of benzodiazepines, barbiturates, and opiates detected. Prior to admission, all patients had been treated with lactulose (orally or by enema) and the majority was also given a total parenteral nutrition (using amino acid solutions with high content of branched chain amino acids). Other treatments of complications of the underlying disease were given according to the individual need of each patient and included antibiotics, diuretics, H_2-receptor blocking agents, and infusions of human albumin or fresh frozen plasma.

Treatment with flumazenil

After selection for the study patients were monitored for at least 8 hours and the actual treatment was continued. During the prestudy period HE did not improve in any of the 20 episodes studied and in most of them even deteriorated. The first 5 patients with 7 episodes of HE were treated with various doses of flumazenil over varying time intervals (one 2mg/15min, each two 5/45 or 15/525, respectively). In the remaining 13 episodes 15 mg flumazenil were infused over 3 hours. Before treatment SEPs were recorded using a NICOLET 2000 (Nicolet Biomedical Instruments, Madison, Wisconsin, USA) by standard techniques (15) and a clinical neurological examination including calculation of the Glasgow coma scale (GCS, 14) was performed. The stage of HE was graded from 0 to IV according to Conn (16) by the treating physician. During administration of flumazenil patients were monitored clinically. At the end of the infusion, recording of SEPs and the clinical neurological examination were repeated. Thereafter patients were monitored clinically at 10 min intervals.

RESULTS

Patients with fulminant liver failure

In 9 patients with fulminant hepatic failure 11 episodes of HE were treated and considerable improvement occurred in 6 episodes (see table 1). 2 patients were treated twice. One patient responded transiently with improvement of HE from grade IV to grade III, of the GCS from 7 to 11, and of SEPs (N3:152 msec to 121, P3 absent to 171 msec, N1-N3: 134 to 103). Subsequently, the patient remained in HE grade IV and further deteriorated with clinical evidence of increased intracranial pressure 41 hours later. A CT-scan of the brain showed marked brain edema. The patient now failed to respond to flumazenil and died without regaining consciousness. A second patient improved following the administration of 5 mg flumazenil and consciousness spontaneously fluctuated (between stage III and IV) during the next 18 hours. A repeated treatment with 5 mg flumazenil was followed by a sustained mental improvement. 4 of the 5 patients not reacting to flumazenil had clinical evidence of increased intracranial pressure by neurological examinations (see table 2). In two of them brain edema was documented by CT. All 4 patients died and brain edema was confirmed by postmortem examination. The remaining nonresponding patient underwent orthotopic liver transplantation and regained full consciousness thereafter.

Patients with cirrhosis

6 out of 9 episodes of HE in 8 patients with cirrhosis treated with flumazenil improved. Out of the 3 patients not reacting to flumazenil one had clinical evidence of increased intracranial pressure. This patient improved after treatment with mannitol twelve hours later. The second patient was only in stage I of HE. The third patient regained consciousness three days after

	Stage of HE		GCS		Time to first response (min)	Duration of improvement after treatment (min)
	before	after	before	after		
FHF	IV	II	10	14	5	80
FHF	IV	III	11	14	10	35
FHFa	IV	III	6	11	5	60
FHFb	III	I	13	15	3	no worsening
FHF	IV	III	7	11	15	90
FHF	III	II	12	15	30	120
Ci	III	II	10	14	60	120
Ci	III	I	14	15	15	120
Ci	III	I	14	15	30	240
Cia	III	I	12	15	20	no worsening
Cib	III	II	12	14	gradual	no worsening
Ci	II	I	13	15	30	no worsening

Table 1: **Changes in the stage of hepatic encephalopathy and in Glasgow coma scale (GCS) in the 12 episodes of hepatic encephalopathy responding to treatment with flumazenil. FHF: fulminant hepatic failure, Ci:cirrhosis**

administration of flumazenil. This patient also had renal failure which improved gradually.

Table 2: Clinical characteristics of patients with HE not reacting to flumazenil treatment (Abbreviations: GCS: Glasgow Coma Scale; normal:15; ND:not done, LTX:orthotopic liver transplantation, FHF: fulminant hepatic failure, Ci:cirrhosis)

			BRAIN EDEMA			OUTCOME
	HE grade	GCS	clinical	CT	post mortem	
FHF	IV	4	+	ND	+	died
FHF	IV	4	+	+	+	died
FHF	IV	3	+	+	+	died
FHF	IV	4	+	ND	+	died
FHF	IV	8	?	ND		LTX
Ci	I	15	-	ND		survived
Ci	IV	4	+	ND		survived
Ci	IV	3	-	ND	-	died

Clinical response to flumazenil and changes in somatosensory evoked potentials

The results in the 12 episodes of improvement of HE after flumazenil are summarized in table 1. The first unequivocal mental improvement occurred within 3 and 60 min. In one patient HE improved very slowly but gradually over three hours. Nine patients (including the two who did not respond) became agitated following administration of flumazenil. The agitation abated in all cases without treatment within few minutes. After stopping treatment there was an exacerbation of HE in 8 patients. The time between the end of the infusion and unequivocal worsening of the encephalopathy varied between 35 and 240 min.

Figure 1 shows the results of cortical SEPs recorded in 14 episodes of HE in 12 patients prior and after treatment with flumazenil. All patients had abnormal SEPs with markedly prolonged latencies of the peaks N3 and P3, and of the N1-N3 interpeak time intervals, respectively. In three patients with evidence of brain edema SEPs were not only highly abnormal but also qualitatively different from those of the other patients by absence of the peaks N3 and P3. Overall, there was a high correlation between the mental condition of the patients and the SEP recordings.

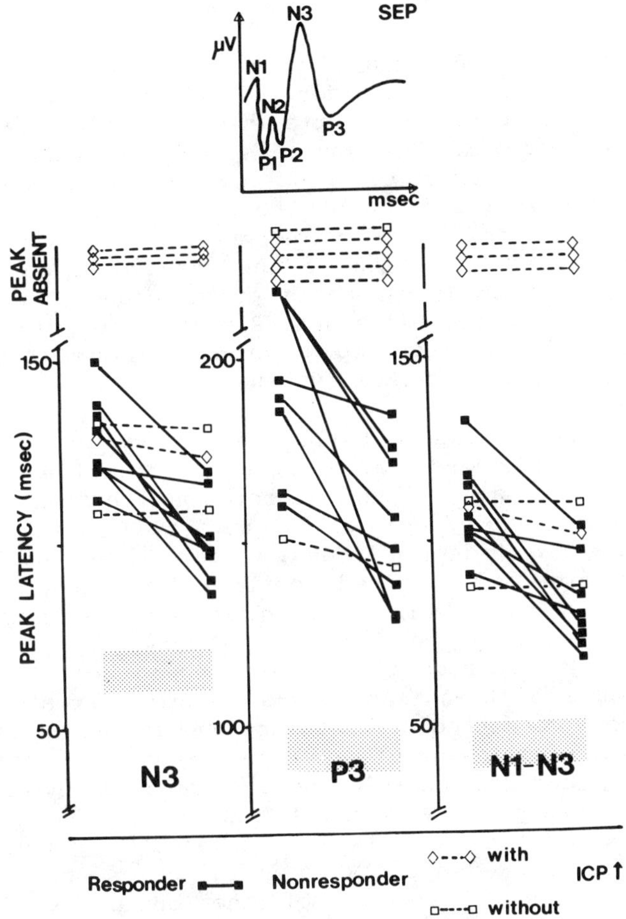

Fig. 1: Changes in SEP's after treatment with flumazenil. A normal cortical SEP record (top) reveals 3 major negative and 3 positive peaks, labelled as peaks N1...N3, P3 consecutively. Peak latencies of N1, N3, P3 and time intervals between N1 and N3 were measured in 14 episodes of HE before and after flumazenil treatment (left and right sides of the corresponding three columns, respectively). The shaded areas show the range of the corresponding peak latencies measured in 40 healthy subjects (N3: 69.5±10 msec, P3: 95.2±11, N1-N3: 49±10.6, x±2.5 SD). ICP = increased intracranial pressure.

FLUMAZENIL TREATMENT IN A PATIENT WITH PORTOSYSTEMIC ENCEPHALOPATHY

In February 1984 in a 42 year old woman a large tumor originating in the gallbladder and infiltrating the right lobe of the liver, the porta hepatis, the transverse colon, the head of the pancreas and the duodenum was found. Surgical treatment included an extended right hemihepatectomy, resection of the biliary confluence, the common duct and the portal vein, subtotal duodenopancreatectomy, partial resection of the transverse colon, and construction of an intrahepatic ducto-jejunostomy and of an end-to-side portacaval shunt. Histologic examination of the resected tissue revealed a chronic inflammatory tumor arising within chronic cholecystitis.

The clinical course following surgery is summarized in figure 2. Six weeks after surgery the patient became encephalopathic and did not improve on treatment with a low protein (40 g/day) diet, neomycin (4g/day), and lactulose (60 ml/day). Indeed, after three weeks she developed coma and remained comatose for one week. When the patient was in coma, venous ammonia concentration was increased (ranging between 214 and 411 µM, normal: 5-40 µM) and electroencephalograms showed delta activity with triphasic transients. The regular treatment of hepatic coma consisted of lactulose enemas and intravenous infusion of L-valine or of ornithine-alpha-ketoglutarate. After she regained consciousness, branched chain amino acids (3x10g/day) were added to the diet. For the next three weeks she remained free of clinical signs of HE. A liver biopsy showed a fatty infiltration of about 30% of the hepatocytes. A CT-scan of the brain revealed no abnormality and a CT-scan of the abdomen showed a liver of normal size and appearence. During the following months she had recurrent fever and multiple episodes of coma. Longterm oral antibiotic treatment with Augmentin^{R} (Beecham, UK) was started and neomycin was stopped. She remained mildly encephalopathic. From September 1985 to February 1986 she had 8 further episodes of hepatic coma. At the last episode of coma the patient woke up 30 seconds after 1 mg of flumazenil was injected intravenously and remained conscious for about 2 hours. In the light of this favourable response, oral treatment with 25 mg flumazenil each morning was started and after two weeks the dose was increased by 25 mg given at noon. Signs of HE as assessed

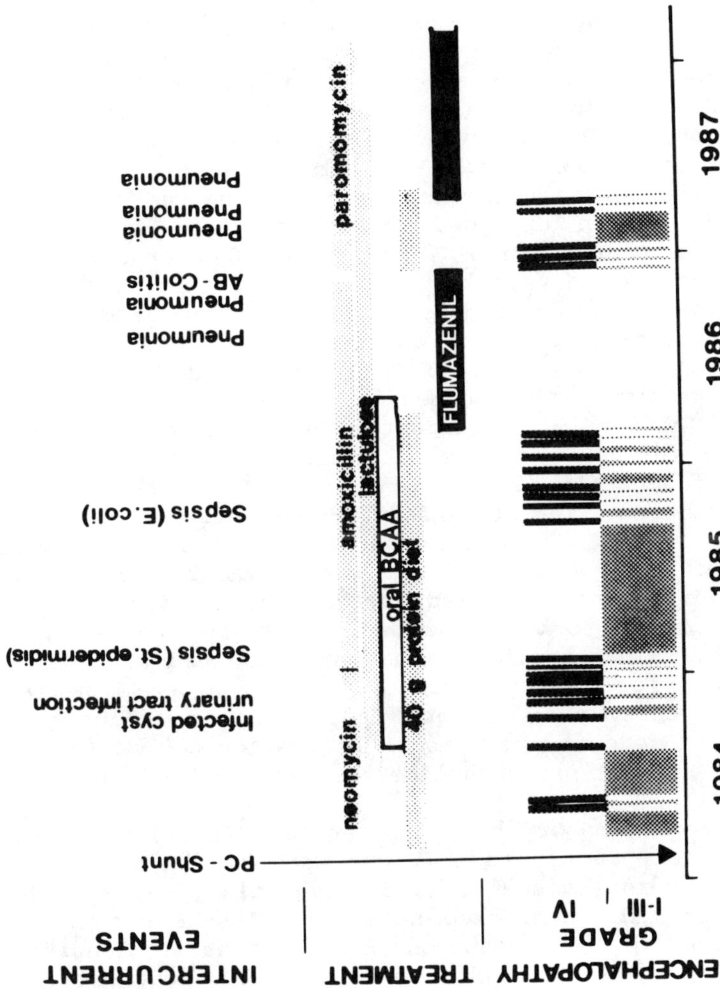

Fig. 2: Clinical course of the patient with PSE following surgery

clinically and by psychometric testing abated. The patient consistently became anxious about 25 to 40 minutes after the intake of the drug and remained anxious for 30 to 60 minutes. Within one month oral protein intake was gradually increased until it was unrestricted. The daily protein intake calculated from the patient's dietary records was between 70 and 90 g. On one day she tolerated 150 g protein without any signs of HE. In July 1986 she became febrile. No apparent reason for the fever was found. On repeated treatment with ceftriaxon the fever disappeared, but recurred shortly after discontinuation of antibiotic treatment. Later also diarrhea developed. As the possibility that the fever was induced by flumazenil could not be excluded, treatment with this drug was stopped on December 2, 1986 and a 40 g protein diet was started. Two days later she became comatous. Clostridium difficile toxin was detected in feces and a diagnosis of antibiotic-associated colitis was made. Following treatment with oral vancomycin fever and diarrhea abated. Oral treatment with Augmentin[R] was followed by paromomycin (2x40 ml/day). The patient remained encephalopathic and had 4 further episodes of coma. In March 1987 treatment with flumazenil (2x25 mg/day) was restarted. The patient again became free of signs of HE and her protein tolerance became normal. In April and in June 1987 recurrences of pneumonia were treated with Ceftriaxon. Since then, the patient has not required further hospitalisation. Indeed, in contrast to when she was not taking the drug, while receiving flumazenil she was able to lead a normal life. Except for short periods, mild hyperammonemia was consistently present, and indices of liver function did not change over the 4-year observation period (bilirubin: 2 to 4 mg/dl, pseudocholinesterase: 1 - 2 kU/l).

Somatosensory evoked potentials (SEPs) were recorded repeatedly in HE grade IV and were highly abnormal. Pattern reversal visual evoked potentials (PR-VEP) and event related auditory evoked potentials (ER-AEP) could not be obtained in the unconscious state. After starting flumazenil treatment in March 1986 serial recordings of PR-VEP's and SEP's in monthly intervals were within normal range.

In May 1987 ER-AEP's were studied after flumazenil was stopped on th preceeding day. Highly abnormal ER-AEP became normal 2 hours after intake of 25 mg of flumazenil.

In January 1988, while on flumazenil treatment SEP's, ER-AEP's and PR-VEP's and the Reitan test were completely normal. Four hours after ingestion of 300 g beef only a moderate prolongation (by less than 10%) of SEP's, PR-VEP's, ER-AEP's was observed.

DISCUSSION

This study represents the largest experience with the benzodiazepine antagonist flumazenil in the treatment of human HE so far. The effects of the drug were documented clinically and by repeated recordings of SEPs. The late components of cortical SEPs (peaks N3 and P3) appear to be highly sensitive indicators of cortical dysfunction in HE (17). The results indicate that this drug may be useful in inducing an amelioration of acute HE complicating both acute and chronic liver failure. Flumazenil treatment was associated with an improvement in 60% of episodes of HE and except in one case improvement occurred within a few minutes of drug administration. The rapidity of these responses is in marked contrast to the intervals of several hours that are typically necessary before improvement of HE can be documented following the institution of conventional therapies.

The 60% improvement rate may even underestimate the potential efficacy of flumazenil in the treatment of HE since most of the patients studied were encephalopathic for many days prior to flumazenil treatment and had failed to respond to conventional therapy. Furthermore, all 5 patients with clinical evidence of increased intracranial pressure did not respond to treatment with flumazenil. One of these patients improved after treatment with mannitol. The remaining four died within three days of flumazenil administration. By excluding these 5 patients the response rate to flumazenil in this study would be 80%.

In 8 of the 12 episodes responding to flumazenil there was an exacerbation of HE 0.5 to 4 hours after stopping treatment. This transient effect of the drug is consistent with its pharmacokinetic properties (18,19,20). In order to achieve a long lasting response continuous administration of the drug over longer time periods may be necessary. Although an unequivocal response to flumazenil treatment was documented in these 12 episodes, it should be stressed that none of these patients regained normal brain function at the end of treatment. The possibility that larger doses or a longer duration of treatment would have achieved a complete improvement seems unlikely since in patients with benzodiazepine intoxication much lower doses are sufficient for complete recovery (20). Alternatively, it could be assumed that an increased GABA-ergic tone is only one of many abnormalities of brain function in patients with liver failure and that correction of this particular abnormality may therefore induce only an incomplete improvement.

Furthermore, in a patient with a surgical portocaval shunt chronic PSE refractory to standard therapies was effectively controlled with orally administered flumazenil. The numerous attacks of hepatic coma in this patient enabled the effects of treatments to be readily assessed both clinically and electrophysiologically. During longterm flumazenil treatment there was no clinical evidence of even mild HE and the results of several sensitive tests for HE such as psychometric tests and serial multimodality EP recordings were normal. HE recurred immediately after stopping the drug and subsided after reinstituting it. Omission of flumazenil intake for 36 hours in the patient was associated with a prolongation of ER-AEPs (P300) to a similar extent as was observed in healthy controls after intake of a small dose of a benzodiazepine (Grimm et al, unpublished observations). Her ER-AEP's improved to normal 2 hours after oral flumazenil administration, further confirming the effects of flumazenil on subclinical HE. Clearly, the efficacy of a new treatment cannot be regarded as proven by observations made in a single case. Although the findings suggest that the improvement was mediated by flumazenil the possibility that other factors improved the condition of the patient cannot be excluded with certanty.

The mechanism by which flumazenil may improve HE is uncertain. Displacement of an endogenous benzodiazepine-like substance from the $GABA_A$-benzodiazepine receptor by flumazenil may account for its favourable effect. The presence of such a substance was suggested in the brains of animals of HE and in cerebrospinal fluid of patients dying with HE (21).

In summary, the findings in this study clearly suggest that the benzodiazepine antagonist flumazenil is an effective treatment modality for HE. Prospective randomized controlled studies should be conducted to establish the effectiveness of benzodiazepine antagonists in reversing HE.

REFERENCES

1. Schafer DF, Pappas SC, Brady LE, Jacobs R, Jones EA. Visual evoked potentials in a rabbit model of hepatic encephalopathy. I: Sequential changes and comparisons with drug induced comas. Gastroenterology 1984; 86:540-545

2. Basile AS, Gammal S, Mullen KD, Jones EA, Skolnick P. Differential responsiveness of cerebellar Purkinje neurons to GABA and benzodiazepine receptor ligands in an animal model of hepatic encephalopathy. J Neurosci 1988; 8:2414-2421

3. Schafer DF, Jones EA. Hepatic encephalopathy and the γ-aminobutyric acid neurotransmitter system. Lancet 1982; 2:18-19

4. Paul SM, Marangos PJ, Skolnick P. The benzodiazepine-GABA-chloride ionophore receptor complex. Common site of minor tranquillizer action. Biological Psych 1981; 16:213-229

5. Baraldi M, Zeneroli ML, Ventura E, et al. Supersensitivity of benzodiazepine receptors in hepatic encephalopathy due to fulminant hepatic failure in the rat: reversal by a benzodiazepine antagonist. Clin Sci 1984; 67:167-175

6. Bassett ML, Mullen KD, Skolnick P, et al. Amelioration of hepatic encephalopathy by pharmacologic antagonism of the $GABA_A$-benzodiazepine receptor complex in a rabbit model of fulminant hepatic failure. Gastroenterology 1987; 93:1069-1077

7. Bansky G, Meier PJ, Ziegler WH. Reversal of hepatic coma by benzodiazepine antagonist (Ro 15-1788). Lancet 1985; I:1324-1325

8. Bansky G, Meier PJ, Riederer E, et al. Effect of a benzodiazepine antagonist in hepatic encephalopathy. Hepatology 1987; 7:1103

9. Scollo-Lavizzari G, Steinmann E. Reversal of hepatic

coma by benzodiazepine antagonist (Ro 15-1788). Lancet 1985; i:1324

10. Burke DA, Mitchell KW, Al Mardini H, Record CO. Reversal of hepatic coma with flumazenil with improvement in visual evoked potentials. Lancet 1988; i:505-506

11. Sutherland LR, Minuk GY. Ro 15-1788 and hepatic failure. Ann Int Med 1988; 108:158

12. Grimm G, Lenz K, Kleinberger G, et al. Ro 15-1788 improves coma in 4 out of 5 patients with fulminant hepatic failure; verification by long latency auditory and somatosensory potentials. J Hepatol 1987; 4(suppl 1):S 21

13. Meier R, Gyr K. Treatment of hepatic encephalopathy (HE) with the benzodiazepine antagonist flumazenil: a pilot study. Eur J Anaesthesiology 1988; Suppl 2:139-146

14. Conn HO, Lieberthal M. The hepatic coma syndromes and lactulose. Williams & Wilkins, Baltimore, 1979, p. 6

15. Teasdale G, Jennett B. Assessment of coma and impaired consciousness. Lancet 1974; i:81-83

16. Frontiers of Clinical Neuroscience, Vol. 3, Evoked Potentials. R.Q. Cracco, I. Bodis-Wolner (Eds.), Alan R. Liss, New York 1986

17. Chu NS, Yang SS. Portal-systemic encephalopathy: alterations in somatosensory and brainstem auditory evoked potentials. J Neurol Sci 1988; 84:41-50

18. Lister R, Greenblatt D, Abernathy D, et al. Pharmacokinetic studies on Ro 15-1788, a benzodiazepine receptor ligand, in the brain of rat. Brain Res 1984; 290:183-186

19. Roncari G, Ziegler WH, Guentert TW. Pharmacokinetics of the new benzodiazepine antagonist Ro 15-1788 in man following intravenous and oral administration. Br J Clin Pharmac 1986; 22:421-428

20. Prischl F, Donner A, Grimm G, et al. Value of flumazenil in benzodiazepine self-poisoning. Med Toxicol 1988; 3:334-339

21. Mullen KD, Martin JV, Mendelson WB, et al. Could an endogenous benzodiazepine ligand contribute to hepatic encephalopathy? Lancet 1988; i:457-459

Subject Index